Properties and Processes of Crustal Fault Zones
Volume I

Edited by
Yehuda Ben-Zion
Antonio Rovelli

Previously published in *Pure and Applied Geophysics* (PAGEOPH), Volume 171, No. 11, 2014

 Birkhäuser

Editors
Yehuda Ben-Zion
Department of Earth Sciences
University of Southern California
Zumberge Hall
CA 90089-0740 Los Angeles
USA

Antonio Rovelli
Istituto Nazionale di Geofisica e Vulcanologia
Via di Vigna Murata 605
00143 Rome
Italy

ISBN 978-3-0348-0876-7   ISBN 978-3-0348-0877-4 (eBook)
DOI 10.1007/978-3-0348-0877-4

Library of Congress Control Number: 2014956919

Springer Basel Heidelberg New York Dordrecht London
© Springer Basel 2015

Cover illustration: Based on Fig. 5 in "Ground Motion Prediction Equations in the San Jacinto Fault Zone - Significant Effects of Rupture Directivity and Fault Zone Amplification" by I. Kurzon, F.L. Vernon, Y. Ben-Zion and G.M. Atkinson.

Cover design: deblik, Berlin.

Printed on acid-free paper

Springer Basel AG is part of Springer Science+Business Media

www.springer.com

Contents

Pure Appl. Geophys. 171 (2014), 2863–2865
© 2014 Springer Basel
DOI 10.1007/s00024-014-0943-3

Properties and Processes of Crustal Fault Zones: Volume I

YEHUDA BEN-ZION[1] and ANTONIO ROVELLI[2]

1. Introduction

Crustal fault zones are complex regions of localized deformation with evolving geometries and altered rheological properties from those of the host material. Seismic ruptures modify the geometry, internal structure, and material properties of fault zones. Conversely, properties of earthquake ruptures, seismic radiation, inter- and post-seismic deformation, and local seismicity patterns are controlled by the fault zone structure. It is also well established that faulting modifies the porosity and permeability of the crust and that faults serve as both conduits and barriers to fluid flow. The geometrical complexity, material heterogeneities and wide range of effective space–time scales make the quantification of fault zone properties and processes highly challenging. Many fundamental questions concerning the structure and evolution of fault zones in relation to earthquake properties and generated ground motion remain unanswered despite considerable research spanning over 100 years.

Recent theoretical developments, acquisitions of large seismic and other data sets, detailed geological studies, and novel laboratory experiments provide new opportunities for advancing the understanding of fault zone and earthquake properties. The state-of-the-art in the field was discussed at the 40th Workshop of the International School of Geophysics titled "Properties and Processes of Crustal Fault Zones", conducted during May 18–24, 2013, in the Ettore Majorana Foundation and Centre for Scientific Culture of Erice, Sicily. The papers in the present and follow-up volumes provide broad perspectives on crustal fault zones based on presentations given at the workshop and additional contributions. Topics covered in this volume include fluids and faulting, characterization of fault zone materials, seismic ground motion, geodetic deformation, seismicity and hazard, imaging fault zone structures, experiments on fault evolution, and damage-based rheologies for shear deformation.

SIBSON provides an overview on fluid overpressures in different fault zone environments. The results indicate that fluid overpressures are more likely to exist in compressional/transpressional crustal regimes, and subduction interface shear zones, compared to extensional/transtensional settings. Whether earthquakes tend to nucleate or just rupture through overpressure regions and the heterogeneity of overpressuring remain to be clarified in future studies. PLACE et al. describe an experimental approach to estimate the normal and tangential compliances of a rough interface by comparing source P and S waves with reflected phases. The ratio of the compliances depends on the type of saturating fluids and effective pressure. Measuring the temporal variations of the ratio may be used to monitor the evolution in the state of the interface. CURREN and BIRD use plane-stress clay experiments to study the evolution of strike-slip faults in the presence of preexisting, obliquely oriented faults and joints. The results indicate that preexisting fault networks can cause deflection from pure strike-slip faulting, suppress the development of a through-going fault and lead to distributed shear deformation. Several corresponding examples from natural faulting areas are discussed.

ROCKWELL et al. use non-central principal component analysis to study chemical alteration in rock

[1] Department of Earth Sciences, University of Southern California, Los Angeles, CA 90089-0740, USA. E-mail: benzion@usc.edu
[2] Istituto Nazionale di Geofisica e Vulcanologia, Seismology and Tectonophysics, Rome, Italy.

Reprinted from the journal

samples from different internal units of the San Jacinto fault zone in southern California. Differences between the fault core and surrounding damage units, combined with XRD results, are interpreted in terms of illite/smectite mineralogical changes during brittle rock damage aided by hot acidic fluids. LINDSEY et al. analyze interseismic geodetic deformation across the San Jacinto fault near Anza. GPS and InSAR data indicate a 2–3 km wide shear zone deforming more than twice as fast as the background strain rate. Model simulations show that reduced shear modulus in the fault zone consistent with tomographic images can explain about 50 % of the elevated strain rate, with a best-fitting locking depth several kilometers less than the depth of seismicity. A deep fault zone with larger shear modulus reduction and deeper locking depth can explain fully the elevated strain rate. Additional possibilities include significant fault creep in the section of deep microseismicity, or reduced yield strength and distributed plastic failure in the interseismic period within the upper fault zone. ZÖLLER and BEN-ZION discuss spatio-temporal seismicity patterns along the San Jacinto fault zone, using a long simulated record generated by a model that reproduces the main statistical properties of the available instrumental and paleoseismic data. Augmenting the limited observed data with the longer simulated record increases significantly the hazard associated with a future large San Jacinto earthquake. The simulations indicate further that the recent observed increasing number of moderate events is a robust precursory statistical feature, and that the hypocenters of the moderate events move progressively toward the location of the future large earthquake.

SCHULTE-PELKUM and MAHAN present receiver function results on imaging faults and shear zones that separate isotropic S or anisotropic P structures. Synthetic calculations clarify features of dipping interfaces that may be used to determine their strike and depth; other interface parameters are subjected to trade-offs. Realistic shear fabric anisotropy is shown to produce signals with amplitudes comparable to that generated by the Moho. The method is applied to image the Wind River thrust fault in the western U.S. as a steeply dipping interface in the middle crust that flattens in the lower crust, and to predict seismic

signatures from microstructural data of exhumed ductile shear zones in Scotland and the western Canadian Shield. ALLAM et al. analyze early P waveforms generated by earthquakes along the Hayward fault, CA, to detect and utilize fault zone head waves that refract along a major bimaterial fault interface. Time differences between the head and direct P arrivals at stations on the slower crustal block vs. propagation distance along the fault are used to calculate average velocity contrasts across different fault sections. The results imply a continuous bimaterial interface in the seismogenic zone spanning the entire ~80 km examined fault length, and structural complexities near the junction between the Hayward and Calaveras faults and the city of Oakland. ZHANG et al. incorporate surface wave dispersion data into double-difference tomography of body wave arrival times, and apply the method to obtain three-dimensional P and S velocity models around the site of the San Andreas Fault Observatory at Depth. An appropriate weighting of the body and surface waves data defines an optimal joint solution. Adding the surface-wave constraints to the body waves inversion provides a smoother S wave velocity model.

BOORE describes the global database of the Pacific Earthquake Engineering Center used to develop regional Ground Motion Prediction Equations, along with relevant parameters and techniques, and discusses properties of seismic ground motion near faults based on these data. The available seismic records near faults have complex distributions of motion amplitude and polarization. Scaling relations of the observed ground motion in the entire data set are consistent on average with simple models of source, path and site effects. KURZON et al. develop Ground Motion Prediction Equations based exclusively on seismograms recorded in close proximity to the San Jacinto fault zone. The results indicate systematic rupture directivities for given fault sections even for small events, which have significant effects on the ground motion. Fault zone amplification within a region several kilometers wide around the fault is also shown to have an important impact on the observed motion. PANZERA et al. analyze directional amplification in ambient seismic noise recorded across and around fault zones in the western flank of Mt. Etna. Near-fault results exhibit systematic motion amplification in a direction close to the fault normal, which is lacking in data several kilometers away and is interpreted in terms of compliance anisotropy of the damage zone. Across-fault variations of the spectral peak frequency may reflect fault damage asymmetry.

LYAKHOVSKY and BEN-ZION present continuum damage-breakage model results on phase transitions between solid and granular states of material during brittle deformation. The developed framework accounts for faulting with fracturing, granulation, and friction. Shear localization and dynamic instability involve solid-granular transition in the failure zone, while material healing between failure episodes produces the reverse granular-solid transition. BONEH et al. study wear and friction in laboratory shear experiments with a high-velocity rotary apparatus and five rock types. An early wear stage of fault roughening is followed by simultaneous reduction of friction and wear-rate, followed by steady-state deformation with low friction and wear-rate when the fault is covered by a gouge layer. The final stage is most relevant to deformation of natural faults with preexisting gouge zones. LYAKHOVSKY et al. uses the damage-breakage framework to develop a model for the wear-rate in the rotary shear experiments as a propagating damage front. Simulation results fit the measured friction and wear-rate in experiments with several types of rocks and loadings. The model predicts that the thickness of the generated gouge depends strongly on the frictional strength. VEVEAKIS and REGENAUER-LIEB provide a theoretical explanation for double-localization of shear deformation, observed under both brittle and ductile conditions, in terms of solid–fluid transitions that lead to a second internal localization. Microstructural evolutionary processes govern the time-scale of the transition between slow background shear and rapid episodic localization instabilities. Additional papers on properties and processes of crustal fault zones will be published in a follow up Volume II.

Acknowledgments

We are grateful to Silvia Nardi for invaluable help with the organization of the 40th Workshop of the International School of Geophysics on Properties and Processes of Crustal Fault Zones. The workshop was funded by the Istituto Nazionale di Geofisica e Vulcanologia, the European Geophysical Union and the National Science Foundation (Grant EAR-1303569). We thank the authors of the papers for their contributions and the referees for critical reviews that improved the scientific quality of the volume. The latter include Mustafa Aktar, Dave Boore, Eran Bouchbinder, Theodore Brzinski, Emily Brodsky, Fatih Bulut, Giovanna Calderoni, Fred Chester, Cristiano Collettini, Carola DiAlessandro, Ashley Griffith, Karen Daniels, Eric Hetland, Fan-Chi Lin, Vladimir Lyakhovsky, Amil Misra, Charles Mueller, David Okaya, Francesca Pacor, Ze'ev Reches, Joerg Renner, Amir Sagy, Charlie Sammis, Heather Savage, Chris Scholz, Robert Shcherbakov, Toshihiro Shimamoto, Norm Sleep, Ondrej Soucck, Toshiro Tanimoto, Clifford Thurber, Joshua West, Shimon Wdowinski, Hongfeng Yang, and Ramon Zuniga.

Pure Appl. Geophys. 171 (2014), 2867–2885
© 2014 Springer Basel
DOI 10.1007/s00024-014-0838-3

Earthquake Rupturing in Fluid-Overpressured Crust: How Common?

RICHARD H. SIBSON[1]

Abstract—Whether or not ruptures nucleate in fluid-overpressured crust ($\lambda_v = P_f/\sigma_v > 0.4$) is important because pore-fluids overpressured above hydrostatic lower fault frictional strength and may also vary through the earthquake cycle, acting as an independent variable affecting fault failure. Containment of fluid overpressure is precarious because pressure-dependent activation of faults and fractures allows drainage from overpressured portions of the crust. Discharge of fluids through activated fault-fracture permeability (fault-valve action) decreases overpressure so that subsequent failure depends on the cycling of both overpressure and frictional strength as well as tectonic stress. Geometric and mechanical considerations suggest that fluid overpressures are more likely to develop and be sustained in compressional/transpressional regimes as opposed to extensional/transtensional tectonic settings. On the basis of geophysical observations and force-balance analyses, subduction interface shear zones appear to be strongly but variably overpressured to near-lithostatic levels ($\lambda_v > 0.9$) over the full depth range of seismogenic megathrusts. Strong overpressuring at seismogenic depths is also documented in active fold-thrust belts and in areas of ongoing compressional inversion (e.g., northern Honshu) where inherited normal faults are reactivated as steep reverse faults, requiring near-lithostatic overpressures ($\lambda_v \to 1.0$) at depths of rupture initiation. Evidence for overpressuring around strike-slip faults is less clear but tends to be strongest in areas of transpression. In areas of extensional tectonics coincident with particularly high fluid discharge, there is some evidence of overpressuring concentrated towards the base of the seismogenic zone. In general, because of the limited resolution of geophysical techniques, it is easier to make the case for rupture propagation through overpressured crust than to make a definitive case for the direct involvement of overpressured fluids in rupture nucleation, though in some instances the circumstantial evidence is compelling. An unresolved related issue is the heterogeneity of overpressuring. Do the active fault zones themselves serve as fluid conduits that are locally overpressured with respect to the surrounding crust?

1. Introduction

This paper reviews evidence as to whether earthquake ruptures nucleate and/or propagate through fluid overpressured crust. Alteration and metamorphic assemblages in fault-rocks suggest that fault structures are generally fluid-saturated in both brittle (PARRY 1998) and ductile (McCAIG 1997) regimes, with aqueous pore-fluids dominant through much deforming crust though CO_2, hydrocarbons, etc., may also be significant in some settings. Fluid pressure, P_f, in pore and fracture space within the crust is usefully referenced to the vertical stress, $\sigma_v = \rho gz$, by the pore-fluid factor, $\lambda_v = P_f/\sigma_v$. Hydrostatic fluid pressures ($\lambda_v \sim 0.4$) prevail where pore and/or fracture space is freely interconnected up to the Earth's surface. Overpressures above hydrostatic (i.e., $\lambda_v > c.\ 0.4$) develop where permeability barriers impede flow in areas of strong fluid release (e.g., compacting sedimentary basins) when fluid pressure may approach lithostatic values (i.e,. $\lambda_v \to 1.0$) on occasion.

Seismic activity is commonly restricted to the upper half of deforming continental crust defining a 10–20 km thick seismogenic zone (e.g., NAZARETH and HAUKSSON 2004. Within this unstable frictional regime, fault strength is likely approximated by a Coulomb-type criterion:

$$\tau = \tau_f = c + \mu_s \sigma_n' = c + \mu_s(\sigma_n - P_f), \quad (1)$$

where τ and σ_n are, respectively, the shear and normal stress acting on the fault, c is residual cohesive strength and μ_s is the coefficient of friction. Fault strength, τ_f, in the frictional regime of the upper crust is, thus, critically dependent on effective normal stress $\sigma_n' = (\sigma_n - P_f)$ and, consequently, on the level of pore fluid pressure (Fig. 1). If P_f is time-varying, then so is fault frictional strength (SIBSON 1992).

---
[1] Department of Geology, University of Otago, P.O. Box 56, Dunedin 9054, New Zealand. E-mail: rick.sibson@otago.ac.nz

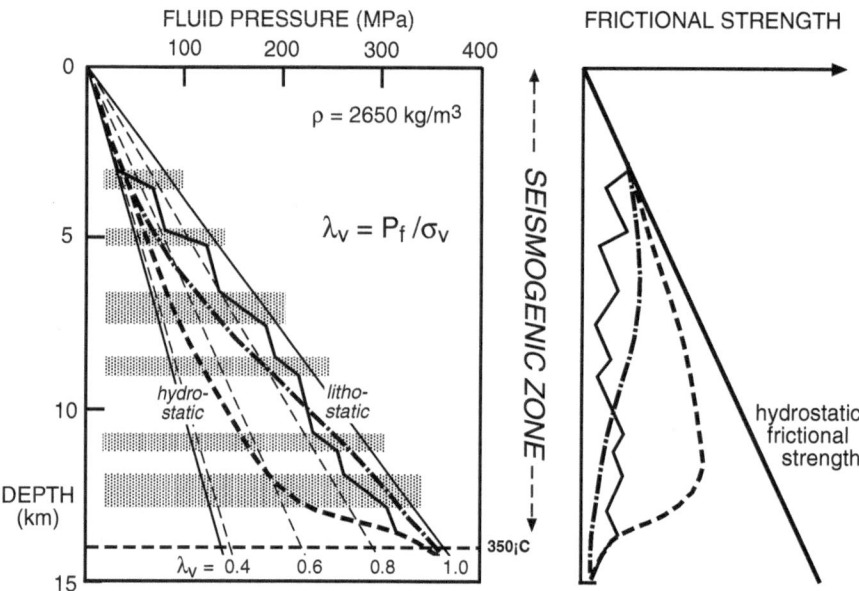

Figure 1

a Hypothetical fluid pressure profiles through seismogenic crust and, **b** their effect on fault frictional strength. Two smooth progressive profiles are shown plus one with overpressure compartments separated by permeability barriers (*shaded bands*). *Dashed horizontal line* defines approximate base to crustal seismogenic zone assumed coincident with 350 °C isotherm

Understanding the physical conditions of seismogenesis at sites of rupture nucleation, therefore, requires knowledge of the fluid pressure state (P_f or λ_v) as well as depth and overburden pressure (z, σ_v), temperature, T, differential stress, ($\sigma_1-\sigma_3$), and strain-rate, $\partial\gamma/\partial t$.

Sedimentary basins are major fluid repositories within the crust and fluid overpressures are known within many deep basins, especially those undergoing active shortening (BERRY 1973; MCPHERSON and GARVEN 1999). Larger earthquake ruptures, however, mostly nucleate in crystalline basement comprising fluid-depleted metamorphic and/or igneous rocks at depths in excess of 5 km (SIBSON 1983). Nevertheless, fault-related hydrothermal vein systems provide evidence that, at least locally, fluid overpressuring also develops within crystalline basement assemblages with fluid inclusions recording evidence of repeated fluid pressure cycling (PARRY and BRUHN 1990; BOULLIER and ROBERT 1992; COX 1995; ROBERT *et al.* 1995; FOXFORD *et al.* 2000). Questions over the involvement of overpressured fluids in faulting raise important issues as to whether fault instability in the lower seismogenic zone arises solely from rate-and-state friction in response to stress accumulation (e.g.

SCHOLZ 1998), or to the cyclical accumulation of overpressure as well as stress (SIBSON 1992; MILLER 2002). A range of failure mechanisms can be envisaged ranging from purely stress-driven failure under constant fluid-pressure, to fluid-driven failure under constant stress levels. Similar issues have been addressed by BYERLEE (1993), MILLER (2002), and COLLETTINI *et al.* (2006). This assessment of the involvement of overpressured fluids in faulting focuses on continental fault zones which are generally better characterized by both geological and geophysical observations. It is important, however, to keep scale considerations in mind. Geological exposures of hydrothermal veining around faults are commonly of the order of 1–100 m, sometimes ranging up to 1 km or so, but the highest resolution of seismic tomographic investigations in the crust is generally of the order of a few kilometres (e.g., OKADA *et al.* 2006).

1.1. Maintenance of Overpressure

Values of rock permeability, k, measured in the laboratory range over 13 orders of magnitude from $10^{-22} < k < 10^{-9}$ m^2 (BRACE 1980). Low permeabilities ($k \ll 10^{-18}$ m^2) are generally considered necessary for the development and maintenance of overpressures in areas of fluid release, for example, from prograde metamorphism at depth (INGEBRITSEN and MANNING 1999). From borehole measurements to depths of several kilometres within intraplate crystalline crust, TOWNEND and ZOBACK (2000) showed that bulk permeability on length scales of 1–10 km ranged from 10^{-17} to 10^{-16} m^2, inhibiting the development of overpressure and maintaining 'hydrostatic-Byerlee' stress states where the differential stress state is governed by the presence of optimally oriented faults that are critically stressed under hydrostatic fluid pressure in the prevailing tectonic regime. Values of permeability so obtained are not far from estimates of 'seismogenic permeability' ($5 \times 10^{-16} < k < 5 \times 10^{-14}$ m^2) derived from the migration of microseismicity triggered by artificially induced or natural fluid pressure fluctuations (TALWANI *et al.* 2007), though these likely represent permeabilities developed after or during brittle failure. The critical issue remains, however, that if permeabilities of the order of those estimated by TOWNEND and ZOBACK (2000) prevail throughout seismically active crust, they preclude the possibility of earthquake rupturing under fluid overpressure and suggest that seismogenic crust is generally hydrostatically pressured. Note, however, that the borehole

stress measurements compiled by TOWNEND and ZOBACK (2000) (including those from the 9.1 km deep KTB borehole) all come from normal fault and/or strike-slip fault stress regimes where active faults and subsidiary fractures are likely to be subvertical, facilitating easy drainage. Fluid overpressure is more easily generated and sustained in compressional regimes where $\sigma_v = \sigma_3$ (SIBSON 2003).

Indeed, as previously discussed, evidence does exist for near-lithostatic overpressures developed in the vicinity of faults developed within basement assemblages (COX 1995; PARRY and BRUHN 1990; ROBERT *et al.* 1995; TUNKS *et al.* 2004). Moreover, hydrothermal fluid ($T < 250$ °C) at significant overpressure ($\lambda_v \approx 0.7$) has been directly encountered at c. 4.2 km depth in a borehole penetrating granitic basement below 3.6 km of sedimentary cover in the Cooper Basin of South Australia where the regional stress state is compressional ($\sigma_v = \sigma_3$) (BAISCH *et al.* 2006). A possibility to be considered is the likelihood of permeability reduction by hydrothermal sealing in the lower seismogenic zone over the temperature range 150 °C $< T <$ 350 °C where solution transfer (particularly from fine-grained fault rocks) and hydrothermal sealing may significantly reduce porosity and permeability (PARRY 1998; SIBSON and SCOTT 1998). In addition, under comparable hydrothermal conditions, MORROW *et al.* (2001) have shown

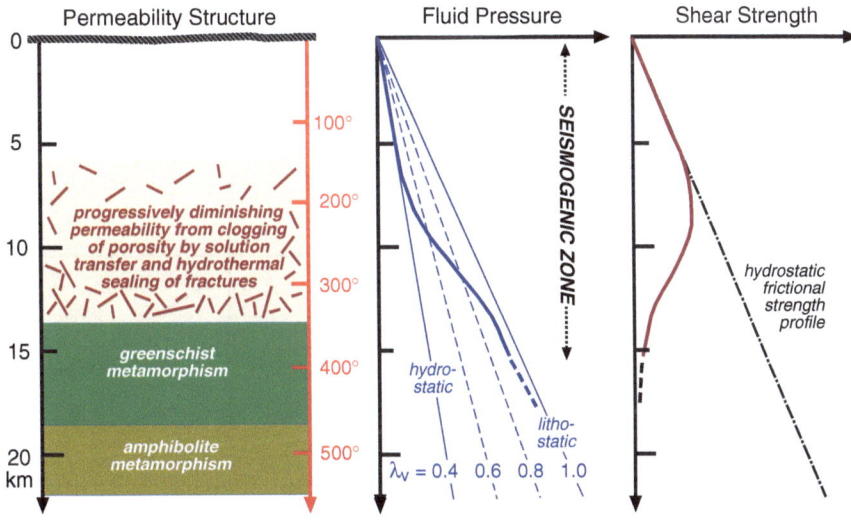

Figure 2
Conceptual model of permeability, overpressure and frictional strength in seismogenic continental crust with a geothermal gradient of 25 °C/km. Solution transfer progressively clogs porosity and heals fractures through the lower seismogenic zone (150 $< T <$ 350 °C)

experimentally that dramatic reductions in fracture permeability within granites may occur over time periods comparable to earthquake recurrence intervals through hydrothermal sealing of fractures during minor cycling of pore pressure. Permeability of quartz-rich gouge material has also been shown to decrease rapidly under hydrothermal conditions (GIGER et al. 2007). Such effects are likely to become increasingly important through the lower seismogenic zone at depths great than 5–7 km, depending on geothermal gradient. Under these circumstances, overpressure development throughout the lower seismogenic zone may have a dramatic effect on the profile of frictional fault strength (Fig. 2).

1.2. Control of Fault and Fracture Orientation by the Tectonic Stress Field

Stress at a point within the earth is characterized by the orientation and relative magnitude of three orthogonal principal compressive stresses ($\sigma_1 > \sigma_2 > \sigma_3$). ANDERSON (1905) pointed out that because of the boundary condition imposed by Earth's free surface, only three basic stress regimes exist in the crust depending whether $\sigma_v = \sigma_1$, σ_2, or σ_3 (Fig. 3).

In classical brittle failure theory (JAEGER and COOK 1979), extension fractures form perpendicular to σ_3 in intact isotropic rock with tensile strength, T_i, in accordance with the hydraulic fracturing criterion:

$$P_f = \sigma_3 + T_i \qquad (2)$$

provided $(\sigma_1 - \sigma_3) < 4T_i$. For the majority of sedimentary rocks, $T_i < 10$ MPa but crystalline rocks may have intact tensile strengths up to c. 50 MPa (LOCKNER 1995). Shear fractures (faults) form along planes containing σ_2 and typically at angles of c. 25–30° to σ_1, in rough accord with the empirical Coulomb criterion:

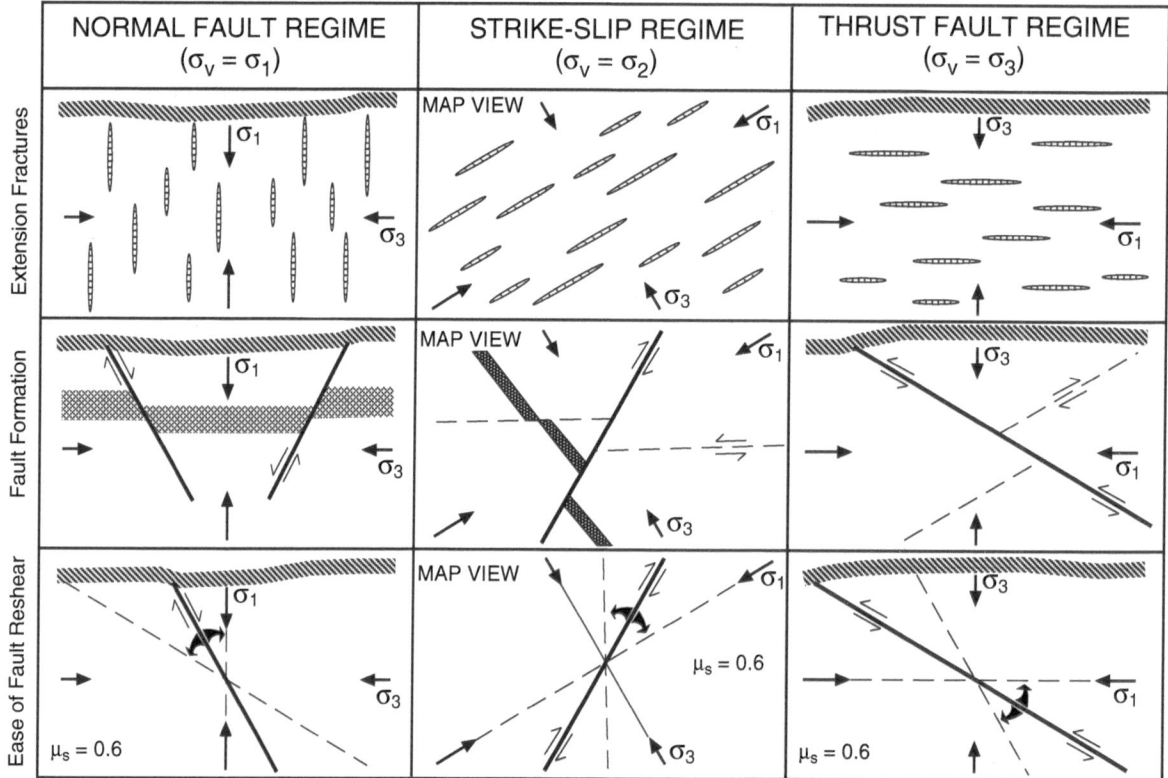

Figure 3

The three 'Andersonian' stress regimes together with the likely orientations of extension fractures and new-forming faults in the different regimes (cross-sections apart from the strike-slip column). Lowest row shows the likely range of fault orientation for reshear ($\mu_s = 0.6$), with reactivation becoming increasingly difficult as orientation departs from optimal

$$\tau = C + \mu_i \sigma_n' = C + \mu_i (\sigma_n - P_f), \qquad (3)$$

where μ_i is the coefficient of internal friction (typically, $0.5 < \mu_i < 1.0$) and C is the cohesive strength of intact rock. A composite brittle failure envelope for intact rock (Fig. 4) can be constructed employing the macroscopic Griffith criterion of parabolic form:

$$\tau^2 = 4(\sigma_n - P_f)T_i + 4T_i^2 \qquad (4)$$

to merge the Coulomb and hydraulic fracture criteria over the interval $4T_i < (\sigma_1 - \sigma_3) < 5.66T_i$ (BRACE 1960; SECOR 1965). Note that this relationship gives rise to the criterion for hydraulic extension fracturing (Eq. 2) on planes of zero shear stress perpendicular to σ_3 and is also consistent with empirical evidence that $C \approx 2T_i$ (JAEGER and COOK 1979).

Reshear of existing faults is governed by Eq. 1 and, provided $\mu_s \sim \mu_i$, is generally easiest when faults are oriented close to their initial formation angle with respect to σ_1, but becomes progressively more difficult as the angle increases or decreases from this optimal value. Frictional lock-up occurs when the fault is oriented at twice the optimal reshear angle (SIBSON 1985). Beyond lock-up, reshear

requires the tensile overpressure condition ($P_f > \sigma_3$) to be met, allowing the possibility of hydraulic extension fracturing.

Systematic orientations of extension fractures and faults (newly-formed or undergoing reshear) developed under quasi-static stress fields within the three 'Andersonian' regimes (ANDERSON 1905; SECOR 1965) are illustrated in Fig. 3. Such systematic stress-controlled fracture sets may develop over broad areas. For example, in the Glenorchy area of the Otago Schist Belt, New Zealand, mapping revealed a set of quartz-filled extension veins (typically ~ 1 m in length) oriented perpendicular to schist foliation over an area $>1,500$ km^2. The veins lies in an Andersonian relationship to normal faults hosting quartz-scheelite lodes, intersecting them at c. 30° (BEGBIE and SIBSON 2006). Attitudes of both the veins and the associated faults have been affected by late-stage regional folding of the schist foliation. Non-systematic fractures derived from dynamic rupture processes may also develop, however, but tend to be restricted to the damage zones flanking the faults (e.g. MITCHELL and FAULKNER 2009).

1.3. Buffering of Overpressure by Activation of Faults and Fractures

In low-porosity rocks it is abundantly clear from hydrothermal vein systems that faults and fractures are the dominant fluid conduits (PARRY 1998). On geometric grounds alone it is evident from Fig. 3 that vertical transport of fluids is facilitated in normal and strike-slip fault regimes because both faults and extension fractures tend to be steep or subvertical. By contrast, in thrust fault regimes extension fractures tend to be subhorizontal and the thrusts themselves are comparatively low-dipping, especially when deflected along flat-lying bedding anisotropy as often occurs in cover sequences (SIBSON 2004).

The mechanics of brittle failure and fault reshear in different stress regimes place additional constraints on the containment of overpressure. Because the criteria for brittle failure of intact rock and fault reshear are all fluid-pressure dependent, drainage of overpressured fluid becomes likely when any kind of brittle failure or reshear occurs to create a high permeability fault and/or fracture conduit.

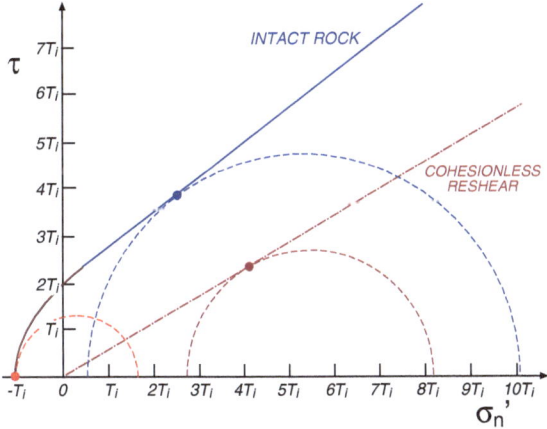

Figure 4

Composite brittle failure envelope on a Mohr diagram of shear stress, τ, against effective normal stress, $\sigma_n' = (\sigma_n - P_f)$ normalized to intact rock tensile strength, T_i (after SECOR 1965). *Red segment* defines the macroscopic Griffith criterion embracing extensional and extensional-shear fracturing when $\sigma_n' < 0$; *blue segment* the Coulomb shear failure criterion for $\mu_i = 0.75$; *dash-dot maroon line* the reshear criterion for $\mu_s = 0.6$. *Stress circles* representing possible failure conditions for shear failure of intact rock, hydraulic extension fracturing, and reshear of an optimally oriented cohesionless fault follow the same *colour* scheme

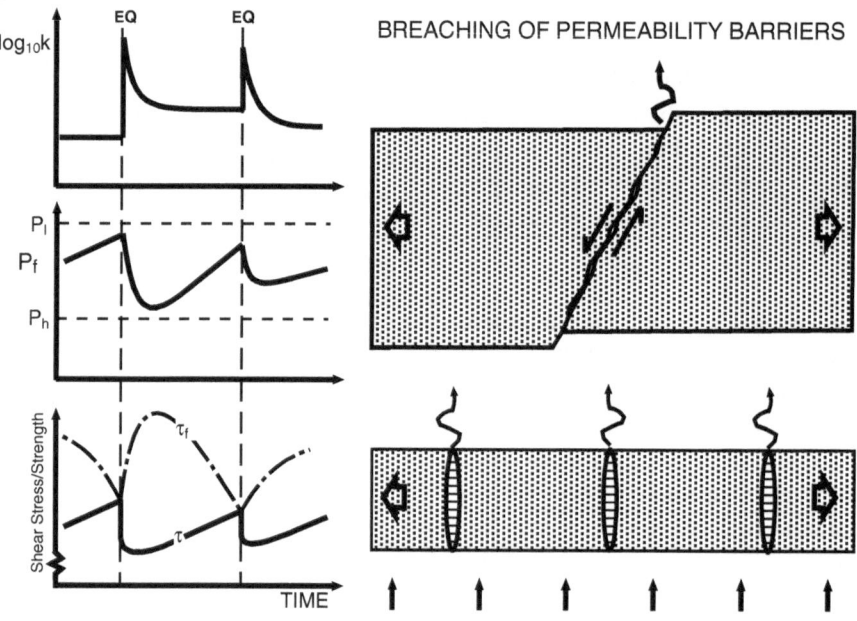

Figure 5

Fault/fracture valving action accompanying earthquake rupture (EQ). Fault and/or fracture systems breach permeability barriers across a vertical gradient in hydraulic head, increasing local permeability (k) and allowing discharge, thereby decreasing fluid-pressure (P_f) and increasing fault frictional strength (τ_f)

This gives rise to various forms of 'valving' action (SIBSON 1992, 2003) involving discharge of overpressured fluid along fault and/or fracture systems cutting across permeability barriers (Fig. 5). The standard failure criteria (Eqs. 1–4) may all be recast in terms of differential stress ($\sigma_1 - \sigma_3$) as a function of effective vertical stress, $\sigma_v' = (\sigma_v - P_f)$ and the material constants (SIBSON 1998). Containment of overpressure for the different 'Andersonian' regimes can then be illustrated on failure mode plots of differential stress, ($\sigma_1 - \sigma_3$), versus the pore-fluid factor, $\lambda_v = P_f/\sigma_v$, constructed for a particular depth (λ_v then being equivalent to values of effective vertical stress, σ_v'). The plots in Fig. 6 are constructed for a rupture nucleation depth of 10 km within a uniform crust with density, $\rho = 2,650$ kg/m^3. 'Generic' failure curves are constructed for intact rock with material properties, $\mu_i = 0.75$ and $T_i = 10$ MPa, and for optimal reshear of existing cohesionless faults with $\mu_s = 0.6$. The failure curves for intact rock are colour-coded: blue segments denote brittle shear failure and red segments the

formation of extensional or extensional-shear fractures likely to promote high-flux flow.

It is apparent that because of the stress configuration, overpressures above hydrostatic are contained over a far greater range of differential stress values in a compressional thrust fault regime than in an extensional normal fault regime, with strike-slip regimes having an intermediate degree of containment depending on the σ_2 level. Overpressure confinement during strike-slip faulting approaches that of a thrust-fault regime as $\sigma_2 \rightarrow \sigma_3$, and that of a normal-fault regime as $\sigma_2 \rightarrow \sigma_1$. Note further that the presence of an optimally oriented cohesionless fault inhibits all other modes of brittle failure. It is also clear that extensional failure requiring ($\sigma_1 - \sigma_3$) < $4T_i$ (ETHERIDGE 1983) is restricted to comparatively low levels of differential stress. Thus, along with the geometric considerations related to fault/fracture orientation, the failure mode plots demonstrate the ease of 'holding-in' overpressure in compressional regimes ($\sigma_v = \sigma_3$), in comparison with the difficulty of sustaining overpressure in extensional regimes ($\sigma_v = \sigma_1$).

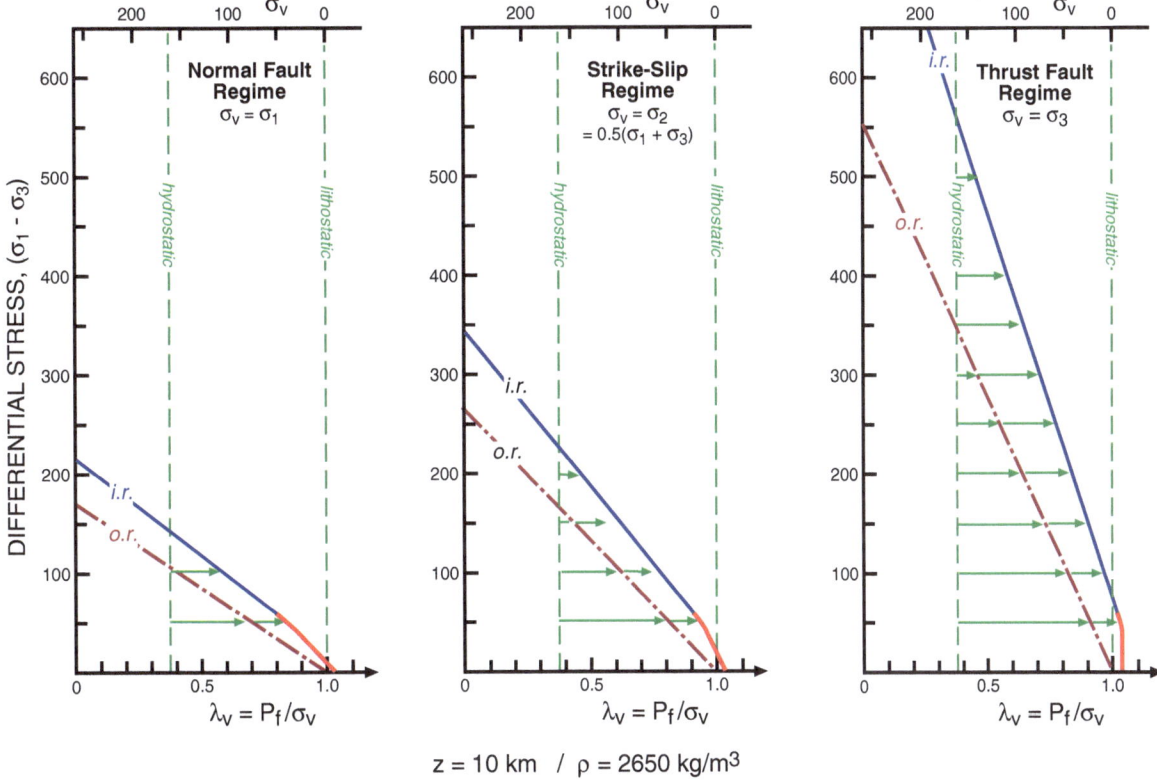

Figure 6

Brittle failure mode envelopes for intact rock (i.r.) with $\mu_i = 0.75$ and $T_i = 10$ MPa, and for reshear of optimally oriented faults (o.r.) with $\mu_s = 0.6$, constructed for a depth of 10 km on plots of differential stress, $(\sigma_1-\sigma_3)$, versus the pore-fluid factor, λ_v, for the three 'Andersonian' stress regimes. *Blue segments* of the failure envelopes for intact rock denote shear failure; *red segments* denote formation of extension or extensional-shear fractures; *maroon dash-dot lines* denote optimal reshear condition. *Green arrows* denote level of overpressure above hydrostatic sustained at failure

1.4. Evidence for Overpressure in Different Seismotectonic Regimes

The employment of seismic and electrical techniques to investigate fluids in and around fault zones at seismogenic depth was reviewed by EBERHART-PHILLIPS *et al.* (1995). Seismic velocities are affected by rock composition (CHRISTENSEN 1996), varying crack densities (as may occur in fault damage zones) and, especially, pore-fluid pressure. Increased pore-fluid pressure lowers V_p and, to a greater extent, V_s, through inhibition of crack closure, thereby increasing the V_p/V_s ratio. The increase in V_p/V_s becomes pronounced as effective confining pressure approaches zero (i.e., $\lambda_v \to 1.0$). Overpressured pore fluids also tend to enhance pore connectivity and lower bulk electrical resistivity. Thus, a particularly strong case for fluid overpressuring may be made

where anomalously high V_p/V_s ratios coincide with regions of high electrical conductivity. Bear in mind, however, that while experimental measurements of rock properties in the laboratory are generally restricted to small specimens of individual rock types (e.g., O'CONNELL and BUDIANSKY 1974), rock volumes sampled by geophysical techniques are likely to be polylithologic with dimensions of 1–10 km.

1.5. Subduction Interface Megathrusts

Fore-arcs to island and mountain arcs are critically organized structures whose metastability is governed by the configuration of boundary stresses and the internal distribution of fluid-pressure in pore/fracture space (DAHLEN 1990). Seismogenic

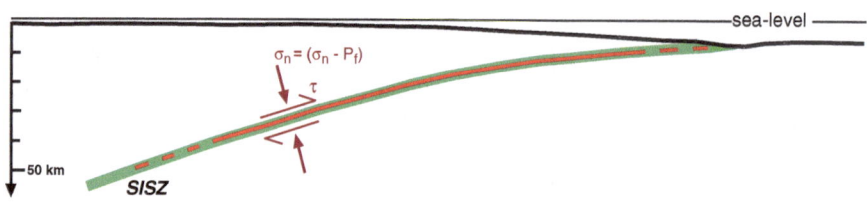

Figure 7
Schematic representation of a subduction megathrust rupture contained within a subduction interface shear zone (modeled on the Tohoku interface in northern Japan). *Solid red line* denotes seismogenic megathrust; *dashed segments* show regions sometimes characterized by episodic slow-slip events

megathrusts underlying fore-arcs may extend from the inner trench wall to depths of 40–50 km, perhaps 200 km inboard from the trench (Fig. 7), with the down-dip limit of the megathrust defined by the onset of crystal-plasticity at temperatures in the range 350–450 °C (HYNDMAN 2007). The megathrusts are hosted within subduction interface shear zones (SISZ) inferred to contain a mélange assemblage of subducted ocean floor sediments (muds, siliceous oozes, etc.) and trench wall sediments (muds, turbidite sands) together with similar material derived from the accretionary prism, and occasional seamount volcanics and slivers of oceanic crust (VON HUENE and SCHOLL 1991). Geophysical imaging reveals a thin (several km-scale) tabular shear zone, locally highly reflective and with anomalously low V_P, and high V_P/V_S (e.g., KODAIRA et al. 2002; ABERS 2005; SONG et al. 2009), thought to represent a fluid-rich but highly impermeable SISZ. The fluids are inferred to be predominantly aqueous (including free water within pore/fracture space and bound water within hydrous minerals) with lesser CO_2 and hydrocarbons.

The case for strong fluid overpressuring within SISZ throughout the full depth of subduction interface megathrusts is founded on: (1) observations and mechanical analyses limiting values of shear resistance along the interface to less than a few tens of MPa; and (2) measured seismological and other geophysical characteristics of the interface consistent with fluid-overpressuring to near-lithostatic values.

1.6. Constraints on Megathrust Shear Resistance

Heat flux generated along a plate interface is the product of average shear resistance and the interplate slip-rate. The lack of evidence for significant shear

heating along subduction megathrust interfaces limits the average shear stress along the interface to at most a few tens of MPa (PEACOCK 2004; WANG et al. 1995; VON HERZEN et al. 2001). Horizontal compressive stress in the forearcs of subduction systems is largely determined by the basal shear traction along the megathrust, which may extend 100–200 km down-dip (Fig. 7). Force balance analyses taking account of surface and Moho topography and the integrated effect of this basal shear traction generally limit depth-averaged shear stress along the subduction interface to c. 10 MPa (rising to as much as 40 MPa in a few instances), corresponding to 'effective' friction coefficients of c. 0.03 (LAMB 2006; SENO 2009; WANG and SUYEHIRO 1999; LUTTRELL et al. 2011). Note that while dynamic processes may operate to lower frictional shear resistance and heat generation during seismic slip episodes (see ROCKWELL and BEN-ZION 2007), this does not affect the estimates of basal shear stress from static force-balance analyses.

Low values of effective friction and shear resistance could be explained either by the presence of extremely low friction material along the interface or by fluid overpressuring. BYERLEE (1978) found sliding friction to be largely independent of rock type and generally restricted to the range, $0.6 < \mu_s < 0.9$. Clay-rich material, however, provides significant exceptions. For example, illite-rich and montmorillonite-rich gouges are characterized by $\mu_s \sim 0.4$ and 0.2, respectively (SAFFER and MARONE 2003), and recent friction experiments on saponite-rich San Andreas fault gouge recovered from 2.7 km depth in the SAFOD borehole suggest $\mu_s \rightarrow 0.1$ (LOCKNER et al. 2011; CARPENTER et al. 2012). However, saponite-montomorillonite remains stable only for $T < 150$ °C, so its weakening effect is likely restricted to the top

Table 1

Values of shear resistance and effective friction calculated from Eq. 5 for a megathrust interface at 20 km depth and an average crustal density of 2,550 kg/m³ for various coefficients of sliding friction, μ_s, and values of the pore-fluid factor, λ_v

	$\mu_s = 0.6$	$\mu_s = 0.4$	$\mu_s = 0.2$	$\mu_s = 0.1$
$\lambda_v = 0.4$ (hydrostatic)	$\tau = 180$ MPa $\mu_s' = 0.36$	$\tau = 120$ MPa $\mu_s' = 0.24$	$\tau = 60$ MPa $\mu_s' = 0.12$	$\tau = 30$ MPa $\mu_s' = 0.06$
$\lambda_v = 0.5$	$\tau = 150$ MPa	$\tau = 100$ MPa	$\tau = 50$ MPa	$\tau = 25$ MPa
$\lambda_v = 0.7$	$\mu_s' = 0.30$ $\tau = 90$ MPa	$\mu_s' = 0.20$ $\tau = 60$ MPa	$\mu_s' = 0.10$ $\tau = 30$ MPa	$\mu_s' = 0.05$ $\tau = 15$ MPa
$\lambda_v = 0.9$	$\mu_s' = 0.18$ $\tau = 30$ MPa	$\mu_s' = 0.12$ $\tau = 20$ MPa	$\mu_s' = 0.06$ $\tau = 10$ MPa	$\mu_s' = 0.03$ $\tau = 5$ MPa
$\lambda_v = 0.95$	$\mu_s' = 0.06$ $\tau = 15$ MPa $\mu_s' = 0.03$	$\mu_s' = 0.04$ $\tau = 10$ MPa $\mu_s' = 0.02$	$\mu_s' = 0.02$ $\tau = 5$ MPa $\mu_s' = 0.01$	$\mu_s' = 0.01$ $\tau = 2.5$ MPa $\mu_s' = 0.005$

half of the seismogenic zone. Normal stress across a flat-lying subduction interface is close to the vertical stress (i.e., $\sigma_n \sim \sigma_v$) so that, neglecting any cohesion, Eq. 1 may be rewritten to approximate the shear resistance along the interface as:

$$\tau_f = \mu_s' \sigma_v' = \mu_s' \sigma_v = \mu_s' \rho g z, \qquad (5)$$

where the 'effective' friction, $\mu_s' = \mu_s(1 - \lambda_v)$. Choosing $z = 20$ km as the average depth of a megathrust interface, values of shear resistance and effective friction along the interface are calculated at that depth for an average crustal density of 2,550 kg/m³ for $\mu_s = 0.6$ (the bottom of the 'Byerlee' range), 0.4 (illite-rich gouge), 0.2 (montmorillonite-rich gouge), and 0.1 (saponite-rich San Andreas gouge) (see Table 1). Even the lowest friction values require some degree of overpressuring to match the stress and effective friction constraints outlined above. With higher friction coefficients, overpressuring to near-lithostatic values (i.e., $\lambda_v > 0.9$) is needed to match the constraints.

1.7. Observational Evidence for Overpressuring within SISZ

A strong observational case can be made for fluid-overpressuring within the tabular SISZ hosting megathrust ruptures, which appear to be localized in the oceanic crust of the descending plate (AUDET *et al.* 2009; SONG *et al.* 2009), and are variously characterized by low P-wave and S-wave velocities, locally high reflectivity and anomalously high V_p/V_s ratios (ABERS 2005; EBERHART-PHILLIPS and REYNERS 1999; KODAIRA *et al.* 2002). For example, along the north Chilean subduction interface the V_p/V_s ratio increases from background values of c. 1.70 to values approaching 1.85 around the SISZ (HUSEN and KISSLING 2001; NIPPRESS and RIETBROCK 2007).

At shallow depth overpressures are generated by compaction of high porosity sediments contained within the SISZ and the accretionary prism occupying the hangingwall, with contributions from progressive metamorphic dewatering, especially the smectite-illite transition, as depth and temperature increase (SAFFER and TOBIN 2011). At greater depths an increasingly high proportion of fluid is derived from progressive metamorphism of the underthrust basaltic oceanic crust and lithosphere (HACKER 2008; PEACOCK *et al.* 2011). Generation and maintenance of near-lithostatic overpressures requires bulk permeabilities $<10^{-21}$ m² (AUDET *et al.* 2009), at least through interseismic periods. Factors contributing to the development of such low permeabilities within SISZ include: (1) the presence of a substantial proportion of fine-grained silicic mudrock within the mélange matrix; (2) the likely development of a phyllosilicate foliation subparallel to the shear zone walls, creating strong permeability anisotropy inhibiting vertical fluid transport; and, (3) an active hydrothermal environment along the lower half of the megathrust

13

interface ($150 < T < 350$ °C) so that solution transfer processes in fine-grained siliceous material clog pore and fracture space and contribute to the development of low-permeability seals (RUTTER 1976; McCLAY 1977; FAGERENG 2011). Thrust décollements are likely to localize at the base of such seals where overpressuring tends to be highest and shear resistance is at its lowest.

Transition zones occur at both the up-dip and down-dip margins to the locked seismogenic portion of the megathrust (Fig. 7) characterized by combinations of slow-slip events (LFE, VLFE, episodic tremor). From their association with low velocity zones, high seismic reflectivities, and high V_P/V_S, these transition regions are thought to be areas of extreme overpressuring to near-lithostatic values ($\lambda_v \to 1.0$) (SHELLY et al. 2006; SONG et al. 2009; PEACOCK et al. 2011). A somewhat lower degree of overpressuring is inferred for the locked central portions of seismogenic megathrusts (SAFFER and TOBIN 2011). Tomographic studies of fore-arc hangingwalls suggest, however, that the degree of overpressuring revealed by the distribution of V_P/V_S anomalies may be quite heterogeneous (HUANG et al. 2011; ZHAO et al. 2011) and may also change from before to after large megathrust ruptures (HUSEN and KISSLING 2001; TSURU et al. 2005).

Any inferred heterogeneity in the distribution of overpressure must, however, be reconciled with estimates from force balance analysis of an average $\lambda_v > 0.90$ for circum-Pacific megathrusts, with $0.95 < \lambda_v < 0.98$ in the case of the Miyagi, Shikoku, Peru, and N. and S. Chile subduction interfaces (LAMB 2006; SENO 2009). It thus seems inescapable that rupturing of subduction megathrusts occurs within SISZ that are strongly, if variably, fluid-overpressured ($\lambda_v \to 1.0$).

1.8. Intracontinental Thrust Faults

The likely involvement of fluid overpressures in facilitating low-angle thrust displacements within foreland fold-thrust belts was invoked first by HUBBERT and RUBEY (1959) but took no account of whether displacements were seismic or aseismic. Nonetheless, several of the active fold-thrust belts considered (e.g., Zagros, Iran; northern Pakistan) are

actively deforming and associated with distributed seismicity.

There are a limited number of seismically active areas where extensive subsurface information from oilfield drilling may be combined with high resolution earthquake locations from dense seismograph networks. One such is the area of the 1983–85 **M** 6.5 Coalinga–**M** 6.1 Kettleman Hills earthquake sequence in California where thrust rupturing initiated at c. 10 km depth below producing oilfields (EBERHART-PHILLIPS 1989; EKSTROM et al. 1992). Mainshock and aftershocks appear to have occurred within a crustal section where variably overpressured Tertiary and Cretaceous Great Valley sedimentary sequences overlie a wedge of the Franciscan metamorphic assemblage which is in turn in overthrust contact with underlying Great Valley sequence. There is direct evidence of overpressuring around the base of the Cenozoic section from boreholes extending to 2–3 km depth around the Coalinga Anticline and from the presence of a pronounced low-velocity zone associated with the contact at c. 10 km where the Franciscan assemblage overthrusts the Great Valley Sequence and the Coalinga mainshock nucleated (EBERHART-PHILLIPS 1989; YERKES et al. 1990). A similar case can be made for overpressuring within the active fold-thrust belt in the Santa Barbara Channel–Ventura Basin area of the western Transverse Ranges (YEATS 1983; SHAW and SUPPE 1994; SIBSON 2004). Further east in the Transverse Ranges, the LARSE seismic line was shot running NNE from the Los Angeles basin across the crystalline massif of the San Gabriel mountains to the San Andreas fault trace and the Mojave Desert (FUIS et al. 2001). Notable at depths of 18–23 km below the San Gabriel Mountains around the base of the seismogenic layer was a brightly reflective, c. 500 m thick, low-velocity zone dipping gently northwards towards the San Andreas fault. This feature has been interpreted as a fluid-rich thrust décollement shear zone replete with flat-lying, fluid-filled macro-cracks overpressured to near-lithostatic values (RYBERG and FUIS 1998), inviting comparison with the low-velocity zone associated with the nucleation of the 1983 **M** 6.5 Coalinga thrust rupture (EBERHART-PHILLIPS 1989). Analogies may also be drawn with mineralized low-dipping thrust faults hosting Au-quartz veins and lying subparallel to arrays of subhorizontal extension veins, which have been interpreted as developing incrementally through episodic 'fault-valve' action under near-lithostatic overpressure (e.g. HARLEY and CHARLESWORTH 1996).

Along the southern margin of the Po Plain region in northern Italy, the close association of active mud volcanoes with seismically active thrusts (in some instances mud eruptions are penecontemporaneous with rupturing) also indicates rupturing through an overpressured cover sequence (BONINI 2009). In New Zealand, overpressures approach lithostatic in the top few kilometres of the cover sequence in a region of active thrust/reverse faulting (again with widespread mud volcanism) along the hanging wall of the Hikurangi subduction margin (SIBSON and ROWLAND 2003). Wedge analysis suggests that $\lambda_v \sim 0.7$ prevails across the seismically active fold-thrust belt of Taiwan (DAVIS *et al.* 1983; DAHLEN 1990) with tomographic studies revealing coincident low P-wave and high Vp/Vs anomalies diagnostic of overpressuring in the focal zone of the 1999 **M** 7.6 Chi-Chi thrust rupture, varying from before, to after, the mainshock rupture (CHEN *et al.* 2001).

1.9. Steep Reverse Faulting During Compressional Inversion

The likelihood of overpressuring at seismogenic depths is also strong in areas of ongoing compressional inversion. For example, in northern Honshu several recent strong (6.0 < **M** < 7.0) earthquakes have involved close-to-pure reverse-slip rupturing on inherited normal faults along the margins of Miocene extensional basins flanking the magmatic arc, both to the east and to the west. On the assumption of horizontal maximum compressive stress, reactivation of these unfavourably oriented structures as steep reverse faults (dips > 45°), requires fluid pressure elevated towards lithostatic levels ($\lambda_v \to 1.0$) within the lower portions of the upper crustal seismogenic zone where the larger ruptures typically nucleate (SIBSON 2012). Geophysical diagnostics support the existence of a fluid-rich ($H_2O \pm CO_2$), variably overpressured mid-crust extending into the lower seismogenic zone, especially around the major fault systems (SIBSON 2009). The combination of

anomalously low P-wave and S-wave velocities with elevated V_p/V_s ratios, bright-spot reflectors and high electrical conductivity makes a particularly strong case for overpressuring, a notable example being the 2004 **M** 6.6 Mid-Niigata earthquake sequence (MATSUMOTO *et al.* 2005; UYESHIMA *et al.* 2005; OKADA *et al.* 2006; WANG and ZHAO 2006; SIBSON 2007; XIA *et al.* 2008). In the case of the Shonai Plain reverse fault system (the source of an **M** ~7 rupture in 1896), the causative structure is defined by a steeply dipping conductor extending through the seismogenic upper crust which may represent a residual zone of interconnected and probably overpressured fracture permeability (ICHIHARA *et al.* 2011). The seismogenic upper crust in NE Honshu, therefore, appears to function as a brittle stress guide under compressional stress, confining overpressured hydrothermal fluids in the mid-crust that are probably derived from a

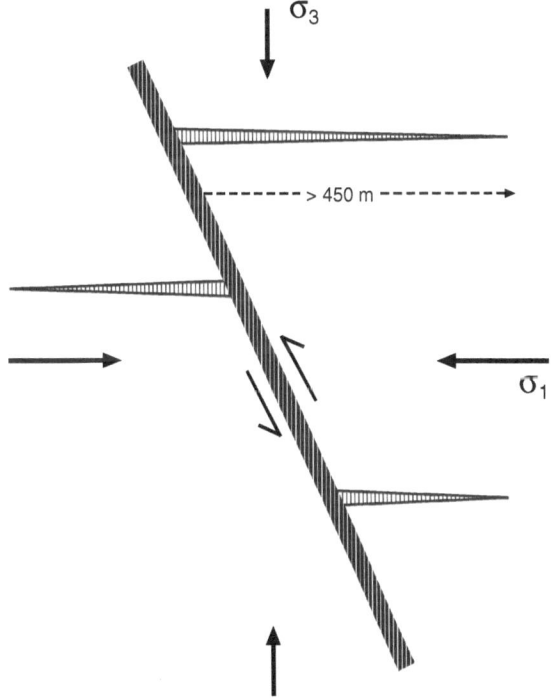

Figure 8
Cartoon representation of flat-lying quartz-infilled extension veins ('flats' that opened vertically) tapering away from a steep reverse shear zone, all inferred to have developed within a compressional stress field ($\sigma_v = \sigma_3$). Diagram is based on Hollinger Mine, Ontario, Canada, where one such 'flat' was mined laterally for 450 m (HALL 1985). Hosting Archean assemblage at Hollinger Mine comprises greenstone metavolcanics cross-cut by felsic intrusives

combination of magmatic intrusion and prograde metamorphism within the crust, as well as from the underlying dewatering slab. The congruence between geological evidence for ~ lithostatic overpressure in the form of flat-lying hydrofracture extension veins associated with exhumed steep reverse faults (as illustrated in Fig. 8), and the afore-mentioned geophysical diagnostics of a fluid-overpressured mid-crust points to direct involvement of overpressured fluids in rupture nucleation in such settings.

Comparable fluid overpressuring within the lower seismogenic zone seems likely in other areas of active compressional inversion such as the north-western South Island of New Zealand (GHISETTI and SIBSON 2006).

1.10. Intracontinental Strike-Slip Faults

The manifest problems of generating and sustaining fluid overpressures in and around subvertical strike-slip faults have been discussed by RICE (1992) and FAULKNER and RUTTER (2001). However, strike-slip faults occur over an enormous scale-range: we consider first the case for fluid involvement in isolated low-displacement structures. LECLÈRE et al. (2012) analysed a swarm of predominantly strike-slip earthquakes that occurred along the border of the Alps in SE France in 2003–2004. The swarm, concentrated at depths of 6–7 km within the Argentera crystalline massif, involved selective dextral reactivation of one of a set of subvertical strike-slip faults lying at a rather high angle (c. 63°) to the regional σ_1 direction. Significant fluid overpressuring was invoked to account for the reactivation of the strike-slip fault at such an unfavourable orientation with respect to the stress field. In southern Italy, BONCIO (2008) employed rheological modelling to invoke moderate overpressuring ($\lambda_v > 0.65$) in the mid-crust, possibly from mantle-derived CO_2 (CHIODINI et al. 2004) to account for a series of E-W striking, dextral **M** 5.7 strike-slip ruptures occurring at depths of 10–25 km in the mid-crust beneath the Apulian foreland of the Apennines, below the usual seismogenic layer. In Japan, tomographic analysis led ZHAO and NEGISHI (1998) to infer an overpressured, fluid-filled rock matrix at c. 18 km depth around the base of the Nojima Fault seismogenic zone and close

to the hypocentre of the 1995 **M** 7.2 Kobe strike-slip rupture.

For large-displacement (>300 km) intracontinental strike-slip faults we concentrate our attention on the San Andreas dextral fault system because it has been intensively studied with high-resolution investigations of its geophysical attributes. WESNOUSKY (2005) has emphasized its overall transpressional character. Through southern and central California the main trace of the San Andreas fault lies at 60–85° to the regional maximum compressive stress (TOWNEND and ZOBACK 2004) and it is flanked by systems of thrust faults which favour containment of overpressured fluids. Indeed, there is widespread evidence for fluid overpressures at depths >2 km or so below the Coast Ranges abutting the San Andreas fault to the northeast (BERRY 1973; YERKES et al. 1990; UNRUH et al. 1992; MCPHERSON and GARVEN 1999) and in the Western Transverse Ranges to the southwest (YEATS 1983; RYBERG and FUIS 1998). It would be surprising, therefore, if the San Andreas seismogenic zone was not, at least locally, overpressured. IRWIN and BARNES (1975) pointed out that those segments of the San Andreas fault system in central and northern California that exhibit creep coupled to high levels of microseismicity correspond to areas where the major strike-slip faults are transecting the Coast Range thrust to the northeast with a hanging-wall rich in impermeable serpentinite from the Coast Range ophiolite. They suggested that CO_2-rich metamorphic fluids (presumably overpressured) derived from the Franciscan assemblage below the Coast Range thrust were infiltrating the San Andreas system at depth and contributing to the anomalous slip behaviour of the creeping segments.

High-resolution geophysical investigations have been widespread, but non-uniform, along the San Andreas fault with attention especially focused on the Parkfield area at the transition between the creeping and locked portions of the fault, the site of the SAFOD scientific drilling experiment which intersected the fault zone at c. 2.7 km depth (ZOBACK et al. 2011). While fluid overpressures were not directly encountered in the SAFOD borehole, overpressures with $\lambda_v \sim 0.7$ were measured at c. 1.5 km depth in the Varian well located approximately 1.5 km NE of the San Andreas surface trace (MALIN et al. 1987). It appears that the SAF acts as a hydrologic barrier between the overpressured region northeast of the fault and the essentially hydrostatically pressured region to the southwest (ZOBACK *et al.* 2011). However, from recovered core from the San Andreas fault zone, the existence of hydrothermal extension veins suggests that transitory overpressuring does sometimes occur within the fault zone (MITTEMPERGHER *et al.* 2011). Moreover, the morphology of pores in ultracataclasite from the SAFOD core led JANSSEN *et al.* (2011) to conclude that they were held open by elevated fluid pressure during deformation.

In the Parkfield area a thin vertical band of microseismicity flanks a low-velocity damage zone c. 200 m wide that has now been imaged to at least 7 km depth (LI and MALIN 2008). Magnetotelluric (MT) transects have identified a tabular subvertical fault zone conductor (FZC) extending to depths of a few kilometres (UNSWORTH *et al.* 1997; UNSWORTH and BEDROSIAN 2004) coincident with the c. 200 m wide damage zone defined by this fault zone wave-guide. Comparable FZC have been identified at other San Andreas localities but possibly extend to c. 6 km depth in the locked segment of the Carrizo Plain (UNSWORTH *et al.* 1999) and to the base of the microseismic zone at c. 10 km in the Hollister area (BEDROSIAN *et al.* 2002). These FZC are interpreted as comprising high-porosity fractured rock saturated with variably saline fluids within the fault damage zone.

In the region south of Hollister, towards the northwestern termination of the creeping segment of the San Andreas fault, THURBER *et al.* (1997) demonstrated the existence of a zone of anomalously low Vp and high Vp/Vs with associated microseismicity extending to 6–7 km depth below the San Andreas trace, with further zones of anomalously low Vp/Vs flanking the Calaveras fault to the northeast. They interpret these Vp/Vs anomalies as fluid-rich regions involved with the faulting process. In the same area, BEDROSIAN *et al.* (2004) showed that the low Vp and high Vp/Vs anomalies coincide with low resistivity anomalies from MT studies that flank the microseismicity below the San Andreas trace to 8–10 km depth, essentially right through the seismogenic zone. Inferred porosities range from 15 % to 35 % in the top 2–3 km to 2–15 % at greater depths below the

San Andreas trace. Coincidental low Vp, high Vp/Vs, and high electrical conductivity anomalies extending through the seismogenic zone to 10 km depth are highly suggestive of fluid overpressuring through at least the lower seismogenic zone. While it is possible to maintain a network of interconnected fractures under hydrostatic fluid pressure in the top few kilometres (SECOR 1965), some degree of fluid overpressuring seems likely to be required at greater depths.

More recently, a larger aperture MT transect near the SAFOD site (BECKEN *et al.* 2008) revealed that, as well as the previously recognized FZC, a broad tabular high conductivity anomaly dips 70° SW from an area northeast of the San Andreas fault extending downward through the base of the San Andreas seismogenic zone at 10–15 km depth to continue into the lower crust and upper mantle. This anomaly is interpreted as a crustal fluid channel and is of particular interest because of the evidence from helium isotope ratios for an influx of mantle fluids (probably CO_2-rich) into the ductile roots of the San Andreas fault zone (KENNEDY *et al.* 1997). Moreover, the inference of near-lithostatic overpressure accompanying tremor activity at 25–30 km depth along the downward projection of the SAF (THOMAS *et al.* 2009) could also be associated with this high conductivity anomaly (BECKEN *et al.* 2011).

Overall, though much of the evidence is circumstantial, the indications are that at least some of the seismically active portions of the San Andreas fault system, especially in the lower seismogenic zone, are fluid overpressured. One may also note the not infrequent occurrence of earthquake bursts throughout southern California which, in some hypotheses, are attributed to migration of overpressured fluid (VIDALE and SHEARER 2006).

1.11. Intracontinental Normal Faults

Fluid overpressures are least likely to be sustained in areas of active tectonic extension unless fluid forcing is extreme because of the prevalence of steep normal faults and subvertical fractures (Figs. 3, 6). For example, in the Taupo Volcanic Zone, New Zealand, which is actively extending NW–SE at c. 10 mm/year, arguments have been advanced that vigorous convection of hydrothermal fluids under hot/cold hydrostatic conditions extends through the full (c. 7 km) depth of the seismogenic layer because the average spacing of upwelling hydrothermal plumes is about twice the layer thickness (BIBBY et al. 1995; SIBSON and ROWLAND 2003). Local overpressuring may develop, however, where hydrothermal flow within fault conduits is locally and probably transitorily impeded by 'throttles' created by mineral precipitation (ROWLAND and SIMMONS 2012). Note, however, that active normal faulting is known in some strongly overpressured though largely aseismic sedimentary basins (e.g. ROBERTS et al. 1996).

Central and southern Italy is a region of particular interest because the belt of active extension which gives rise to moderate-to-large normal fault ruptures southwest of the main divide of the Apennines coincides with the northeastern boundary of an area of abnormally high (4–13 Mt/year), mantle-derived CO_2 flux (CHIODINI et al. 2004; FREZZOTTI et al. 2009). Throughout this region, there is considerable evidence for the involvement of overpressured CO_2-rich fluids in the nucleation of normal fault ruptures. The progression of earthquakes in the 1997 Umbria-Marche earthquake sequence has been modelled in terms of a pulse of overpressured CO_2 fluids migrating upwards and along strike from initial rupture across a sealing horizon of Triassic evaporates (MILLER et al. 2004; ANTONIOLI et al. 2005). On the basis of local time-dependent variations in seismic velocity, LUCENTE et al. (2010) put the case for involvement of overpressured fluid in the nucleation of the 2009 **M** 6.3 L'Aquila normal fault rupture. Likewise, TERAKAWA et al. (2010) argue for heterogeneous overpressuring at 7–10 km depth to account for the reactivation of unfavorably oriented slip planes during the aftershock sequence. CHIARALUCE et al. (2007) also invoke the accumulation of overpressured CO_2-rich fluids to account for microseismic reactivation, further to the west, of the very low-angle (c. 15°) NE-dipping Alto Tiberina extensional fault where fluid overpressures have been documented in the footwall of the fault structure by borehole measurements. Evidence for significant fluid-pressure cycling from sub-lithostatic towards hydrostatic values has also been recorded from fluid

inclusions in hydrothermal veining developed within the exhumed footwalls of active normal faults in the western US (PARRY and BRUHN 1990).

2. Discussion

Because of the lack of comparable high-resolution geophysical data on oceanic fault systems aside from subduction megathrusts, this review has focused on intracontinental fault systems. Critical indicators of local fluid-overpressure include: (1) anomalously low seismic P-wave and S-wave velocities; (2) anomalously high Vp/Vs ratios; (3) bright-spot reflectors in compressional stress regimes; (4) zones of anomalously high electrical conductivity; and, where the local stress field is well defined, (5) evidence of persistent reactivation of unfavourably oriented faults. The case for overpressuring is strongest where two or more of these indicators are congruent, but each case has to be evaluated on its merits. In general agreement with geometric and mechanical considerations, it does seem that fluid overpressures are more likely to develop and be sustained in seismically active compressional/transpressional tectonic settings compared with extensional/transtensional regimes.

On the basis of geophysical observations and force-balance analyses, SISZ appear overpressured to near-lithostatic levels ($\lambda_v > 0.9$) over the full depth of seismogenic megathrusts. Accepting the estimate of PACHECO and SYKES (1992) that megathrust interfaces within SISZ give rise to >90 % of global seismic moment release, it follows that the bulk of seismic moment release comes from rupture nucleation and propagation within fault zones that are highly fluid-overpressured. Strong overpressuring at depths of rupture nucleation is also inferred for active fold-thrust belts and in areas of ongoing compressional inversion (e.g., northern Honshu) where the reactivation of inherited normal faults as steep reverse faults requires near-lithostatic overpressure ($\lambda_v \to 1.0$) at depths of rupture initiation. Certainly, field evidence for significant overpressuring in and around rupture nucleation sites is strongest in areas of active compressional tectonics, though higher resolution of geophysical anomalies will be needed to make a definitive case for fluid involvement in rupture nucleation. Evidence for overpressuring around strike-slip faults is generally less clear but tends to be stronger in areas of transpression with overpressuring concentrated in the lower portions of the upper crustal seismogenic layer. In areas of extensional tectonics coincident with high levels of fluid forcing (e.g., Italy), there are inferences of overpressuring also concentrated towards the base of the seismogenic zone where the larger ruptures tend to nucleate.

Nonetheless, cycling of fluid overpressure and frictional strength through some form of fault-valve action seem likely to be widespread, especially in areas of active compressional/transpressional tectonics. Recurrence and nucleation of earthquake ruptures in such areas are, thus, likely to be affected by the cycling of overpressure and frictional strength as well as tectonic stress. An unresolved issue is the heterogeneity of overpressuring; the extent to which active fault zones themselves serve as fluid conduits that are locally overpressured with respect to the surrounding crust? In mesozonal Au-quartz vein systems inferred to have developed by fault-valve action on steep reverse faults (BOULLIER and ROBERT 1992; COX 1995; SIBSON and SCOTT 1998) the habitual tapering of flat-lying extension veins away from the faults in which they root suggests that the fault zones themselves are preferentially overpressured above the surrounding crust (Fig. 8). This tends to accord with heterogeneity of overpressuring inferred around active reverse fault systems from tomographic analyses (OKADA et al. 2006; SIBSON 2007).

Acknowledgments

Sincere thanks to the organisers, Yehuda Ben-Zion and Antonio Ravolli for making it possible for me to participate in the 40th Workshop of the International School of Geophysics—Properties and Processes of Crustal Fault Zones.

REFERENCES

ABERS, G.A. (2005), Seismic low-velocity layer at the top of subducting slabs: observations, predictions, and systematics, Phys Earth Planet Int 149, 7–29.

ANDERSON, E.M. (1905), The dynamics of faulting, Trans Edin Geol Soc 8, 387–402.

ANTONIOLI, A., PICCININI, D., CHIARALUCE, L. and COCCO, M. (2005), Fluid flow and seismicity pattern: evidence from the 1997 Umbria-Marche (central Italy) seismic sequence, Geophys Res Lett 32, L10311, doi:10.1029/2004GL022256.

AUDET, P., BOSTOCK, M.G., CHRISTENSEN, N.I. and PEACOCK, S.M. (2009), Seismic evidence for overpressured subducted ocean crust and megathrust fault sealing, Nature 457, 76–78.

BAISCH, S., WEIDLER, R., VÖRÖS, R, WYBORN, D. and DE GRAAF, L. (2006), Induced seismicity during the stimulation of a geothermal HFR reservoir in the Cooper Basin, Australia, Bull Seism Soc Am 96, 2242–2256.

BECKEN, M., RITTER, O., Bedrosian, P.A. and Weckmann, U. (2011), Correlation between deep fluids, tremor and creep along the central San Andreas fault, Nature 480, 87–90.

BECKEN, M., RITTER, O., PARK, S.K., BEDROSIAN, P.A., WECKMANN, U. and WEBER, M. (2008), A deep crustal fluid channel into the San Andreas Fault system near Parkfield, California, Geophys J Int 173, 718–732.

BEDROSIAN, P.A., UNSWORTH, M.J. and EGBERT, G. (2002), Magnetotelluric imaging of the creeping segment of the San Andreas Fault near Hollister, Geophys Res Lett 29, 1–4.

BEDROSIAN, P.A., UNSWORTH, M.J., EGBERT, G.D. and THURBER, C.H. (2004), Geophysical images of the creeping segment of the San Andreas fault: implications for the role of crustal fluids in the earthquake process, Tectonophysics 385, 137–158.

BEGBIE, M.J. and SIBSON, R.H. (2006), Structural controls on the development of fault-hosted scheelite-gold mineralisation in Glenorchy, Northwest Otago, In, Christie, A. and Brathwaite, R. (eds.), Mineral Deposits of New Zealand, Aust Inst Min Metall Mon 25, 291–298.

BERRY, F.A.F. (1973), High fluid potentials in California Coast Ranges and their tectonic significance, AAPG Bulletin 57, 1219–1249.

BIBBY, H.M., CALDWELL, T.G., DAVEY, F.J. and WEBB, T.H. (1995), Geophysical evidence on the structure of the Taupo volcanic zone and its hydrothermal circulation, J Volc Geotherm Res 68, 29–58.

BONCIO, P. (2008), Deep-crust strike-slip earthquake faulting in southern Italy aided by high fluid pressure: insights from rheological analysis, Geol Soc Lond Spec Publ 299, 195–210.

BONINI, M. (2009), Mud volcano eruptions and earthquakes in the Northern Apennines and Sicily, Italy, Tectonophysics 474, 723–735.

BOULLIER, A.-M. and ROBERT, F. (1992), Paleoseismic events recorded in Archean gold-quartz vein networks, Val d'Or, Abitibi, Quebec, Canada, J Struct Geol 14, 161–179.

BRACE, W.F. (1960), An extension of the Griffith theory of fracture to rocks, J Geophys Res 65, 3477–3480.

BRACE, W.F. (1980), Permeability of crystalline and argillaceous rocks, Int J Rock Mech Min Sci 17, 241–251.

BYERLEE, J.D. (1978), Friction of rocks, Pure Appl Geophys 116, 615–626.

BYERLEE, J.D. (1993), Model for episodic flow of high-pressure water in fault zones before earthquakes, Geology 21, 303–306.

CARPENTER, B.M., SAFFER, D.M. and MARONE, C. (2012), Frictional properties and sliding stability of the San Andreas fault from deep drill core, Geology 40, 759-762.

CHEN, C-H., WANG, W-H. and TENG, T-L. (2001), 3D velocity structure around the source area of the 1999 Chi-Chi Taiwan *earthquake before and after the mainshock*, Bull Seism Soc Am *91*, 1013–1027.

CHIARALUCE, L., CHIARABBA, C., COLLETTINI, C., PICCININI, D. and COCCO, M. (2007), *Architecture and mechanics of an active low-angle normal fault: Alto Tiberina Fault, northern Apennines, Italy*, J Geophys Res *112*, B10310, doi:10.1029/2007JB005015.

CHIODINI, G., CARDELLINI, C., AMATO, A., BOSCHI, E., CALIRO, S., FRONDINI, F. and VENTURA, G. (2004), *Carbon dioxide Earth degassing and seismogenesis in central and southern Italy*, Geophys Res Lett *31*, L07615, doi:10.1029/2004GL019480.

CHRISTENSEN, N.I. (1996), *Poisson's ratio and crustal seismology*, J Geophys Res *101*, 3139–3156.

COLLETTINI, C., DE PAOLA, N. and GOULTY, N.R. (2006), *Switches in the minimum compressive stress direction induced by overpressure beneath a low-permeability fault zone*, Terra Nova *18*, 224–231.

COX, S.F. (1995), *Faulting processes at high fluid pressures: an example of fault-valve behavior from the Wattle Gully Fault, Victoria, Australia*, J Geophys Res *100*, 12,841–12,859.

DAHLEN, F.A. (1990), *Critical taper model of fold-and-thrust belts and accretionary wedges*, Ann Rev Earth Planet Sci *18*, 55–99.

DAVIS, D., SUPPE, J. and DAHLEN, F.A. (1983), *Mechanics of fold-and-thrust belts and accretionary wedges*, J Geophys Res *88*, 1153–1172.

EBERHART-PHILLIPS, D. (1989), *Active faulting and deformation of the Coalinga anticline as interpreted from three-dimensional velocity structure and seismicity*, J Geophys Res *94*, 15,565–15,586.

EBERHART-PHILLIPS, D., STANLEY, W.D., RODRIGUEZ, B.D. and LUTTER, W.J. (1995), *Surface seismic and electrical methods to detect fluids related to faulting*, J Geophys Res *100*, 12,919–12,936.

EBERHART-PHILLIPS, D. and REYNERS, M. (1999), *Plate interface properties in the northeast Hikurangi subduction zone, New Zealand, from converted seismic waves*, Geophys Res Lett *26*, 2565–2568.

EKSTROM, G., STEIN, R.S., EATON, J.P. and EBERHART-PHILLIPS, D. (1992), *Seismicity and geometry of a 110 km long blind thrust fault: 1. The 1985 Kettleman Hills, California earthquake*, J Geophys Res *97*, 4843–4864.

ETHERIDGE, M.A. (1983), *Differential stress magnitudes during regional deformation and metamorphism: upper bound imposed by tensile fracturing*, Geology *11*, 231–234.

FAGERENG, A. (2011), *Geology of the seismogenic subduction thrust interface*, Geol Soc Lond Spec Publ *359*, 55–76.

FAULKNER, D.R. and RUTTER, E.H. (2001), *Can the maintenance of overpressured fluids in large strike-slip fault zones explain their apparent weakness?*, Geology *29*, 503–506.

FOXFORD, K.A., NICHOLSON, R., POLYA, D.A. and HEBBLEWHITE, R.P.B. (2000), *Extensional failure and hydraulic valving at Minas da Panasqueira, Portugal: evidence from vein spatial distributions, displacements, and geometries*, J Struct Geol *22*, 1065–1086.

FREZZOTTI, M., PECCERILLO, A. and PANZA, G. (2009), *Carbonate metasomatism and CO_2 lithosphere-asthenosphere degassing beneath the Western Mediterranean: an integrated model arising from petrological and geophysical data*, Chem Geol *262*, 108–120.

FUIS, G., RYBERG, T., GODFREY, N.J., OKAYA, D.A. and MURPHY, J.M. (2001), *Crustal structure and tectonics from the Los Angeles basin to the Mojave Desert, southern California*. Geology *29*, 15–18.

GHISETTI, F. and SIBSON, R.H. (2006), *Accommodation of compressional inversion in northwestern South Island (New Zealand): old faults versus new?* J Struct Geol *28*, 1994–2010.

GIGER, S.B., TENTOREY, E., COX, S.F. and FITZGERALD, J.D. (2007), *Permeability evolution in quartz fault gouge under hydrothermal conditions*, J Geophys Res *112*, B07202, doi:10.1029/2006 JB004828.

HACKER, B.R (2008), *H_2O subduction beyond arcs*, Geochem Geophys Geosyst 9, Q03001:doi:10.1029/2007GC001707.

HALL, D.J. (1985), *Flat veins at the Hollinger Mine, Timmins, Ontario*, unpublished MSc dissertation, Queen's University, Kingston, Ontario, Canada, pp. 93.

HARLEY, M. and CHARLESWORTH, E.G. (1996), *The role of fluid pressure in the formation of bedding-parallel, thrust-hosted gold deposits, Sabie-Pilgrim's Rest goldfield, eastern Transvaal*, Precambrian Research *79*, 125–140.

HUANG Z., ZHAO, D. and WANG, L. (2011), *Seismic heterogeneity and anisotropy of the Honshu arc from the Japan Trench to the Japan Sea*, Geophys J Int *184*, 1428–1444.

HUBBERT, M.K. and RUBEY, W.W. (1959), *Role of fluid pressure in mechanics of overthrust faulting*, Geol Soc Am Bull *70*, 115–166.

HUSEN, S. and KISSLING, E. (2001), *Postseismic fluid flow after the large subduction earthquake of Antofagasta, Chile*, Geology *29*, 847–850.

HYNDMAN, R.D. (2007), *The seismogenic zone of subduction thrust faults: what we know and what we don't know*, In, Dixon, T.H. and Moore, J.C. (eds.), *The Seismogenic Zone of Subduction Thrust Faults*, Columbia University Press, New York, pp. 15–40.

ICHIHARA, H., UYESHIMA, M., SAKANAKA, S, OGAWA, T., MISHINA, M., OGAWA, Y., NISHITANI, T., YAMAYA, Y., WATANABE, A., MORITA,. Y., YOSHIMURA, R. and USUI, Y. (2011), *A fault-zone conductor beneath a compressional inversion zone, northeastern Honshu, Japan*, Geophys Res Lett *38*, L09301, doi:10.1029/2011GL047382.

INGEBRITSEN, S. and MANNING, C. (1999), *Geological implications of a permeability depth curve for the continental crust*, Geology *27*, 1107–1110.

IRWIN, W.P. and BARNES, I. (1975), *Effect of geologic structure and metamorphic fluids on seismic behavior of the San Andreas fault system in central and northern California*, Geology *3*, 713–716.

JAEGER, J.C. and COOK, N.G.W. (1979), *Fundamentals of Rock Mechanics (3rd edn)*, Chapman and Hall, London, pp. 593.

JANSSEN, C., WIRTH, R., REINICKE, A., RYBACKI, E., NAUMANN, R., WENK, H-R. and DRESEN, G. (2011), *Nanoscale porosity in SAFOD core samples (San Andreas Fault)*, Earth Planet Sci Lett *301*, 179–189.

KENNEDY, B.M., KHARAKA, Y.K., EVANS, W.C., ELLWOOD, A., DEPAOLO, D.J., THORDSEN, J., AMBATS, G. and MARINER, R.H. (1997), *Mantle fluids in the San Andreas fault system, California*, Science *278*, 1278–1281.

KODAIRA, S., KURASHIMO, E., PARK, J.-O., TAKAHASHI, N., NAKANISHI, A., MIURA, S., IWASAKI, T., HIRATA, N., ITO, K., and KANEDA, Y. (2002), *Structural factors controlling the rupture process of megathrust earthquake at the Nankai trough seismogenic zone*, Geophys J Int *149*, 815–835.

LAMB, S. (2006), *Shear stress on megathrusts: Implications for mountain building behind subduction zones*, J Geophys Res *111*, B07401, doi:10.1029/2005JB003916.

LECLÈRE, H., FABBRI, O., DANIEL, G. and CAPPA, F. (2012), *Reactivation of a strike-slip fault by fluid overpressuring in the southwestern French-Italian Alps*, Geophys J Int *189*, 29–37.

LI, Y.-G. and MALIN, P.E. (2008), *San Andreas Fault damage at SAFOD viewed with fault-guided waves*, Geophys Res Lett *35*, L08304, doi:10.1029/2007GL032924.

LOCKNER, D.A. (1995), *Rock failure*, In: Ahrens, T.J. (ed.), *Rock Physics and Phase Relations: A Handbook of Physical Constants*, AGU Ref Shelf *3*, 127–147.

LOCKNER, D.A., MORROW, C., MOORE, D. and HICKMAN, S. (2011), *Low strength of deep San Andreas fault gouge from SAFOD core*, Nature *472*, 82–86.

LUCENTE, F.P., DE GORI, P., MARGHERITI, L., PICCININI, D., DI BONA, M., CHIARABBA, C. and AGOSTINETTI, N.P. (2010), *Temporal variation of seismic velocity and anisotropy before the 2009 $M_w6.3$ L'Aquila earthquake*, Italy. Geology *38*, 1015–1018.

LUTTRELL, K.M., TONG, X., SANDWELL, D.T., BROOKS, B.A. and BEVIS, M.G. (2011), *Estimates of stress drop and crustal tectonic stress from the 7 February 2010 Maule, Chile, earthquake: implications for fault strength*, J Geophys Res *116*, B11401, doi:10.1029/2011JB008509.

MCCAIG, A.M. (1997), *The geochemistry of volatile fluid flow in shear zones*, In: *Deformation-Enhanced Fluid Transport in the Earth's Crust and Upper Mantle*, (M.B. Holness, Ed.), pp. 227-266, Chapman and Hall, London.

MCCLAY, K.R. (1977), *Pressure solution and Coble creep in rocks and minerals: a review*, J Geol Soc Lond *134*, 57–50.

MCPHERSON, B.J.O. and GARVEN, G. (1999), *Hydrodynamics and overpressure mechanisms in the Sacramento Basin, California*, Am J Sci *299*, 429–466.

MALIN, P., MCEVILLY, T. and MOSES, T. (1987), *Cooperative UCSB-UCB-USGS recovery and instrumentation of the 5150 ft Varian A-1 well as a geophysical observatory*, EOS Trans Am Geophys Union *68*, 1357.

MATSUMOTO, S., IIO, Y., UEHIRA, K. and SHIBUTANI, T. (2005), *Imaging of S-wave reflectors in and around the hypocentral area of the 2004 mid-Niigata Prefecture earthquake (M6.8)*, Earth Planets Space *57*, 557–561.

MILLER, S.A. (2002), *Inferring fault strength from earthquake rupture properties and the tectonic implications of high pore pressure faulting*, Earth Planets Space *54*, 1173–1179.

MILLER, S.A., COLLETTINI, C., CHIARALUCE, L., COCCO, M., BARCHI, M. and KAUS, B.J.P. (2004), *Aftershocks drive by a high-pressure CO_2 source at depth*, Nature *427*, 724–727.

MITCHELL, T.M. and FAULKNER, D.R. (2009), *The nature and origin of off-fault damage surrounding strike-slip fault zones with a wide range of displacements: A field study from the Atacama fault system, northern Chile*, J Struct Geol. *31*, 802–816.

MITTEMPERGHER, S., DI TORO, G., GRATIER, J.P., HADIZADEH, J., SMITH, S.A.F. and SPIESS, R. (2011), *Evidence of transient increases of fluid pressure in SAFOD phase III cores*, Geophys Res Lett *38*, L03301, doi:10.1029/2010GL046129.

MORROW, C.A., MOORE, D.E., and LOCKNER, D.A. (2001), *Permeability reduction in granite under hydrothermal conditions*, J Geophys. Res *106*, 30,551–30,560.

NAZARETH, J. and HAUKSSON, E. (2004), *The seismogenic thickness of the southern California crust*, Bull Seism Soc Am *94*, 940–960.

NIPPRESS, S.E.J. and RIETBROCK, A. (2007), *Seismogenic zone high permeability in the Central Andes inferred from relocations of micro-earthquakes*, Earth Planet Sci Lett *263*, 235–245.

O'CONNELL, R.J. and BUDIANSKY, B. (1974), *Seismic velocities in dry and saturated cracked solids*, J Geophys Res *79*, 5412–5735.

OKADA, T., YAGINUMA, T., UMINO, N., MATSIZAWA, T., HASEGAWA, A., ZHANG, H. and THURBER, C.H. (2006), *Detailed imaging of the fault planes of the 2004 Niigata-Chuetsu, central Japan, earthquake sequence by double-difference tomography*. Earth Planet Sci Lett *244*, 32–43.

PACHECO, J.F. and SYKES, L.R. (1992), *Seismic moment catalog of large shallow earthquakes, 1900–1989*, Bull Seism Soc Am *82*, 1306–1349.

PARRY, W.T. (1998), *Fault-fluid compositions from fluid-inclusion observations and solubilities of fracture-sealing minerals*, Tectonophysics, *290*, 1–26.

PARRY, W.T. and BRUHN, R.L. (1990), *Fluid pressure transients on seismogenic normal faults*, Tectonophysics *179*, 335–344.

PEACOCK, S.M. (2004), *Thermal structure and metamorphic evolution of subducting slabs*, In, Eiler, J. (ed.), *Inside the Subduction Factory*, Geophysical Monograph *138*, American Geophysical Union, Washington, D.C., pp. 7–22.

PEACOCK, S.M., CHRISTENSEN, N.I., BOSTOCK, M.G. and AUDET, P. (2011), *High pore pressures and porosity at 35 km depth in the Cascadia subduction zone*, Geology *39*, 471–474.

RICE, J.R. (1992), *Fault stress states, pore pressure distributions, and the weakness of the San Andreas fault*, In: Evans, B. and Wong, T.F. (eds.) *Earthquake Mechanisms and Transport Properties of Rocks*, London, Academic Press, pp. 475–503.

ROBERT, F., BOULLIER, A.-M. and FIRDAOUS, K. (1995), *Gold-quartz veins in metamorphic terranes and their bearing on the role of fluids in faulting*, J Geophys Res *100*, 12,861–12,879.

ROBERTS, S.J., NUNN, J.A., CATHLES, L. and CIPRIANI, F.-D. (1996), *Expulsion of abnormally pressured fluids along faults*, J Geophys Res *101*, 28,231–28,252.

ROCKWELL, T.K. and BEN-ZION, Y. (2007), *High localization of primary slip zones in large earthquakes from paleoseismic trenches: Observations and implications for earthquake physics*, J Geophys Res *112*, B10304, doi:10.1029/2006JB004764.

ROWLAND, J.V. and SIMMONS, S.F. (2012), *Hydrologic, magmatic, and tectonic controls on hydrothermal flow, Taupo Volcanic Zone, New Zealand: Implications for the formation of epithermal vein deposits*, Econ Geol *107*, 427–457.

RUTTER, E.H. (1976), *The kinetics of rock deformation by pressure solution*, Phil Trans Roy Soc *A283*, 203–219.

RYBERG, T. and FUIS, G. (1998), *The San Gabriel Mountains bright reflective zone: possible evidence of young mid-crustal thrust faulting in southern California*. Tectonophysics *286*, 31–46.

SAFFER, D.E. and MARONE, C. (2003), *Comparison of smectite- and illite-rich gouge frictional properties: application to the up-dip limit of the seismogenic zone along subduction megathrusts*, Earth Planet Sci Lett *215*, 219–235.

SAFFER, D.M. and TOBIN, H.J. (2011), *Hydrology and mechanics of subduction zone forearcs: fluid flow and pore pressure*, Ann Rev Earth Planet Sci *215*, 219–235.

SCHOLZ, C.H. (1998), *Earthquakes and friction laws*, Nature *391*, 37–42.

SECOR, D.T. (1965), *Role of fluid pressure in jointing*, Am J Sci *263*, 633–646.

SENO, T. (2009), *Determination of the pore-fluid pressure ratio at seismogenic megathrusts in subduction zones: implications for strength of asperities and Andean-type mountain building*, J Geophys Res *114*, B05405, doi:10.1029/2008JB005889.

SHAW, J.H. and SUPPE, J. (1994), *Active faulting and fold growth in the eastern San Barbara Channel, California*, Geol Soc Am Bull *106*, 606–626.

SHELLY, D.R., BEROZA, G., IDE, S. and NAKAMULA, S. (2006), *Low-frequency earthquakes in Shikoku, Japan, and their relationship to episodic tremor and slip*, Nature 442, 188191.

SIBSON, R.H. (1983), *Continental fault structure and the shallow earthquake source*, J Geol Soc Lond 140, 741–767.

SIBSON, R.H. (1985), *A note on fault reactivation*, J Struct Geol 7, 751–754.

SIBSON, R.H. (1992), *Implications of fault-valve behavior for rupture nucleation and recurrence*, Tectonophysics 18, 1031–1042.

SIBSON, R.H. (1998), *Brittle failure mode plots for compressional and extensional tectonic regimes*, J Struct Geol 20, 655–660.

SIBSON, R.H. (2003), *Brittle failure controls on maximum sustainable overpressure*, Am Assoc Petrol Geol Bull 87, 901–908.

SIBSON, R.H. (2004), *Frictional mechanics of seismogenic thrust systems in the upper continental crust: implications for fluid overpressures and redistribution*, In, McClay, K.R. (ed.), *Thrust Tectonics and Hydrocarbon Systems*, AAPG Mem 82, 1–17.

Sibson, R.H. (2007), *An episode of fault-valve behavior during compressional inversion?: The 2004 $M_j6.8$ Mid-Niigata Prefecture, Japan, earthquake sequence*, Earth Planet Sci Lett 257, 188–199.

SIBSON, R.H. (2009), *Rupturing in overpressured crust during compressional inversion: the case from NE Honshu, Japan*, Tectonophysics 473, 404–416.

SIBSON, R.H. (2012), *Reverse fault rupturing: competition between non-optimal and optimal fault orientations*, In, Healy, D., Butler, R.W.H., Shipton, Z.K. and Sibson, R.H. (eds.), *Faulting, Fracturing, and Igneous intrusion in the Earth's Crust*, Geol Soc Lond Spec Publ 367, 39–50.

SIBSON, R.H. and ROWLAND, J.V. (2003), *Stress, fluid pressure, and structural permeability in seismogenic crust, North Island, New Zealand*, Geophys J Int 154, 584–594.

SIBSON, R.H. and SCOTT, R.H. (1998), *Stress/fault controls on the containment and release of overpressured fluids: examples from gold-quartz vein systems in Juneau, Alaska; Victoria, Australia; and Otago, New Zealand*, Ore Geol Rev 13, 293–306.

SONG, T.-R.A., HELMBERGER, D.V., BRUDZINSKI, M.R., CLAYTON, R.W., PÉREZ-CAMPOS, X. and SINGH, S.K. (2009), *Subducting slab ultra-slow velocity layer coincident with silent earthquakes in southern Mexico*, Science 324, 502–506.

TALWANI, P., CHEN, L. and GAHALAUT, K. (2007), *Seismogenic permeability, k_s*, J. Geophys Res 112, B07309, doi:10.1029/2006JB004665.

TERAKAWA, T., ZOPOROWSKI, A., GALVAN, B. and MILLER, S.A. (2010), *High-pressure fluid at hypocentral depths in the L'Aquila region inferred from earthquake focal mechanisms*, Geology 38, 995–998.

THOMAS, A.M., NADEAU, R.M. and BÜRGMANN, R. (2009), *Tremor-tide correlations and near-lithostatic pore pressure on the deep San Andreas fault*, Nature 462, 1048–1051.

THURBER, C., ROECKER, S., ELLSWORTH, W., CHEN, Y., LUTTER, W. AND SESSIONS, R. (1997), *Two-dimensional seismic image of the San Andreas fault in the Northern Gabilan Range, central California: evidence for fluids in the fault zone*, Geophys Res Lett 24, 1591–1594.

TOWNEND, J. and ZOBACK, M.D. (2000), *How faulting keeps the crust strong*, Geology 28, 399–402.

TOWNEND, J. and ZOBACK, M.D. (2004), *Regional tectonic stress near the San Andreas fault in central and southern California*, Geophys Res Lett 31, L15S11, doi:10.1029/2003GL018918.

TSURU, T., PARK, J.-O., KIDO, Y., ITO, A, KANEDA, Y., YAMADA, T., SHINOHARA, M. and KANAZAWA, T. (2005), *Did expanded porous patches guide rupture propagation in 2003 Tokachi-oki earthquake?*, Geophys Res Lett 32, L20310, doi:10.1029/2005GL023753.

TUNKS, A.J., SELLEY, D., ROGERS, J.R., and BRABHAM, G. (2004), *Vein mineralization at the Damang Gold Mine, Ghana, controls on mineralization*, J Struct. Geol 26, 1257–1273.

UNRUH, J.R., DAVISSON, M.L., CRISS, R.E. and MOORES, E.M. (1992), *Implications of perennial saline springs for abnormally high fluid pressures and active thrusting in western California*, Geology 20, 431–434.

UNSWORTH, M.J., EGBERT, G.D. and BOOKER, J.R. (1999), *High resolution imaging of the San Andreas fault in Central California*, J Geophys Res 104, 1131–1150.

UNSWORTH, M.J., MALIN, P.E., EGBERT, G.D. and BOOKER, J.R. (1997), *Internal structure of the San Andreas fault at Parkfield, California*, Geology 25, 359–362.

UNSWORTH, M.J. and BEDROSIAN, P.A. (2004), *Electrical resistivity structure at the SAFOD site from magnetotelluric exploration*, Geophys Res Lett 31, L12S05, doi:10.1029/2003GL019405.

UYESHIMA, M., OGAWA, Y., HONKURA, Y., KOYAMA, S., UJIHARA, N., MOGI, T., YAMAYA, Y., HARADA, M., YAMAGUCHI, S., SHIOZAKI, I., NOGUCHI, T., KUWABA, Y., TANAKA, Y., MOCHIDO, Y., MANABE, N., NISHIHARA, M., SAKA, M. and SERIZAWA, M. (2005), *Resistivity imaging across the source area of the 2004 Mid-Niigata prefecture earthquake (M6.8), central Japan*, Earth Planets Space 57, 441–446.

VIDALE, J.E. and SHEARER, P.M. (2006), *A survey of 71 earthquake bursts across southern California: Exploring the role of pore fluid pressure fluctuations and aseismic slip as drivers*, J Geophys Res 111, B05312, doi:10.1029/2005JB004034.

VON HERZEN, RUPPEL, C., MOLNAR, P., NETTLES, M., NGIHARA, S. and EKSTRÖM, G. (2001), *A constraint on the shear stress at the Pacific-Australian plate boundary from heat flow and seismicity at the Kermadec forearc*, J Geophys Res 106, 6817–6833.

VON HUENE, R. and SCHOLL, D.W. (1991), *Concerning sediment subduction, subduction erosion, and the growth of continental crust*, Rev Geophys 29, 279–316.

WANG, K., MULDER, T., ROGERS, G.C. and HYNDMAN, R.D. (1995), *Case for very low coupling stress on the Cascadia subduction fault*, J Geophys Res 100, 12,907–12,918.

WANG, K. and SUYEHIRO, K. (1999), *How does plate coupling affect crustal stresses in Northeast and Southwest Japan?*, Geophys Res Lett 26, 2307–2310.

WANG, Z. and ZHAO, D. (2006), *Seismic images of the source area of the 2004 Mid-Niigata prefecture earthquake in Northeast Japan*, Earth Planet Sci Lett 244, 16–31.

WESNOUSKY, S.G. (2005), *The San Andreas and Walker Lane fault systems, western North America: transpression, transtension, cumulative slip and the structural evolution of a major transform plate boundary*, J Struct Geol 27, 1505–1512.

YEATS, R.S. (1983), *Large-scale Quaternary detachments in Ventura Basin, southern California*, J Geophys Res 88, 569–583.

YERKES, R.F., LEVINE, P. and WENTWORTH, C.M. (1990), *Abnormally high fluid pressures in the region of the Coalinga earthquake sequence and their significance*, US Geol Surv Prof Pap 1487, 235–257.

XIA, S., ZHAO, D. and QIU, X. (2008), *The 2007 Niigata earthquake: effect of arc magma and fluids*, Phys Earth Planet Int *166*, 153–166.

ZHAO, D. and NEGISHI, H. (1998), *The 1995 Kobe earthquake: seismic image of the source zone and its implications for the rupture nucleation*, J Geophys Res *103*, 9967–9986.

ZHAO, D., HUANG, Z., UMINO, N., HASEGAWA, A., and KANAMORI, H., (2011), *Structural heterogeneity in the megathrust zone and mechanism of the 2011 Tohoku-Oki earthquake (M_w9.0)*, Geophys Res Lett *38*, L17308, doi:10.1029/2011GL048408.

ZOBACK, M.D., HICKMAN, S., ELLSWORTH, W. and the SAFOD SCIENCE TEAM (2011), *Scientific drilling into the San Andreas Fault Zone: an overview of SAFODS's first five years*, Scientific Drilling *11*, 14–28.

(Received September 11, 2013, revised March 13, 2014, accepted March 15, 2014, Published online April 4, 2014)

Pure Appl. Geophys. 171 (2014), 2887–2897
© 2014 Springer Basel
DOI 10.1007/s00024-014-0805-z

Wet Fault or Dry Fault? A Laboratory Approach to Remotely Monitor the Hydro-Mechanical State of a Discontinuity Using Controlled-Source Seismics

Joachim Place,[1,2] Oshaine Blake,[1] Daniel Faulkner,[1] and Andreas Rietbrock[1]

Abstract—Stress variation and fluid migration occur in deformation zones, which are expected to affect seismic waves reflected off or propagating across such structures. We developed a basic experimental approach to monitor the mechanical coupling with respect to seismic coupling across a single discontinuity between a granite sample in contact with a steel platen. Piezoceramics located on the platen were used to both generate and record the P and S wave fields reflected off the discontinuity at normal incidence. This way, normal (B_n) and tangential (B_t) compliances were calculated using Schoenberg's linear slip theory (Schoenberg, J Acoust Soc Am 68:1516–1521, 1980) when the roughness, the effective pressure (P_{eff}, up to 200 MPa), and the nature of the filling (gas or water) vary. We observe that increasing the effective pressure decreases B_n and B_t, which is interpreted as the effect of the closure of the voids at the interface, permitting more seismic energy to be transmitted across the interface. Values of B_n are significantly higher than those of B_t at low P_{eff} (<60–80 MPa) in dry conditions, and significantly drop under water-saturated conditions. The water filling the voids therefore helps to transmit the seismic energy of compressional waves across the interface. These results show that the assumption $B_n \approx B_t$ commonly found in some theoretical approaches does not always stand. The ratio B_n/B_t actually reflects the type of saturating fluids and the effective pressure, in agreement with other experimental studies. However, we illustrate that only the relative variations of this ratio seem to be relevant, not its absolute value as suggested in previous studies. Consequently, the use of B_n against B_t plots may allow effective pressure variation and the nature of the pore fluid to be inferred. In this respect, this experimental approach at sample scale helps to pave the way for remotely monitoring in the field the hydro-mechanical state of deformation zones, such as seismogenic faults, fractured reservoirs, or lava conduits.

Key words: Laboratory measurements, seismic, fault, discontinuity, fracture compliance, linear slip theory, hydro mechanical monitoring, effective pressure, fluid identification.

1. Introduction

The propagation of seismic waves is known to be sensitive to various physical properties of rocks. Seismic energy incident on an interface separating two rocks with different velocity and/or density undergoes a partitioning between reflected and transmitted wave fields (e.g. Aki and Richards 1980). Such partitioning has been extensively studied, especially when the interface is welded, to find applications in conventional reservoirs, for instance. However, the effect of mechanical decoupling occurring at partially welded interfaces can dramatically alter the behaviour of seismic waves, as does the occurrence of fluids at such interfaces. Such systems are commonly found in nature, especially when the fluid storage and transfer properties of the rocks are mainly related to fault and fracture networks. For example, fluids can be involved in large earthquakes which often occur in crystalline rocks. Basement rocks can also host geothermal and hydrocarbon resources, and lava flow is also channelized in conduits at depth before reaching the surface. Therefore, there is a need to detect changes in physical properties of discontinuities containing any kind of fluid, and subsequently interpret them in terms of fluid-pressure variation, crack opening/closure, mechanical coupling of the two compartments, healing processes, etc. Such information is essential to understand fluid-assisted deformation within fault zones, to manage naturally fractured reservoirs, and to monitor volcanic hazard (e.g. Kelly et al. 2013). Seismic methods have great potential to monitor the hydro-mechanical state of structures both remotely and non-destructively, with no need for drilling through the structures, and at higher resolution in space and time

---
[1] Department of Earth, Ocean and Ecological Sciences, School of Environmental Sciences, University of Liverpool, 4 Brownlow Street, Herdman Building, Liverpool L69 3GP, UK. E-mail: joachim.place@geo.uu.se
[2] Department of Earth Sciences, Uppsala University, Villavägen 16, 752 36 Uppsala, Sweden.

than other methods such as magnetotellurics. In this paper, we therefore address the transport properties of seismic waves across a discontinuity, focussing on rock response to stresses (through confining and fluid pressure) and the nature of the pore fluid.

Traction and displacement fields related to elastic-wave propagation are continuous across a perfectly welded interface. However, non-welded discontinuities imply a mechanical defect; they allow small displacements, such as those related to elastic waves, to be discontinuous across the interface. This led SCHOENBERG (1980) to introduce the "linear-slip condition" in order to replace the condition of continuous displacement. This theory uses the fracture "stiffness", which is a parameter describing how easy it is to transmit small displacements (e.g. CARCIONE and PICOTTI 2012; LI et al. 2012; MILLER 1978; PYRAK-NOLTE et al. 1990). Conversely, the fracture compliance, which is used in the present paper, was introduced in order to facilitate some analytical developments.

Fracture compliance is usually considered along two components: normal and tangential compliance related to the displacement perpendicular and along the interface plane, respectively. Therefore, the dynamic normal compliance B_n can be derived using normally incident compressional waves, whereas normally incident shear waves can be used to characterize the dynamic tangential compliance B_t (LUBBE et al. 2008; SCHOENBERG 1980; VAN DER NEUT et al. 2008).

Theoretical models have typically assumed that $B_n \approx B_t$ due to the scarcity of experimental data or for convenience (e.g. SCHOENBERG and SAYERS 1995; ZHU et al. 2011). However, geometrical parameters of compliant interfaces, such as the surface roughness or the distribution of microcracks, imply $B_n \neq B_t$. This discrepancy has been addressed in various theoretical approaches and calculations (HAUGEN and SCHOENBERG 2000; LEE and LUNEVA 2012; ZHU and SNIEDER 2002) but only in a few experimental studies (LUBBE et al. 2008; PYRAK-NOLTE et al. 1990). Regarding the influence of fluids, theoretical approaches argue that the viscosity and compressibility of the fluid that is present in the fracture voids also dramatically determines the partitioning between transmitted, reflected, and dissipated energy from the incident seismic

waves. In particular, B_n is predicted to be more sensitive to the presence of fluids than B_t, making the ratio B_n/B_t a good indicator of fluid presence (CARCIONE and PICOTTI 2012; LIU et al. 2000; VAN DER NEUT et al. 2008). Laboratory estimates indeed observed a reduction of the magnitude of B_n/B_t after saturating a discontinuity with a fluid (LUBBE et al. 2008; PYRAK-NOLTE et al. 1990). However, we are aware of only these two experiments carried out on individual fractures, in limestone (up to 60 MPa) and quartz monzonite (at 85 MPa), using honey or water as saturating fluid. Further work is therefore required to address other rock types and settings that relate to deeper crustal conditions.

In this paper, we investigate the seismic transport properties of a discontinuity where the materials on either side of the discontinuity are different from each other, with various degrees of roughness at the contact, under dry and water-saturated conditions, and within a wide range of effective pressure (up to 200 MPa, ca. 8 km depth under lithostatic pressure). Most of the experimental compliance measurements of an individual fracture or a set of fractures have been published based on the transmitted wavefield (ACOSTA-COLON et al. 2009; HOBDAY and WORTHINGTON 2012; LUBBE and WORTHINGTON 2006; PYRAK-NOLTE et al. 1990). Such an approach up-scaled to the field for either reservoir or earthquake studies would likely require instrumented boreholes to be drilled across or on either side of the fault. Aside from extra costs, this may disturb the behaviour of the fault due to fluid circulation through the hole. Thus, in the same way as LUBBE et al. (2008), we prefer focusing our work on the waves reflected off the discontinuity, which could still be applicable to vertical seismic profiling (VSP).

2. Experimental Setup

We used a triaxial deformation apparatus, as described in BLAKE et al. (2012), for instance, to perform the experiments and focused on the interface between a sample of granite and a steel platen (Fig. 1). Confining (P_{conf}) and fluid (P_f) pressures were servo-controlled; P_{conf} was held constant at various levels up to 200 MPa in dry experiments,

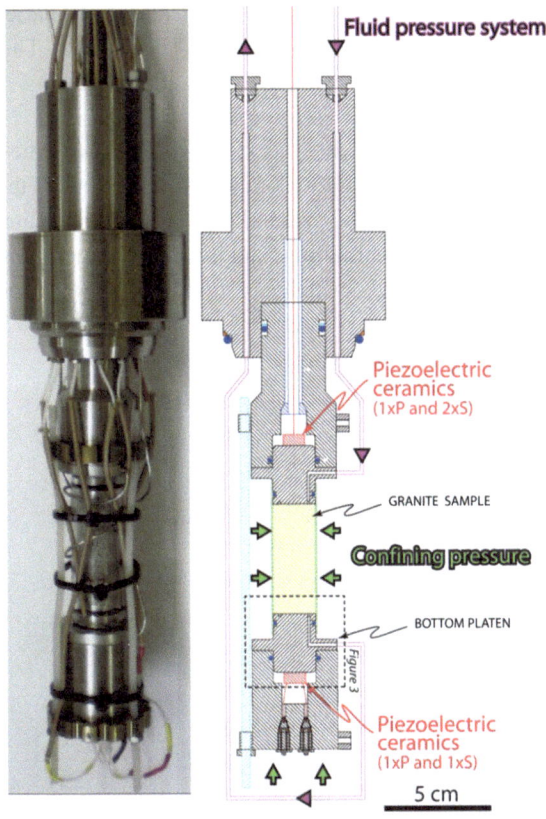

Figure 1
Photograph and schematic diagram of the experimental sample
setup [modified after BLAKE (2011)]. The sample is placed in a PVC
jacket and held between two platens equipped with piezoceramics
to study seismic wave propagation

whereas it was set at 200 MPa in water-saturated
experiments, with P_f increasing up to 195 MPa in
order to set the desired value of effective pressure
(P_{eff}) and, therefore, the effective normal stress at the
interface. Samples were placed in a PVC jacket so
that they were kept separate from the confining
medium and so that de-ionized water could be
injected into the sample to set the fluid pressure
(Fig. 1). The fluid pressure P_f was kept at least
5 MPa below P_{conf} to prevent any water invasion into
the confining oil system. Experiments were carried
out at room temperature, and processes involving
pressure variation were assumed to be adiabatic.

Westerly granite was used because of its small
average grain size (~ 0.75 mm), negligible anisotropy, and wide use in rock deformation experiments.
Samples were cored to a diameter of 20 mm and cut

to a length of 50 mm. Circular faces were squared
with a grinding machine (so that the faces were
parallel to a tolerance of ± 0.02 mm) in order to
obtain a standard roughness. Such a surface was
preserved on one sample and is referred to as the
"rough" surface. Another surface called "smooth"
was obtained on the other circular face of that sample
after polishing it for an hour with emery paper of
gradually finer grain size (P800, P1200, and P2400,
according to the ISO/FEPA grit designation). A second sample, named "fractured", was heated to
500 °C at a rate of 3 °C/min and kept at 500 °C for
15 h before being rapidly quenched in a bath of de-
ionized water at room temperature. Due to the differential thermal expansion of the minerals present in
the granite (COOPER and SIMMONS 1977), the thermal
treatment produces rougher surfaces.

Seismic waves were generated and recorded using
an ultrasonic measurement system. Piezoelectric
ceramics were located either side of the sample,
attached with silver-loaded–epoxy resin to the reverse
faces of the platens in contact with the rock (Fig. 1).
Two ceramics are present at the bottom of the sample
assembly (P and S) and three at the top (P and two S,
perpendicularly polarized). Ceramics polarized in the
direction of the sample axis were used for P wave
propagation, whereas S waves were studied using
ceramics polarized normal to the direction of propagation. This way, our results were obtained at normal
incidence for the contact between the platen and the
rock sample. The two ceramics at the bottom of the
sample assembly were preferentially used in reflection mode because of their bigger size and therefore
larger power than that of the ceramics on top (Fig. 1).
Seismic records were bandpass-filtered (0.8–3 MHz)
around the dominant frequency of the source which
was around 1.5 MHz.

A pulser/receiver (JSR DPR300) was used to
excite the ceramics in either reflection or transmission mode. Data were collected as an amount of
received seismic energy, measured as a voltage, with
a 10^{-8} s sampling interval, using a 300 MHz bandwidth digital oscilloscope (Tektronix TDS 32B).
Each record is the result of 512 signals stacked in
order to increase the signal-to-noise ratio. Further
details regarding the apparatus can be found in BLAKE
(2011) and BLAKE et al. (2012).

3. Data Analysis

3.1. Raw Data Analysis

As stated in the introduction, the mechanical properties of a non-welded interface (such as the surface between our samples and the sample assembly) is expected to influence the propagation of seismic waves interacting with it. Figure 2a shows a set of S-wave records obtained under water-saturated conditions. The amplitude of the seismic signal varies as a function of fluid pressure from about 16 μs. This value is consistent with the propagation time of S waves within the steel platen [(0.025 m × 2)/ 3,070 m.s^{-1} ≈ 1.6 × 10^{-5} s two-way–travel-time], indicating the arrival time of the first reflection off the contact between the platen and the sample. Later arrivals, also exhibiting amplitude changes, are observed at times which are multiples of the first

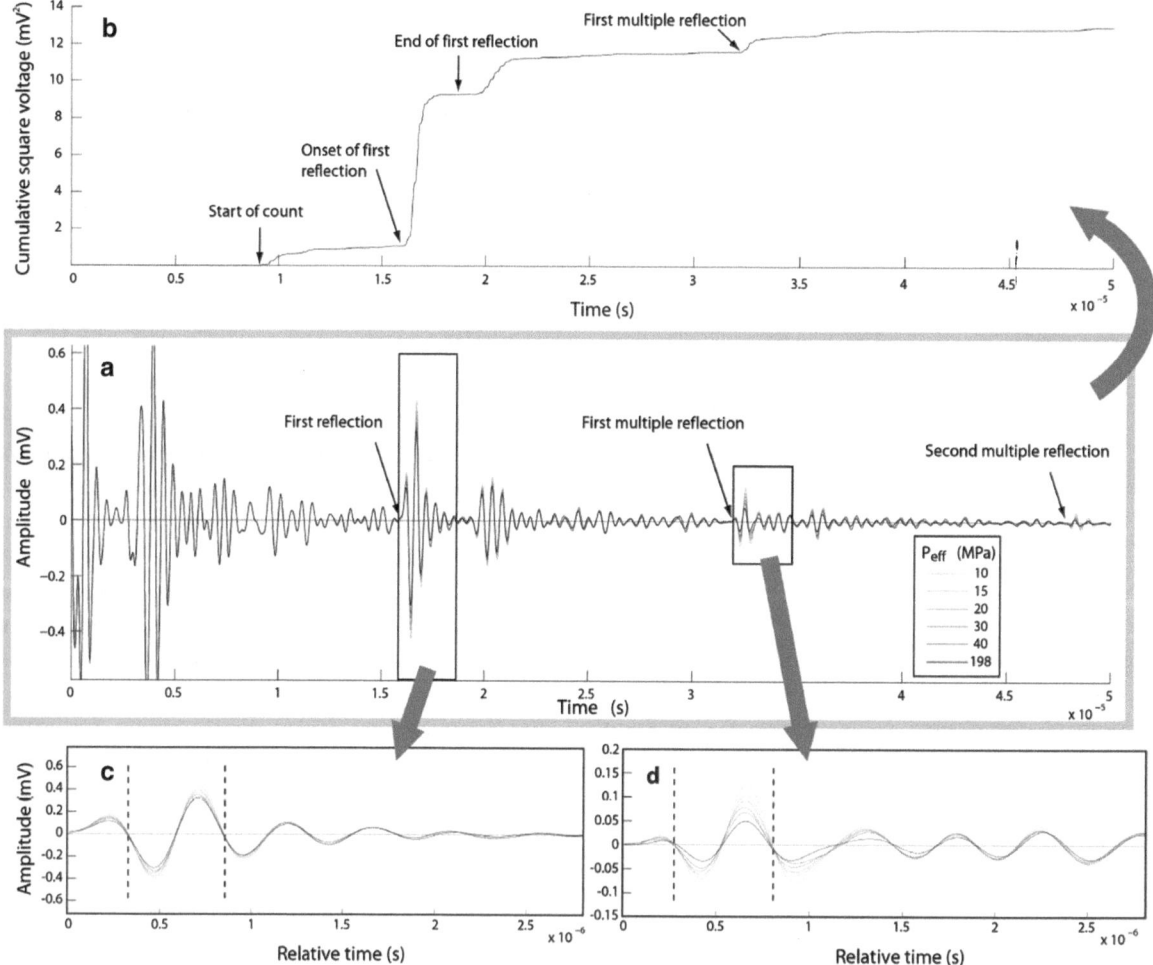

Figure 2

S wave records exhibiting the first S–S reflection and subsequent multiples reflected off the platen–sample interface (rough surface, water-saturated). The confining pressure was 200 MPa and the signal was recorded at various values of fluid pressure between 2 and 190 MPa, resulting in effective pressure from 198 to 10 MPa, respectively. **a** Dataset after bandpass filtering (only part of the dataset is represented for clarity). **b** Example of cumulative energy plot used for picking the first reflection (effective pressure was 10 MPa). For clarity, the cumulative energy is computed after the source-related noise has significantly reduced. **c** First reflection signal. **d** First multiple signal, obtained after cross-correlating the first reflection (**c**) with the entire record (**a**). Note the amplitude variations in (**a**), (**c**), and (**d**) related to changes in effective pressure for the first reflection (from ~1.6 × 10^{-5} s) and the first multiple (from ~3.2 × 10^{-5} s). *Dotted lines* in (**c**) and (**d**) indicate the windowed signals used to compute the reflection coefficients

reflection arrival time, denoting their multiple reflection nature (Fig. 2). Prior to the first reflection, all signals can be interpreted as noise, which is confirmed by the persistence of the later signals whatever the values of P_{eff}. Assuming no variation in attenuation within the steel, the changes in the magnitude of the reflections must be related to changes at the interface and/or within the granite sample. Such amplitude variations of the reflected waves are observed in any experiment involving water-saturated or dry samples: the higher the effective pressure, the weaker the reflected P or S waves.

3.2. Application of the Linear-Slip Theory

The displacement discontinuity theory (or linear-slip theory) of SCHOENBERG (1980), which has been validated by WORTHINGTON and LUBBE (2007) in both laboratory and field approaches, is applied to our dataset. Fracture compliance can be inferred from the transmitted wave field as well as the reflected field (SCHOENBERG 1980). Our records in transmission mode have not been considered for this purpose because they are noisier, especially at low P_{eff}, and transmitted waves are affected by two discontinuities (at either circular face of the rock sample); also, intrinsic attenuation occurs within the sample, and the difficulties related to its measurement and subsequent correction would have significantly impinged the reliability of the results. In the case of the reflected wave field, Schoenberg's theory states:

$$R = \frac{Z_2 - Z_1 + i\omega B Z_1 Z_2}{Z_2 - Z_1 - i\omega B Z_1 Z_2} \qquad (1)$$

where R is the measured reflection coefficient, Z is the impedance (the product of the density and the velocity of P or S waves) of the materials either side of the discontinuity (the subscript 1 refers to the material where the propagation occurs), ω is the angular frequency, and B is the compliance. After some manipulation, B_n and B_t can be derived from the measurement of P and S reflection coefficients, respectively, as follows:

$$B_n = \frac{\sqrt{|R^P|^2(Z_1^P + Z_2^P)^2 - (Z_2^P + Z_1^P)^2}}{\omega Z_1^P Z_2^P \sqrt{1 - |R^P|^2}} \qquad (2a)$$

$$B_t = \frac{\sqrt{|R^S|^2(Z_1^S + Z_2^S)^2 - (Z_2^S + Z_1^S)^2}}{\omega Z_1^S Z_2^S \sqrt{1 - |R^S|^2}} \qquad (2b)$$

where the superscripts P and S indicate whether P or S waves are considered, respectively. The compliance B is the inverse of the stiffness and is expressed in m/Pa. The higher the compliance, the lower the amount of seismic energy can be transmitted across the interface. It is worth noting that Eqs. (2a) and (2b) simplify the form used by LUBBE et al. (2008) when the two materials either side of the discontinuity have identical impedances.

3.3. Reflection Coefficient Measurements

In order to obtain B_n and B_t, the reflection coefficients are required (Eqs. 2a and 2b) and, therefore, have to be measured. The crystals used in the sample assembly are not calibrated, which means that the relationship between voltage and displacement is unknown. Our calculations of the reflection coefficients are therefore restricted to a basic approach, typically based on a signal considered both before and after reflection off an interface. However, in our experiments, as the source signal delivered by the ceramics is not exactly known, the reflection coefficients cannot be estimated from the source signal and the first reflection; the first multiple reflection whose incident wave is the first reflection was used instead (Fig. 3). For each experiment, the onset and end of first reflections were detected using cumulative energy plots against time (Fig. 2b) as, this way, the noise level and seismic arrivals can be easily distinguished. As the noise does not depend on the pressure conditions but the amplitude of reflected waves does, reflected arrivals are prominent on complementary standard-deviation plots computed over several records at various P_{eff} values. Once picked, the waveform of the first reflection (Fig. 2c) was cross-correlated with the full record (Fig. 2a) in order to pick the multiples. Reflection coefficients were calculated after the energy definition of reflection coefficients (STEIN and WYSESSION 2003): defining A as the square root of the energy contained in the first distinguishable period of a signal, the ratio of A for the first multiple and the first reflection,

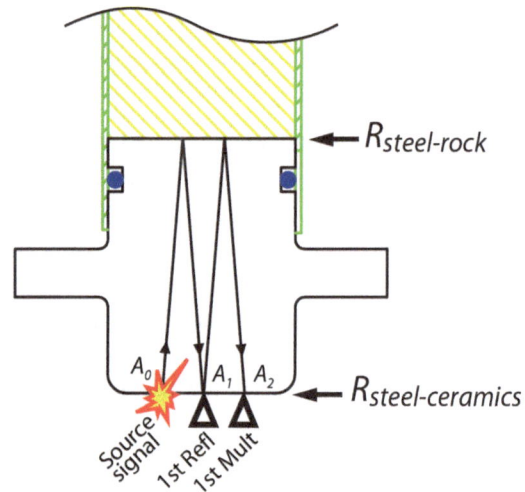

$R_{steel\text{-}rock}$

$R_{steel\text{-}ceramics}$

A_0 A_1 A_2

Source signal   1st Refl   1st Mult

Figure 3
Detail of the *bottom* loading platen schematically illustrating the
ray paths and A notations (square root of the energy contained in
the first distinguishable period of a signal, Fig. 2). As the signal is
generated and recorded by the same ceramic, the actual ray paths
occur along the same *line*, at normal incidence

(Fig. 2d, c respectively) normally provides the
reflection coefficient. However, as the first reflected
waves lose some amplitude when reflected off the
side of the platen where the ceramics are positioned
(Fig. 3), and in order to compensate for possible
attenuation within the steel, a correction factor called
"$R_{steel\text{-}ceramics}$" had to be used: the calibration is
simply obtained by the ratio of A for the first
reflection and the first multiple with no rock sample,
so the reflection off the free interface is perfect. To
sum up, the reflection coefficient (in P or S mode)
used from now on in this study is given by:

$$R_{steel-rock} = \frac{A_2}{A_1 \times R_{steel-ceramics}} \quad (3)$$

where A_1 and A_2 refer to the first reflection and first
multiple, respectively (see Fig. 3 for a sketch of the
propagation paths). Reflection coefficients calculated
using ratios between amplitudes of peaks or troughs
appeared less stable as a function of the pressure than
those obtained over a full period, hence our choice to
work on the latter (Fig. 2). Reflection coefficients
have not been derived from higher-order multiples, as
their signal-to-noise ratio becomes too low due to
coda of earlier arrivals, converted phases, and other
sources of noise.

Other than the reflection coefficients, impedance
values were also needed to use Eqs. (2a) and (2b).
Variations in density and length of the sample and the
platens as a function of pressure were calculated
using their respective bulk moduli. Wave travel times
within the steel were read from the first reflections
(such as in Fig. 2); when subtracted from the travel
times of waves transmitted from the ceramics at the
bottom platen to the ceramics on top (Fig. 1), the
travel time within the sample can be derived. Thus,
using the length values, velocities can be calculated.
Changes in effective pressure affect the porosity of
the samples, and to a much lesser extent, their length
and density (BLAKE *et al.* 2012). Impedance variations
are therefore expected in our experiments. Bulk
modulus and velocity changes are taken into account
in Eqs. (2a) and (2b).

By estimating the maximum uncertainty in all the
parameters involved (voltage, time windows where
the amplitudes are computed, density, and velocity
values), we estimate the error to be no more than
$\pm 1\%$ regarding R and $\pm 5\%$ regarding B.

4. Results and Interpretations

4.1. Dry Conditions (Gas-Saturated)

Both the effect of the roughness and the normal
stress across the interface (provided by the effective
pressure) were investigated in this study. Regarding
the normal compliance B_n, the rougher the sample
surface, the higher the compliance of the discontinuity at any P_{eff} value (Fig. 4). This trend can also be
identified to a lesser extent in the tangential compliance values, although the magnitudes fall within the
range of uncertainty (Fig. 4). This observation means
that the seismic energy of P waves (and marginally of
S waves) is transmitted less effectively when the
contact area (number of contacts and/or their respective surface area) is small, which is in agreement with
previous studies (e.g. MISRA and MARANGOS 2008,
2011; LIU *et al.* 2000; LUBBE *et al.* 2008).

In addition, an increase in the effective pressure
results in a decrease of the compliance (Fig. 4). The
discontinuity therefore becomes stiffer, presumably
because of an increasing number of contacts and/or

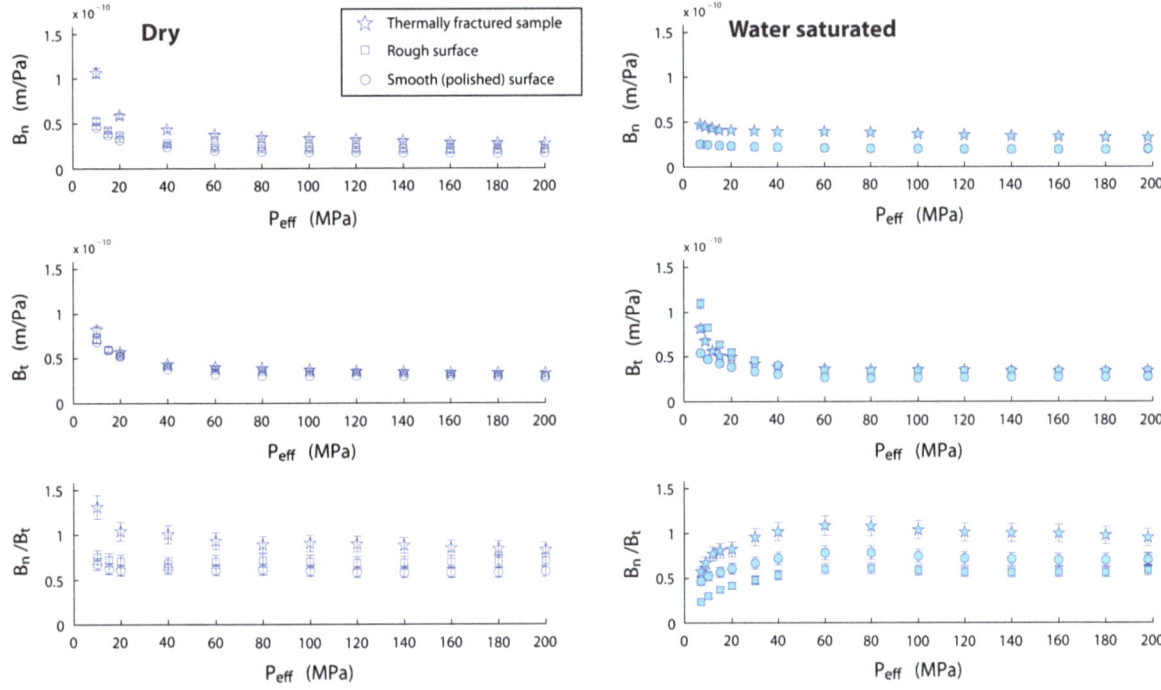

Figure 4

Compliance measured in all experiments as a function of the effective pressure and the presence of pore fluid. General decrease of the normal B_n or tangential B_t compliance is observed with an increase of the effective pressure P_{eff}, denoting a better mechanical coupling at the interface. Increase or decrease of the ratio B_n/B_t with P_{eff} depends on the saturation state of the voids at the interface

the increase in surface area of existing contacts under the effect of the normal stress across the interface (see models from MISRA and MARANGOS 2008, 2011). This trend is particularly apparent at low values of P_{eff} [in quite similar ranges as in PYRAK-NOLTE et al. (1990, Fig. 1) and LUBBE et al. (2008)]; however, our results extend to much higher pressure and exhibit no further dramatic change above 60–80 MPa. We expect this plateau effect to stand until significant changes in the contact area occur, related, for example, to reaching a phase transition or the plastic yield strength.

For all experiments, and at any value of P_{eff}, B_t varies less than B_n. Thus, the ratio B_n/B_t reflects the surface roughness. It decreases from low values of P_{eff} up to about 60–80 MPa, after which a plateau is observed.

4.2. Water-Saturated Conditions

Curves of B_n appear flatter in water-saturated conditions than in dry conditions (Fig. 4). Relative

variations of B_n as a function of P_{eff} are less obvious under water-saturated conditions than under dry conditions. At low effective pressure, B_t values are significantly lower than in dry conditions. The water filling the voids at the interface makes the discontinuity stiffer, which helps in transmitting the seismic energy of the compressional waves. Results obtained at higher P_{eff} are quite similar to those measured in dry conditions, which means that the closure of the voids is significant enough for the presence of water to have a marginal effect. This effect is not as dramatic regarding B_t, as its magnitudes are closer to those measured under dry conditions. Nevertheless, slight variations from the dry case studies are observed, as at each P_{eff} value, the magnitude of B_n no longer reflects the roughness. Nevertheless, the weaker influence of the water on B_n than on B_t can be explained by the inability of shear waves to be transmitted through liquids and gas, in contrast to P waves. In addition, a general decrease of B_n and B_t is observed with increasing P_{eff}, reflecting a better mechanical coupling with an increase of the normal stress.

31

The ratio B_n/B_t does not exhibit a clear dependence upon the roughness, as B_t values measured with the smooth and rough surfaces are extremely close. However, results from these two surfaces are distinguishable from those obtained from the fractured sample. But the most important feature is the obvious increase of B_n/B_t with the effective pressure, in contrast to the observation made in dry conditions. This behaviour is mainly related to the decrease of B_n due to the presence of water. This highlights the sensitivity of the ratio B_n/B_t to the effective pressure (therefore, the fluid pressure) and the nature of the pore fluid.

5. Discussion

5.1. Requirements for Sample Assemblies

As the contact between the sample and the platen is critical in such experiments, special care was taken regarding the way these pieces were held together. The bottom platen was fixed to the upper one with studs (in white on the photograph, Fig. 1) for the sole purpose of preventing it from falling by gravity before the application of confining pressure. Upward movement of the sample and the bottom platen were permitted on the application of pressure, as well as a slight non-parallelism of their respective longitudinal axes. This setup allowed remaining squaring inaccuracies (see above) to be accommodated. In other words, our sample assembly ensured the rock and the steel would be in full contact under the effect of the confining pressure.

Loss of transmitted seismic energy at a compliant interface is frequency dependent (PYRAK-NOLTE et al. 1990; SCHOENBERG 1980). As our results illustrate the compliant behaviour of the interface between platens and sample, it is recommended that the piezoceramics are in direct contact with the samples in any laboratory experiments requiring frequency spectra measurements, especially when the stresses applied to the system may vary.

5.2. Use of the B_n/B_t Ratio

Our results illustrate that the ratio B_n/B_t depends upon both the state of stress and the pore fluid at an

interface: normal stress can alter the roughness at the contact in terms of number of contacts, shapes of indentation, and the contact area, whereas the void space may contribute to the mechanical coupling depending on what is present. This is in agreement with theoretical models (MISRA and MARANGOS 2008, 2011) and previous experimental studies: filling a fracture or a set of fractures with honey makes the B_n/B_t ratio drop significantly (HSU and SCHOENBERG 1993; LUBBE et al. 2008). Conversely, in a field study, VERDON and WÜSTEFELD (2013) measured an increase in the B_n/B_t ratio whilst injecting proppant in a gas-saturated fracture network.

However, our study illustrates that the B_n/B_t ratio is relevant at low effective pressure (<80 MPa), as in such conditions the compliance varies significantly. Higher normal stress allows less differential displacement to occur and also closes the voids, which results in a better mechanical coupling. The effect of repeated loading and unloading cycles has not been directly investigated, but anelastic deformation, especially relative displacement of the grains or yielding of the steel surface near the indenters, is likely to occur [see YOSHIOKA and IWASA (2006) and references therein]. An increase of the surface area is therefore expected, which would be reflected as a general decrease of the compliance with the number of loading and unloading cycles.

In addition, several authors have argued that the absolute value of the ratio B_n/B_t can reflect the nature of the pore fluid, with $B_n/B_t \approx 1$ for dry fractures and $B_n/B_t \approx 0$ when saturating the voids with a relatively incompressible fluid (see, for example, LIU et al. 2000; SAYERS 2002; SAYERS and KACHANOV 1995; SCHOENBERG and SAYERS 1995; WHITE 1983). However, LUBBE et al. (2008) and PYRAK-NOLTE et al. (1990) have shown experimentally that B_n/B_t is more likely to vary between 0.25 and 0.5 under dry conditions. Considering the variety of fault-zone architectures and conditions described in the field, such as large planar fault planes as opposed to sets of non-parallel and non-planar slip surfaces delimiting lenses and unevenly distributed fluid overpressures [see a synthesis in FAULKNER et al. (2010)], a wider variation of the B_n/B_t ratio might be expected. Based on our experimental results obtained on a simple case, we argue that, in first instance, the absolute value of B_n/B_t does not have to be regarded as an accurate indicator of the pore fluid. If similar experiments were to be repeated in time for monitoring purpose, the relative variation of B_n/B_t over time seems to be a more relevant parameter to characterise fluid type and pressure changes at a non-welded interface.

5.3. Towards Field Applications: Up-Scaling of the Experimental Approach

To investigate further the meaning of the B_n/B_t ratio, and envision possible field applications of the approach detailed in this paper, values of B_n were plotted against those of B_t in Fig. 5. Data from the plateaus (Fig. 4) were excluded for clarity, so the plot is restricted to the values obtained at $P_{\text{eff}} \leq 60$ MPa for the fractured and smooth samples in saturated conditions, and $P_{\text{eff}} \leq 80$ MPa for all other experiments. The curves exhibit distinctive patterns as their gradients depend on the pore fluid: gradients calculated locally between two successive dots present a minimum value of 0.39 in dry conditions and a maximum of 0.39 in wet conditions. Therefore, an increase of both B_n and B_t reflects a drop in effective pressure (and vice versa), and the gradient of the curves relative to 0.39 indicates if the interface is wet or dry. To sum up, it is observed that only the relative variation of the ratio B_n/B_t can reflect the saturation state of the interface (along with changes in the normal stress), but not the absolute value of the ratio. The value of 0.39 is likely to depend on the experimental settings, especially the materials involved at the interface and the range of roughness; further studies should be carried out to identify and characterize the parameters governing such a threshold. Nevertheless, it is anticipated that similar analysis of B_n and B_t in field studies could allow hydro-mechanical variations within deformation zones to be characterized at a distance, with no need to measure the fluid pressure through a borehole intersecting a fault, for example.

Fracture compliance can be inferred from the transmitted wave field as well as the reflected field (SCHOENBERG 1980). In field studies, recording the wave field transmitted across a fault is likely to require an observation borehole to be drilled through

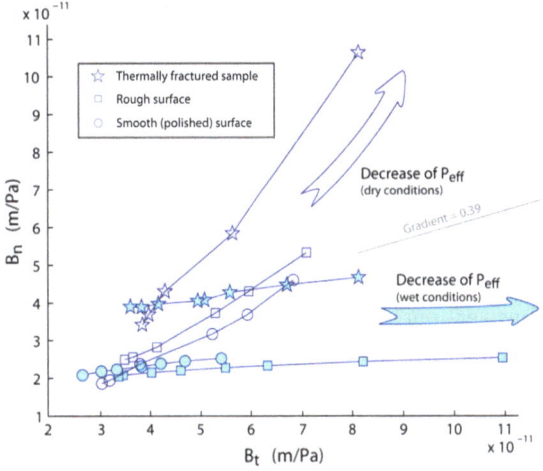

Figure 5

Normal compliance B_n against tangential compliance B_t plot exhibiting distinctive relationships in air or water-saturated conditions as a function of P_{eff}: the gradient of the curves reflects the nature of the pore fluid

this structure (especially in the case of VSP). In this respect, our approach is focused on the reflected field in order to anticipate field applications with only one borehole drilled in the hanging wall of the structure to be monitored. In our experiments, at low effective pressure, the transmitted waves were within the noise level, preventing the velocity within the sample to be measured. Therefore, the impedance and the compliance could not be estimated. However, in field studies, fluid flow concentrated within a fracture or fault zone, or within parts of them, would, in short term, not affect the compartments they separate. Therefore, their properties, especially elastic wave velocities, could be assumed to be in a steady state to some extent. This way, the compliance measurements could be derived from the reflected wave field only, with no need to record transmitted waves.

6. Conclusion

This study describes a simple method to measure the normal and tangential compliance at an interface after Schoenberg's linear-slip theory. Our experimental results illustrate that a non-welded fracture becomes stiffer when the normal stress (or the effective pressure) increases. B_n can reflect the roughness of the contact, especially in dry conditions. The presence of water at the interface allows the compressional waves to be transmitted more easily, whereas the changes observed regarding shear wave propagation are less dramatic. Therefore, and in agreement with previous studies, the ratio B_n/B_t is sensitive to the type of saturating fluid. In addition, our study illustrates the sensitivity of this parameter to the effective pressure, hence the fluid pressure. However, our results suggest that not the absolute magnitude of the B_n/B_t ratio but only its relative variations should be used to characterise the hydro-mechanical state of a fracture; we have shown that plotting B_n against B_t seems to be a simple and accurate way to detect an increase or decrease of the effective pressure at a discontinuity as well as identify if it is air- or water-saturated. Even if the analogy of our experimental settings with a geological case study can be significantly improved, the approach and results presented in this paper show that spatial variations or non-seismogenic phenomenon such as partial rupture or healing processes, as well as fluid migration, could be detected in the field using repeated active seismic surveys.

Acknowledgments

We would like to thank members of the University of Liverpool, especially Betty Mariani, Julia Behnsen, Dan Tatham, John Wheeler, Steve Hicks and Richard Holme, as well as Clarisse Bordes (Université de Pau et des Pays de l'Adour) for their technical help and constructive discussions. The paper benefited from constructive comments from two anonymous reviewers and Antonio Rovelli, Editor. Kay Lancaster (University of Liverpool) is acknowledged for improvement of the figures.

REFERENCES

ACOSTA-COLON, A., L. J. PYRAK-NOLTE, and D. D. NOLTE (2009), *Laboratory-scale study of field of view and the seismic interpretation of fracture specific stiffness*, Geophysical Prospecting, *57*(2), 209–224.

AKI, K., and P. G. RICHARDS (1980), *Quantitative Seismology, Theory and Methods*, University Science Books.

BLAKE, O. O. (2011), *Seismic transport properties of fractured rocks*, PhD of the University of Liverpool, pp. 151.

BLAKE, O. O., D. R. FAULKNER, and A. RIETBROCK (2012), *The Effect of Varying Damage History in Crystalline Rocks on the P- and S-Wave Velocity under Hydrostatic Confining Pressure*, Pure and Applied Geophysics, 1–13.

CARCIONE, J., and S. PICOTTI (2012), *Reflection and transmission coefficients of a fracture in transversely isotropic media*, Studia Geophysica et Geodaetica, *56*(2), 307–322.

COOPER, H. W. and SIMMONS, G. (1977), *The effect of cracks on the thermal expansion of rocks*, Earth and Planetary Science Letters, *36*, 404–412.

FAULKNER, D. R., C. A. L. JACKSON, R. J. LUNN, R. W. SCHLISCHE, Z. K. SHIPTON, C. A. J. WIBBERLEY, and M. O. WITHJACK (2010), *A review of recent developments concerning the structure, mechanics and fluid flow properties of fault zones*, Journal of Structural Geology, *32*(11), 1557–1575.

HAUGEN, G. U., and M. A. SCHOENBERG (2000), *The echo of a fault or fracture*, Geophysics, *65*(1), 176–189.

HOBDAY, C., and M. H. WORTHINGTON (2012), *Field measurements of normal and shear fracture compliance*, Geophysical Prospecting, *60*(3), 488–499.

HSU, C., and M. SCHOENBERG (1993), *Elastic waves through a simulated fractured medium, Geophysics, 58*(7), 964–977.

KELLY, C. M., A. RIETBROCK, D. R. FAULKNER, R. M. NADEAU (2013), *Temporal changes in attenuation associated with the 2004 M6.0 Parkfield earthquake*, Journal of Geophysical ResearchB: Solid Earth, *118*(2), 630–645.

LEE, J., and M. LUNEVA (2012), *Variations of reflection and transmission coefficients at the boundaries of different media and types of contact between media*, Geosciences Journal, *16*(1), 91–104.

LI, J. C., H. B. LI, G. W. MA, and J. ZHAO (2012), *A time-domain recursive method to analyse transient wave propagation across rock joints*, Geophysical Journal International, *188*(2), 631–644.

LIU, E., J. A. HUDSON, and T. POINTER (2000), *Equivalent medium representation of fractured rock*, Journal of Geophysical Research: Solid Earth, *105*(B2), 2981–3000.

LUBBE, R., and M. H. WORTHINGTON (2006), *A field investigation of fracture compliance*, Geophysical Prospecting, *54*(3), 319–331.

LUBBE, R., J. SOTHCOTT, M. H. WORTHINGTON, and C. MCCANN (2008), *Laboratory estimates of normal and shear fracture compliance*, Geophysical Prospecting, *56*(2), 239–247.

MILLER, R. K. (1978), *The effects of boundary friction on the propagation of elastic waves*, Bulletin of the Seismological Society of America, *68*(4), 987–998.

MISRA, A., and MARANGOS, O. (2011), *Rock-joint micromechanics: relationship of roughness to closure and wave propagation*, International Journal of Geomechanics, *11*(6), 431–439.

MISRA, A., and MARANGOS, O. (2008), *Micromechanical model of rough contact between rock blocks with application to wave propagation*, Acta Geophysica, *56*(4), 1109–1128.

PYRAK-NOLTE, L. J., L. R. MYER, and N. G. W. COOK (1990), *Transmission of seismic waves across single natural fractures*, J. Geophys. Res., *95*(B6), 8617–8638.

SAYERS, C. M. (2002), *Fluid-dependent shear-wave splitting in fractured media*, Geophysical Prospecting, *50*(4), 393–401.

SAYERS, C. M., and M. KACHANOV (1995), *Microcrack-induced elastic wave anisotropy of brittle rocks*, Journal of Geophysical Research: Solid Earth, *100*(B3), 4149–4156.

Schoenberg, M. (1980), *Elastic wave behavior across linear slip interfaces*, The Journal of the Acoustical Society of America, *68*(5), 1516–1521.

Schoenberg, M., and C. Sayers (1995), *Seismic anisotropy of fractured rock*, Geophysics, *60*(1), 204–211.

Stein, S., and M. Wysession (2003), An introduction to seismology, earthquakes, and Earth structure, *Blackwell Publishing Ltd.*, pp. 498.

Van Der Neut, J., M. K. Sen, and K. Wapenaar (2008), *Seismic reflection coefficients of faults at low frequencies: a model study*, Geophysical Prospecting, *56*(3), 287–292.

Verdon, J. P., and A. Wüstefeld (2013), *Measurement of the normal/tangential fracture compliance ratio (ZN/ZT) during hydraulic fracture stimulation using S-wave splitting data*, Geophysical Prospecting, *61*, 461–475.

White, J. E. (1983), Underground sound - Application of seismic waves, *Elsevier, New York*.

Worthington, M. H., and R. Lubbe (2007), The scaling of fracture compliance, *Geological Society, London, Special Publications*, *270*(1), 73-82.

Yoshioka N., K. Iwasa (2006), *A laboratory experiment to monitor the contact state of a fault by transmission waves*. Tectonophysics, *413*, 221–238.

Zhu, J. B., A. Perino, G. F. Zhao, G. Barla, J. C. Li, G. W. Ma, and J. Zhao (2011), *Seismic response of a single and a set of filled joints of viscoelastic deformational behaviour*, Geophysical Journal International, *186*(3), 1315–1330.

Zhu, Y., and R. Snieder (2002), Reflected and transmitted waves from fault zones, *SEG International Exposition and 72nd Annual Meeting*, pp. 14.

(Received August 14, 2013, revised February 12, 2014, accepted February 19, 2014, Published online July 11, 2014)

Pure Appl. Geophys. 171 (2014), 2899–2918
© 2014 Springer Basel
DOI 10.1007/s00024-014-0826-7

▌Pure and Applied Geophysics

Formation and Suppression of Strike–Slip Fault Systems

IVY S. CURREN[1] and PETER BIRD[1]

Abstract—Strike–slip faults are a defining feature of plate tectonics, yet many aspects of their development and evolution remain unresolved. For intact materials and/or regions, a standard sequence of shear development is predicted from physical models and field studies, commencing with the formation of Riedel shears and culminating with the development of a throughgoing fault. However, for materials and/or regions that contain crustal heterogeneities (normal and/or thrust faults, joints, etc.) that predate shear deformation, kinematic evolution of strike–slip faulting is poorly constrained. We present a new plane-stress finite-strain physical analog model developed to investigate primary deformation zone evolution in simple shear, pure strike–slip fault systems in which faults or joints are present before shear initiation. Experimental results suggest that preexisting mechanical discontinuities (faults and/or joints) have a marked effect on the geometry of such systems, causing deflection, lateral distribution, and suppression of shears. A lower limit is placed on shear offset necessary to produce a throughgoing fault in systems containing preexisting structures. Fault zone development observed in these experiments provides new insight for kinematic interpretation of structural data from strike–slip fault zones on Earth, Venus, and other terrestrial bodies.

Key words: Strike–slip faults, analog modeling, wet kaolin, brothers fault zone, SISZ, venus.

1. Introduction

On Earth, regions dominated by toroidal (shear) motion are characterized by faults with heave parallel to their strike (BATES and JACKSON 1987), which are known as strike–slip in continental settings or transform faults in oceanic settings. They are responsible for accommodating motion between plates and between offset spreading ridges, rotating blocks, and within compressional orogens. Unlike normal and thrust faults along which area is gained or lost by extension or compression, respectively, strike–slip faults conserve area by sliding horizontally along approximately vertical fault planes (O'CONNELL *et al.* 1991; SLEEP 1992). Their lateral motion produces horizontal offsets that are easily identified in plan view; as a result, strike–slip faults are ideal candidates for identifying and reconstructing tectonic processes and histories on terrestrial planets and satellites (SLEEP 1994; YIN 2012).

Physical analog models (CLOOS 1928; RIEDEL 1929; EMMONS 1969; TCHALENKO 1970; WILCOX *et al.* 1973; HILDEBRAND-MITTLEFEHLDT 1979; ODONNE and VIALON 1983; HEMPTON and NEHER 1986; NAYLOR *et al.* 1986; RECHES 1988; RICHARD *et al.* 1995; LAZARTE and BRAY 1996; CASAS *et al.* 2001; SCHÖPFER and STEYRER 2001; ATMAOUI *et al.* 2006; LE GUERROUÉ and COBBOLD 2006; SCHWARZ and KILFITT 2008; MISRA *et al.* 2009) and field studies (BARTLETT *et al.* 1981; AYDIN and RECHES 1982; SEGALL and POLLARD 1983; MARTEL 1990; CRIDER and PEACOCK 2004) have revealed that strike–slip faults nucleate and develop in undeformed materials by a predictable sequence of events that begins with formation of conjugate Riedel shears (R and R′) and culminates with development of a throughgoing fault, sometimes referred to as the "Y-shear" (MORGENSTERN and TCHALENKO 1967). Although the sequence of intermediate shear development is dictated by the properties of the material being sheared, Riedel experiments invariably result in the formation of some or all of the following shear components before the development of a throughgoing fault: conjugate R (synthetic) and R′ (antithetic), conjugate P (synthetic) and P′ (synthetic), and X (antithetic; BARTLETT *et al.* 1981) shears.

Although Riedel experiments have provided much insight into the development of fault zones in pristine (undeformed) materials, real planetary surfaces are rarely left intact before shear deformation. Instead, the crust often contains preferentially oriented joints,

---
[1] Department of Earth, Planetary, and Space Sciences, University of California, Los Angeles, CA 90095-1567, USA. E-mail: iscurren@epss.ucla.edu; pbird@epss.ucla.edu

thrust, and/or normal faults resulting from episodes of compression and/or extension (THOMPSON and BURKE 1974; ANGELIER 1994; JOHNSON 1995; TUCKWELL and GHAIL 2003; CRIDER and PEACOCK 2004). The effects of preexisting structures on new faulting have been studied extensively with regard to fault reactivation in extensional and compressional settings (SASSI et al. 1993; FACCENNA et al. 1995; TONG and YIN 2011); studies investigating the effect of preexisting structures on shear zone development have been fewer. FREUND (1974) performed experiments in which "faults" were cut into plasticine, which was then shortened to investigate nonrotational strain processes. ARCH et al. (1988) studied the effects of primary anisotropic fabrics on micro-shear zone geometry. ROMÁN-BERDIEL et al. (1997), CORTI et al. (2005), and MAZZARINI et al. (2010) examined the effect of dike and magma emplacement on synchronous shearing. LET-OUZEY and SHERKATI (2004) and KOYI et al. (2008) investigated the effects of salt diapirs on localized strike–slip faulting, and HOLOHAN et al. (2008) studied the effect of caldera collapse in strike–slip tectonic regions. Similarly, DOOLEY and SCHREURS (2012; Sect. 7) studied the effects of a single weak zone oriented oblique to their basement fault. RICHARD and KRANTZ (1991) and VIOLA et al. (2004) investigated the effects of a solitary dip–slip basement fault on the development of strike–slip faulting with regard to reactivation of the preexisting fault; both studies found that preexisting faults were often reactivated and responsible for localized strike–slip faulting.

In this contribution, we introduce a new plane-stress finite-strain physical analog model of the Riedel experiment variety, developed to investigate the effect of preexisting joints and faults on the development of pure strike–slip faulting (fault plane parallel to displacement vector) in a material (clay) deformed by simple shear. We build upon the work of previous Riedel shear modelers, adding experiments containing preexisting mechanical discontinuities that were cut into the clay surface before shear deformation.

2. The Analog Model

Our physical analog model follows the basic design of previous workers' Riedel shear analog models (RIEDEL 1929; TCHALENKO 1970; WILCOX et al. 1973), but differs in three significant ways. First, we removed basal shear traction by injecting high-density $ZnCl_2$ solution beneath the clay. This caused layering (as may be appropriate for terrestrial bodies) and decoupled the clay layer and strong base plates, similar to the decoupling of the Earth's crust from the potentially-strong upper mantle in continental zones of normal or elevated heat flow (BIRD 1978a, b, 1979; KIRBY 1980). Second, we left the wet kaolin to viscously and hydrostatically relax before shearing, which concurrently established a thin (≤ 0.002 m) semidry (damp, but unsaturated) layer. We suspect the critical changes during this time were compaction and relaxation by diffusion of super-hydrostatic pore pressures in the bulk of the clay, and that partial drying of the thin surface layer was less critical. Finally, our model focuses on the effect of preexisting structures in subsequent shear zone development.

2.1. Model Design

All experiments in this study were performed on a simple shear deformation table constructed by the Modeling and Educational Demonstrations Laboratory (MEDL) at the University of California Los Angeles. The apparatus consisted of two base plates, each 1 m-long by 0.5 m-wide, and a geared-down motor that drove one plate in a straight line, parallel to heave, past the other plate at a fixed velocity; relative motion of the plates subsequently defined a sinistral (left–lateral) shear zone. A split frame (interior dimensions 0.36 m × 0.35 m) was affixed to and centered over the contact between the base plates, and ten 0.02 m-wide scalene triangular teeth were attached to its inside edges to transmit shear tractions into the clay and prevent lateral slip along the boundaries (Fig. 1). Experiments were assembled by pouring homogenous wet kaolin into the frame, smoothing its surface (taking special care to prevent anisotropy), and leaving time for relaxation and equilibration (~ 168 h). Kaolin has a linear dry shrinkage of 5.8 %, which caused the drying clay to pull away from the frame; to compensate for the loss of volume a dilute plaster-of-Paris solution was piped into the void between the clay and frame. To avoid adding strength along the shear plane, a small gap

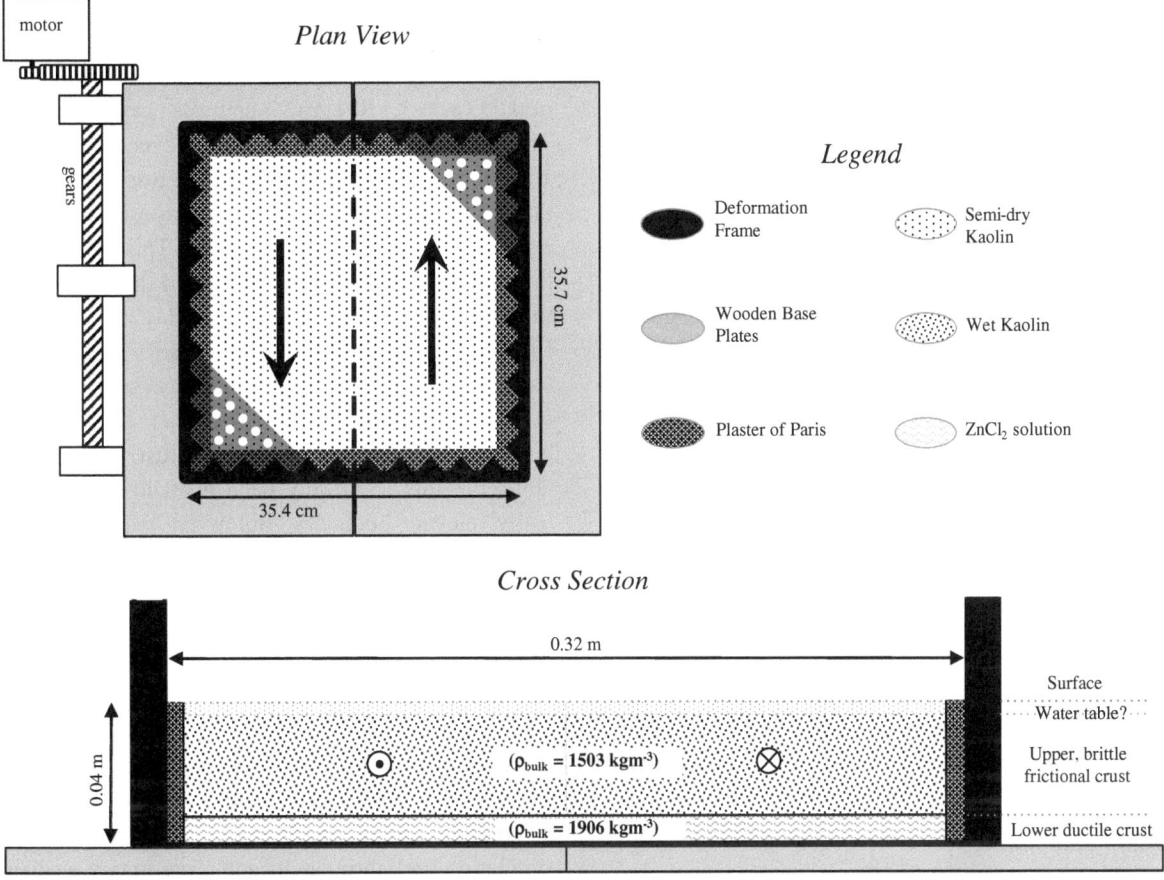

Figure 1

Plan and cross-sectional views of the physical analog model presented here. *Large black triangles with white dots* in the *upper right* and *lower left* corners of plan view represent nail combs used to apply additional tension to the clay in the extensional–diagonal corners of the deformation frame. Nail combs are omitted from the cross-sectional view. Relationship of layered modeling materials to nature is shown to the *right* of the cross section

was left in the plaster at the contact between the base plates. Although the plaster-of-Paris probably added some strength to the analog model, it did not seem to affect the primary deformation zone (PDZ).

Before shearing, the clay surface was left intact for control experiments. For other experiments, cuts meant to mimic deformed crust were incised into the clay. Faults and joints were hand-cut with a wetted blade to a depth of ∼ 0.01 m and overlapped the contact between the two base plates by 0.03 m on either side. Because of the difficulty associated with cutting dip angles into wet kaolin, we opted to cut all preexisting faults vertically. Although this does not represent dip seen in true normal faults, the vertical incisions represent mechanical discontinuities in the crust, which orthogonally (or obliquely) abut strike–slip faults produced by horizontal shear. If we visualize this in 3D space with the x- and y-axes horizontal and z-axis vertical, the shear plane (x-axis) and preexisting fault plane (y-axis) are always normal or subperpendicular (depending on the experiment). Because the x and y-axes are both horizontal, shear stress applied in the horizontal direction resulting in strike–slip faulting should be most affected by crosscutting in the horizontal plane. Variations in dip angle (z-axis) should not be as critical to the development of strike–slip faulting, except in the sense of defining where the two faults will converge first. For a normal fault, new strike–slip faults would encounter the dipping fault plane at depth first, causing deflection and/or suppression to occur at depth initially, then move upward; for a thrust fault, we would expect the opposite.

Early trials showed that the underlying $ZnCl_2$ solution intruded between the kaolin and frame, which prevented exertion of deviatoric tension along the appropriate stress axis and resulted in a transpressive state. We eliminated the transpress by placing triangular wooden blocks containing evenly spaced blunt-tipped nails in the extensional-diagonal corners of the frame (Fig. 1). Although the nail combs created highly localized non-analogous effects in the extensional-diagonal corners of the experiment, the effects did not seem to affect the PDZ and can therefore be ignored for the purposes of this model.

Fault-zone development was monitored by use of time-lapse digital photography (1 and 5-min intervals for experiments with velocities 2.03×10^{-5} and 3.83×10^{-6} m s^{-1}, respectively). Low-angle lighting was used to accentuate the appearance of faults and topography. A passive marker grid on the clay surface was used to map progressive fault deformation. Shear offset of the deformation frame was measured at 5-min intervals by use of a fixed ruler. Real-time observation was used to establish the offset at which a throughgoing fault formed; this offset was later confirmed or adjusted by analysis of map-view digital imagery.

2.2. Materials

The primary modeling material used in this study, Edgar Plastic Kaolin from Edgar, Florida, is a clay consisting of small grains (~ 1.36 μm) which has low plasticity and viscoelasticity when mixed with water (ATMAOUI et al. 2006; COOKE and VAN DER ELST 2012). Shear and cohesive strengths of kaolin can be varied easily by increasing or reducing the water content (KRYNINE 1947; TERZAGHI and PECK 1948; BAIN and BEEBE 1954; KEMPER and ROSENAU 1984; EISENSTADT and SIMS 2005). Wet kaolin approximately 50–60 % water by mass ($\rho = 1,600-1,630$ kgm^{-3}) has low cohesive strength (40–65 Pa; SIMS 1993; EISENSTADT and SIMS 2005) appropriate for analog modeling. Although wet kaolin is not brittle, but a complex visco-elastic–plastic material, Mohr–Coulomb failure and bi-viscous creep are observed; as a result, it develops discrete faults (velocity discontinuities and strain singularities) that mimic natural strain-weakening features in the Earth's crust (ODE 1960; OERTEL

1965; ARCH et al. 1988; RECHES 1988; EISENSTADT and SIMS 2005; COOKE and VAN DER ELST 2012).

Our wet kaolin had an initial water-to-EPK ratio of 0.55 ($\rho = 1,170$ kgm^{-3}) and was nearly saturated. The wet kaolin was left to relax over a period of 1 week (~ 168 h), which resulted in loss of ~ 5 % of the water (by mass) through the exposed diffusive/evaporative surface during drying. This produced a thin unsaturated layer and depressed the water table to a depth 0.002 m below the clay surface. We suspect there was only slight disparity in frictional strength between the semidry and wet clay, because of lack of pore pressure in the former; an analogous boundary layer exists in Earth's crust at the water table, but this has rarely been invoked as a tectonically relevant boundary. Below the water table we assume retention of the original water content (55 wt%) and saturation >80 %, and hence assume a hydrostatic pressure gradient (increasing linearly with depth) for this portion of the wet kaolin. The bulk density of the kaolin after the drying period was 1,503 kg m^{-3}.

Although it has been demonstrated that basal traction can affect surface and volumetric strains, kinematics, and geometry in physical analog models simulating fold–thrust belts and accretionary wedges, basal boundary conditions are often overlooked in models of strike–slip faulting (LIU et al. 1992; GUTSCHER et al. 1996; NILFOROUSHAN et al. 2007). KLINKMÜLLER (2011) showed that in a 0.04 m-thick analog model comprising granular materials and a variety of basal surfaces, the angle of internal friction at the base can be as much as 20° lower than the angle of internal friction of the bulk material at peak strength, indicating that small differences in boundary conditions can have large effects on frictional behavior. AN and SAMMIS (1996) placed a layer of plastic wrap sprinkled with water beneath their gravity sliding model, but this is problematic, because it could result in water absorption by the clay and cause strain hardening. To minimize basal shear tractions in our model, we utilized a high-density (1,906 kg m^{-3}) zinc chloride solution that was injected between two thin plastic bags beneath the wet kaolin. The plastic bags, pleated parallel to the direction of maximum compression (which enabled pleats to open in the direction of greatest extension), were placed inside the deformation frame to prevent mixing and/or absorption of water from the wet kaolin into the underlying $ZnCl_2$ solution. Although the strength of the plastic bag at the base of the clay may have caused subtle mechanical effects, it is important to note that pleats (and free unfolding of those pleats into the underlying fluid layer) prevented the plastic bag from carrying any regional deviatoric stress. The clay floated above the 0.01 m-thick layer of $ZnCl_2$ solution; consequently, basal traction was eliminated and a free surface was created at the base of the model, providing the added benefit of free isostatic compensation, similar to the layered sand–honey model of DAUTEUIL and MART (1998).

2.3. Scaling Issues

HUBBERT (1937) showed that for proper scaling of a brittle-deformation experimental model, two conditions must be satisfied. First, the model and natural materials must have similar coefficients of friction, μ. Second, the model must scale linearly to nature with regard to cohesion, density, gravity, and length. Although scaling is necessary to achieve accurate representation of natural systems, models are often limited by poorly constrained natural and model properties. Nevertheless, analog models with only geometric similarity or partial scaling to natural settings have been invaluable in interpreting tectonic structures and processes (RIEDEL 1929; TCHALENKO 1970; TAPPONNIER et al. 1986).

The model presented in this study is qualitative; however, like other qualitative analog models, our model has geometric similarity to nature in the sense that the strong layer (clay) is planar (sometimes prefaulted) similar to real crust in simple shear, and that boundary conditions are comparable. In addition, direct-shear experiments by SIMS (1993) identified the angle of internal friction, ϕ, for wet clay (50 % water by weight) as 28°, which corresponds to $\mu = 0.53$. The coefficient of friction of intact crustal rock, μ_E, is estimated to be 0.85 (HANDIN 1966; JAEGER and COOK 1976; BYERLEE 1978), but there is much uncertainty in this value for regions of pervasive and/or active faulting, and strain weakening could cause large drops in effective friction to values as low as 0.03 (BRACE

and KOHLSTEDT 1980; BIRD and KONG 1994; CHESTER 1995; BIRD 1996, 1998; KONG and BIRD 1996; JIMÉNEZ-MUNT et al. 2001; LIU and BIRD 2002; BIRD et al. 2008; GOLDSBY and TULLIS 2011; TONG and YIN 2011). If we assume an average μ_E for the crust from these end-member values, we satisfy the first condition set by HUBBERT (1937). Because our modeling material is likely to have a coefficient of friction similar to rock and because the most natural geometric scale is according to the ratio of thicknesses of the strong layers, it may be possible to make very tentative length-scale evaluations between the model and natural systems. For the purposes of this model, we define the length scale, $\tilde{\lambda}$, on the basis of the thickness of wet kaolin, $\lambda_m = 0.04$ m, and approximate depth to the bottom of Earth's seismogenic crust, $\lambda_E = 10^4$ m. Thus, we derive a length scale $\tilde{\lambda} = \lambda_m/\lambda_E = 4.0 \times 10^{-6}$, indicating that our scaled clay surface represents an area of approximately 8×10^9 m^2.

Because clay has complex material properties, there are several failures of our model to reproduce natural settings. One distinct failure of the model is the apparent lack of earthquakes (inertia-limited slip events). Natural earthquakes may exert both transient and lasting stress concentrations on crack tips (DUNHAM 2007) that could control their propagation (SLEEP and HAGIN 2008), whereas our model is apparently slow and quasi-static at all times. A second notable failure is the development of features with no analogous natural counterparts, for example Mode-I (opening) cracks parallel to the most compressive stress (σ_1). This is probably because the cohesive strength is too high to accurately represent natural crust where extension usually results in normal faulting. Third, although shear stress may be independent of depth during the early stages of faulting, resulting in fault initiation at the surface where strength is the lowest (LOCKNER et al. 1991; SCHULSON et al. 1999), it is also plausible that some faults form at depths that are initially blind. Slip at depth imposes strain on a shallow lid above a blind fault, which eventually leads to frictional failure if static stress is high (SLEEP 2013). Because wet kaolin has viscous properties, our model does not allow for lid failure and self-organization.

Although our model is not dynamically scaled to nature, similar statements could be made with regard

Table 1

Results from the model

Experiment	Total offset at first shear formation (m)	Mean departure from applied shear direction (degrees)					Offset necessary for throughgoing faulting (m)	R-shear fault spacing (m)	PDZ width (m)	Surface-to-total offset ratio
		R-shear	R′-shear	P-shear	P′-shear	X-shear				
P-1	0.0100	10.8	75.9	166.7	47.7	121.6	0.0222	0.007	0.045	0.765
P-2	0.0080	11.3	–	–	–	–	0.0242	0.006	0.039	0.735
P-3	0.0070	–	–	–	–	–	0.0259	–	–	0.816
P-4 (slow)	0.0025	11.0	–	–	–	–	0.014	0.008	0.051	0.865
BF-1	0.0110	13.2	73.8	152.6	54.5	104.1	>0.049[a]	0.010	0.102	0.788
BF-2	0.0090	15.5	71.3	155.0	33.1	104.7	>0.043[a]	0.015	0.100	0.834
BF-3	0.0083	19.7	79.0	154.5	38.7	99.6	>0.052[a]	0.008	0.090	0.826
MD-1	0.0075	20.3	77.2	130.6	44.9	104.7	>0.047[a]	–	0.101	0.805
MD-2	0.0110	15.5	72.6	166.3	27.1	114.6	>0.039[a]	–	0.100	0.785
CJ-1	0.0060	13.0	79.3	156.5	38.1	105.5	>0.048[a]	–	0.170	0.819
CJ-2	0.0120	13.2	74.0	162.7	34.3	99.2	>0.043[a]	–	0.203	0.811

[a] Denotes lower limit of total offset required to form a throughgoing fault for experiments in which throughgoing faulting was not achieved

to analog models using different materials (e.g., sand with high shear dilatancy; MANDL 1988) and numerical models that often, as a result of unconstrained natural properties, have more variables than constraints. Accordingly, the limitations of all models simulating natural environments should be interpreted with both an open mind and a watchful eye.

3. Experimental Results

Results of the eleven analog experiments performed in this study are summarized in Table 1. Experiments are organized into four types based on the structural similarity of their preexisting structures:

1 Pristine (control);
2 Bookshelf faults;
3 Multidirectional faults; and
4 Columnar joints.

All shear angles, α, are reported as departure from the applied shear direction.

3.1. Pristine (Control)

During these experiments (P-1–P-4), the surface of the clay was left intact before shear deformation. As expected, a discrete fault zone composed of en

échelon faults and folds developed with progressive strain (Fig. 2, Table 1; RIEDEL 1929; TCHALENKO 1970; WILCOX et al. 1973; NAYLOR et al. 1986; DOOLEY and SCHREURS 2012). Synthetic R-shears began to form after 0.01 m of offset at angle $\alpha = 8°–17°$. JAEGER and COOK (1976) determined that α can be expressed as a function of the angle of internal friction, ϕ, or coefficient of friction for a material, μ, such that $\alpha = \frac{1}{2}\phi = \frac{1}{2}\tan^{-1}\mu$; for wet kaolin with $\mu = 0.5–0.6$ (SIMS 1993; HENZA et al. 2010), α should be approximately $13°–15°$, which is in general agreement with our results.

After 0.005 m of additional offset, syntectonic conjugate R′-shears began to form at $\alpha = 65°–83°$. P-shears were few in number, formed at $\alpha = 164°–170°$, and were concentrated near the leading and trailing edges of clay. Because of their azimuthal similarity, it is difficult to distinguish between R′- and P′-shears, however a set of shears ($\alpha = 35°–57°$) we assume to be P′-shears appeared during later stages of the experiment. Antithetic X-shears ($\alpha = 123°–134°$) formed along the contact with the deformation frame and may be artifact structures. Rotation and linkage of shears into a throughgoing fault occurred within the primary deformation zone (PDZ) after 0.024 m (avg.) of total (base plate) offset. The average width of the PDZ was 0.045 m, and average spacing between R-shears within the PDZ was 0.007 m (Fig. 2). En échelon folds perpendicular to σ_1 formed

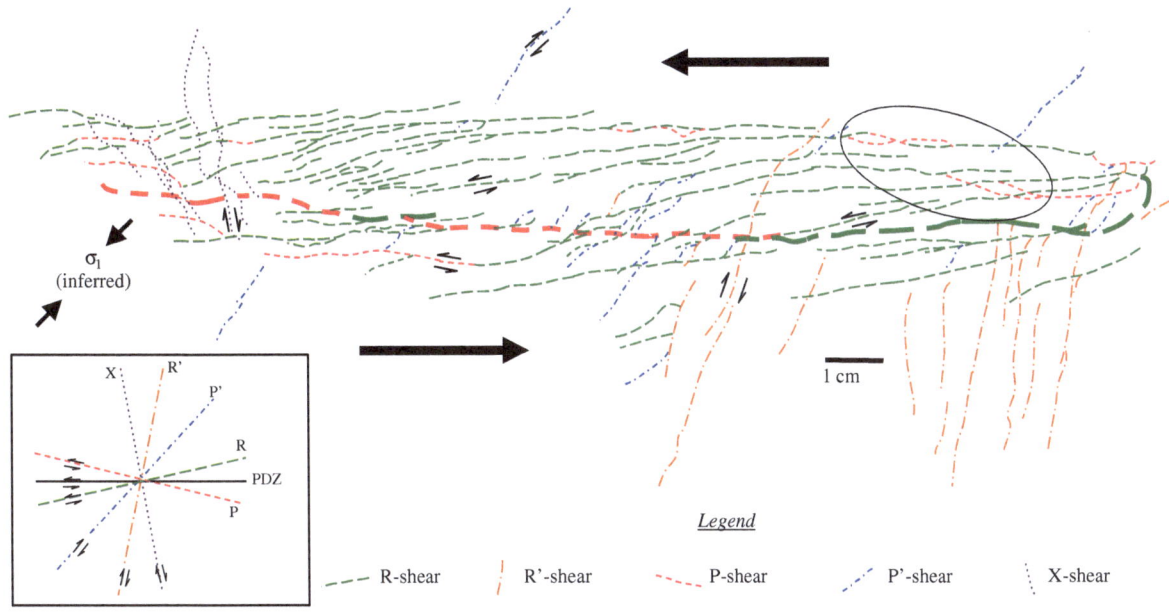

Figure 2

Map view line drawing of strike–slip faults developed during sinistral shearing in wet kaolin with an originally intact surface (total relative shear offset of base plates = 37 mm). Classic Riedel fault assemblage (*inset*) developed during pristine experiments. *Small arrows* indicate the inferred and observed shear direction for each shear type. The *bold line* is throughgoing fault (Y-shear). The *oval* indicates pop-up and pull-apart basin location. See text for detailed discussion

and interacted with faults, resulting in the development of a variety of structures such as along-strike pop-ups, orthogonal pull-apart basins (oval in Fig. 2), tension fractures, and coalesced and/or reactivated faults, all of which have commonly been reported in Riedel experiments (CLOOS 1928; RIEDEL 1929; TCHALENKO 1970; WILCOX *et al.* 1973; AYDIN and SCHULTZ 1990; ATMAOUI *et al.* 2006). Most of the coalesced shears produced right-stepping compound faults that formed restraining bends.

Maximum clay displacement was measured as the average offset of the most displaced row of passive markers relative to a fixed point. With the exception of one low-velocity experiment (3.83×10^{-6} m s^{-1}), the maximum accommodation of total offset at the clay surface was 77.2 %. The shear array produced by the low-velocity pristine experiment was geometrically similar to classic Riedel shear array (Fig. 2, inset); however, a higher percentage (86 %) of the total offset was accommodated at the surface and a throughgoing fault appeared after less total offset. It is unclear if this is indicative of temporal dependence of PDZ evolution in the wet kaolin.

3.2. Bookshelf Faulting

Experiments containing preexisting structures classified as bookshelf faulting consisted of five parallel to sub-parallel linear incisions oriented normal (BF-1), 45° (but antithetic; BF-2), or parallel (BF-3) to the inferred principal stress direction (σ_1), equivalent to 135°, normal, or 45° transects of the base plate contact and applied shear direction, respectively (Fig. 3; Table 1). Experiments resulted in distributed shear zones, wide PDZs and fault spacing, and progressively rotated R-shears with decreasing obliquity of preexisting structures with regard to σ_1 (Table 1). No throughgoing faults were produced during bookshelf experiments, although fault traces were often 0.05–0.08 m long. Preexisting faults showed signs of rotation and offset, and shears formed simultaneously with preexisting fault reactivation manifested reverse drag, similar to that described by RECHES and EIDELMAN (1995). The average accommodation of total offset (81.6 %) was larger than for pristine experiments, implying that deviatoric stress was probably lower in bookshelf experiments.

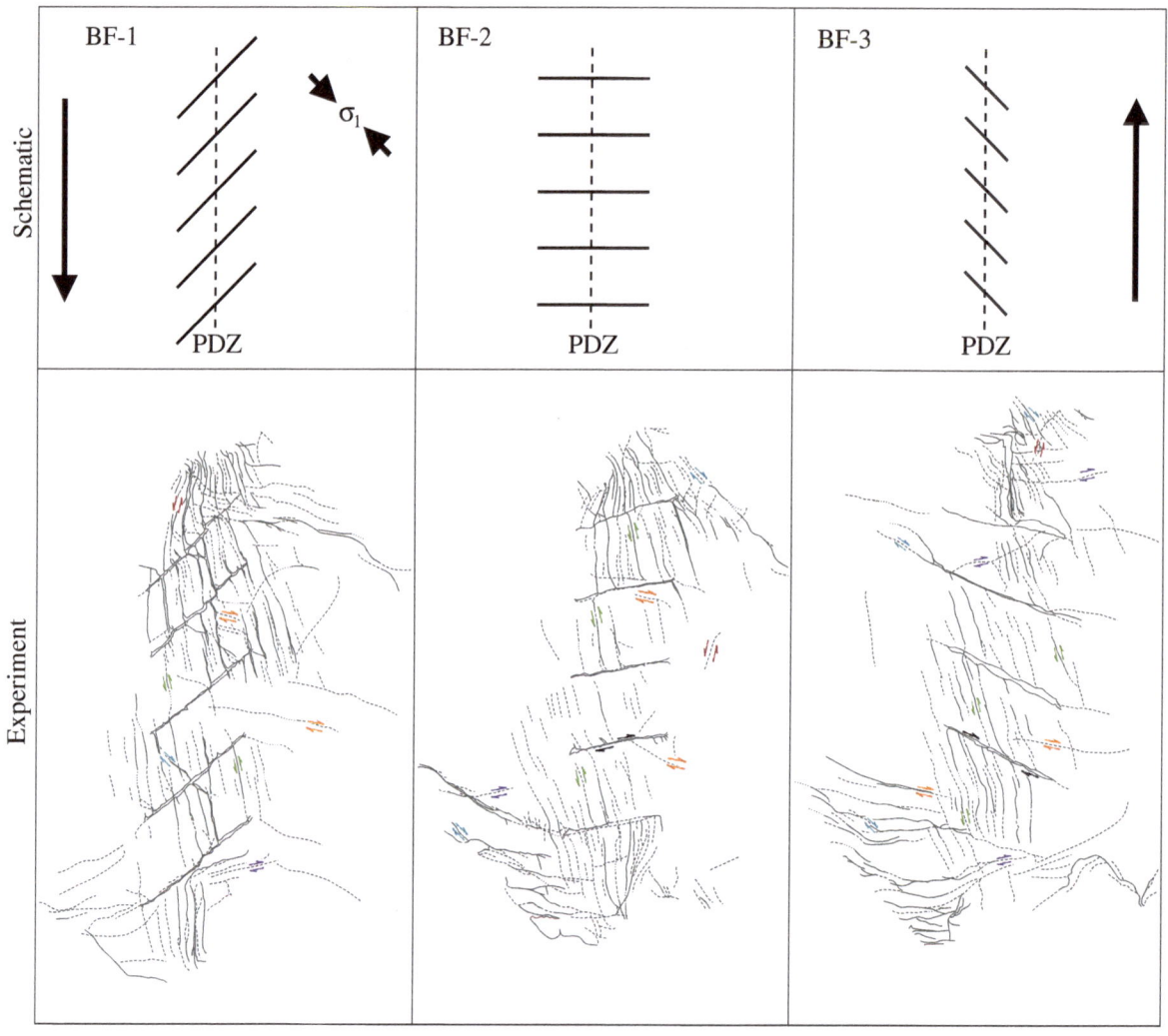

Figure 3

Map view of wet kaolin containing preexisting bookshelf faults oriented approximately perpendicular (BF-1), at 45° (BF-2), and parallel (BF-3) to the inferred principal stress direction (σ_1). *The top three diagrams* are schematic representations of faults that were pre-cut into the wet kaolin. *The bottom three figures* are line drawings of faults that formed after 37 mm of sinistral base plate offset in the analog model. *Large arrows* denote the applied shear direction, and *small arrows* on fault maps indicate inferred and observed relative motion of individual shears. *Arrow colors* indicate shear type: R-shears (*green*), R′- shears (*orange*), P-shears (*red*), P′-shears (*blue*), X-shears (*purple*). *Solid lines* indicate distinct fault traces, *dashed lines* indicate inferred faults (visible but not fully ruptured at the surface), *dotted lines* indicate concealed fault segments

3.3. Multi-directional Faulting

Multi-directional fault experiments contained networks of preexisting intersecting linear features meant to simulate a highly faulted/fractured crust. The first experiment (MDF-1) consisted of incisions normal, at 45°, and parallel to σ_1, essentially representing a combination of the bookshelf faulting scenarios (discussed in the section "Bookshelf Faulting"). A second experiment (MDF-2) consisted of a network of linear features cut into the clay surface transecting the base plate contact along planes oriented at 60°, normal, and 120° (Fig. 4). These experiments produced fault zones that were significantly reduced in both number and length of shears; as a result, throughgoing faults did not develop (Table 1). Shears that did develop were delimited by preexisting faults and were rotated away from the applied shear direction at larger angles than in

MDF-1   MDF-2

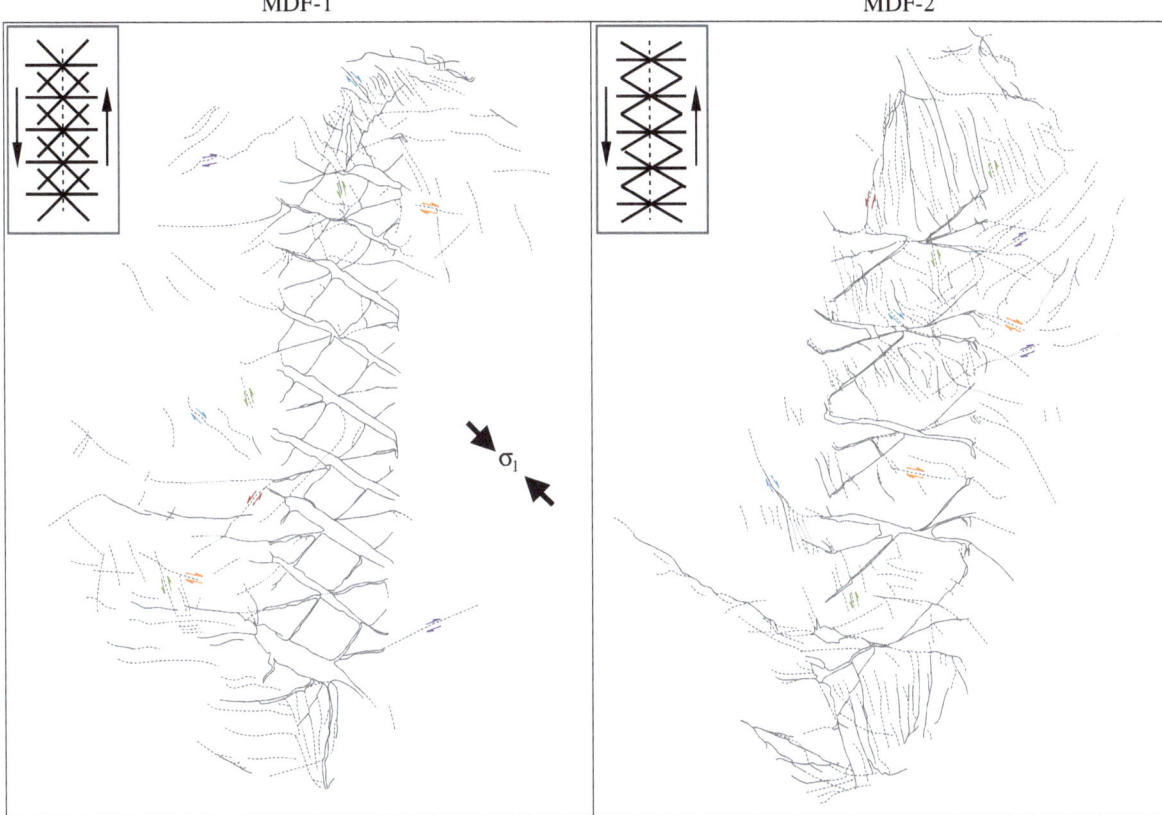

Figure 4

Map view of sinistral shearing in wet kaolin containing preexisting multi-directional faults, oriented perpendicular, at 45°, and parallel (MDF-1) to the inferred principal stress direction (σ_1) and at 60°, perpendicular, and 120° to the applied shear direction (MDF-2), after 37 mm of base plate offset. *Insets* are schematic representations of faults that were pre-cut into the wet kaolin. *Arrows in insets* denote the applied shear direction, and *small arrows on fault maps* indicate observed and inferred relative motion of individual shears. *Arrow colors* indicate shear type: R-shears (*green*), R′-shears (*orange*), P-shears (*red*), P′-shears (*blue*), X-shears (*purple*). *Solid lines* indicate distinct fault traces, *dashed lines* indicate inferred faults (visible but fully ruptured at the surface), *dotted lines* indicate concealed fault segments

pristine experiments. The minimum PDZ width in these experiments was affected by the width of the preexisting fault network. Average surface accommodation of the total offset was 79.5 %.

3.4. Columnar Jointing

Columnar jointing experiments comprised an interconnected hexagonal pattern (CJ-1) and a mesh-like network of equilateral triangles (CJ-2; Fig. 5). Because of the difficulty of cutting intricate patterns into the clay surface, "columnar joints" in these models are not true-to-scale with natural columnar joints; rather, they are meant to mimic the regional effects of basaltic or andesitic cover that are known to commonly manifest extensive columnar

jointing (TOMKEIEFF 1940). Although columnar joints do not behave as crustal-scale tectonic structures in nature, it is possible that during the early stages of an incipient shear zone, the shear stress may be independent of depth in the upper crust. In such a scenario, faults could nucleate near the surface and rupture downward; thus, columnar joints at the surface could have a deeper effect by controlling the location and geometry of fault nucleation.

Both columnar jointing experiments produced chaotic shear deformation. Shears were short, fault linkage did not occur, and throughgoing faults did not develop. Mean R-shear direction, $\bar{\alpha} = 13°$, was within the predicted range for R-shears in pristine materials; however, shears did not resemble the en échelon pattern developed during pristine experiments. Faults

CJ-1   CJ-2

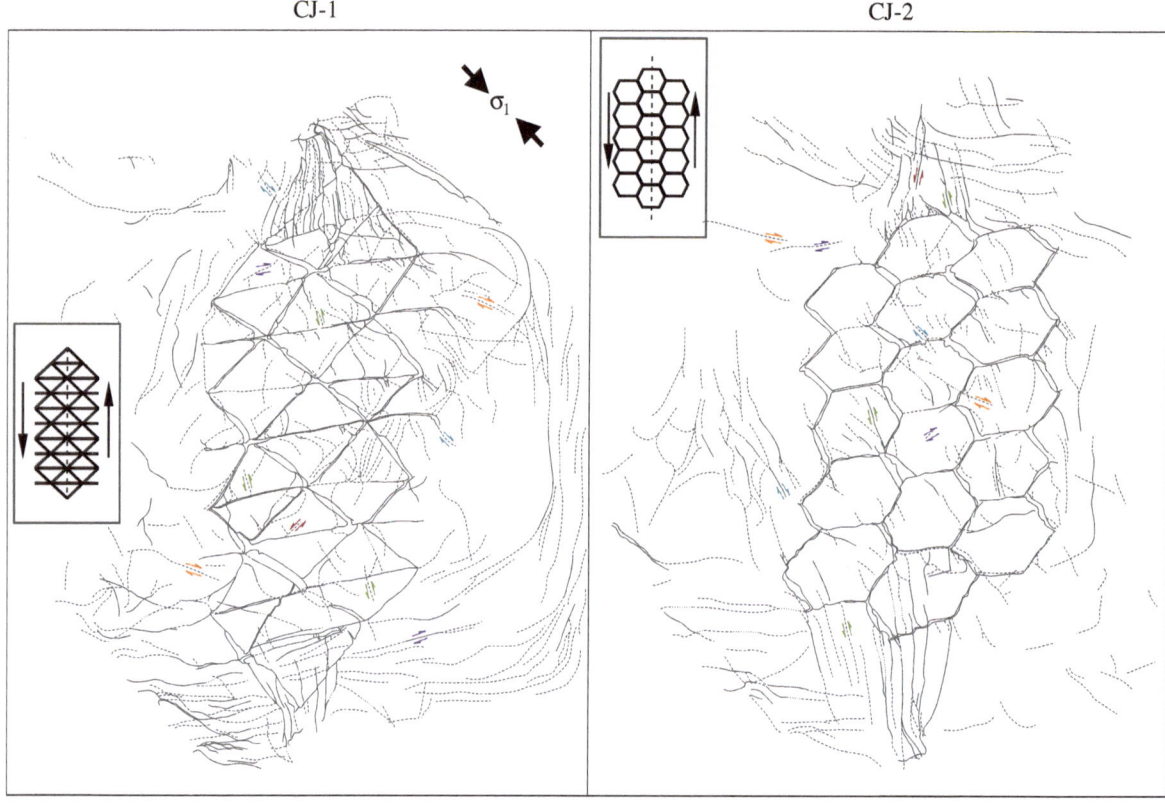

Figure 5

Map view of sinistral shearing in wet kaolin containing preexisting columnar joint structures, represented by an equilateral triangle mesh (CJ-1) or a hexagonal mesh (CJ-2), after 37 mm of base plate offset. *Insets* are schematic representations of faults that were pre-cut into the wet kaolin. *Arrows in insets* denote the applied shear direction, and *small arrows on fault maps* indicate observed and inferred relative motion of individual shears. *Arrow colors* indicate shear type: R-shears (*green*), R′-shears (*orange*), P-shears (*red*), P′-shears (*blue*), X-shears (*purple*). *Solid lines* indicate distinct fault traces, *dashed lines* indicate inferred faults (visible but not fully ruptured at the surface), *dotted lines* indicate concealed fault segments

were restricted to the interior and/or exterior of preexisting structures and did not usually intersect preexisting linear traces. Counterclockwise rotation of the modeling material, dilation on portions of preexisting structures parallel or sub-parallel to σ_1, and compression along portions of preexisting features approximately normal to σ_1 are apparent in both experiments. PDZs were widely distributed (>0.17 m-wide) and extended past preexisting structures. The average accommodation of base plate motion was 81.9 and 81.1 % in CJ-1 and CJ-2, respectively (Table 1).

4. Discussion

We find that mechanical discontinuities (faults/fractures) that predate shear deformation affect the

evolution and geometry of an incipient shear zone by means of fault suppression and dispersion. Figure 6 plots the relationship of experiment type (i.e., pristine, bookshelf, etc.), required offset for throughgoing fault formation (filled squares), or total offset if no throughgoing faults formed (open squares), and the accommodation of base total offset in the clay surface. Pristine experiments produced throughgoing faults after ≤0.025 m total offset, whereas experiments containing preexisting structures did not produce throughgoing faults in the range of offsets available to us. Because of structural limits, we were required to terminate shear at total offsets of 0.05 m; we can, therefore, only assign a lower limit to the total offset necessary to form throughgoing faults in systems containing preexisting structures. It seems that more than twice as much total offset (with regard

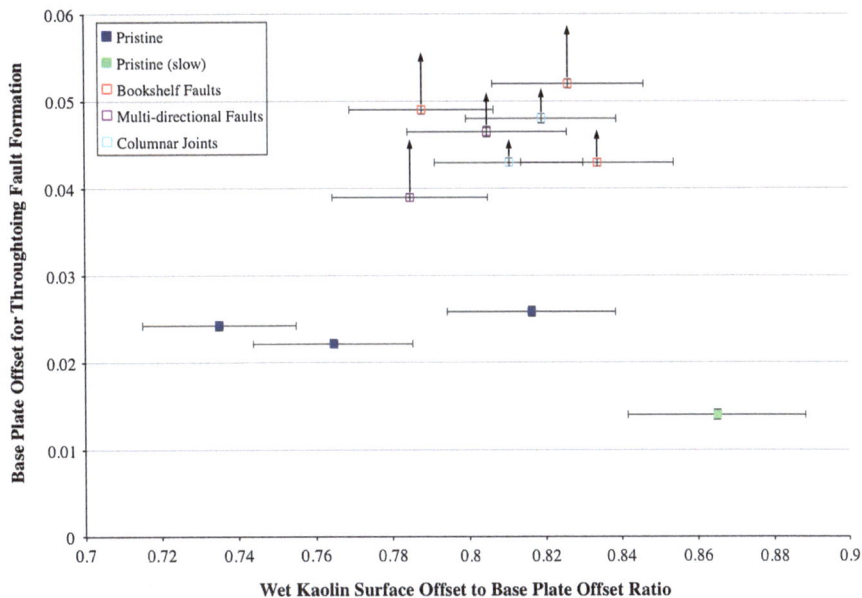

Figure 6
Plot illustrating the relationship between base plate offset necessary to form throughgoing faults and surface offset as a fraction of base plate offset for each experiment. *Open squares* indicate experiments in which throughgoing faults did not form. *Arrows* imply a minimum value of offset necessary for throughgoing faults to develop

to pristine experiments) is required to develop throughgoing strike–slip faults in systems containing preexisting faults, and it may be that the ratio is much higher. No statistically significant relationship exists between accommodation of total offset by the clay surface and experiment type.

Our conclusions agree with theoretical predictions for strike–slip fault response to obliquely oriented preexisting faults and joints. In our model, preexisting structures normal or oblique to applied shear produced distributed and less ordered PDZs; shears were fragmented and were often deflected or unable to cut preexisting structures. It has been reported a strike–slip fault may develop secondary features with reversed-polarity (with respect to large-scale slip) at its fault tips (BEN-ZION *et al.* 2012); if this is true for our model, then reversed polarity tips could have an inhibiting effect on shear propagation as shears approach an oblique and/or orthogonal structure in the horizontal plane, resulting in observed deflection or arrest.

Preexisting "normal faults" were reactivated to slip opposite the applied shear direction. Additional strain resulted in counterclockwise block rotation defined by the location of preexisting faults, dilation

normal to σ_1 of incisions parallel to σ_1, and compression parallel to σ_1 of incisions normal to σ_1 (Figs. 3, 4). These observations are consistent with crustal block rotation theory, which states that because the stress tensor is symmetric in any coordinate system ($\sigma_{xy} \equiv \sigma_{yx}$), shear stress on a linear shear zone and on preexisting high-angle faults perpendicular to the linear shear zone are always equal. Subsequently, regional stress should be regulated by slipping on high-angle faults in the sense opposite to the bulk shear, which would, in theory, delay throughgoing faulting and produce rotated crustal blocks (NUR *et al.* 1986; RON *et al.* 2001).

While crack dilation along preexisting "faults" has no natural analog, we assume these might be equated with slip on normal faults, which also form by extension perpendicular to the strike. If a preexisting structure forms an extensional crack in the model, then we may infer that the fault is reactivated in the normal sense and that slip would occur along the weakest plane (in our model, these planes are vertical, but in nature fault dip would determine the direction of slip); if a preexisting fault is compressed in the model, we would infer that the fault is reactivated in a reverse sense, with slip occurring along the predefined fault plane. These faults may also show some component of horizontal slip along strike as a result of horizontal shear, making these faults dip–slip. Although we cannot observe what happens at depth in our model, we predict that preexisting fault geometry will greatly affect reactivated fault geometry. From global and regional studies of strike–slip fault systems, MANN (2007) and MANN et al. (2007) proposed that the formation of releasing, restraining, and paired bend formation is often the result of propagating fault interaction with preexisting crustal structures, for example ancient rift basins. Dilation of preexisting faults with strike oriented parallel to σ_1 and compression of those with strike oriented oblique to σ_1 in our model suggests that this may be at least partly correct, although it does not require that all strike–slip fault bends form in this particular fashion.

Columnar joints are commonly formed by thermal contraction in flood basalts and andesite (AYDIN and DEGRAFF 1988; DEGRAFF and AYDIN 1993). Although individual joints do not generally exceed a diameter of 2 m (RYAN and SAMMIS 1978; DEGRAFF and AYDIN 1987; GROSSENBACHER and MCDUFFIE 1995), they can form expansive colonnades—collections of adjacent equant polygonal columns (TOMKEIEFF 1940). Assuming that colonnades act mechanically as a unified structure larger than its individual constituents, we may surmise that columnar jointing will suppress strike–slip faulting in the following ways:

1 Shears aligned with the applied shear direction will be deflected or dampened when they encounter hexagonal columnar joints, and/or
2 Columnar joint networks will tend to dilate under in-plane shear resulting in the opening of some joints.

If lateral boundary conditions are resistant to dilation, then horizontal compressive stress will increase, adding frictional strength to the deformed region and subsequently suppressing any faulting instability (SLEEP 1997). Although we did not replicate the scale of individual joints in our model, preexisting equilateral triangle (CJ-1) and hexagonal (CJ-2) patterns we used presumably mimic larger-scale colonnades found in many flood basalts, for example the Columbia river basalt group (LONG and WOOD 1986). Results from our model show dilation

under in-plane shear that caused joints parallel to σ_1 to open (Fig. 6), as predicted by SLEEP (1997). Shear deflection was not overly apparent in experiment CJ-1, but discrete strike–slip fault formation along the PDZ was almost completely subdued or absent.

4.1. Natural Examples

Our model is beset with caveats and uncertainties with regard to natural settings; however, if we adopt a geometric similarity length-based approximation for which the thickness of our wet kaolin corresponds to a seismogenic crust of 10 km, then the offset at which throughgoing faults formed in pristine surfaces corresponds to 3.5–6.5 km in nature. Many regions follow the deformation evolution of our pristine experiments, forming throughgoing shear zones after only relatively little offset. Among these regions are the Calico and West Calico faults in the southern Mojave Desert (WESNOUSKY 1986; PETERSEN and WESNOUSKY 1994) and the Dasht-e-Bayaz sinistral fault in Iran (TCHALENKO 1970; WALKER et al. 2004).

Using global plate model PB2002 (BIRD 2003) that combines GPS, focal mechanisms, and neotectonic modeling to produce a continuous global plate boundary map, it was possible to identify several regions on Earth that are dominated by toroidal motion, but without extensive and/or discrete strike–slip faulting as predicted by the model (Fig. 7). For example, the 70 km-long and 20–60 km-wide South Icelandic seismic zone (SISZ), a seismically active transform boundary between the NE–SW-trending Western and Eastern volcanic zones, has been shearing at ~ 19 mm a^{-1} (BIRD 2003) for ~ 2 Ma (SAEMUNDSSON 1974). Net offset for the SISZ is therefore ~ 38 km. Although earthquakes generated in the SISZ manifest strike–slip focal mechanisms and GPS measurements confirm sinistral motion at >11 mm a^{-1} and extension (LAFEMINA et al. 2005), virtually no expected E–W trending sinistral faults are present (Fig. 8; EINARSSON et al. 1981; HACKMAN et al. 1990; BERGERAT and ANGELIER 2000; BERGERAT 2001; GUDMUNDSSON 2007). Instead, the SISZ, which is located partly in Upper Pliocene–Pleistocene rocks and partly in Holocene lava flows, comprises conjugate NNE–SSW, NE–SW, and ENE–WSW-striking fractures that form as a result of crustal spreading

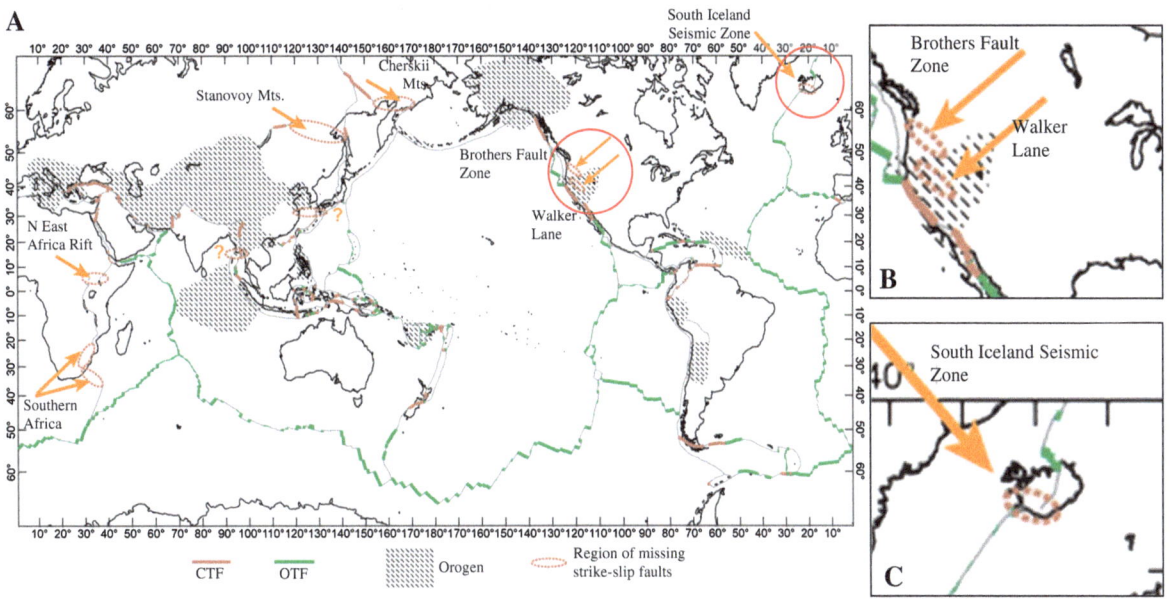

Figure 7

a Strike–slip plate boundaries of the world, from the PB2002 global plate model of BIRD (2003). The model is known to be incomplete in shaded regions of complex (orogenic) geometry. *Dashed ellipses* highlight regions where fault systems are underdeveloped or missing. *CTF* continental transform fault; *OTF* oceanic transform fault. *Red circles* indicate the locations of insets (**b**) and (**c**). **b** Walker Lane, California contains sets of preexisting high-angle bookshelf faults sub-perpendicular to the shear direction and Brothers fault zone, Oregon contains columnar-jointed basaltic and andesitic cover and high-angle faults sub-perpendicular to the shear direction. **c** The South Iceland seismic zone contains high-angle faults sub-perpendicular to the shear direction and regional cover of flood basalts

(BERGERAT and ANGELIER 2000). NE–SW-trending normal faults are dated to an age similar to the Upper Pliocene–Pleistocene lava pile they are contained within, whereas NNE–SSW dextral and ENE–WSW and NE–SW sinistral fractures are younger (GUDMUNDSSON and BRYNJOLFSSON 1993; BERGERAT and ANGELIER 2000; GUDMUNDSSON 2007). Net offset in the SISZ corresponds to a total offset of 0.15 m in our model. Experiments MDF-1 and MDF-2, which best represent the complex fabric of the pre-fractured basaltic crust of the SISZ, show very little to almost no deformation in the form of strike–slip faults parallel to applied shear after 0.15 m total offset. GUDMUNDSSON (2007) inferred a principal stress direction for the region (Fig. 8) which agrees with E–W transform motion, and went further to suggest that E–W strike–slip faults have not formed because the structure of South Iceland (e.g., the Hreppar anticline, mechanically dissimilar materials, and oblique-trending normal faults) is unfavorable for fault propagation. Results from our model suggest that preexisting faults and/or fractures have resulted

in mechanical suppression of throughgoing E–W trending strike–slip faults and strain accommodation by dextral slip on NNE–SSW-trending fractures, which is in general agreement with GUDMUNDSSON (2007).

We will also discuss two regions in the broad, dextral Pacific–North America boundary zone: the Brothers fault zone in Oregon and Walker Lane in California and Nevada. We focus on these relatively inland portions of the developing Pacific–North America plate boundary because they are probably less strongly controlled by the Mendocino triple-junction singularity than are the more proximal strike–slip fault of the Coast Ranges and Klamath Mountains. Nevertheless, it is true that both may have involved an element of southeast-to-northwest propagating development, and therefore do not exactly match our experimental conditions of shear initiating simultaneously in all locations of the future shear zone.

Brothers fault zone (BFZ) is a NW–SE-trending 280-km-long, 30-km-wide zone of poorly developed

Map of the South Iceland Seismic Zone

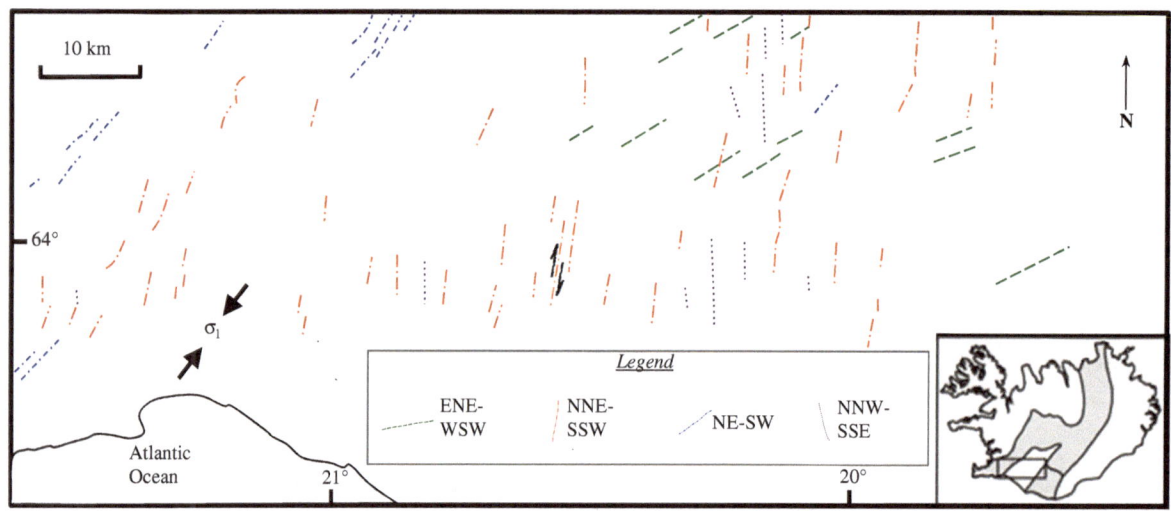

Figure 8

Fault map of the South Iceland seismic zone (SISZ) and its surroundings (modified from JOHANNESSON and SAEMUNDSSON 1998). The NE–SW-trending fractures (*blue dashed-dotted lines*) are primarily high-angle normal, NNE–SSW-trending fractures (*orange long-dashed-dotted lines*) are mostly dextral (as noted by *directional arrows*), ENE–WSW-trending fractures (*green long-dashed lines*) are primarily sinistral, and NNW–SSE-trending fractures (*purple dotted lines*) are of an "unknown type" (BERGERAT *et al.* 1998). Most compressive stress direction is from GUDMUNDSSON (2007). *Grey area in inset map* is the Mid-Atlantic ridge system as it is expressed on Iceland

NW–SE-trending dip–slip faults in southeast Oregon (Figs. 7, 9; HIGGINS and WATERS 1967; WALKER 1969, 1977; LAWRENCE 1976; WALKER and MACLEOD 1991; CHRISTIANSEN and YEATS 1992; HOOPER *et al.* 2002b). The onset of faulting is younger than the Early Miocene Columbia River basalt group covering much of the region, and probably coincides with the opening of the Gulf of California (12–6 Ma). The BFZ is intersected by a set of NE–SW-trending high-angle basin and range extensional normal faults, including Steens Mountain fault, the largest normal fault in Oregon (LAWRENCE 1976). Steens Mountain fault is a long-lived horst structure that predates basin and range extension (MINOR *et al.* 1987a, b; LANGER 1991; HOOPER *et al.* 2002a; SCARBERRY *et al.* 2010) and is responsible for expansive ($\leq$60,000 km^2) and thick (450–1,000 m) basalt flows that erupted approximately 16.6–15.3 Ma (CARLSON and HART 1987; SWISHER *et al.* 1990; LANGER 1991; BRUESEKE and HART 2000; HOOPER *et al.* 2002a; CAMP *et al.* 2003; BONDRE *et al.* 2004; BONDRE and HART 2006; BRUESEKE *et al.* 2007). The neotectonic model of BIRD (2009) and GPS data of HAMMOND and THATCHER (2005) predict and confirm a dextral offset rate of 2–3 mm a^{-1}, implying a net offset of 12–36 km. On

the basis of the simple approximations from our model, we would expect real shear zones containing preexisting faults to deform by at least 13 km (possibly much more) without developing new throughgoing faults, which agrees with the observations in the BFZ, and also suggests that the formation of integrated strike–slip fault systems is quite slow where the upper crust is predominantly composed of basalt flows and/or contains preexisting faults oriented normal and/or oblique to shear.

The Walker Lane (WL) in the western Great Basin of North America (Fig. 7) is a case where net offset is quite large, but throughgoing strike–slip faulting is sparse. The WL has been shearing at rates up to 12 mm a^{-1} (ARGUS and GORDON 2001) probably since the opening of the Gulf of California at 6 Ma, so its net offset may be as much as 72 km. It shows numerous NW-striking dextral strike–slip faults that cut previous thrusts and normal faults, but that are not integrated into a throughgoing fault (WESNOUSKY 2005). These very approximate numbers may confirm the mechanical importance of preexisting faults in suppressing or delaying the appearance of discrete, throughgoing strike–slip faults, and suggest that even higher displacement than the lower limit determined

Map of Brothers Fault Zone, Oregon

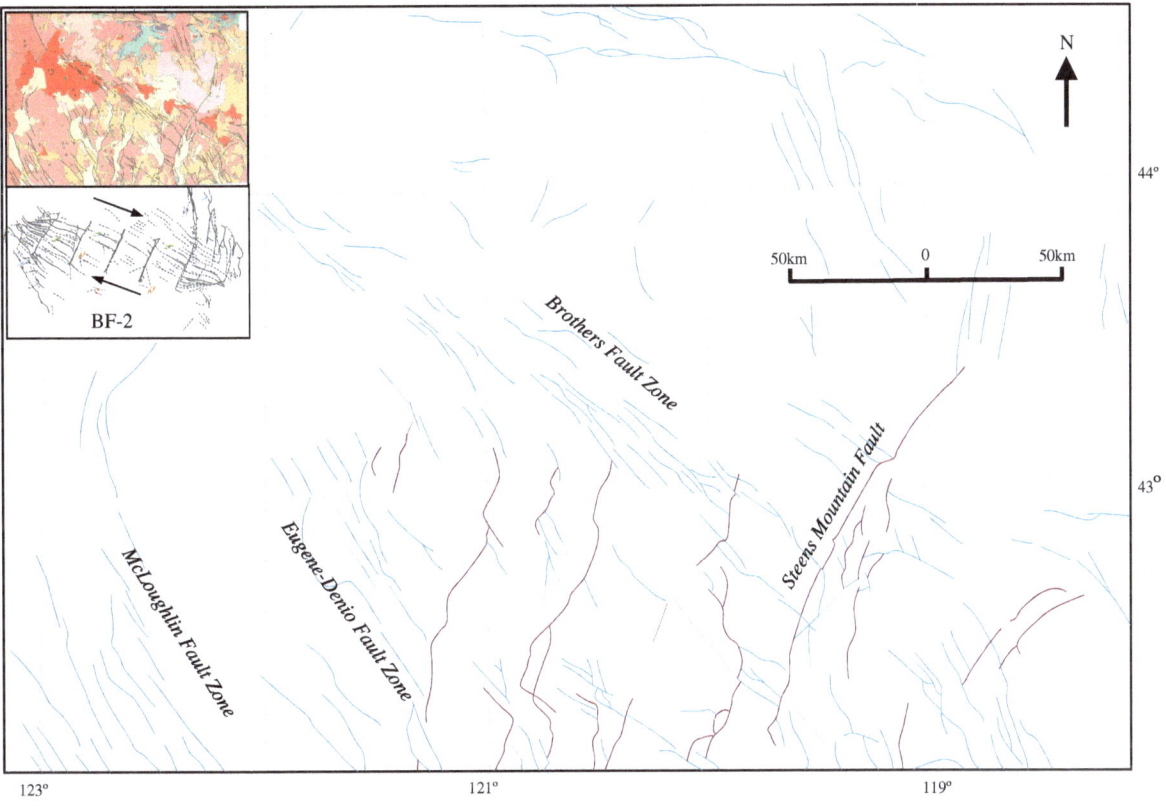

Figure 9

Fault map of SE Oregon. *Blue lines* denote NW–SE trending dextral dip-slip faults (LAWRENCE 1976; HAMMOND and THATCHER 2005) and *red lines* denote NNE-SSW trending steeply dipping normal faults corresponding to Steens Mountain basalt and andesite (WALKER and KING 1969; LANGER 1991; BRUESEKE *et al.* 2007). The *top inset* is a geologic map of SE Oregon from which the fault map was developed (WALKER and KING 1969), showing primarily basaltic and/or andesitic crust (*pink* and *red hues*). The *bottom inset* is the fault map from experiment BF-2 of this study, mirrored and rotated to correspond to the tectonic setting of SE Oregon

in this model may be necessary for throughgoing faults to develop where faults and/or joints exist before shear deformation.

SLEEP (1994) suggested that when searching for plate tectonics on terrestrial planets, we should search for transform boundaries because they reveal the relatively rigid motion of plate interiors. Our results indicate, however, that strike–slip and transform fault zones may be subdued or highly distributed in specific settings, making their identification difficult. As an example, the basaltic crust of Venus has been subjected to multiple episodes of deformation (CRUMPLER *et al.* 1986; RAITALA 1994; KOENIG and AYDIN 1998; PHILLIPS and HANSEN 1998; GHENT and HANSEN 1999; TUCKWELL and GHAIL 2003), which has left its crust riddled with normal and thrust faults, sinuous deformation fabrics called ribbon-tessera, cross-cutting linea, and many other features. Regions assumed to contain strike–slip faulting, for example Lavinia Planitia, (volcanic plains) and Thetis and Ovda Regios (upland areas), contain highly distributed shear zones that are often blunted at their ends by obliquely oriented fractures (SOLOMON *et al.* 1992; SQUYRES *et al.* 1992; KOENIG and AYDIN 1998; TUCKWELL and GHAIL 2003; ROMEO *et al.* 2005). Perhaps these shear zones are forming in geometries and on timescales defined by preexisting structures. More detailed map analysis is needed to establish the mechanics of faulting in these regions.

5. Conclusions

Preexisting fault networks affect fault zone development both geometrically and systematically by shear dispersion and suppression of throughgoing faults. Whether preexisting structures in our model correspond to natural deformation patterns or to the mechanical properties of regional flood basalts is still not completely clear. However, the results of the model match the theoretical interactions of strike–slip and obliquely oriented faults and joints. In addition, our model seems to fit select regions on Earth and could possibly correspond to regional shear deformation on other terrestrial planets, suggesting that it may, at least in part, be insightful in respect of shear zone development. Finally, it is well documented that faults rarely occur as isolated structures, but rather as complex zones of fault interaction that accommodates local strain, which suggests that preexisting structures are probably involved in the evolution of most new fault zones.

Acknowledgments

We would like to thank Y. Ben-Zion, N. H. Sleep, and two anonymous reviewers for their detailed criticism and comments, which improved this paper. We also thank Z. Reches for bringing relevant literature on analog modeling to our attention. This study was supported in part by the Committee on Research at the University of California Los Angeles and in part by the Southern California Earthquake Center. SCEC is supported in part by the National Science Foundation under Cooperative Agreement EAR-1033462 and United States Geological Survey under Cooperative Agreement G12AC20038. This report is SCEC contribution 1898. Any opinions, findings, conclusions, or recommendations expressed in this material are those of the authors and do not necessarily reflect the views of the National Science Foundation, the US Geological Survey, or the Southern California Earthquake Center.

REFERENCES

AN, L.-J., and SAMMIS, C.G. (1996), *Development of strike–slip faults: shear experiments in granular materials and clay using a new technique*, J. Struc. Geol., *18*, 1061–1077.

ANGELIER, J., *Fault slip analysis and paleostress reconstruction*, In Continental Deformation (ed., Hancock, P.L.) (Pergamon Press, Oxford 1994), pp. 53–100.

ARCH, J., MALTMAN, A.J., and KNIPE, R.J. (1988), *Shear-zone geometries in experimentally deformed clays: the influence of water content, strain rate and primary fabric*, J. Struc. Geol., *10*, 91–99.

ARGUS, D.F., and GORDON, R.G. (2001), *Present tectonic motion across the Coast Ranges and San Andreas fault system in central California*, Geo. Soc. Am. Bull., *113*, 1580–1592.

ATMAOUI, N., KUKOWSKI, N., STÖCKHERT, B., and KÖNIG, D. (2006), *Initiation and development of pull-apart basins with Riedel shear mechanism: insights from scaled clay experiments*, Int. J. Earth Sci., *95*, 225–238.

AYDIN, A., and DEGRAFF, J.M. (1988), *Evolution of polygonal fracture patterns in lava flows*, Science, *239*, 471–476.

AYDIN, A., and RECHES, Z. (1982), *Number and orientation of fault sets in the field and in experiments*, Geol., *19*, 107–112.

AYDIN, A., and SCHULTZ, R.A. (1990), *Effect of mechanical interaction on the development of strike–slip fault with en échelon patterns*, J. Struc. Geol., *12*, 123–129.

BAIN, G.W., and BEEBE, J.H. (1954), *Scale model reproduction of tension faults*, Amer. J. Sci., *252*, 745–754.

BARTLETT, W.L., FRIEDMAN, M., and LOGAN, J.M. (1981), *Experimental folding and faulting of rocks under confining pressure, Part IX. Wrench faults in limestone layers*, Tectonophys., *79*, 255–277.

BATES, R.L., and JACKSON, J.A., *Glossary of geology—3rd edition* (Amer. Geol. Inst., Alexandria, VA 1987), 788 p.

BEN-ZION, Y., ROCKWELL, T.K., ZHEQIANG, S., SHIQING, X. (2012), *Reversed-polarity secondary deformation structures near fault stepovers*, J. Appl. Mech., *79*, 031025.

BERGERAT, F. (2001), *Historical seismicity in Iceland: geological aspects, environmental and social impacts. Examples in the South Iceland Seismic Zone*, Comp. Rend. l'Acad. Sci. Ser. IIA Earth. Planet. Sci., *333*, 81–92.

BERGERAT, F., and ANGELIER, J. (2000), *The South Iceland Seismic Zone: tectonic and seismotectonic analyses revealing the evolution from rifting to transform motion*, J. Geodyn., *29*, 211–231.

BERGERAT, F., GUDMUNDSSON, A., ANGELIER, J., RÖGNVALDSSON, S.T. (1998), *Seismotectonics of the central part of the South Iceland Seismic Zone*, Tectonophys., *298*, 319–335.

BIRD, P. (1978a), *Finite element modeling of lithosphere deformation: The Zagros collision orogeny*, Tectonphys., *50*, 307–336.

BIRD, P. (1978b), *Initiation of intracontinetal subduction in the Himalaya*, J. Geophys. Res. *83*, 4975–4987.

BIRD, P. (1979), *Continental delamination and the Colorado plateau*, J. Geophys. Res., *84*, B13.

BIRD, P. (1996), *Computer simulations of Alaskan neotectonics*, Tectonics, *15*, 225–236.

BIRD, P. (1998), *Testing hypotheses on plate-driving-mechanism with global lithosphere models including topography, thermal structure, and faults*, J. Geophys. Res., *103*, 10115–10129.

BIRD, P. (2003), *An updated digital model of plate boundaries*, Geochem., Geophys., Geosys., *4*, 1027.

BIRD, P. (2009), *Long-term fault slip rates, distributed deformation rates, and forecast seismicity in the western United States from fitting of community geologic, geodetic, and stress direction datasets*, J. Geophys. Res., *114*, B11403.

BIRD, P., and KONG, X. (1994), *Computer simulations of California tectonics confirm very low strength of major faults*, Geo. Soc. Am. Bull., *106*, 159–174.

BIRD, P., LIU, Z., and RUCKER, W.K. (2008), *Stresses that drive the plates from below: Definitions, computational path, model optimization, and error analysis*, J. Geophys. Res., *113*, B11406.

BONDRE, N.R., DURAISWAMI, R.A., and DOLE, G. (2004), *A brief comparison of lava flows from the Deccan Volcanic Province and the Columbia-Oregon Plateau flood basalts: Implications for models of flood basalt emplacement*, J. Earth Sys. Sci., *113*, 809–817.

BONDRE, N.R., and HART, W.K. (2006), *Morphological and textural diversity of the Steens Basalt lava flows, Southwest Oregon, USA: implications for emplacement style and nature of eruptive episodes*, Bull. Volcan., *70*, 999–1019.

BRACE, W.F., and KOHLSTEDT, D.L. (1980), *Limits on lithospheric stress imposed by laboratory experiments*, J. Geophys. Res., *85*, 6248–6252.

BRUESEKE, M.E., and HART, W.K. (2000), *Reevaluation and new age constraints on the eruptive history of the mid-Miocene Steens basalt, southeastern Oregon*, Geol. Soc. Amer. Abs. Prog., *32*, A-142.

BRUESEKE, M.E., HEIZLER, M.T., HART, W.K., and MERTZMAN, S.A. (2007), *Distribution and geochronology of Oregon Plateau (U.S.A.) flood basalt volcanism: The Steens Basalt revisited*, J. Volcan. Geotherm. Res., *161*, 187–214.

BYERLEE, J.D. (1978), *Friction of rocks*, Pure App. Geophys., *116*, 615–626.

CAMP, V.E., ROSS, M.E., and HANSON, W.E. (2003), *Genesis of flood basalts and Basin and Range volcanic rocks from Steens Mountain to the Malheur River Gorge, Oregon*, Geol. Soc. Am. Bull., *115*, 105–128.

CARLSON, R.W., and HART, W.K. (1987), *Crustal genesis on the Oregon Plateau*, J. Geophys. Res., *92*, 6191–6206.

CASAS, A.M., GAPAIS, D., NALPAS, T., BESNARD, K., and ROMÁN-BERDIEL, T. (2001), *Analogue models of transpressive systems*, J. Struc. Geol., *23*, 733–743.

CHESTER, F.M. (1995), *A rheologic model for wet crust applied to strike–slip faults*, J. Geophys. Res., *100*, 13033–13044.

CHRISTIANSEN, R.L., and YEATS, R.S., *Post-Laramide geology of the U.S. Cordilleran region*, In The Cordilleran Orogen: Conterminous U.S. (eds. Burchfiel, B.C., Lipman, P.W., and Zoback, M.L.) (Geol. Soc. Amer., Boulder, CO 1992), pp. 261–406.

CLOOS, H. (1928), *Experimente zur inneren Tektonik*, Zen. Min., Geol. Pal., 1928B, 609–621.

COOKE, M.L., and VAN DER ELST, N.J. (2012), *Rheologic testing of wet kaolin reveals frictional and bi-viscous behavior typical of crustal materials*, Geophys. Res. Lett., *39*, L01308.

CORTI, G., MORATTI, G., and SANI, F. (2005), *Relations between surface faulting and granite intrusions in analogue models of strike–slip deformation*, J. Struc. Geol., *27*, 1547–1562.

CRIDER, J.G., and PEACOCK, D.C.P. (2004), *Initiation of brittle faults in the upper crust: a review of field observations*, J. Struc. Geol., *26*, 691–707.

CRUMPLER, L.S., HEAD, J.W., and CAMPBELL, D.B. (1986), *Orogenic belts on Venus*, Geol., *14*, 1031–1034.

DAUTEUIL, O., and MART, Y. (1998), *Analogue modeling of faulting pattern, ductile deformation, and vertical motion in strike–slip fault zones*, Tectonics, *17*, 303–310.

DEGRAFF, J.M., and AYDIN, A. (1987), *Surface morphology of columnar joints and its significance to mechanics and direction of joint growth*, Geo. Soc. Am. Bull., *99*, 605–617.

DEGRAFF, J.M., and AYDIN, A. (1993), *Effect of thermal regime on growth increment and spacing of contraction joints in basaltic lava*, J. Geophys. Res.: Solid Earth, *98*, 6411–6430.

DOOLEY, T.P., and SCHREURS, G. (2012), *Analogue modelling of intraplate strike–slip tectonics: A review and new experimental results*, Tectonophys., 574–575, 1–71.

DUNHAM, E.M. (2007), *Conditions governing the occurrence of supershear ruptures under slip-weakening friction*, J. Geophs. Res., *112*, B07302.

EINARSSON, P., BJÖRNSSON, S., FOUGLER, G., STEFANSSON, R., and SKAFTADOTTIER, T., *Seismicity pattern in the South Iceland Seismic Zone*, In Earthquake Prediction: An International Review (eds. Simpson, D.W., and Richards, P.G.) (AGU Maurice Ewing Series 4, Washington D.C. 1981), pp. 141–151.

EISENSTADT, G., and SIMS, D. (2005), *Evaluating sand and clay models: Do rheological differences matter?*, J. Struc. Geol., *27*, 1399–1412.

EMMONS, R.C. (1969), *Strike–slip rupture patterns in sand models*, Tectonophys., *7*, 71–87.

FACCENNA, C., NALPAS, T., BRUN, J.P., DAVY, P., and BOSI, V. (1995), *The influence of pre-existing thrust faults on normal fault geometry in nature and in experiments*, J. Struc. Geol., *17*, 1139–1149.

FREUND, R. (1974), *Kinematics of transform and transcurrent faults*, Tectonophys., *21*, 93–134.

GHENT, R.R., and HANSEN, V.L. (1999), *Structural and Kinematic Analysis of Eastern Ovda Regio, Venus: Implications for Crustal Plateau Formation*, Icarus, *139*, 116–136.

GOLDSBY, D.L., and TULLIS, T.E. (2011), *Flash heating leads to low frictional strength of crustal rocks at earthquake slip rates*, Science, *14*, 216–218.

GROSSENBACHER, K.A., and MCDUFFIE, S.M. (1995), *Conductive cooling of lava: Columnar joint diameter and stria width as functions of cooling rate and thermal gradient*, J. Volcan. Geotherm. Res., *69*, 95–103.

GUDMUNDSSON, A. (2007), *Infrastructure and evolution of ocean-ridge discontinuities in Iceland*, J. Geodyn., *43*, 6–29.

GUDMUNDSSON, A., and BRYNJOLFSSON, S. (1993), *Overlapping rift-zone segments and the evolution of the South Iceland Seismic Zone*, Geophys. Res. Lett., *20*, 1903–1906.

GUTSCHER, M.A., KUKOWSKI, N., MALAVIELLE, J., and LALLEMAND, S. (1996), *Cyclical behavior of thrust wedges; insights from high basal friction sandbox experiments*, Geol., *24*, 135–138.

HACKMAN, M.C., KING, G.C.P., and BILHAM, R. (1990), *The mechanics of the South Iceland Seismic Zone*, J. Geophys. Res., *95*, 17339–17351.

HAMMOND, W.C., and THATCHER, W. (2005), *Northwest Basin and Range tectonic deformation observed with the 1456 Global Positioning System, 1999–2003*, J. Geophys. Res., *110*, B10405.

HANDIN, J., *Strength and ductility*, In Handbook of Physical Constants—Memoir 97 (ed. Clark, S.P., Jr.) (Geol. Soc. Amer., Boulder, CO 1966), pp. 223–289.

HEMPTON, M.R., and NEHER, K. (1986), *Experimental fracture, strain and subsidence patterns over en échelon strike–slip faults: implications for the structural evolution of pull-apart basins*, J. Struc. Geol., *8*, 597–605.

HENZA, A.A., WITHJACK, M.O., and SCHLISCHE, R.W. (2010), *Normal-fault development during two phases of non-coaxial extension: An experimental study*, J. Struc. Geol., *9*, 573–584.

HIGGINS, M.W., and WATERS, A.C. (1967), *Newberry Caldera, Oregon: A preliminary report*, Ore Bin, *29*, 37–60.

HILDEBRAND-MITTLEFEHLDT, N. (1979), *Deformation near a fault termination, part I: A fault in a clay experiment*, Tectonophys., *57*, 131–150.

HOLOHAN, E.P., VAN WYK DE VRIES, B., and TROLL, V.R. (2008), *Analogue models of caldera collapse in strike–slip tectonic regimes*, Bull. Volcan., *70*, 774–796.

HOOPER, P.R., BINGER, G.B., and LEES, K.R. (2002a), *Ages of the Steens and Columbia River flood basalts and their relationship to extension-related calc-alkalic volcanism in eastern Oregon*, Geo. Soc. Am. Bull., *114*, 43–50.

HOOPER, P.R., JOHNSON, J.A., and Hawkesworth, C.J. (2002b), *A model for the origin of the western Snake River Plain as an extensional strike–slip duplex, Oregon-Idaho*, In Tectonic and magmatic evolution of the Snake River Plain volcanic province (eds. Bonnichsen, B., White, C.M., and McCurry, M.), Idaho Geol. Sur. Bull. *30*, 59–67.

HUBBERT, M.K. (1937), *Theory of scale models as applied to the study of geologic structures*, Geo. Soc. Am. Bull., *48*, 1459–1519.

JAEGER, J.C., and COOK, N.G.W., *Fundamentals of Rock Mechanics* (John Wiley and Sons, New York 1976).

JIMÉNEZ-MUNT, I., BIRD, P. and FERNÀNDEZ, M. (2001), *Thin-shell modeling of neotectonics in the Azores-Gibraltar Region*, Geophys. Res. Lett., *28*, 1083–1086.

JOHANNESSON, H., and SAEMUNDSSON, K., *Geologic map of Iceland, 1:500,000, Tectonics* (Iceland Inst. Nat. Hist., Reykjavik 1998).

JOHNSON, A.M. (1995), *Orientations of faults determined by pre-monitory shear zones*, Tectonophys., *247*, 161–238.

KEMPER, W.D., and ROSENAU, R.C. (1984), *Soil cohesion as affected by time and water content*, Soil Sci. Soc. Amer. J., *48*, 1001–1006.

KIRBY, S.H. (1980), *Tectonic stresses in the lithosphere: Constraints provided by the experimental deformation of rocks*, J. Geophys. Res., *85*, 6353–6363.

KLINKMÜLLER, M., *Properties of analogue materials, experimental reproducibility and 2D/3D deformation quantification techniques in analogue modelling of crustal-scale processes* (Ph.D. dissertation) (University of Bern 2011).

KOENIG, E., and AYDIN, A. (1998), *Evidence for large-scale strike–slip faulting on Venus*, Geol., *26*, 551–554.

KONG, X., and BIRD, P., *Neotectonics of Asia: Thin-shell finite-element models with faults*, In Tectonic Evolution of Asia (eds. Yin, A., and Harrison, T.M.) (Cambridge University Press, New York 1996), pp. 18–34.

KOYI, H., GHASEMI, A., HESSAMI, K., and DIETL, C. (2008), *The mechanical relationship between strike–slip faults and salt diapirs in the Zagros fold-thrust belt*, J. Geol. Soc., *165*, 1031–1044.

KRYNINE, D.P., *Soil Mechanics. Its Principal and Structural Applications* (McGraw-Hill, New York 1947).

LAFEMINA, P.C., DIXON, T.H., MALSERVISI, R., ARNADOTTIR, T., STURKELL, E., SIGMUNDSSON, F., and EINARSSON, P. (2005), *Geodetic GPS measurements in southern Iceland: Strain accumulation and partitioning in a propagating rift system*, J. Geophys. Res., *110*, B11405.

LANGER, V., *Geology and petrologic evolution of the silicic to intermediate volcanic rocks underneath Steens Mountain basalt, southeastern Oregon* (M.S. thesis) (Oregon State Univ., Corvallis 1991).

LAWRENCE, R.D. (1976), *Strike–slip faulting terminates the Basin and Range province in Oregon*, Geo. Soc. Am. Bull., *87*, 846–850.

LAZARTE, C.A., and BRAY, J.D. (1996), *A study of strike–slip faulting using small-scale models*, ASTM Geotech. Test. J., *19*, 118–129.

LE GUERROUÉ, E., and COBBOLD, P.R. (2006), *Influence of erosion and sedimentation on strike–slip fault systems: Insights from analogue models*, J. Struc. Geol., *28*, 421–430.

LETOUZEY, J., and SHERKATI, S., *Salt movement, tectonic events, and structural style in the central Zagros fold and thrust belt (Iran)*, In Salt-Sediment Interactions and Hydrocarbon Prospectivity: Concepts, Applications, and Case Studies for the 21st Century, 24th Ann. Res. Conf. (eds. Post, P.J., Olson, D.L., Lyons, K.T., Palmes, S.L., Harrison, P.F., and Rosen, N.C.) (SEPM Foundation 2004), pp. 444–463.

LIU, H., MCCLAY, K.R., and POWELL, D., *Physical models of thrust wedges*, In Thrust Tectonics (ed., McClay, K.R.) (Springer Netherlands 1992), pp. 71–81.

LIU, Z., and BIRD, P. (2002), *Finite element modeling of neotectonics in New Zealand*, J. Geophys. Res.: Solid Earth, *107*, ETG-1.

LOCKNER, D.A., BYERLEE, J.D., KUKSENKO, V., PONOMAREV, A., and SIDORIN, A. (1991), *Quasi-static fault growth and shear fracture energy in granite*, Nature, *350*, 39–42.

LONG, P.E., and WOOD, B.J. (1986), *Structures, textures, and cooling histories of the Columbia River basalt flows*, Geo. Soc. Am. Bull., *97*, 1144–1155.

MANDL, G., *Mechanics of tectonic faulting* (Elsevier, Amsterdam, Netherlands 1988), p. 407.

MANN, P. (2007), *Global catalog, classification and tectonic origins of restraining- and releasing bends on active and ancient strike–slip fault systems*, Geol. Soc. London Spec. Pub., *290*, 13–142.

MANN, P., DEMETS, C., and WIGGINS-GRANDISON, M. (2007), *Toward a better understanding of the Late Neogene strike–slip restraining bend in Jamaica: geodetic, geological, and seismic constraints*, Geol. Soc. London Spec. Pub., *290*, 239–253.

MARTEL, S. J. (1990). *Formation of compound strike–slip fault zones at Mount Abbot quadrangle, California.*, J. Struc. Geol., *12*, 869–882.

MAZZARINI, F., MUSUMECI, G., MONTANARI, D., and CORTI, G. (2010), *Relations between deformation and upper crustal magma emplacement in laboratory physical models*, Tectonophys., *484*, 139–146.

MINOR, S.A., PLOUFF, D., ESPARZA, L.E., and PETERS, T.J. (1987a), *Mineral resources of the High Steens and Little Blitzen Gorge Wilderness study area, Harney County, Oregon*, USGS.

MINOR, S.A., RYTUBA, J.J., GRUBENSHY, M.J., VANDER MEULEN, D.B., GOELDNER, C.A., and TEGTMEYER, K.J. (1987b), *Geologic map of the High Steens and Little Blitzen Gorge Wilderness study area, Harney County, Oregon*, USGS Misc. Field Stud. Map MF-1876, scale 1:24,000.

MISRA, S., MANDAL, N., and CHAKRABORTY, C. (2009), *Formation of Riedel shear fractures in granular materials: Findings from analogue shear experiments and theoretical analyses*, Tectonophys., *471*, 253–259.

MORGENSTERN, N.R., and TCHALENKO, J.S. (1967), *Microscopic structures in kaolin subjected to direct shear*, Geotechnique, *17*, 309–328.

NAYLOR, M.A., MANDL, G., and SIJPESTEIJN, C.H.K. (1986), *Fault geometries in basement-induced wrench faulting under different initial stress states*, J. Struc. Geol., *8*, 737–752.

NILFOROUSHAN, F., KOYI, H.A., SWANTESSON, J., and TALBOT, C.J. (2007), *Effect of basal friction and volumetric strain in models of convergent settings measured by laser scanner*, J. Struc. Geol., *30*, 366–379.

NUR, A., RON, H., and SCOTTI, O. (1986), *Fault mechanics and the kinematics of block rotations*, Geol., *14*, 746–749.

O'CONNELL, R.J., GABLE, C.W., and HAGER, B.H., *Toroidal-poloidal partitioning of lithospheric plate motions*, In Glacial Isostasy, Sea-Level and Mantle Rheology (eds. Sabadini, R., Lambeck, K., and Boschi, E.) (Kluwer Academic, Norwell, MA 1991), pp. 535–551.

ODE, H. (1960), *Faulting as a velocity discontinuity in plastic deformation*, Mem. Geol. Soc. Am., *79*, 293–321.

ODONNE, F., and VIALON, P. (1983), *Analogue models of folds above a wrench fault*, Tectonophys., *99*, 31–46.

OERTEL, G. (1965), *The mechanism of faulting in clay experiments*, Tectonophys., *2*, 343–393.

PETERSEN, M.D., and WESNOUSKY, S.G. (1994), *Fault slip rates and earthquake histories for active faults in southern CA*, Bull. Seismo. Soc. Amer., *84*, 1608–1649.

PHILLIPS, R.J., and HANSEN, V.L. (1998), *Geological evolution of Venus: A geodynamical and magmatic framework*, Science, *279*, 1492–1497.

RAITALA, J. (1994), *Crustal tectonic zone on Venus*, Earth. Moon. Planets., *64*, 155–164.

RECHES, Z. (1988), *Evolution of fault patterns in clay experiments*, Tectonophys., *145*, 141–156.

RECHES, Z., and EIDELMAN, A. (1995), *Drag along faults*, Tectonophys., *247*, 145–156.

RICHARD, P.D., and KRANTZ, R.W. (1991), *Experiments on fault reactivation in strike–slip mode*, Tectonophys., *188*, 117–131.

RICHARD, P.D., NAYLOR, M.A., and KOOPMAN, A. (1995), *Experimental models of strike–slip tectonics*, Petrol. Geosci., *1*, 71–80.

RIEDEL, W. (1929), *Zur Mechanik geologisher Brucherscheinungen*, Zen. Min. Geol. Pal., *1929B*, 354–368.

ROMÁN-BERDIEL, T., GAPAIS, D., and BRUN, J.P. (1997), *Granite intrusion along strike–slip zones in experiment and nature*, Amer. J. Sci., *297*, 651–678.

ROMEO, I., CAPOTE, R., and ANGUITA, F. (2005), *Tectonic and kinematic study of a strike–slip zone along the southern margin of Central Ovda Regio, Venus: Geodynamical implications for crustal plateaux formation and evolution*, Icarus, *175*, 320–334.

RON, H., BEROZA, G., and NUR, A. (2001), *Simple model explains complex faulting, Eos*, Trans. Amer. Geophys. Union, *82*, 125–129.

RYAN, M.P., and SAMMIS, C.G. (1978), *Cyclic fracture mechanisms in cooling basalt*, Geo. Soc. Am. Bull., *89*, 1295–1308.

SAEMUNDSSON, K. (1974), *Evolution of the axial rift zone in northern Iceland and the Tjornes fracture zone*, Geo. Soc. Am. Bull., *85*, 495–504.

SASSI, W., COLETTA, B., BALÉ, P., and PAQUEREAU, T. (1993), *Modelling of structural complexity in sedimentary basins: the role of pre-existing faults in thrust tectonics*, Tectonophys., *226*, 97–112.

SCARBERRY, K.C., MEIGS, A.J., and GRUNDER, A.L. (2010), *Faulting in a propagating continental rift: Insight from the late Miocene structural development of the Abert Rim fault, southern Oregon, USA*, Tectonophys., *488*, 71–86.

SCHÖPFER, M.P.J., and STEYRER, H.P., *Experimental modeling of strike–slip faults and the self- similar behavior*, In Tectonic Modeling: A Volume in Honor of Hans Ramberg (eds. Koyi, H.A., and Mancktelow, N.S.) (Geol. Soc. Amer. Mem. 193, Boulder, Colorado 2001), pp. 21–27.

SCHULSON, E.M., ILIESCU, D., and RENSHAW, C.E. (1999), *On the initiation of shear faults during brittle compressive failure: A new mechanism*, J. Geophys. Res.: Solid Earth, *104*, 695–705.

SCHWARZ, H.-U., and KILFITT, F.-W. (2008), *Confluence and intersection of interacting conjugate faults: A new concept based on analogue experiments*, J. Struc. Geol., *30*, 1126–1137.

SEGALL, P., and POLLARD, D. D. (1983). *Nucleation and growth of strike slip faults in granite*, J. Geophys. Res., *88*, 555–568.

SIMS, D. (1993), *The rheology of clay: A modeling material for geologic structures, Eos*, Trans. Amer. Geophys. Union, *74*, 569.

SLEEP, N.H. (1992), *Archean plate tectonics: What can be learned from continental geology?*, Canadian J. Earth Sci., *29*, 2066–2071.

SLEEP, N.H. (1994), *Martian plate tectonics*, J. Geophys. Res., *99*, 5639–5655.

SLEEP, N.H. (1997), *Application of a unified rate and state friction theory to the mechanics of fault zones with strain localization*, J. Geophys. Res., *102*, 2875–2896.

SLEEP, N.H. (2013), *Self-organization of elastic moduli in the rock above blind faults*, Geochem., Geophys., Geosys., *14*, 733–750.

SLEEP, N.H., and HAGIN, P. (2008), *Nonlinear attenuation and rock damage during strong seismic ground motions*, Geochem., Geophys., Geosys., *9*, Q10015.

SOLOMON, S.C., SMREKAR, S.E., BINDSCHADLER, D.L., GRIMM, R.E., KAULA, W.M., MCGILL, G.E., PHILLIPS, R.J., SAUNDERS, R.S., SCHUBERT, G., SQUYRES, S.W., and STOFAN, E.R. (1992), *Venus tectonics: An overview of Magellan observations*, J. Geophys. Res., *97*, 13199–13255.

SQUYRES, S.W., JANKOWSKI, D.G., SIMONS, M., SOLOMON, S.C., HAGER, B.H., and MCGILL, G.E. (1992), *Plains tectonism on Venus: the deformation belts of Lavinia Planitia*, J. Geophys. Res., *94*, 13579–13599.

SWISHER, C.C., ACH, J.A., and HART, W.K. (1990), *Laser fusion $^{40}Ar/^{39}Ar$ dating of the type Steens Mountain Basalt, southeastern Oregon and the age of the Steens geomagnetic polarity transition, Eos*, Trans. Amer. Geophys. Union, *71*, 1296.

TAPPONIER, P., PELTZER, G., and ARMIJO, R. (1986), *On the mechanics of the collision between India and Asia*, Geol. Soc. Lon. Spec. Pub., *19*, 115–157.

TCHALENKO, J.S. (1970), *Similarities between Shear Zones of Different Magnitudes*, Geo. Soc. Am. Bull., *81*, 1625–1640.

TERZAGHI, K., and PECK, R.B., *Soil Mechanics in Engineering Practice* (Wiley, New York 1948).

THOMPSON, G.A., and BURKE, D.B. (1974), *Regional geophysics of the Basin and Range province*, Ann. Rev. Earth. Planet. Sci., *2*, 213–238.

TOMKEIEFF, S.I. (1940), *The basalt lavas of the Giant's Causeway district of Northern Ireland*, Bull. Volcan., *6*, 89–143.

TONG, H., and YIN, A. (2011), *Reactivation tendency analysis: A theory for predicting the temporal evolution of preexisting weakness under uniform stress state*, Tectonophys., *503*, 195–200.

TUCKWELL, G.W., and GHAIL, R.C. (2003), *A 400-km-scale strike–slip zone near the boundary of Thetis Regio, Venus*, Earth. Planet. Sci. Lett., *211*, 45–55.

VIOLA, G., ODONNE, F., and MANCKTELOW, N.S. (2004), *Analogue modelling of reverse fault reactivation in strike–slip and transpressive regimes: application to the Giudicarie fault system, Italian Eastern Alps*, J. Struc. Geol., *26*, 401–418.

WALKER, G.W. (1969), *Geology of the High Lava Plains province, Mineral and Water Resources of Oregon*, Oregon Dept. of Geol. and Min. Ind. Bull., *64*, 77–79.

WALKER, G.W. (1977), *Geologic map of Oregon east of the 121st meridian*, USGS Misc. Geol. Inv. Map I-902, scale 1:500,000.

WALKER, G.W., and KING, P.B. (1969), *Geologic map of Oregon*, USGS Misc. Geol. Inv. Map I- 595, scale 1:2,000,000.

WALKER, G.W., and MACLEOD, N.S. (1991), *Geologic map of Oregon*, USGS special map, scale 1:500,000.

WALKER, R., JACKSON, J., and BAKER, C. (2004), *Active faulting and seismicity of the Dasht-e- Bayaz region, eastern Iran*, Geophys. J. Int., *157*, 265–282.

WEIJERMARS, R., JACKSON, M.P.A., and VENDEVILLE, B. (1993), *Rheological and tectonic modeling of salt provinces*, Tectonophys., *217,* 143–174.

WESNOUSKY, S.G. (1986), *Earthquakes, quaternary faults, and seismic hazard in California*, J. Geophys. Res., *91*, 12587–12631.

WESNOUSKY, S.G. (2005), *Active faulting in the Walker Lane*, Tectonics, *24*, TC3009.

WILCOX, R.E., HARDING, T.P., and SEELY, D.R. (1973), *Basic wrench tectonics*, AAPG Bull., *57*, 74–96.

YIN, A. (2012*), Structural analysis of the Valles Marineris fault zone: Possible evidence for large-scale strike–slip faulting on Mars*, Lithos., *4*, 286–330.

(Received July 23, 2013, revised March 7, 2014, accepted March 7, 2014, Published online March 30, 2014)

Pure Appl. Geophys. 171 (2014), 2919–2935
© 2014 Springer Basel
DOI 10.1007/s00024-014-0851-6

A Statistical Framework for Calculating and Assessing Compositional Linear Trends Within Fault Zones: A Case Study of the NE Block of the Clark Segment, San Jacinto Fault, California, USA

BRIAN G. ROCKWELL,[1] GARY H. GIRTY,[1] and THOMAS K. ROCKWELL[1]

Abstract—Utilizing chemical data derived from the various fault zone architectural components of the Clark strand of the San Jacinto fault, southern California, USA, we apply for the first time non-central principal component analysis to calculate a compositional linear trend within molar A–CN–K space. In this procedure A–CN–K are calculated as the molar proportions of Al_2O_3 (A), CaO^* + Na_2O (CN), and K_2O (K) in the sum of molar Al_2O_3, Na_2O, CaO^*, and K_2O. CaO^* is the molar CaO after correction for apatite. We then derive translational invariant chemical alteration intensity factors, t, for each architectural component through orthogonal projection of analyzed samples onto the compositional linear trend. The chemical alteration intensity factor t determines the relative change in composition compared to the original state (i.e., the composition of the altered wall rocks). It is dependent on the degree of intensity to which the process or processes responsible for the change in composition of each architectural component has been active. These processes include shearing, fragmentation, fluid flow, and generation of frictional heat. Non-central principal component analysis indicates that principal component 1 explains 99.7 % of the spread of A–CN–K data about the calculated compositional linear trend (i.e., the variance). The significance level for the overall one-way analysis of variance (ANOVA) is 0.0001. Such a result indicates that at least one significant difference across the group of means of t values is different at the 95 % confidence level. Following completion of the overall one-way ANOVA, the difference in means t test indicated that the mean of the t values for the fault core are different than the means obtained from the transition and damage zones. In contrast, at the 95 % confidence level, the means of the t values for the transition and damage zones are not statistically distinguishable. The results of XRD work completed during this study revealed that the <2 μm fraction is composed primarily of illite/smectite with ~15 % illite in the damage zone, of illite/smectite with ~30 % illite in the transition zone, and of discreet illite with very minor smectite in the fault core. These changes parallel the increasing values of the chemical alteration intensity factors (i.e., t). Based on the above results, it is speculated that when fault zones are derived from tonalitic wall rocks at depths of ~400 ± 100 m, the onset of the illite/smectite to illite conversion will occur when t values exceed

0.20 ± 0.12, the average chemical alteration intensity factor calculated for the transition zone. Under such conditions during repeated rupturing events, frictional heat is produced and acidic fluids with elevated temperatures (≥ ~125 °C) are flushed through the fault core. Over time, the combination of shearing, fragmentation, and frictionally elevated temperatures eventually overcomes the kinetic barrier for the illite/smectite to illite transition. Such settings and processes are unique to fault zones, and as a result, they represent an underappreciated setting for the development of illite from illite/smectite. The success of non-central principal component analysis in this environment offers the first statistically rigorous methodology for establishing the existence of compositional linear trends in fault zones. This method also derives quantifiable alteration intensity factors that could potentially be used to compare the intensity of alteration at different segments of a fault, as well as offer a foundation to interpret the potential driving forces for said alteration and differences therein.

Key words: San Jacinto fault zone, fault zone architecture, linear compositional trends, alteration intensity factors, illite/smectite-illite transition, non-central principal component analysis.

1. Introduction

Fault zones consist of a fault core or cores and adjacent damage zones (e.g., CHESTER and LOGAN 1986; SCHULZ and EVANS 1998; FAULKNER et al. 2003, 2010; MORTON et al. 2012; COLBY and GIRTY 2013). The fault core or cores are discreet narrow tabular zones, commonly centimeters in width, where grain size has been reduced relative to adjacent areas through fragmentation, abrasion, and sliding during multiple rupture events (e.g., CHESTER et al. 1993; CAINE et al. 1996; CHESTER and CHESTER 1998). Adjacent mechanically linked damage zones contain rocks that are not as severely affected by the comminution process, but are generally areas of high fracture density. With increasing distance from the fault core, fracture density within such zones

---
[1] Department of Geological Sciences, San Diego State University, San Diego, CA 92182, USA. E-mail: brockwell090@gmail.com; ggirty@mail.sdsu.edu; trockwell@mail.sdsu.edu

decreases to that characteristic of the bedrock system (e.g., CHESTER and LOGAN 1986; SCHULZ and EVANS 1998, 2000; KIM et al. 2004; MITCHELL and FAULKNER 2009; FAULKNER et al. 2010).

There is considerable evidence for focused fluid flow within fault zones (e.g., EVANS and CHESTER 1995; CAINE et al. 1996; SCHULZ and EVANS 1998; FAULKNER et al. 2010; MORTON et al. 2012; COLBY and GIRTY 2013). Incongruent reactions resulting from invading fluids commonly result in the formation of clays, and the leaching and introduction of various ions out of or into the fault zone, respectively (e.g., MORTON et al. 2012; COLBY and GIRTY 2013). When viewed as a whole, the resulting pattern of chemical alteration can be ascribed to a compositional linear trend, i.e., a trend that describes the direction and intensity of chemical alteration within the fault zone.

An important and necessary property of any statistical procedure established for calculating and assessing the probabilities associated with a compositional linear trend is that it must address the closed-sum format of chemical data, and the fact that such data are not amenable to standard parametric statistical procedures (AITCHISON 1986; WORONOW and LOVE 1990; AGUE 1994; AGUE and VAN HAREN 1996; WARREN and GIRTY 1999; VON EYNATTEN et al. 2003; VON EYNATTEN 2004; PAWLOWSKY-GLAHN and BUCCIANTI 2011; BUCCIANTI et al. 2006; and the many references therein). As described in a later section of this paper, much recent work has shown that the log ratio transformation of AITCHISON (1986) circumvents this problem, and when combined with non-central principal component analysis can be used to calculate and assess the probability associated with a compositional linear trend developed within a fault zone.

Below we present a case study of the above concepts and ideas as they apply to the main fault core and adjacent damage zone of the NE block of the Clark segment of the San Jacinto fault zone, SE of Anza, California, USA. The results of our work suggest that a compositional linear trend with an associated high probability can be established for the NE block of the Clark segment, and that this trend is the direct result of invading relatively hot fluids reacting with the fragmented bedrock system,

dissolving and removing selected ions, and converting mixed-layer illite/smectite to illite.

Much of what follows involves the evaluation of chemical data and how it changes across a fault zone. In the context of compositional data (e.g., AITCHISON 1986; BUCCIANTI et al. 2006; PAWLOWSKY-GLAHN and BUCCIANTI 2011), our evaluation of these changes utilizes multivariate and small-sample statistical procedures including non-central principal component analysis, one-way analysis of variance, and the Student's t difference in means test. For those readers who are not familiar with these techniques, we refer them to VON EYNATTEN et al. (2003) and references therein for non-central principal component analysis, and DAVIS (2002) for all of the other statistical methods referred to in this paper.

2. The Clark Segment

2.1. General Features

The ~250 km long San Jacinto fault zone, a major branch of the San Andreas fault system, is characterized by a series of northwest-trending segments that accumulated 21–26 km of total dextral slip (Fig. 1) (SHARP 1967). Southeast of Anza most fault activity is focused on the Clark segment, the main strand of the San Jacinto fault, where interseismic creep is not known to occur (SAHAKIAN et al. 2011) and offset geomorphic features indicate dextral slip of ~3 m during the most recent event (SALISBURY et al. 2012). Within Horse Canyon, the Clark segment strikes ~N50W and juxtaposes the Horse Canyon and Cahuilla Valley plutons along an ~1 km long fault contact (Fig. 1c). Erosion and mass wasting within Horse Canyon has exposed portions of the fault cores, damage zones, and wall rocks. As discussed below, the presence of a Tertiary erosion surface above the exposures at Horse Canyon indicates an exhumation depth of ~400 ± 100 m.

2.2. Age, Slip-Rate, and Depth of Exhumation

The Horse Canyon exposure of the damage zone and fault core of the Clark strand lies 300–500 m below a broad upland surface that includes Anza

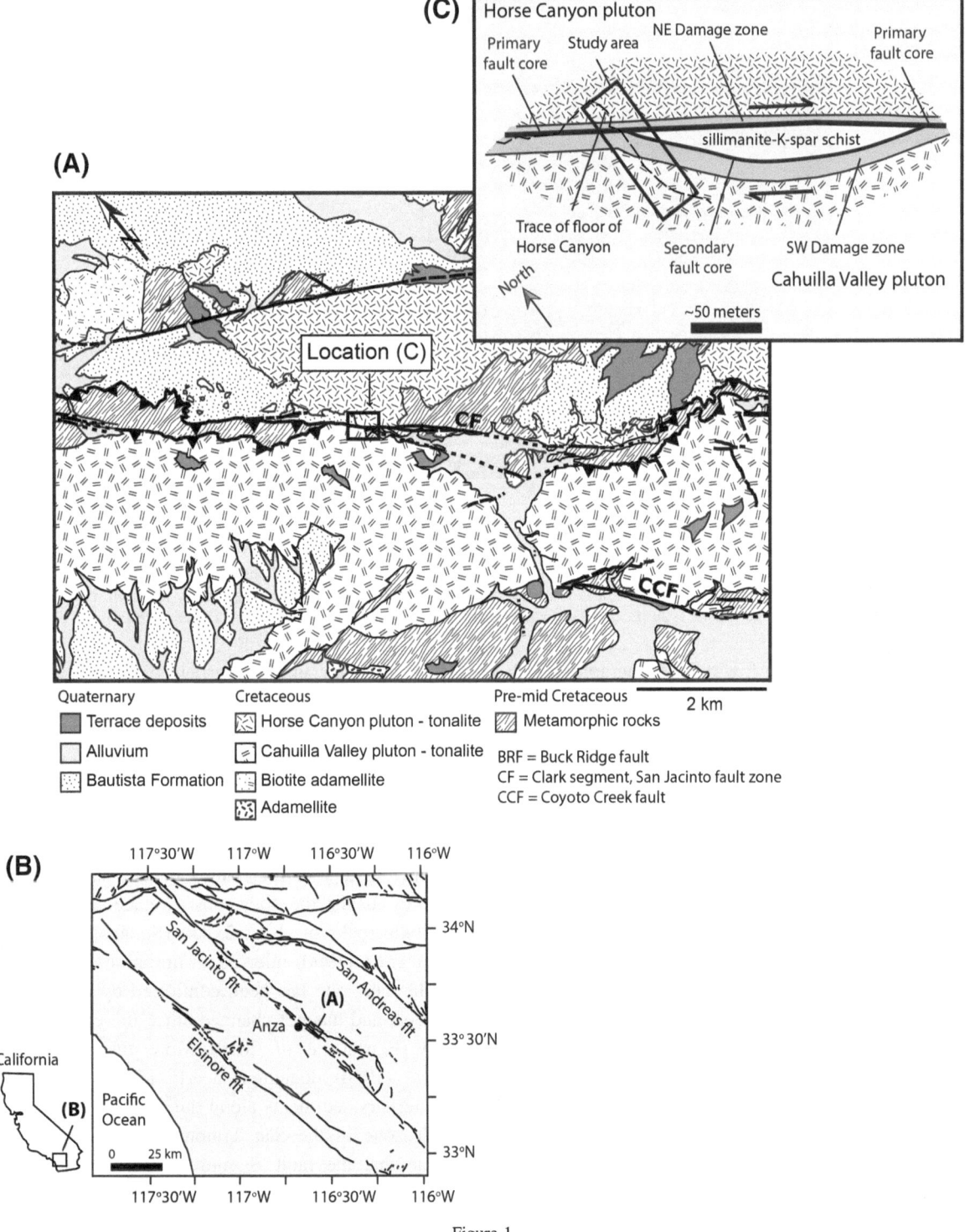

Figure 1

a Regional map after SHARP (1967) showing location of San Jacinto fault in southern California, USA. **b** Generalized map of southern California showing locations of Elsinore, San Jacinto, and San Andreas fault zones, and location of **a**. **c** Sketch map of the study area, showing Clark segment of the San Jacinto fault juxtaposing the Horse Canyon pluton against the Cahuilla Valley pluton, and distribution of fault zone architectural elements. **c** is based on map from Google Earth© image in MORTON *et al.* (2012) and field observations by authors

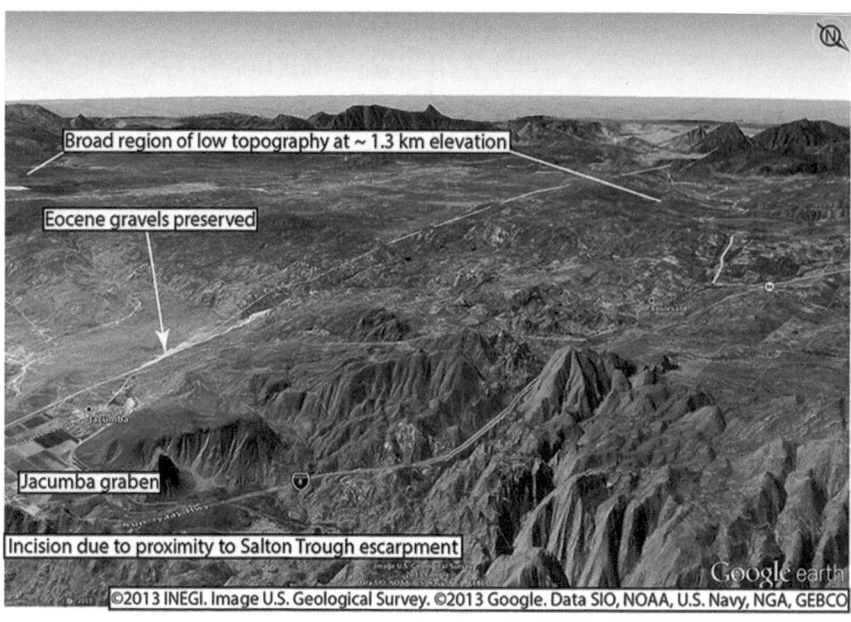

Figure 2
Gently undulating, broad erosional surface in southernmost California, USA and northern Baja California, Mexico cuts across Cretaceous plutonic rocks. Eocene and Miocene sedimentary and volcanic rocks are found at various locations atop this broad surface, and latest Cretaceous marine rocks are cut into the plutonic rocks near the coast. The erosion surface for nearly a century has been interpreted to be Tertiary in age (SAUER 1929; ECKIS 1930), with uplift being Pliocene and younger (MUELLER *et al.* 2009)

Valley and surrounding low-relief topography developed in plutonic rocks of the southern California batholith. The presence of this extensive and ancient, broad erosional surface has been known for nearly a century and was first recognized by SAUER (1929) and ECKIS (1930). The broad upland terrain is particularly well developed in southern San Diego County and extends southward into Baja California, Mexico (Fig. 2), where remnants of an Eocene fluvial system are still locally preserved on the upland surface. The Eocene fluvial gravels are remnants of a large river system (Ballena River) that crossed the Peninsular Ranges prior to their uplift and are sourced almost entirely from Sonora, Mexico (ABBOTT and SMITH 1978, 1989; STEER and ABBOTT 1984). The lack of locally derived materials within these deposits has been used to argue that the region had eroded to low relief, consistent with the preserved remnants of this surface today. In Anza Valley, early Quaternary lacustrine deposits contain the 780 ka Bishop Tuff (SHARP 1981), providing a minimum age for the antiquity of the landscape above Horse Canyon.

The San Jacinto fault has been inferred to be entirely Quaternary in age. For example, south of

Horse Canyon in the Borrego Badlands, JANECKE *et al.* (2010 and references therein) have inferred an initiation age as young as 1.1–1.3 Ma based on the presence of an unconformity that they believe marks the beginning of folding associated with early fault motion. In the Anza area, assuming that the late Quaternary slip rate of about 12 mm/year is valid for the life history of the fault and back-calculating its age by using the total post-Cretaceous offset on plutonic rocks of about 25 km (SHARP 1967), ROCKWELL *et al.* (1990) inferred an initiation age of about 2 Ma. This rate has been confirmed by more recent studies, and has not changed since the early Quaternary (BLISNIUK *et al.* 2010). More directly, MORTON *et al.* (1986) document nearly identical offset of Quaternary sediments along the northern third of the fault zone in the San Timoteo Badlands and also argue that the fault is entirely Quaternary in age. KENDRICK *et al.* (2002) argue for a slip rate >20 mm/year in the San Timoteo Badlands, which combined with total offset, would yield an age of initiation similar to that of JANECKE *et al.* (2010).

In summary, the age of initiation of the San Jacinto fault is likely Quaternary, and there is no

Figure 3
View (to the southeast) of Horse Canyon and the San Jacinto fault (indicated by *arrows*). Horse Canyon is incised ∼400 m into a broad, regional scale erosional surface, which is still preserved in Anza Valley and surrounding areas. The Quaternary incision is easily seen by the deep and steep gorges into the more gentle regional topography. At Horse Canyon, the 400 m estimated depth of exhumation assumes that there was no proto-Horse Canyon, so actual exhumation may be slightly less

evidence that an earlier version of the fault existed prior to this time. Taken in context with the antiquity of the Tertiary upland erosion surface, where not dissected by faulting or eroded by the retreating escarpment bounding the Salton Trough, the above observations imply that all fault zone characteristics associated with the Clark strand are Quaternary in age and are exhumed only as deeply as any exposure is preserved below the Tertiary erosion surface.

The difference in elevation between Anza Valley and the exposure of the Clark strand in Horse Canyon is ∼300 m. We utilize this as our minimum estimate. In contrast, the difference in elevation between the highest nearby terrain developed in the Horse Canyon pluton and the exposure in Horse Canyon is about 500 m. As Horse Canyon likely eroded in an area of lower topography than the highest elements of the local terrain, we use 400 ± 100 m as a best estimate of the depth of total incision of Horse Canyon at the fault exposure. This interpretation is consistent with the relative elevations of the Horse Canyon exposures below the broad upland surface (Fig. 3), and implies that all fault zone architectural elements discussed below are exhumed from this approximate depth.

2.3. Characteristics of Fault Zone Architectural Elements

In this paper we use the term fault core to refer to that part or parts of the fault zone where the greatest grain size reduction has taken place by comminution, and the term damage zone to refer to the adjacent rock material where evidence of more widely spaced grain-size reduction and fracturing is evident. The above components commonly grade one into another across a narrow transition zone (e.g., Fig. 4).

In classifying rocks characteristic of the fault core, transition zone, and damage zone, we utilized the general classification scheme of Sibson (1977) with the addition of foliated gouge. Operationally, cataclasite, gouge, and foliated gouge are both composed of more than 70 % matrix (i.e., material that does not contain grains visible to the naked eye)

61

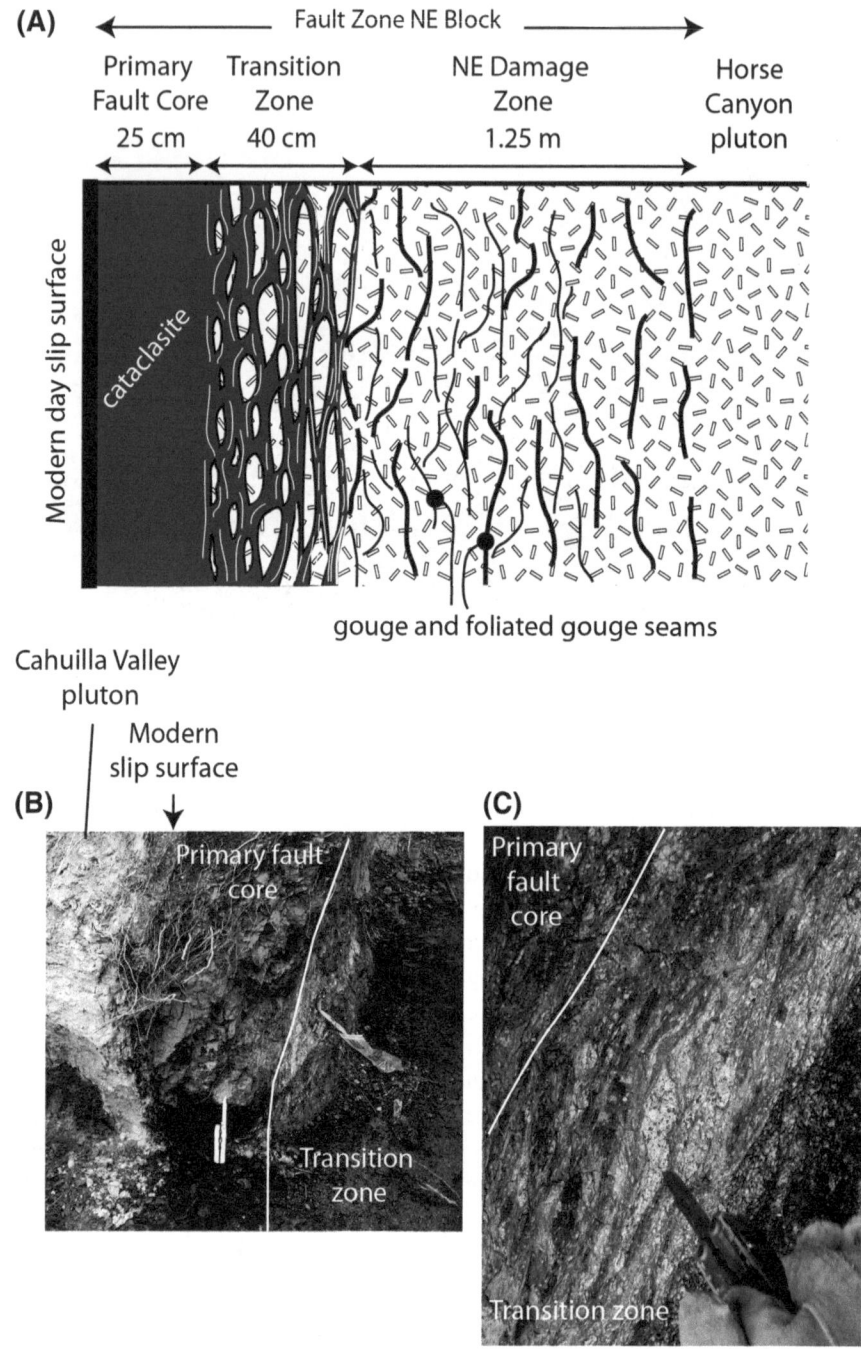

Figure 4
a Idealized diagram of the fault zone architecture of the NE block of the Clark strand of the San Jacinto fault zone. **b** Photograph showing modern slip surface, primary fault core, and inner part of transition zone. **c** Close-up view of contact between primary fault core and transition zone

and <30 % visible framework components (coarser grained rock and mineral fragments), and differ only in degree of consolidation and the present or absence of a visible foliation. Cataclasite, though fractured, is sufficiently consolidated that extraction of a sample requires a hammer, while blocks or pieces of both foliated and non-foliated gouge can be extracted with a putty knife or similarly designed tool.

Within Horse Canyon at 33°30′41.39″N latitude and 116°33′6.32″W longitude, the main trace of the Clark segment bifurcates around a lensoid block of sillimanite-K-feldspar mica schist (Fig. 1c). From the point of bifurcation to the southeast, the ~0.25 m thick primary fault core lies along the NE side of the lensoid block of schist, while an ~0.18 to ~0.3 m thick secondary fault core lies along the SW side. Well defined damage zones, characterized by a fracture density greater than that observed in the adjacent wallrocks, are present along both the NE margin of the primary and the SW margin of the secondary fault cores. Within both the ~1.25 m wide NE and the ~9.0 m wide SW damage zones (Fig. 1c), fractures are filled with dark, fine-grained gouge. Separating the primary fault core and the NE damage zone is a narrow ~0.4 m thick transition zone (Fig. 4). A similar interval separates the secondary fault core and SW damage zone. Across both transition zones, the proportion of gouge surrounding centimeter to millimeter sized lensoid fragments of the wall rocks increases from ~5 to 15 % near the contact with the damage zone to >50 % near the contact with the fault core. The primary fault core is composed of moderately indurated cataclasite (Fig. 4) while the secondary fault core is composed of foliated gouge. The principal slip surface occurs along the SW boundary of the primary fault core (Fig. 4), and appears to have accommodated the most recent slip events as it follows the current geomorphic expression of the fault trace (SHARP 1967; MORTON et al. 2012).

Thin section samples from the various architectural components described above were point counted to determine the proportions of the various mineral phases. This commonly used method for quantifying the mineralogical proportions present within a rock involves identifying and counting the mineral phase falling under the cross-hairs of the microscope as a thin section is moved mechanically across the stage at ~1 mm intervals (e.g., HUTCHISON 1974).

Point-count data derived from the various architectural components of the NE block are summarized graphically in Fig. 5. Note that the highly quartzofeldspathic character of the tonalitic wallrock system is systematically modified to a mostly quartz-clast cataclasite in the primary fault core. This change

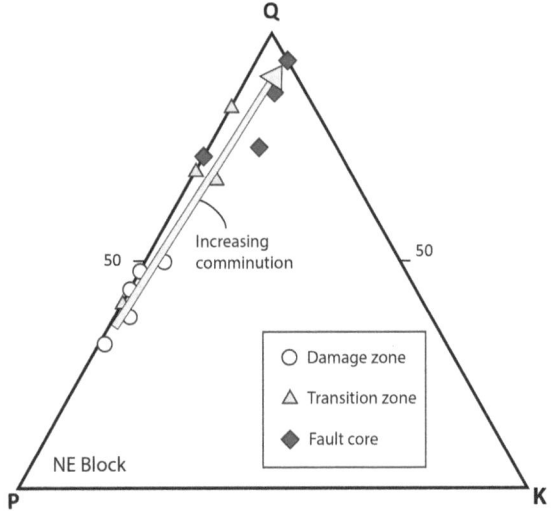

Figure 5
Ternary plot of point count data derived from samples analyzed from the northeast block of the Clark strand, San Jacinto fault zone. Data are from MORTON et al. (2012). Q quartz, P plagioclase, K K-feldspar

parallels a reduction in grain size from ~1 to 2 mm in the wallrock system to <0.005 mm in the primary fault core (MORTON et al. 2012). Moreover, XRD analysis of the <4 μm sized fraction separated from samples collected across the damage zone, transition zone, and primary fault core by MORTON et al. (2012) consistently indicated that the core is dominated by illite while samples from the transition and damage zones appear to contain both illite and mixed-layer illite/smectite. In order to refine the clay mineralogy study of MORTON et al. (2012), an XRD study of the <2 μm fraction derived from a series of samples collected across the NE block of the fault zone was completed. The <2 μm fraction represents a more standard size fraction for analyzing clay minerals than that used by MORTON et al. (2012), and leads to a more precise characterization of mixed-layer clays (e.g., MOORE and REYNOLDS 1997).

3. Clay Mineralogy

Approximately 130–140 g of samples characteristic of the damage zone, transition zone, and fault core of the NE block were powdered with a mortar and pestle. The resulting powdered sample was then placed within a 500 mL plastic bottle which was subsequently filled with deionized water and ~160 mL of 0.1 % Calgon (Na hexametaphosphate), a commonly used dispersing agent (MOORE and REYNOLDS 1997). The container was sealed and placed on an innOva 2100 platform shaker where it was agitated for 8 h. Utilizing Stokes Law and normal clay mineralogy extraction practices (MOORE and REYNOLDS 1997), the <2 μm fraction was then separated from this mixture. The resulting <2 μm fractions were smeared onto glass slides, and were then analyzed for their clay mineralogy utilizing a Philips X'Pert multipurpose diffractometer with copper Kα radiation (1.5405 Å), and 45 kV and 40 mA settings. In the case of foliated gouge, particles within the <2 μm fraction are derived primarily from clay or aggregates of such material, while in the sample from the cataclasite, the <2 μm fraction may include physical pieces of the very fine-grained cataclasite matrix. Each sample was analyzed following air drying, ethylene glycol solvation, heat saturation at 400 °C for 1 h, and heat saturation at 550 °C for 1 h. Unfortunately, during heat saturation treatments at 550 °C, the glass slide prepared for the damage zone sample warped. Hence, no data are available for this treatment for the damage zone. Typical scans ranged from 2° to 35° 2θ at speeds of 0.08°/s for damage and transition zone samples, and at 0.01°/s for specimens from the fault core (e.g., MOORE and REYNOLDS 1997).

3.1. Damage Zone

In the <2 μm fraction, a broad peak at 15.4 Å is present in the air dried sample derived from the damage zone (Fig. 6a). Upon ethylene glycol solvation, this peak expands to 16.8 Å and small, broad peaks appear at ~5.6 and 4.23 Å. These results indicate the presence of mixed-layer illite/smectite, and are consistent with ~15 % illite in illite/smectite (MOORE and REYNOLDS 1997). Following heat saturation treatment at 400 °C (Fig. 6a), the 16.8 Å peak collapsed into the 10 Å peak, while the 8.4, 5.6, and 4.23 Å peaks disappeared. In contrast, after such treatment, the 10, 5, and 3.35 Å peaks characteristic of illite are evident (MOORE and REYNOLDS 1997).

In addition to the above, in the air-dried sample, a low intensity peak at 7.1 Å is not affected by ethylene glycol solvation or heat saturation treatment to 400 °C. Due to the lack of the distinctive 14.1 and 4.7 Å chlorite peaks (for an example, see Fig. 6b), this peak is most likely derived from kaolinite. However, due to the destruction of the slide at 550 °C, the presence of kaolinite cannot be conclusively confirmed (MOORE and REYNOLDS 1997). Nonetheless, data presented in MORTON et al. (2012) provide unambiguous evidence for the presence of kaolinite in another split derived from the sample (njfz06) analyzed from the damage zone.

Fragments of plagioclase are also present in the <2 μm fraction. For example, a low intensity peak at 3.19 Å is not affected by any of the above treatments and is likely derived from this mineral (Fig. 6a) (MOORE and REYNOLDS 1997).

3.2. Transition Zone

In the <2 μm fraction derived from the transition zone sample, a broad peak at 14.26 Å is present in the air-dried sample (Fig. 6b). Upon ethylene glycol solvation, this peak expands to 16.8 Å, a possible ~8.4 Å peak occurs on the high 2θ shoulder of the illite peak, and a small, broad peak appears at ~5.54 Å. Though the 4.23 Å peak may be masked by the 4.25 Å peak attributed to quartz, this result is generally consistent with the presence of illite/smectite with ~30 % illite in illite/smectite (MOORE and REYNOLDS 1997). Following heat saturation treatment at 400 °C, the 16.8 Å peak collapses into the 10 Å peak, and the collapse of the broad, high 2θ shoulder possibly containing the 8.4 Å peak, as well as the 5.54 Å illite/smectite peak, reveals the 14.1, 7.1, and 4.7 Å peaks of chlorite, along with the 10, 5, and 3.34 Å peaks of illite. Following heat saturation treatment at 550 °C, these peaks remain relatively unchanged (Fig. 6b) (MOORE and REYNOLDS 1997).

Moderate intensity peaks at 7.1 and 3.54 Å are not affected by ethylene glycol solvation or heating to 400 °C, but are destroyed upon heating to 550 °C, and are probably derived from kaolinite (Fig. 6b). In addition, the low intensity peaks at ~14 and 4.7 Å described above are probably derived from chlorite (MOORE and REYNOLDS 1997).

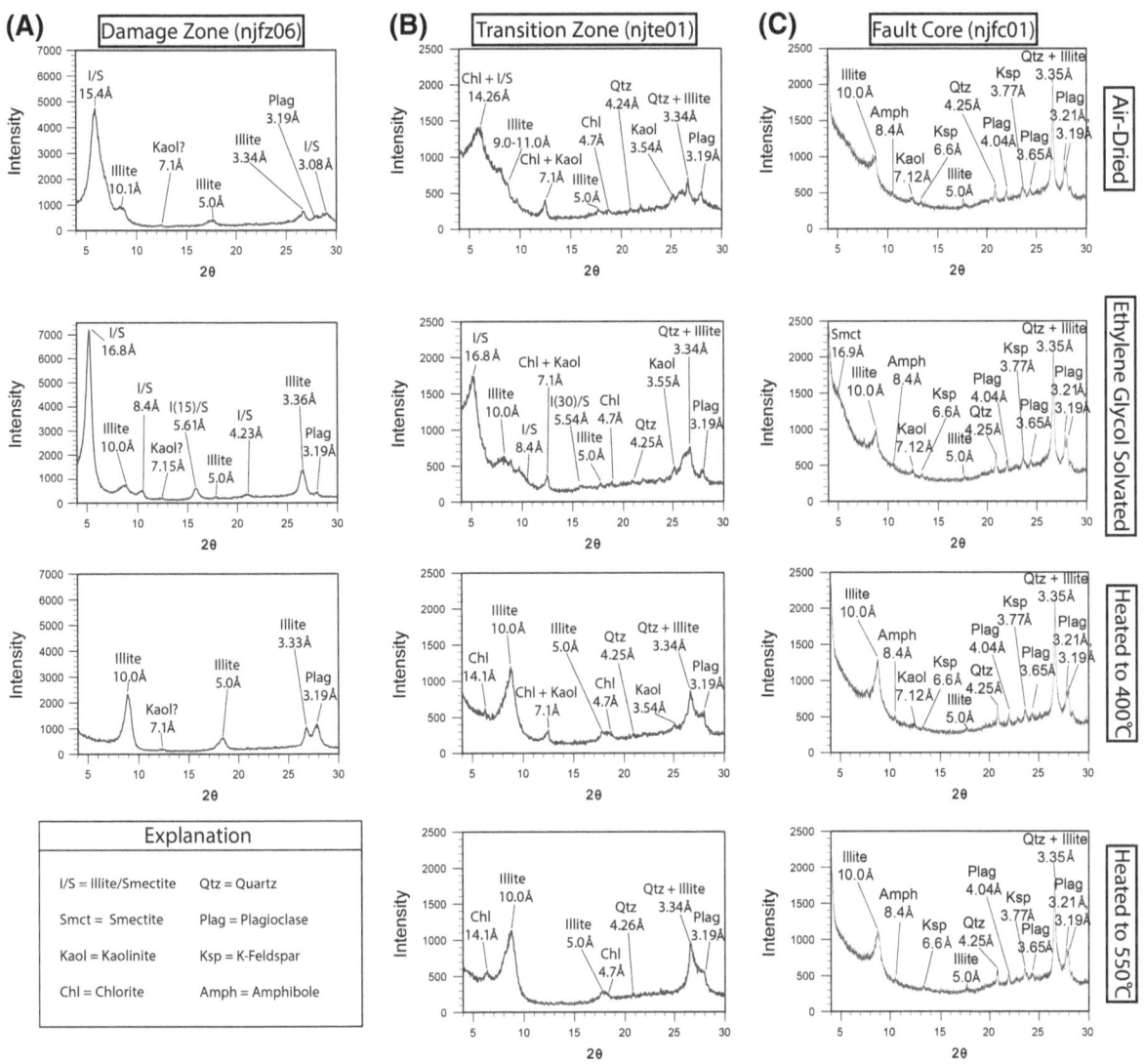

Figure 6
Results of <2 μm XRD analysis of representative samples from the **a** damage zone, **b** transition zone, and **c** primary fault core

Low to very low intensity peaks at 3.19 and 4.25 Å are not affected by any of the above procedures and are attributable to plagioclase and quartz, respectively (Fig. 6b) (MOORE and REYNOLDS 1997).

3.3. Fault Core

In the <2 μm fraction derived from the fault core sample, a high intensity peak at 10 and 3.35 Å as well as a low intensity peak at 5 Å persist through ethylene glycol solvation and heat saturation to 400

and 550 °C, indicating discreet illite (Fig. 6c) (MOORE and REYNOLDS 1997). A very low intensity peak at 16.9 Å appears upon ethylene glycol solvation but collapses into the 10 Å peak upon heat saturation treatments at 400 and 550 °C (Fig. 6c). This peak is likely derived from small amounts of smectite (MOORE and REYNOLDS 1997).

Moderate intensity peaks at 4.04, 3.65, 3.21, and 3.19 Å that are not affected by ethylene glycol solvation and heat saturation treatments are indicative of moderate amounts of plagioclase (Fig. 6c) (MOORE and REYNOLDS 1997). In addition, the moderate peak

Table 1

Major element XRF data for NE block

Zone	Sample	SiO$_2$	Al$_2$O$_3$	FeO$_{(t)}$	CaO	MgO	K$_2$O	Na$_2$O	MnO	TiO$_2$	P$_2$O$_5$	LOI	Total
Wallrock	nj01	64.6	16.5	4.87	4.34	1.69	1.94	3.27	0.06	0.80	0.19	0.59	98.8
	nj02	65.9	16.4	4.00	4.47	1.39	1.63	3.40	0.06	0.68	0.16	0.52	98.6
	nj03	65.8	16.0	4.60	4.14	1.59	1.97	3.16	0.06	0.79	0.18	0.57	98.9
	nj04	66.9	15.6	4.33	4.17	1.47	1.76	3.09	0.06	0.71	0.17	0.67	98.9
	nj05	65.5	16.3	4.60	4.45	1.58	1.85	3.25	0.06	0.77	0.18	0.55	99.1
	nj06	66.4	16.8	3.66	4.07	1.47	1.82	3.22	0.08	0.72	0.05	0.64	98.9
	nj07	67.1	16.6	3.02	4.09	1.50	1.71	3.33	0.06	0.62	0.05	0.74	98.8
	nj08	65.5	17.6	2.77	4.28	1.83	1.77	3.56	0.06	0.58	0.07	0.74	98.8
	nj09	62.5	18.6	3.54	4.54	1.77	2.03	3.70	0.07	0.69	0.08	0.88	98.4
Damage zone	njdz01	65.9	16.5	3.56	4.04	1.44	1.94	3.34	0.06	0.74	0.27	0.99	98.8
	njdz02	65.5	16.4	3.90	3.91	1.48	2.03	3.41	0.06	0.71	0.25	1.14	98.8
	njdz03	65.2	16.6	3.60	3.94	1.47	2.00	3.23	0.06	0.75	0.26	1.33	98.4
	njdz04	68.7	15.6	3.21	3.73	1.22	1.66	3.33	0.05	0.56	0.13	0.67	98.8
	njdz05	67.8	15.6	3.81	3.89	1.30	1.67	3.15	0.05	0.69	0.16	0.65	98.7
	njdz06	65.2	16.6	4.15	4.04	1.44	1.85	3.31	0.06	0.76	0.20	0.88	98.5
Transition zone	njt01	63.6	15.4	4.16	3.61	1.78	1.83	2.71	0.07	0.68	0.18	5.09	99.1
	njt02	62.9	15.0	4.27	3.87	1.77	1.68	2.60	0.07	0.69	0.18	5.55	98.6
	njt03	62.5	14.7	5.63	3.27	1.97	1.90	2.36	0.09	0.78	0.19	6.20	98.6
	njt04	61.9	14.4	5.09	3.12	2.13	1.99	2.12	0.10	0.82	0.19	6.81	98.6
	njt05	61.7	14.8	4.44	3.18	1.97	1.86	2.80	0.08	0.76	0.19	6.67	98.4
Fault core	njfc01	65.3	13.0	4.79	2.65	2.08	2.04	1.75	0.08	0.77	0.15	4.84	97.5
	njfc02	66.6	12.7	4.74	2.37	2.05	2.00	1.59	0.08	0.72	0.15	3.60	96.5
	njfc03	65.6	13.5	4.79	2.88	1.95	1.87	2.03	0.09	0.77	0.17	4.20	97.8
	njfc04	66.2	13.1	4.65	2.74	1.94	1.89	1.73	0.09	0.75	0.16	3.67	97.0
	njfc05	64.8	13.9	4.62	3.12	1.91	1.81	1.91	0.08	0.78	0.17	3.78	96.9

XRF analysis completed by Morton. See MORTON *et al.* (2012) for description of analytical methods

at 4.25 Å and the high intensity peak at 3.35 Å remain after heat saturation treatments at 550 °C, and are likely derived from quartz. Low to moderate intensity peaks at 6.6 and 3.77 Å are likely the result of small quantities of K-feldspar (MOORE and REYNOLDS 1997). A minor peak at 8.4 Å that persists through all stages of treatment indicates the presence of small amounts of amphibole (Fig. 6c).

Data presented above suggests that across the damage and transition zones adjacent to the primary fault core the proportion of illite in illite/smectite progressively increases, and within the primary fault core is converted nearly completely to illite. This transformation is paralleled by a systematic decrease in grain size, dissolution of plagioclase, and enrichment in quartz toward the primary fault core (MORTON *et al.* 2012). In order to assess the effect of these processes on major element chemistry, samples from each fault zone architectural element were collected by Morton, Carrasco, and Girty and were analyzed for major element chemistry in the GeoAnalytical

Laboratory of Washington State University following procedures outlined in JOHNSON *et al.* (1999). Resulting data, including Loss On Ignition (LOI), are provided in Table 1. LOI values represent the weight percent of water and other volatiles contained in each analyzed rock sample.

4. The A–CN–K Diagram

The A–CN–K ternary diagram is commonly used in investigations involving the alteration of shallow crustal rocks by solutions (e.g., NESBITT and YOUNG 1984, 1989; GIRTY *et al.* 2008, 2013, 2014). To plot data on the diagram, major element data are first converted to molar proportions. From such data A, CN, and K are calculated as the proportions of Al$_2$O$_3$, CaO* + Na$_2$O, and K$_2$O in the sum of Al$_2$O$_3$, Na$_2$O, CaO*, and K$_2$O, respectively. Note that CaO* represents the molar proportion of CaO held only in silicate minerals. In this study it was corrected for

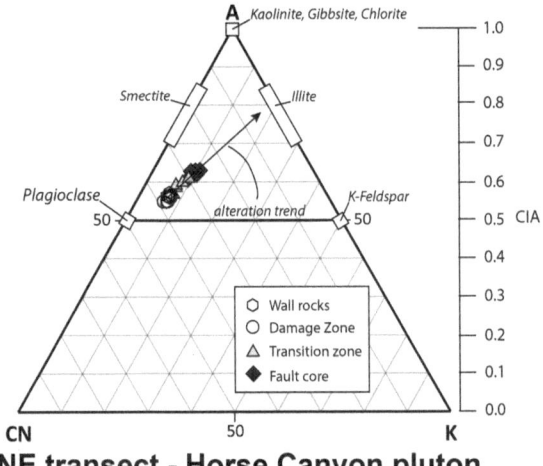

NE transect - Horse Canyon pluton

Figure 7
Ternary plot of A–CN–K data for the northeast block, Clark segment, San Jacinto fault zone

apatite as follows: CaO* = CaO − (3.33 × P_2O_5) (e.g., MCLENNAN 1993). The proportion of Al_2O_3 in the sum of Al_2O_3, Na_2O, CaO*, and K_2O is referred to as the Chemical Index of Alteration (CIA) (NESBITT and YOUNG 1984, 1989). For fresh unaltered granitoids, CIA ranges from ~0.45 to ~0.55. In contrast, the same ratio in secondary minerals such as illite is around 0.80 (NESBITT and YOUNG 1984, 1989).

As shown in Fig. 7, in A–CN–K subspace geochemical data derived from the NE block trend from the compositions of the wallrock system toward the general composition of illite. Moreover, along the alteration trend, samples from the fault core plot closer to the general composition of illite than do samples from the damage or transition zones, a result that supports the earlier findings of MORTON et al. (2012) and is consistent with the clay mineralogy described above. However, chemical compositions are vectors composed of non-negative components that sum to either 1 or 100 % [e.g., A + CN + K = 1 or 100(A + CN + K) = 100 %] (AITCHISON 1986; BUCCIANTI et al. 2006; PAWLOWSKY-GLAHN and BUCCIANTI 2011). They, therefore, convey only relative information, and are not free to range from −∞ to +∞ as do unconstrained variables (PAWLOWSKY-GLAHN and EGOZCUE 2006). As a result of the closed sum format, problems concerning correlation and trend analysis of compositional data as well as other

statistical issues are commonly raised (e.g., AITCHISON 1986; WORONOW and LOVE 1990; AGUE 1994; AGUE and VAN HAREN 1996). However, the general methodology of AITCHISON (1986) as espoused by VON EYNATTEN et al. (2003) and VON EYNATTEN (2004) addresses these issues, and was used to assess whether or not geochemical data like that depicted in Fig. 7 yield a statistically significant linear compositional trend. If the results of such an assessment are positive, then as described below, translational invariant chemical alteration intensity factors can be calculated through orthogonal projection of altered samples onto the compositional linear trend.

5. The Statistics of Calculating Linear Compositional Trends and Alteration Intensity Factors

As originally recognized by AITCHISON (1986), the sample space for D-part compositional data is the simplex, S^D, rather than D-dimensional real space $\Re^D$. Though D can be as large as the total number of elements analyzed in a given composition (e.g., the 10 common major elements shown in Table 1), it is commonly reduced to a sub-compositional space that has some interest to the investigator. For example, if $D = 3$, then the simplex is graphically represented by the ternary diagram with apices of A, CN, and K, the sub-compositional space of interest in this investigation (PAWLOWSKY-GLAHN and EGOZCUE 2006).

The simplex has a vector space structure (AITCHISON 1986; PAWLOWSKY-GLAHN and EGOZCUE 2006; MATEU-FIGUERAS et al. 2011). Within this context, a compositional linear trend as defined by VON EYNATTEN et al. (2003) is determined by an initial composition $\mathbf{a} = [a_1,..., a_D]$, and a unitary composition vector $\mathbf{p} = [p_1,..., p_D]$ that defines the direction of the trend. Any composition $\mathbf{y}$ on the linear trend is perfectly determined by a scalar t ($t \in \Re$) such that

$$\mathbf{y} = \mathbf{a} \oplus (t \otimes \mathbf{p}) = C[a_1 p_1^t,...,a_D p_D^t], \quad (1)$$

where the symbols $\oplus$ and $\otimes$ represent perturbation and powering, respectively, and C represents the closure operator, i.e., each component of the vector is divided by the sum of all of its components and multiplied by a constant, usually 1 or 100. In Eq. (1), the scalar t is a measure of the change in composition relative to the original state (i.e., composition $\mathbf{a}$) as a result of some process or processes.

As noted by VON EYNATTEN (2004), describing compositional linear trends in the simplex is similar to the vectorial expression of straight lines in real space, e.g., $\mathbf{y} = \mathbf{b} + \lambda\mathbf{u}$. By analogy, the position vector $\mathbf{b}$ corresponds to the initial composition $\mathbf{a}$ in Eq. (1), and the scalar values λ and t indicate the position of a specific point on the straight line or a composition on the compositional linear trend, respectively. In terms of a process, t determines the relative change in composition compared to the original state (i.e., composition $\mathbf{a}$). It is dependent on the degree of intensity to which the process or processes responsible for the change in composition has been active (VON EYNATTEN et al. 2003; VON EYNATTEN 2004). In this study, t represents the chemical alteration intensity factor associated with the development of each fault zone architectural element as defined in A, CN, K subspace.

An important property of t is that the difference in intensity of the alteration process between $y(t)$ and $y(t + \Delta t)$ depends on Δt, and not on the particular value of t (VON EYNATTEN 2004). In other words, the amount of the alteration process leading from $\mathbf{a}$ to $y(1)$ equals the amount of the process from, say, $y(3)$ to $y(4)$ because $\Delta t = 1$ in both cases (i.e., t is a translation invariant measure of alteration intensity) (VON EYNATTEN et al. 2003).

Utilizing the general algorithm outlined in VON EYNATTEN et al. (2003), we calculated the compositional linear trend that explained as much as possible of the total variability of the data set characteristic of the NE wallrocks, damage zone, transition zone, and primary fault core. In doing so, the calculated compositional linear trend was forced to pass through the initial composition $\mathbf{a}$ (unaltered wall rocks), with a direction defined by the unitary perturbation vector $\mathbf{p}$. The geometric mean of the unaltered wallrock samples were used as the initial composition $\mathbf{a}$ (Table 2). Non-central principal component analysis (PCA) was used to define the unitary perturbation vector $\mathbf{p}$ (Table 2). Below a brief explanation of non-central PCA is provided. For additional discussion, the

Table 2

*Perturbation vector, **p** and initial composition, **a***

Variable	A	CN	K
$\mathbf{p}$ (perturbation vector)	0.2892	0.1391	0.5718
$\mathbf{a}$ (wallrock geometric mean)	0.3310	0.3670	0.3019

interested reader is referred to VON EYNATTEN et al. (2003) and VON EYNATTEN (2004) and references therein. Much of what follows is a summary of material covered in these papers.

Adjusting a linear trend with known starting point $\mathbf{a}$ to a set of observations $\mathbf{x}_1,..., \mathbf{x}_n$ in S^D consists in finding an axis going through $\mathbf{a}$ which explains as much as possible of the total simplicial variability $\sum|x_i \oplus \mathbf{a}^{-1}|^2_\alpha$ of $\mathbf{x}_1,..., \mathbf{x}_n$ with respect to $\mathbf{a}$ (VON EYNATTEN et al. 2003). This type of problem can be dealt with through the use of non-central principal component analysis, a multivariate statistical technique that finds the unitary eigenvectors $\nu_1,..., \nu_D$ and corresponding eigenvalues $\lambda_1,..., \lambda_D$ of the $D \times D$ product matrix $\mathbf{Z}'\mathbf{Z}$, where $\mathbf{Z}$ is the $n \times D$ matrix whose n rows are the vectors clr $\mathbf{x}_1$ − clr $\mathbf{a}$,..., clr $\mathbf{x}_n$ − clr $\mathbf{a}$. Note that the prime stands for transpose (VON EYNATTEN et al. 2003). The term clr signifies the centered log-ratio transformation and is given by clr $\mathbf{x} = [\log(x_1/g(\mathbf{x})),..., \log(x_D/g(\mathbf{x}))]$ for each $\mathbf{x}$ in S^D, where $g(\mathbf{x})$ is the geometric mean of the components of the composition $\mathbf{x}$ (e.g., AITCHISON 1986; PAWLOWSKY-GLAHN and EGOZCUE 2006; MATEU-FIGUERAS et al. 2011; and references therein). The inverse operation (i.e., clr^{-1}) is defined as follows: clr^{-1} (clr($\mathbf{x}$)) $= C[\exp(\text{clr}(\mathbf{x}_1)),..., \exp(\text{clr}(\mathbf{x}_D))]$ (AITCHISON 1986; PAWLOWSKY-GLAHN and EGOZCUE 2006; MATEU-FIGUERAS et al. 2011).

Given that the eigenvalues are arranged in decreasing order of magnitude, i.e., $\lambda_1 \geq \cdots \geq \lambda_D$, then the vectors $\mathbf{e}_1 = \text{clr}^{-1}\nu_1,...,$ $\mathbf{e}_{D-1} = \text{clr}^{-1}\nu_{D-1}$ explain the variability of $\mathbf{x}_1,..., \mathbf{x}_n$ with respect to $\mathbf{a}$ in decreasing order. Hence, the variability explained by $\mathbf{e}_j$ equals λ_j for $j = 1,..., D-1$, and the variability of $\mathbf{x}_1,..., \mathbf{x}_n$ with respect to $\mathbf{a}$ is given by the sum $\lambda_1 + \cdots + \lambda_{D-1}$ (VON EYNATTEN et al. 2003). In other words, the linear trend with known starting point $\mathbf{a}$, which adjusts the set of compositional observations $\mathbf{x}_1,..., \mathbf{x}_n$ will be L_a (**a**;

Table 3

Alteration intensity factors, t, and associated centered A, CN, K compositions

Zone	Sample	t	A	CN	K	Average
Damage zone	njdz01	0.0850	0.3324	0.3450	0.3224	$t = 0.06 \pm 0.06^a$
	njdz02	0.1209	0.3328	0.3363	0.3307	
	njdz03	0.1265	0.3328	0.3350	0.3320	
	njdz04	−0.0103	0.3309	0.3682	0.3007	
	njdz05	−0.0014	0.3311	0.3660	0.3027	
	njdz06	0.0450	0.3319	0.3547	0.3133	
Transition zone	njt01	0.1438	0.3329	0.3309	0.3361	$t = 0.20 \pm 0.12^a$
	njt02	0.0643	0.3322	0.3500	0.3177	
	njt03	0.2567	0.3327	0.3044	0.3627	
	njt04	0.3415	0.3316	0.2852	0.3831	
	njt05	0.2004	0.3330	0.3175	0.3494	
Fault core	njfc01	0.4794	0.3282	0.2551	0.4165	$t = 0.42 \pm 0.13^a$
	njfc02	0.5446	0.3259	0.2415	0.4324	
	njfc03	0.3425	0.3316	0.2850	0.3833	
	njfc04	0.4159	0.3300	0.2688	0.4011	
	njfc05	0.2996	0.3322	0.2946	0.3730	

[a] 95 % confidence interval

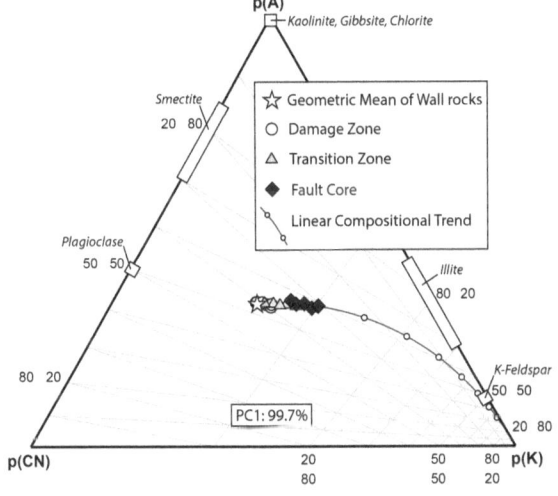

Figure 8

Centered calculated compositional linear trend and results from non-central principal component analysis. Principal component 1 (PC1) explains 99.7 % of the total variance

$clr^{-1}\boldsymbol{v}_1$), and the proportion of the total variability of $\mathbf{x}_1,\ldots, \mathbf{x}_n$ with respect to **a** retained by this trend, will be given by the quotient $\lambda_1/(\lambda_1 + \cdots + \lambda_{D-1})$. This proportion is used to judge the goodness of fit of L_a $(a; clr^{-1}\boldsymbol{v}_1)$ to $\mathbf{x}_1,\ldots,\mathbf{x}_n$, and in this paper is provided as the proportion of the variability defined by principal component 1 (PC1). The alteration intensity factor, t_i for each fault zone sample, $\mathbf{x}_i$, was then derived by orthogonal projection onto the calculated

linear trend. For this projection, the inner product $t_i = \langle x_i \oplus a^{-1}, p \rangle_a$ was used (VON EYNATTEN *et al.* 2003; VON EYNATTEN 2004).

All of the above procedures were implemented in the C# program, Statistical Modeling of Compositional Data (SMCD) written by the second author. Numerical and graphical results output by SMCD are presented in Tables 2 and 3, and Fig. 8, respectively.

6. Results

Data derived from the above calculations were centered following the procedures discussed in VON EYNATTEN *et al.* (2002) and PAWLOWSKY-GLAHN and EGOZCUE (2006). Centering is a special case of perturbation where all of the data are perturbed by the same element, which in the case of Fig. 8, was the inverse of the center of the data set derived from samples analyzed from the wall rock system, damage zone, transition zone, and primary fault core of the NE block (e.g., PAWLOWSKY-GLAHN and EGOZCUE 2006). Following the recommendations of VON EYNATTEN *et al.* (2002), centered data are referred to as $p(A)$, $p(CN)$, and $p(K)$.

Note that in Fig. 8, PC1 explains 99.7 % of the variance, and that the resulting linear compositional trend is oriented toward the $p(K)$-apex and away from

Table 4

Omnibus results for one-way ANOVA

t values PV1	Sum of squares	df	Mean square	F	Significance
Between groups	0.346	2	0.173	22.043	0.000
Within groups	0.102	13	0.008		
Total	0.448	15			

the $p(CN)–p(A)$ join. Moreover, samples derived from the various architectural components of the NE block spread relatively tightly about the calculated linear compositional trend, with samples derived from each architectural component clustering in different groups that reflect a generally increasing degree of alteration toward the fault core. For example, results from calculating the chemical alteration intensity factors (i.e., $t_i = \langle x_i \oplus a^{-1}, p \rangle_a$) (Table 3) suggest that samples from the damage zone (average $t = 0.06 \pm 0.06$) are less intensely altered than are those from the transition zone ($t = 0.20 \pm 0.13$), which in turn are less intensely altered than are samples analyzed from the primary fault core ($t = 0.42 \pm 0.12$) (Fig. 8). These results are consistent with the physical properties, petrologic characteristics, and clay mineralogy of each architectural component noted earlier in this paper, and are likely the result of the increasing reaction capacity of fluids as they migrated through damage zone, transition zone, and fault core.

A one-way Analysis Of Variance (ANOVA) in SPSS© was used to test whether or not the t values for each architectural component was statistically significant at the 95 % confidence level. The omnibus (Table 4), or overall result, provided a significance level <0.05 ($p = 0.0001$). Therefore, the null hypothesis that there is not at least one significant difference across the group of means is rejected. Following normal statistical protocol, after completion of the omnibus, post hoc tests (Table 5) included both Turkey highly significant difference (HSD) and Scheffe tests. When the damage and transition zones are compared, significance levels are 0.052 and 0.064, respectively; hence, at the 95 % confidence level the difference between these two groups is

marginally insignificant. In contrast, when the damage zone or transition zone are compared to the primary fault core, the significance levels for both the Turkey HSD and Scheffe tests are well under 0.05, indicating that the alteration intensity within the fault core is statistically significant (Table 5).

7. Discussion

Results obtained during this study show that the proportion of illite in mixed-layer illite/smectite in the damage and transition zones progressively increases toward the primary fault core where a nearly complete transformation to discreet illite occurs. Following in parallel, chemical alteration intensity factors vary from $t = 0.06$ (± 0.06) in the damage zone to $t = 0.42$ (± 0.12) in the primary fault core. Given that most currently published models for the illite/smectite to illite transformation involve some form of a fluid phase (e.g., POLLASTRO 1993; ESSENE and PEACOR 1995; HAINES and VAN DER PLUIJM 2012; MORTON et al. 2012), the above data and observations imply that fluids were focused through the primary fault core. Moreover, the progressive dissolution of plagioclase from the damage zone to the primary fault core further implies that solutions were likely acidic (e.g., NESBITT and YOUNG 1984, 1989), and, as documented in MORTON et al. (2012), is paralleled by a progressive loss of Ca and Na mass. While there is large uncertainty as to the temperature range over which the conversion of illite/smectite to illite occurs, most authors, depending on kinetic factors (POLLASTRO 1993), place the minimum temperature at about 125–150 °C (POLLASTRO 1993; ESSENE and PEACOR 1995; HAINES and VAN DER PLUIJM 2012; MORTON et al. 2012). Given that recent field and experimental studies suggest that significant frictional heat is produced during high velocity slip events (BRANTUT et al. 2008; COLLETTINI et al. 2013), a characteristic feature of the Clark strand, we speculate that during repeated rupturing, frictional heating and the flushing of high temperature acidic fluids through the primary fault core may have resulted in transient temperatures ≥125 °C. Over time these factors eventually overcame the kinetic barrier for the illite/smectite to illite transition (e.g., HAINES and VAN DER PLUIJM 2012). Such processes are unique to fault zones,

Table 5

Results from SPSS post hoc tests

I type	J type	Mean difference (I − J)	Standard error	Significance	95 % confidence interval	
					Lower bound	Upper bound
Turkey HSD						
Damage zone	TZ	−0.1403	0.0537	0.052	−0.2821	0.0013
	FC	−0.3555*	0.0537	0.000	−0.4971	−0.2138
Transition zone	DZ	0.1404	0.0536	0.052	−0.0013	0.2821
	FC	−0.2151*	0.0560	0.005	−0.3631	−0.0671
Fault core	DZ	0.3555*	0.0537	0.000	0.2138	0.4971
	TZ	0.2151*	0.0560	0.005	0.0671	0.3631
Scheffe						
Damage zone	TZ	−0.1404	0.0537	0.064	−0.2884	0.0076
	FC	−0.3555*	0.0537	0.000	−0.5035	−0.2074
Transition zone	DZ	0.1404	0.0537	0.064	−0.0076	0.2884
	FC	−0.2151*	0.0560	0.007	−0.3697	−0.0605
Fault core	DZ	0.3555*	0.0537	0.000	0.2074	0.5035
	TZ	0.2151*	0.0560	0.007	0.0605	0.3697

* The mean difference is significant at the 0.05 level

and as a result, they represent an underappreciated setting for the development of discreet illite from illite/smectite. Furthermore, the success of non-central principal component analysis in this environment offers the first statistically rigorous methodology for establishing the existence of compositional linear trends in fault zones. This method also derives quantifiable alteration intensity factors that could potentially be used to compare the intensity of alteration at different segments of a fault, as well as offer a foundation to interpret the potential driving forces for said alteration and differences therein.

Acknowledgments

Work completed during this study was funded by National Science Foundation Grant EAR 0908515 to T. Rockwell and G. H. Girty. T. Shimamoto and C. Collettini provided constructive reviews of an earlier version of this paper, and an anonymous third reviewer helped us to focus our approach.

REFERENCES

ABBOTT, P.L., and SMITH, T.E. (1978), *Trace-element comparison of clasts in Eocene conglomerates, southwestern California and northwestern Mexico*, J. Geol. *86*, 753–762.

ABBOTT, P.L., and SMITH, T.E. (1989), *Sonora, Mexico, source for the Eocene Poway Conglomerate of southern California*, Geol. *17*, 329–332.

AGUE, J.J. (1994), *Mass Transfer during Barrovian Metamorphism of Pelites, South-Central Connecticut; I, Evidence for Changes in Composition and Volume*, Am. J. Sci. *294*, 989–1057.

AGUE, J.J., and VAN HAREN, J.M. (1996), *Assessing Metasomatic Mass and Volume Changes Using the Bootstrap, with Application to Deep Crustal Hydrothermal Alteration of Marble*, Econ. Geol. *91*, 1169–1182.

AITCHISON, J., The Statistical Analysis of Compositional Data (Chapman and Hall, London 1986).

BLISNIUK, K., ROCKWELL, T., OWEN, L.A., OSKIN, M., LIPPINCOTT, C., CAFFEE, M.W., and DORTCH, J. (2010), *Late Quaternary Slip Rate Gradient Defined using High-resolution Topography and ^{10}Be Dating of Offset Landforms on the Southern San Jacinto Fault Zone, California*, J. Geophys. Res. *115*, B08401.

BRANTUT, N., SCHUBNEL, A., ROUZAUD, J.-N., BRUNET, F., and SHIMAMOTO, T. (2008), *High-Velocity Frictional Properties of a Clay-bearing Fault Gouge and Implications for Earthquake Mechanics*, Jour. J. Geophys. Res. *113*, B10401.

BUCCIANTI, A., MATEU-FIGUERAS, G. and PAWLOWSKY-GLAHN, V. (2006), *Compositional Data Analysis in the Geosciences: From Theory to Practice*, Geol. Soc. *264*, 1–10.

CAINE, J.S., EVANS, J.P., and FORSTER, C.B. (1996), *Fault Zone Architecture and Permeability Structure*, Geol. *24*, 1025–1028.

CHESTER, F.M. and CHESTER, J.S. (1998), *Ultracataclasite Structure and Friction Processes of the Punchbowl Fault, San Andreas System, California*, Tectonophysics *295*, 199–221.

CHESTER, F.M., and LOGAN, J.M. (1986), *Implications for Mechanical Properties of Brittle Faults from Observations of the Punchbowl Fault Zone, California*. Pure Appl. Geophys. *124*, 79–106.

CHESTER, F.M., EVANS, J.P., and BIEGEL, R.L. (1993), *Internal Structure and Weakening Mechanisms of the San Andreas Fault*, J. Geophys. *98*, 771–786.

Colby, T.A., and Girty, G.H. (2013), *Determination of Volume Loss and Element Mobility Patterns Associated with the Development of the Copper Basin Fault, Picacho State Recreation Area, SE California, USA*, J. Struct. Geol. *51*, 14–37.

Collettini, C., Viti, C., Tesei, T., and Mollo, S. (2013), *Thermal Decomposition Along Natural Carbonate Faults During Earthquakes*, Geology *41*, 927–930.

Davis, J.C. Statistics and Data Analysis in Geology (John Wiley & Sons, New York 2002).

Eckis, R. (1930), The Geology of the Southern Part of the Indio Quadrangle, California, MS thesis, California Institute of Technology, 25 p.

Essene, E. J., and Peacor, D.R. (1995), *Clay Mineral Thermometry—a Critical Perspective*, Clays and Clay Min. *43*, 540–553.

Evans, J.P., Chester, F.M. (1995), *Fluid-Rock Interaction in Faults of the San Andreas System: Inferences from San Gabriel Fault Rock Geochemistry and Microstructures*, J. Geophys. Res. *100*, 13007–13020.

Faulkner, D. R., Lewis, A. C., and Rutter, E. H. (2003), *On the Internal Structure and Mechanics of Large Strike-Slip Fault Zones: Field Observations of the Carboneras Fault in Southeastern Spain*, Tectonophysics *367*, 235–251.

Faulkner, D.R., Jackson, C.L., Lunn, R.J., Schlische, R.W., Shipton, Z.K., Wibberley, C.J., Withjack, M.O. (2010), *A Review of Recent Developments Concerning the Structure, Mechanics and Fluid Flow Properties of Fault Zones*, J. Struct. Geol. *32*, 1557–1575.

Girty, G.H., Biggs, M.A., Berry, R.W. (2008), *An Unusual Occurrence of Probable Pleistocene Corestone Within a Cretaceous Dioritic Enclave, Peninsular Ranges, California*, Catena *74*, 43–57.

Girty, G.H., Colby, T.A., Rayburn, J.Z., Parizek, J.R., Voyles, E.M. (2013), *Biotite-controlled Linear Compositional Trends in Quartz Dioritic to Tonalitic Saprock, Santa Margarita Ecological Reserve, Southern California, USA*, Catena *105*, 40–51.

Girty, G.H., Gebhart, K.L., Replogle, C.T., Morton, N., Purcell, J.W., Parizek, J.R., Carrasco, R., Errthum, R., and Groover, K. (2014), *Reconstructing the Regolith from Erosionally Exhumed Corestone and Saprock Derived from the Cretaceous Val Verde Tonalite, Peninsular Ranges, Southern California, USA: A Case Study*, Catena, *113*, 150–164.

Haines, S.H., and van der Pluijm, B.A. (2012), *Patterns of Mineral Transformations in Clay Gouge, with Examples from Low-angle Normal Fault Rocks in the Western USA*, J. Struct. Geol. *43*, 2–32.

Hutchison, C.S., Laboratory Handbook of Petrographic Techniques (John Wiley & Sons, New York 1974).

Janecke, S.U., Dorsey, R.J., Forand, D., Steely, A.N., Kirby, S.M., Lutz, A.T., Housen, B.A., Belgarde, B., Langenheim, V.E., Rittenour, T.M. (2010), *High Geologic Slip Rates Since Early Pleistocene Initiation of the San Jacinto and San Felipe Fault Zones in the San Andreas Fault System, Southern California, USA*, Geol. Soc. of Am. Spc. Paper *475*, 48 p.

Johnson, D.M., Hooper, P.R., Conrey, R.M. (1999), *XRF Analysis of Rocks and Minerals for Major and Trace Elements On a Single Low Dilution Li-tetraborate Fused Bead*. Adv. X-ray Anal. *41*, 843–867.

Kendrick, K. J., Morton, D.M., Wells, S.G., and Simpson, R. W. (2002), *Spatial and Temporal Deformation Along the Northern San Jacinto Fault, Southern California: Implications for Slip Rates*, Bull. Seismol. Soc. Am. *92*, 2782–2802.

Kim, Y.S., Peacock, D.C.P., and Sanderson, D.J. (2004), *Fault Damage Zones*, J. Struct. Geol. *26*, 503–517.

Mateu-Figueras, G., Pawlowsky-Glahn, V., and Egozcue, J.J. (2011), The Principle of Working on Coordinates. In: Pawlowsky-Glahn, V., and Buccianti, A., (eds), Compositional data analysis, John Wiley & Sons, Ltd., 31–42.

McLennan, S. M. (1993), *Weathering and Global Denudation*, J. Geol., *101*, 295–303.

Mitchell, T.M., Faulkner, D.R. (2009), *The Nature and Origin of Off-Fault Damage Surrounding Strike-Slip Fault Zones With a Wide Range of Displacements; a Field Study From the Atacama Fault System, Northern Chile*, J. Struct. Geol. *31*, 802–816.

Morton, N., Girty, G.H., and Rockwell, T.K. (2012), *Fault Zone Architecture of the San Jacinto Fault Zone in Horse Canyon, Southern California: A Model for Focused Post-Seismic Fluid Flow and Heat Transfer in the Shallow Crust*, Earth Planet Sc. Lett. *329–330*, 71–83.

Morton, D.M., Matti, J.C., Miller, F.K., and Repenning, C.A. (1986), *Pleistocene conglomerate from the San Timoteo badlands, southern California: Constraints on Strike Slip Displacements on the San Andreas and San Jacinto Faults (Abstracts with Programs)*, Geol. Soc. Am. *18*, 161.

Moore, D.M., and Reynolds, R.C., X-ray Diffraction and the Identification and Analysis of Clay Minerals (Oxford University Press, Oxford 1997).

Mueller, K., Kier, G., Rockwell, T., and Jones, C.G. (2009), *Quaternary rift flank uplift of the Peninsular Ranges in Baja and southern California by removal of mantle lithosphere*, Tectonics *28*, 1–17.

Nesbitt, H.W., and Young, G.M. (1984), *Prediction of Some Weathering Trends of Plutonic and Volcanic Rocks Based on Thermodynamic and Kinetic Considerations*, Geochim. Cosmochim. Ac. *48*, 1523–1534.

Nesbitt, H.W., and Young, G.M. (1989), *Formation and Diagenesis of Weathering Profiles*, J. Geol. *97*, 129–147.

Pawlowsky-Glahn, V., and Egozcue, J.J. (2006), Compositional Data and their Analysis: an Introduction: In: Buccianti, A., Mateu-Figureeras, G., & Pawlowsky-Glahn, V. (eds) Compositional Data Analysis in the Geosciences: From Theory to Practice, Geological Society London, Special Publications, *264*, 1–10.

Pawlowsky-Glahn, V., and Buccianti, A., Compositional Data Analysis: Theory and Applications (John Wiley & Sons, Ltd., New York 2011).

Pollastro, R.M. (1993), *Considerations and Applications of the Illite/Smectite Geothermometer in Hydrocarbon-Bearing Rocks of Miocene to Mississippian Age*, Clays and Clay Min. *41*, 119–133.

Rockwell, T.K., Loughman, C., and Merifield, P., (1990), *Late Quaternary Rate of Slip Along the San Jacinto Fault Zone Near Anza, Southern California*, J. Geophys. Res. *95*, 8593–8605.

Sahakian, V.J., Fialko, Y., Bock, Y., Rockwell, T. (2011), Space Geodetic Investigation of Interseismic Deformation Due to the San Jacinto Fault Near Anza, CA. Abstr. T43I-05, AGU Fall Meet.

Salisbury, J.S., Rockwell, T.K., Middleton, T.J., Hudnut, K.W. (2012), *LiDAR and Field Observations of Slip Distribution for the Most Recent Surface Ruptures along the Central San Jacinto Fault*, Bull. Seis. Seismo. Soc. Am. *102*, 598–619.

Sauer, C. (1929), *Land Forms in the Peninsular Ranges of California as Developed about Warner's Hot Springs and Mesa Grande*, U. Calif. Pub. Geog. *3*, 199–290.

Schulz, S.E., and Evans, J.P. (1998), *Spatial Variability in Microscopic Deformation and Composition of the Punchbowl Fault, Southern California: Implications for Mechanisms, Fluid-Rock Interaction, and Fault Morphology*, Tectonophysics 295, 223–244.

Schulz, S.E., and Evans, J.P. (2000), *Mesoscopic Structure of the Punchbowl Fault, Southern California, and the Geologic and Geophysical Structure of Active Strike Slip Faults*, J. Struct. Geol. *22*, 913–930.

Sharp, R.V. (1967), *San Jacinto Fault Zone in the Peninsular Ranges of Southern California*. Geol. Soc. Am. Bull. *78*, 705–730.

Sharp, R. V. (1981), *Variable Rates of Late Quaternary Strike Slip on the San Jacinto Fault Zone, Southern California*, J. Geophys. Res. *86*, 1754–1762.

Sibson, R.H. (1977), *Fault Rocks and Fault Mechanics*, J. Geol. Soc. Lond., *133*, 191–213.

Steer, B. L., and Abbott P. L. (1984), *Paleohydrology of the Eocene Ballena Gravels, San Diego County, California*, Sed. Geol. *38*, 181–216.

von Eynatten, H., Pawlowsky-Glahn, V., and Egozcue, J.J. (2002), *Understanding Perturbation on the Simplex: a Simple Method to Better Visualize and Interpret Compositional Data in Ternary Diagrams*, Math. Geol. *34*, 249–257.

von Eynatten, H., Barcelo-Vidal, C., Pawlowsky-Glahn, V. (2003), *Modeling Compositional Change: the Example of Chemical Weathering of Granitoid Rocks*, Math. Geol., *35*, 231–251.

von Eynatten, H. (2004), *Statistical Modeling of Compositional Trends in Sediments*, Sed. Geol. *171*, 79–89.

Warren, N.H., and Girty G.H. (1999), *A Matlab 5 Program for Calculating the Statistics of Mass Change*, J. Geosci. Edu., *47*, 313–320.

Woronow, A., and Love, K.M. (1990), Quantifying and Testing Differences Among Means of Compositional Data Suites, Math. Geol. 22, 837–852.

(Received August 12, 2013, revised April 11, 2014, accepted April 15, 2014, Published online May 20, 2014)

Reprinted from the journal

Pure Appl. Geophys. 171 (2014), 2937–2954
© 2013 Springer Basel
DOI 10.1007/s00024-013-0753-z

Interseismic Strain Localization in the San Jacinto Fault Zone

ERIC O. LINDSEY,[1] VALERIE J. SAHAKIAN,[1] YURI FIALKO,[1] YEHUDA BOCK,[1] SYLVAIN BARBOT,[2] and
THOMAS K. ROCKWELL[3]

Abstract—We investigate interseismic deformation across the San Jacinto fault at Anza, California where previous geodetic observations have indicated an anomalously high shear strain rate. We present an updated set of secular velocities from GPS and InSAR observations that reveal a 2–3 km wide shear zone deforming at a rate that exceeds the background strain rate by more than a factor of two. GPS occupations of an alignment array installed in 1990 across the fault trace at Anza allow us to rule out shallow creep as a possible contributor to the observed strain rate. Using a dislocation model in a heterogeneous elastic half space, we show that a reduction in shear modulus within the fault zone by a factor of 1.2–1.6 as imaged tomographically by ALLAM and BEN-ZION (Geophys J Int 190:1181–1196, 2012) can explain about 50 % of the observed anomalous strain rate. However, the best-fitting locking depth in this case (10.4 ± 1.3 km) is significantly less than the local depth extent of seismicity (14–18 km). We show that a deep fault zone with a shear modulus reduction of at least a factor of 2.4 would be required to explain fully the geodetic strain rate, assuming the locking depth is 15 km. Two alternative possibilities include fault creep at a substantial fraction of the long-term slip rate within the region of deep microseismicity, or a reduced yield strength within the upper fault zone leading to distributed plastic failure during the interseismic period.

Key words: Fault zone, dislocation model, compliant zone, San Jacinto fault, Anza.

1. Introduction

The San Jacinto fault is historically the most seismically active fault in southern California, with

Electronic supplementary material The online version of this article (doi:10.1007/s00024-013-0753-z) contains supplementary material, which is available to authorized users.

[1] Institute of Geophysics and Planetary Physics, Scripps Institution of Oceanography, University of California San Diego, La Jolla, CA 92093, USA. E-mail: elindsey@ucsd.edu

[2] Earth Observatory of Singapore, Nanyang Technological University, Singapore 639798, Singapore.

[3] Department of Geological Sciences, San Diego State University, San Diego, CA 92182, USA.

nine major (M 6–7) earthquakes over the past 120 years. The fault segment near Anza has not ruptured for more than 200 years (ROCKWELL *et al.* 2006), and is considered to represent a "seismic gap" (THATCHER *et al.* 1975; SANDERS and KANAMORI 1984). As a result, the Anza segment of the San Jacinto fault (SJF) has been a subject of numerous geologic, geodetic, and seismic studies (ROCKWELL *et al.* 1990; LISOWSKI *et al.* 1991; ALLAM and BEN-ZION 2012).

Early trilateration surveys and electronic distance measurements (EDM) suggested an unusually high strain rate across this segment (LISOWSKI *et al.* 1991; JOHNSON *et al.* 1994). These observations were interpreted as requiring a locking depth of just 5–6 km assuming a homogeneous elastic model, significantly shallower than the observed 14–18 km depth extent of microseismicity in the area (e.g., SANDERS 1990; LIN *et al.* 2007; HAUKSSON *et al.* 2012). LISOWSKI *et al.* (1991) proposed that the apparent disagreement could be explained by a compliant fault zone with significantly reduced shear modulus, but could not rule out alternatives such as shallow fault creep, which might affect near-field measurements. More recently, WDOWINSKI (2009) proposed that the deep microseismicity may be due to the propagation of "brittle creep" into the seismogenic zone, and, therefore, the locking depth is in fact much shallower than the bottom of the seismogenic zone.

Shallow creep on the Anza segment of the SJF would offset geodetic monuments on either side of the fault and increase the apparent near-fault strain rate. Shallow creep is predicted by laboratory observations, which show that poorly consolidated rocks in the top few kilometers of the Earth's crust exhibit velocity-strengthening behavior (MARONE and SCHOLZ 1988; MARONE 1998), and by rate- and state-dependent frictional models in which low normal stresses

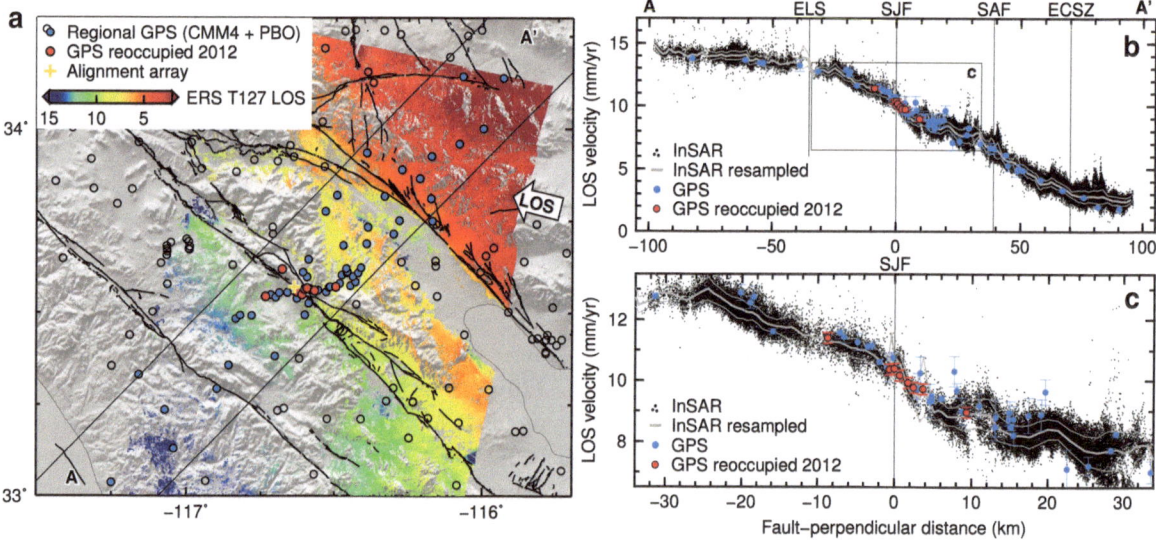

Figure 1
Geodetic data used in this study. **a** Map showing InSAR line-of-sight (LOS) velocities from ERS track 127 (MANZO *et al.* 2011), locations of regional GPS sites, and the re-surveyed Mitchell road alignment array. **b** Selected InSAR LOS velocities and GPS velocities projected onto the radar LOS for comparison. **c** Zoom of (**b**) near the San Jacinto fault (SJF)

near the surface result in stable sliding, even if the frictional properties are velocity-weakening (e.g., TSE and RICE 1986; LAPUSTA *et al.* 2000; KANEKO *et al.* 2013).

A compliant fault zone with a reduced effective shear modulus would have a similar effect on the surface strain rate even if the fault is locked near the surface. Zones of this type have been inferred in several locations, both geodetically (RYBICKI and KASAHARA 1977; FIALKO *et al.* 2002; CHEN and FREY-MUELLER 2002; FIALKO 2004; HAMIEL and FIALKO 2007; BARBOT *et al.* 2009; JOLIVET *et al.* 2009; CAKIR *et al.* 2012) and seismically (SPUDICH and OLSEN 2001; BEN-ZION *et al.* 2003; COCHRAN *et al.* 2009). Observations of trapped waves along the SJF near Anza have suggested such a zone may be present there as well, with inferred reductions in S-wave velocity of up to 50 % (LI and VERNON 2001; LEWIS *et al.* 2005). Recently, ALLAM and BEN-ZION (2012) conducted a high-resolution tomographic study of the velocity structure surrounding the SJF, yielding clear evidence for a low velocity fault zone extending to a depth of several kilometers.

In this study, we present new high-density GPS and InSAR observations (Fig. 1) and numerical models incorporating recent tomographic results (ALLAM and BEN-ZION 2012) to evaluate the

mechanisms responsible for the anomalous interseismic strain rate at Anza. We find that shallow creep is negligible at this location, based on occupations with GPS of an alignment array installed in 1990 (ROCKWELL *et al.* 1992). Next, we show that a compliant fault zone with properties inferred by ALLAM and BEN-ZION (2012) does play a significant role in the localization of strain, affecting inferences of slip rate, locking depth and ultimately seismic hazard. However, the observed elastic heterogeneity is insufficient to fully reconcile the "geodetic" and "seismic" locking depths on the Anza section of the SJF. Instead, we find that the data require either the deep seismogenic zone to be sliding stably at a significant fraction of the long-term rate, or the shallow fault zone to be deforming inelastically, possibly due to a lower yield strength compared to the ambient crust.

2. Alignment Array and Shallow Creep

Shallow creep has been observed on a number of faults in Southern California, including the Coachella segment of the SAF (SIEH and WILLIAMS 1990; RYMER 2000; RYMER *et al.* 2002; LYONS and SANDWELL 2003), the Imperial fault (GOULTY *et al.* 1978; LYONS *et al.*

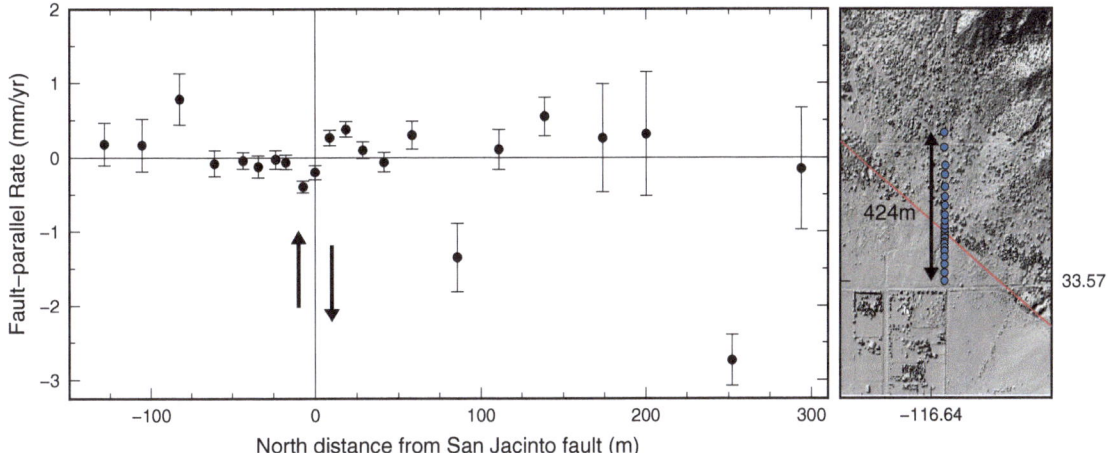

Figure 2

Mitchell road alignment array. Fault-parallel rates computed as total offset divided by the total time, detrended. *Arrows* indicate expected sense of motion on the fault if creep is occurring. *Error bars* reflect the sum in quadrature of GPS and Total Station measurement uncertainties. Map shows monument locations relative to the fault trace in *red*

2002), and the Superstition Hills and Coyote Creek segments of the SJF to the south (SHARP *et al.* 1986; WEI *et al.* 2009, 2011). If shallow creep of this type is occurring on the SJF at Anza, it would offset nearby geodetic monuments and affect the apparent near-field deviatoric strain rate. Here we take advantage of a ~400 m-long alignment array installed in 1990 across the SJF at Anza to quantify the near-field interseismic deformation. The Mitchell road array consists of 22 monuments in a line oriented north–south and centered on the fault trace at −116.640°W, 33.572°N (Fig. 2). Initial surveys in 1990 and 1991 recorded the azimuth and distance from each monument to the central base station (ROCKWELL *et al.* 1992). In August 2010 and March 2011 we occupied the array with survey-mode GPS. By computing the relative changes in the position of each site, we are able to evaluate the presence of shallow creep at a sub-millimeter per year level.

During each GPS campaign, the monuments were occupied with a tripod-mounted chokering antenna and dual-frequency receiver recording at one sample per second for two separate 15 min intervals; the base station on the fault trace was occupied continuously for the duration of the 2-day campaign. We then computed the instantaneous positions at each epoch relative to the nearby continuous station AZRY, using the Real Time Dynamics software package. This rapid-static approach is justified by error

analyses of the instantaneous positioning method for very small baselines (BOCK *et al.* 2000); in this case we were able to determine the baselines between each site with a one-sigma uncertainty of 2–3 mm in the horizontal direction.

The initial optical surveys determined the azimuth from the base station to each monument at a precision of 10^{-4} degrees, translating to an uncertainty in relative position of 1–15 mm with increasing distance from the base station. In principle, the relative station motions computed over the 20-year period could be used to determine the overall shear strain rate near the fault in addition to any localized creep. However, the expected rotation of the array due to tectonic strain accumulation is on the order of 10^{-5} degrees over the 20-year period, while the initial alignment of the array was subject to a much larger compass error of up to 1°. Thus, we subtracted the best-fitting rotation from the differential motion of the benchmarks and focus on near-surface fault creep only. Future GPS surveys of the array should make possible an absolute determination of the local shear strain rate.

In Fig. 2, we show the de-trended offsets in the fault-parallel direction, divided by the total time between observations. The reported uncertainties for each site are the sum in quadrature of the GPS and optical survey errors. These uncertainties are somewhat smaller than overall data scatter, suggesting some non-tectonic motion of individual sites,

77

possibly related to initial settling and subsequent minor disturbance of the monuments. Nevertheless, no systematic right-lateral offset is visible across the fault trace or anywhere within the array. Formally, comparison of the data with a model of surface creep weighted by the relative measurement uncertainties implies a creep rate of <0.2 mm/year with 95 % confidence.

The evident lack of creep on this segment of the SJF is unsurprising; earlier surveys of another alignment array a few kilometers to the north of the Mitchell road array showed no evidence for surface creep over a period of 7 years between 1977 and 1984 at a level of 1 mm/year (LOUIE et al. 1985), although the results were ambiguous owing to motions attributed to the initial settling of the monuments (KELLER et al. 1978; SANDERS and KANAMORI 1984). Our measurements confirm the absence of shallow creep on the Anza section of the San Jacinto fault at a significantly increased precision. In contrast, models of rate- and state-dependent friction on faults predict ubiquitous creep within a shallow layer at a rate that is on the order of ~ 10 % of the fault slip rate if the top 2–3 km of the fault are velocity strengthening (e.g., TSE and RICE 1986; LAPUSTA et al. 2000). Given the long-term slip rate of the SJF at Anza of 10–15 mm/year (ROCKWELL et al. 1990; BLISNIUK et al. 2010, also see below) the predicted rate of shallow creep exceeds the observational constraints by nearly an order of magnitude. Possible explanations for this discrepancy include velocity-weakening behavior of the uppermost seismogenic layer (e.g., KANEKO et al. 2013), or distributed yielding (see "Discussion").

3. Elastic Half-Space Models

The analysis of geodetically determined interseismic motion near an active fault requires a model, including an assumption about the rheology of Earth's lithosphere. The most common models are a dislocation in an elastic half-space (SAVAGE and BURFORD 1973), or an elastic layer over a viscoelastic half-space (NUR and MAVKO 1974; SAVAGE and PRESCOTT 1978). In the context of infinitely long strike-slip faults, it was shown that the two models can explain

interseismic deformation equally well (SAVAGE 1990; FAY and HUMPHREYS 2005). Recent simulations that take into account laboratory-constrained rheologies and long-term strain evolution (TAKEUCHI and FIALKO2012) show a considerable strain localization in the ductile substrate, at least for mature faults such as those of the SAF-SJF system. Thus, in the following, we choose to interpret the data with an elastic dislocation model, in which the fault is fully locked above the locking depth D, and slipping at the full long-term rate below. Simple dislocation models have shown remarkable success at predicting fault slip rates and locking depths, without significant bias relative to seismic and geologic observations (e.g., SMITH-KONTER et al. 2011; LINDSEY and FIALKO 2013). The stress singularity arising from the sudden transition from locked to creeping at the locking depth is nonphysical; in reality, the slip rate must increase gradually with depth to the full rate (e.g., SAVAGE 2006). The implications of a more gradual variation in slip rate with depth are considered in Sect. 4.

3.1. Regional GPS Data

The primary dataset used to infer slip rates on the SAF-SJF system consists of GPS velocities from the Southern California Earthquake Center (SCEC) Crustal Motion Map version 4 (CMM4) GPS velocity solution (SHEN et al. 2011). We use a combination of the CMM4 dataset and continuous GPS velocities from the UNAVCO Plate Boundary Observatory (PBO) network, rotated into the North America Fixed (NAFD) reference frame (Tom Herring, pers. commun., 2011). This is the same GPS dataset used for determination of fault slip rates in the Uniform California Earthquake Rupture Forecast, Version 3 (UCERF3) project, and contains 2,663 GPS velocities distributed throughout California and other parts of North America. We selected 70 sites located along a 30 km-wide fault-perpendicular profile centered on the SJF at Anza (116.6321W, 33.5679N) and projected the horizontal velocities onto the fault azimuth of 313°. This azimuth was chosen to minimize the trend in the residual fault-perpendicular velocities, which would represent unmodeled fault-normal compression; we note that the results are not sensitive to small changes in this value.

Table 1

Updated GPS velocities and one-sigma uncertainties for sites reoccupied in September 2012 and April 2013, in mm/year

Site	Longitude	Latitude	V_E	V_N	V_U	σ_E	σ_N	σ_U
CARY	243.2645	33.5454	−33.43	13.94	−0.45	0.22	0.18	1.00
ANZC	243.3694	33.5578	−31.23	11.93	−0.24	0.21	0.22	1.05
G114	243.3871	33.5502	−31.65	11.57	1.44	0.33	0.23	1.07
TOME	243.3200	33.6190	−30.99	11.36	1.09	0.22	0.21	1.08
G120	243.3970	33.5646	−30.61	10.77	−0.52	0.48	0.42	1.99
RCUT	243.4044	33.5675	−30.14	10.58	0.31	0.29	0.19	1.05
0821	243.4294	33.5613	−30.33	10.21	−0.54	0.19	0.20	1.03
D138	243.5019	33.5711	−28.46	9.03	−0.03	0.31	0.22	1.24

Reference frame is ITRF2008

The dataset, shown in Fig. 1a, has a very dense spacing of GPS sites in the vicinity of Anza: along the entire SJF there are 55 sites located within 10 km of the fault trace, and 26 of these are within our selected profile at Anza. However, many of these are campaign sites with velocities of relatively low accuracy, occupied four to 12 times over the period 1992–2002 when the GPS constellation, orbits, and receivers, especially in the earlier epochs, were less favorable. There were no reported observations in the past 10 years. As a result, the projected velocities show considerable scatter (Fig. 1b). To improve the accuracy of the near-field GPS velocities, in September 2012 and April 2013 we re-occupied eight of the sites near Anza, highlighted in red in Fig. 1. The sites were occupied for 2–4 days at 30-s sampling, and the data were processed in combination with the CMM4 results using the GAMIT/GLOBK software package (HERRING *et al.* 2010). The long timespan between surveys allowed us to determine each velocity with a 1σ uncertainty of 0.2–0.5 mm/year.

Although the intervening time period spans several regional earthquakes (notably, the 1994 M6.7 Northridge, 1999 M7.1 Hector Mine, and 2010 M7.2 El Mayor-Cucapah events), the relatively small size of the network compared with the distance to the events means that coseismic and postseismic offsets do not vary significantly between sites, so they should not significantly affect the relative velocities. This is supported by comparison of continuous GPS time-series within the affected region. To further mitigate the effect of these earthquakes, we subtracted the estimated coseismic offsets for each event based on a dislocation model during the GLOBK analysis. The results are summarized in Table 1 in the ITRF2008 reference frame.

The updated fault-parallel velocities are compared with the regional GPS data in Fig. 1b and c; the best-fitting fault-parallel shear strain rate is 0.60 ± 0.10 μrad/year for the five sites within 3 km of the fault trace [1 μrad/year = 1 (mm/year)/km]. Considering the full 18 km span from CARY to D138, the strain rate is 0.40 ± 0.05 μrad/year, in good agreement with USGS two-color EDM results which covered a similar span across the fault and recorded a rate of 0.42 ± 0.05 μrad/year (http://earthquake.usgs.gov/monitoring/edm/socal/).

3.2. InSAR Data

We also incorporate InSAR data from ERS-1/2 descending track 127, processed with the small baseline subset (SBAS) method in combination with a sparse set of five continuous GPS velocities to constrain the longest-wavelength deformation signals and remove orbital and other systematic errors (MANZO *et al.* 2011). The resulting map of surface deformation may retain unmodeled vertical motion or residual errors at long wavelengths, which can lead to biases in slip rate inversions (e.g., FIALKO 2006; LUNDGREN *et al.* 2009; LINDSEY and FIALKO 2013). Therefore, we adopted the remove–restore method proposed by WEI *et al.* (2010) to constrain the long wavelengths using additional continuous GPS data. Using all available continuous GPS data in the area within and surrounding the radar scene, we interpolated the GPS velocities onto each radar pixel using natural neighbor interpolation, then projected these velocities into the radar LOS. The resulting smooth velocity map was subtracted from the InSAR, and the result was high-pass filtered with a 40 km cutoff wavelength. The interpolated GPS velocities were then added back, resulting in a velocity field that agrees with the GPS at large spatial wavelengths but retains the short-wavelength information provided by the InSAR. Sensitivity testing indicated that the results are not strongly dependent on the interpolation method or the filter cutoff wavelength; for details see WEI et al. (2010) or TONG et al. (2013).

The final InSAR-derived velocity map is shown in Fig. 1. The western portion of the scene is heavily forested, resulting in poor C-band radar correlation. As a result, there are far fewer available data points there than in the eastern portion of the scene where C-band correlation is excellent. Thus, some resampling of the data is needed to avoid over-fitting the eastern part of the selected profile at the expense of points to the west. We considered uniform resampling, or resampling at a rate proportional to a measure of the model sensitivity (normalized derivative of the model predictions with respect to each of the parameters), but found that a strain-based resampling provided the most faithful reproduction of the original dataset and the smallest resulting parameter uncertainties for a given number of resampled data points. We used LOESS local regression (CLEVELAND 1979; CLEVELAND and DEVLIN 1988) to fit a smooth curve to the data, then divided the profile into bins with a width proportional to the derivative of the smooth curve. The mean and standard deviation of the LOS velocities were computed within each bin; we ultimately recovered 190 velocity estimates with standard deviations ranging from 0.2 to 0.9 mm/year, with the highest sample rate in regions of high strain (see Supplementary Figure S1).

The resampled InSAR profile across the fault is shown in Fig. 1b and c along with LOS-projected horizontal GPS velocities. In the figures, the velocities are referenced to a constant incidence angle of 23.5° for plotting purposes; we used the exact incidence angle for each pixel when comparing the data to a model. The data suggest the region of high strain across the fault is approximately 3 km wide and offset to the east of the surface trace of the fault, with a best-fitting strain rate of 0.79 μrad/year, slightly higher than the rate determined above from only GPS.

3.3. Heterogeneous Material Properties

High resolution double-difference tomography by ALLAM and BEN-ZION (2012) indicates the presence of a significant low velocity anomaly along the SJF near Anza. The result is consistent with earlier observations of fault-zone-trapped waves in the area (LI and VERNON 2001; LEWIS et al. 2005), and suggests the presence of an extensive compliant fault zone with a reduced elastic shear modulus, extending several kilometers deep and up to several kilometers from the fault.

We used the seismic velocity model of ALLAM and BEN-ZION (2012) to compute the shear modulus along a 2D fault-perpendicular transect at the location of the geodetic profile, using an empirical relation to compute density from V_p (LUDWIG et al. 1970; BROCHER 2005). The result is shown in Fig. 3a. We then computed the elastic Green functions for anti-plane dislocations representing each fault at a range of locking depths, using the method of fictitious body forces (BARBOT et al. 2009) implemented in a parallel finite-difference framework. As an example, the surface velocity due to a dislocation at 10 km depth is compared for the homogeneous and heterogeneous models in Fig. 3b. The strain rate near the fault is increased by up to 50 %, and the velocities are

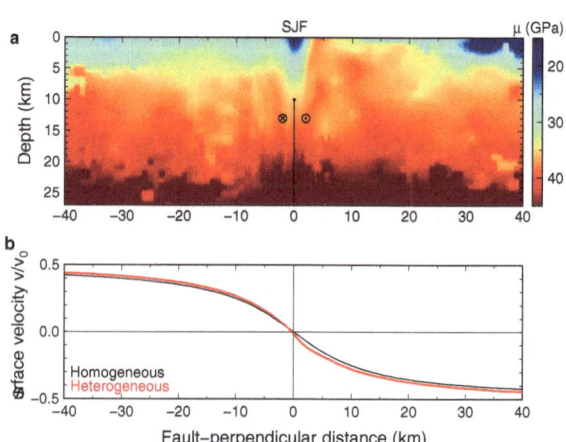

Figure 3

a Fault-perpendicular cross-section showing shear modulus values inferred from tomography of ALLAM and BEN-ZION (2012), centered on the fault trace at Anza (116.632W, 33.568N). **b** Predicted fault-parallel surface velocity for a unit dislocation slipping freely below 10 km, in a homogeneous (*black*) and heterogeneous (*red*) elastic domain

somewhat asymmetric across the fault trace, in qualitative agreement with the geodetic data.

3.4. Inverse Method

Each fault is represented by a semi-infinite dislocation at horizontal position ξ_i slipping at a rate v_i below depth D_i. In this case, the modeled surface velocities are described by

$$\mathbf{m}_j = k_j \sum_i \left(\frac{v_i}{\pi}\right) \tan^{-1}\left(\frac{\mathbf{x} - \xi_i}{D_i}\right) \quad (1)$$

The coefficients k_j represent the projection from fault-parallel velocity to the radar LOS for each pixel; for GPS sites, the model is compared directly to the fault-parallel projection of the observed velocities so that $k_{\mathrm{GPS}} = 1$.

In the heterogeneous case, we use the pre-computed elastic Green functions for each locking depth $\mathbf{G}_i(D_i, \xi_i)$, and the forward model is expressed as a linear combination of these functions:

$$\mathbf{m}_j = k_j \sum_i v_i \mathbf{G}_i(D_i, \xi_i). \quad (2)$$

We conducted the inversions in a Bayesian framework, which has the advantage of providing a simultaneous estimate of the parameters, their uncertainties, and their correlations (e.g., MACKAY 2003; TARANTOLA 2005; MINSON et al. 2013). Bayes' theorem states that the posterior probability distribution function (PDF) $p(\mathbf{m}|\mathbf{d})$ is given by

$$p(\mathbf{m}|\mathbf{d}) \propto p(\mathbf{m})p(\mathbf{d}|\mathbf{m}), \quad (3)$$

where $p(\mathbf{m})$ is the prior PDF and $p(\mathbf{d}|\mathbf{m})$ is the likelihood function. The prior PDF represents our knowledge or assumptions about the model; in this case we assume a uniform prior that requires only right-lateral slip and a positive locking depth. The likelihood function is a measure of goodness of fit, defined by

$$p(\mathbf{d}|\mathbf{m}) = \frac{A}{\sqrt{|\mathbf{C}|}} \exp\left(\frac{1}{2}(\mathbf{m}-\mathbf{d})^T \mathbf{C}^{-1}(\mathbf{m}-\mathbf{d})\right), \quad (4)$$

where A is a normalization constant, $\mathbf{m}$ and $\mathbf{d}$ represent the modeled and observed GPS and

InSAR velocities, and $\mathbf{C}$ is the data covariance matrix.

The data covariance matrix $\mathbf{C}$ represents our best information regarding the uncertainties in the input measurements, an accurate measure of which is critical for the Bayesian method to be valid. For some continuous GPS velocities, the reported one-sigma uncertainties can be very small, <0.1 mm/year. These uncertainties are well justified by analyses considering a combination of white noise and flicker noise (WILLIAMS et al. 2004). However, as noted elsewhere (e.g., PLATT and BECKER 2010), attempting to fit a simplified tectonic model within these uncertainties can lead to over-fitting of the velocities, due to the presence of unmodeled tectonic or non-tectonic sources of deformation. For the diagonal entries representing the GPS uncertainties in $\mathbf{C}$, we, therefore, imposed a minimum value of 0.25 mm/year to avoid over-fitting continuous GPS sites with very low formal uncertainties. For the case of spatially correlated InSAR data, one may consider more realistic non-diagonal models of data covariance (e.g., LOHMAN and SIMONS 2005; SUDHAUS and JONSSON 2008); however, the strain-based resampling of the InSAR significantly reduces the number of samples in non-deforming regions where unmodeled data covariance is typically a problem, so this effect should be limited. Therefore, we computed uncertainties for the resampled InSAR profile as simply the standard deviation of all data points within each resampled bin.

To ensure accurate weighting between the GPS and InSAR datasets, we further modified $\mathbf{C}$ by rescaling the relative GPS and InSAR uncertainties so that the two datasets make equal contributions to the total misfit value. Typically this is done by assigning empirically determined weighting factors to the two datasets (e.g., SIMONS et al. 2002; FIALKO 2004); here we make use of the Bayesian formulation of the problem to determine the ideal weights exactly. In general, any part of Eq. (4) may be treated as a free parameter in the inversion, including parts of the matrix $\mathbf{C}$. For the case of two datasets containing N_1 and N_2 data points, and with unknown weights γ_1 and γ_2, we can decompose Eq. (4) as the product of two likelihood functions:

$$p(\mathbf{d}|\mathbf{m}) \propto \frac{1}{\gamma_1^{N_1}} \exp\left(\frac{1}{2\gamma_1^2} (\mathbf{m_1} - \mathbf{d_1})^T \mathbf{C_1}^{-1} (\mathbf{m_1} - \mathbf{d_1}) \right)$$
$$\times \frac{1}{\gamma_2^{N_2}} \exp\left(\frac{1}{2\gamma_2^2} (\mathbf{m_2} - \mathbf{d_2})^T \mathbf{C_2}^{-1} (\mathbf{m_2} - \mathbf{d_2}) \right).$$

$$(5)$$

If γ_1 and γ_2 are then treated as free parameters, the best-fitting values will be those for which the two datasets contribute an equal amount to the overall likelihood function. In addition, the $\chi^2/d.o.f$ (χ^2 per degree of freedom) statistic will be close to unity, ensuring that the reported parameter uncertainties are scaled appropriately (although this limits the use of this statistic for determining which model provides a better fit to the data). This is similar to the method of FUKUDA and JOHNSON (2010), except that we do not require assumptions regarding the analytic form of the likelihood function for linear parameters. For the GPS and InSAR datasets, we found the best-fitting weights in a preliminary inversion (2.3 and 0.7, respectively), then fixed them at these values for all models. In fact, inversions using each dataset independently result in best-fitting fault parameters that are compatible within the uncertainty, so that this relative weighting is not critical to the values of the inferred parameters.

Several parameters in Eqs. (1) and (2) enter in a nonlinear manner, prohibiting a straightforward solution by least squares. Instead, we seek a distribution of models that is proportional to the posterior PDF (3). Sample models were generated using the "slice sampling" Monte Carlo technique, a computationally efficient relative of Gibbs sampling (NEAL 2003; MACKAY 2003). As with all Markov chain Monte Carlo methods, a single random walk may become trapped if there are several widely separated misfit minima, so we combined the results of a large number of walks to form the final distribution. From this distribution we may compute the means, standard deviations, and covariances between the model parameters, as well as any desired marginal posterior PDFs.

3.5. Fault Geometry

Models of this part of the SAF system suggest that, in addition to the SAF and SJF, contributions from

other major faults, such as the Elsinore fault (ELS) and Eastern California Shear zone (ECSZ) need to be accounted for to fit the geodetic data. These faults are subparallel to the main two, and trend northwest at an average azimuth of 310°–320°. In this area, 2D geodetic models assuming infinitely long faults (e.g., FAY and HUMPHREYS 2005; PLATT and BECKER 2010) have inferred slip rates and locking depths in good agreement with more complex 3D models (e.g., MEADE and HAGER 2005; SMITH-KONTER et al. 2011). Our preliminary modeling of the San Jacinto fault geometry suggested that 3D effects such as a gradually changing locking depth along strike may produce a small fault-normal component of motion, but do not produce a measurable change in the inferred locking depth or peak fault-parallel strain rate across the fault, which is our primary concern here.

Therefore, we chose a simplified model with four dislocations in a 2D elastic half-space, as shown in Fig. 1b. Previous work has shown that dislocation models of this type are extremely sensitive to the assumed fault geometry (LINDSEY and FIALKO 2013), so we initially included the horizontal positions ξ_i as free parameters in the model, then fixed them at the best-fitting locations to prevent trade-off with the other parameters. This approach is similar to the iterative procedure adopted by PLATT and BECKER (2010).

In the final model geometry, the best-fitting SAF location is offset by 6 km to the northeast of the surface trace, suggesting that the SAF dips northeast at 50°–60°, in good agreement with earlier modeling results (FIALKO 2006; LINDSEY and FIALKO 2013) as well as with seismic and other geophysical observations (LIN et al. 2007; FUIS et al. 2012). The best-fitting SJF location is offset 2 km to the northeast from its surface trace, suggesting it may also be steeply dipping to the northeast at 80°–85°. This is in good agreement with the asymmetric location of seismicity at depth and with the geometry of inferred seismic waveguide structures (LI and VERNON 2001; LEWIS et al. 2005) as well as with geologic observations at nearby surface exposures of the fault (DOR et al. 2006). Because of the very high strain rate across the SJF and the high density of geodetic data, the model is unusually sensitive to the location of this fault. We found that placing the dislocation directly

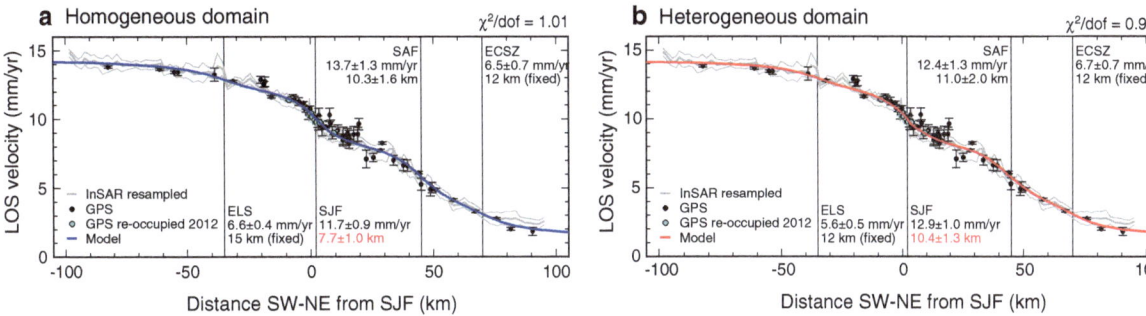

Figure 4
Dislocation model predictions compared to regional geodetic data, projected onto the radar LOS. **a** Best-fitting model in a homogeneous domain. **b** Best-fitting model in the heterogeneous domain shown in Fig. 3a. This case predicts a slightly deeper SJF locking depth, although the two models are visually indistinguishable

below the mapped fault trace can lead to an instability in the Monte Carlo inversion procedure which results in an unreasonably small SJF velocity (<10 mm/year), compensated by a large velocity on the SAF (as high as 30 mm/year, the upper limit of our uniform prior PDF). This is similar to the findings of FUKUDA and JOHNSON (2010), who reported a best-fitting velocity as high as 37 mm/yr on the Coachella segment of the SAF, more than twice the geologic slip rate—a result which we suggest may have been caused by their assumptions regarding the SJF geometry.

3.6. Results

In the case of a homogeneous elastic half-space, the best-fitting model is shown in Fig. 4a. The inferred slip rates are 11.7 ± 0.9 mm/year for the SJF, 13.7 ± 1.3 mm/year for the SAF, 6.6 ± 0.4 mm/year for the Elsinore, and 6.5 ± 0.7 mm/year for the ECSZ. These rates, although not the primary focus of this study, are in excellent agreement with geologic slip rate estimates for each of these faults or fault zones (ROCKWELL *et al.* 1990; PETERSEN and WESNOUSKY 1994; VAN DER WOERD *et al.* 2006; BEHR *et al.* 2010; OSKIN *et al.* 2007). The best-fitting SAF locking depth is 10.3 ± 1.6 km, in good agreement with the depth of seismicity on that fault. The inferred SJF locking depth of 7.7 ± 1.0 km is shallow compared to the 14–18 km maximum depth of seismicity near Anza, although it is in good agreement with previous geodetic models of the area (LISOWSKI *et al.* 1991; BECKER *et al.* 2005; LUNDGREN *et al.* 2009; PLATT and BECKER 2010); the small

locking depth is required to fit the high near-fault strain rate. Although this parameter trades off strongly with the SJF slip rate (correlation −0.92) as well as with the other model parameters (Fig. 5), it appears unlikely that a bias in these values is leading to under-estimation of the SJF locking depth because the other parameters are in good agreement with geologic, seismic, and previous geodetic results.

In the heterogeneous case, we inverted the geodetic data using the Green functions computed for the domain shown in Fig. 3a, using the same fault geometry as in the homogeneous half-space. The results are shown in Fig. 4b; the model fit to the data is visually indistinguishable from the homogeneous case. The best-fitting parameters deviate only slightly from the homogeneous case, except for the SJF locking depth which is increased by 35 % to 10.4 ± 1.3 km, still significantly smaller than the depth of seismicity. This value exceeds that of the homogeneous case by more than two standard deviations, so it is clear that the elastic properties of the crust have a significant effect on this parameter. Part of this increase may be attributed to the layered nature of the rigidity structure, which causes the surface deformation pattern to become narrower compared to the homogeneous case (SAVAGE 1998). For comparison, we conducted an inversion using Green functions computed in a laterally homogeneous, horizontally averaged rigidity structure, in which case the best-fitting SJF locking depth was 8.5 ± 1.1 km. Thus, the layered nature of the rigidity structure accounts for approximately 25 % of the total increase in the inferred locking depth, with the

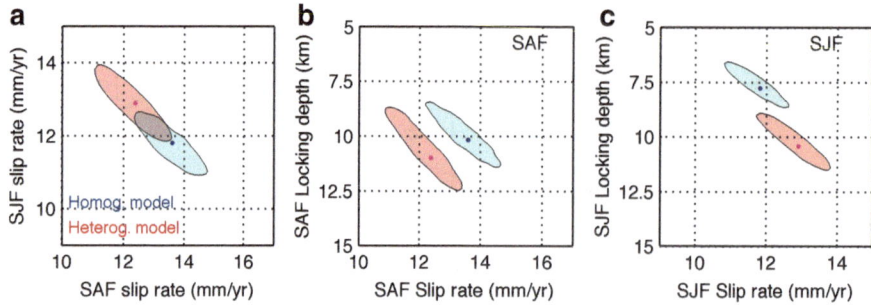

Figure 5
1σ confidence intervals of the two-dimensional marginal posterior probability distributions showing tradeoffs between **a** SAF and SJF slip rates, **b** SAF slip rate and locking depth, and **c** SJF slip rate and locking depth, for a homogeneous (*blue*) and heterogeneous (*red*) domain

remaining 75 % attributed to the presence of the shallow compliant zone.

4. Variation in Fault Locking with Depth

Geodetic locking depths in Southern California have been found to agree with the maximum depth of microseismicity in nearly all cases (SMITH-KONTER et al. 2011; LINDSEY and FIALKO 2013), despite the nonphysical nature of an abrupt transition from locked to sliding in dislocation models. Thus, while the true slip rate likely varies more smoothly with depth, it appears that the transition is typically centered near the bottom of the seismogenic zone, which is the depth where frictional properties transition from unstable velocity-weakening behavior to stable velocity-strengthening.

It has been proposed that the anomalous surface strain rate across the SJF (and small inferred locking depth) represents an upward extension of creep into the nominally unstable region, so that the maximum depth of microseismicity is not a good indicator of the depth of locking (WDOWINSKI 2009). This may be possible if the frictional properties of the fault interface are highly heterogeneous, with a large fraction of the interface undergoing stable creep. The observed microseismicity could then occur within small patches or lineaments of unstable velocity-weakening material, similar to the behavior observed on the creeping section of the SAF near Parkfield (RUBIN et al. 1999; WALDHAUSER et al. 2004), where it occurs at the boundary of the seismogenic zone (BARBOT et al. 2012). A hallmark of this scenario is the presence of repeating earthquakes, a set of seismically similar

events that occur on isolated locked asperities within the creeping fault and have a characteristic magnitude and recurrence interval (NADEAU et al. 1995; NADEAU and Johnson 1998; CHEN and LAPUSTA 2009). Preliminary evidence suggests such events may be occurring at a depth of 12–15 km along the SJF (Taira and Burgmann, pers. comm.), so that low fault coupling within the zone of active microseismicity may be an important contributor to the surface strain rate.

To quantify the effect of creep within the seismogenic zone on the geodetic data, we consider two models that relax the assumption of a sharp transition between locked and freely sliding. As above, we use the Green functions $\mathbf{G}(D)$ determined for the heterogeneous elastic structure shown in Fig. 3a.

Rather than assuming a single locking depth D, a more realistic model would allow the slip rate to increase gradually from zero from the full rate v_0 between depths D_1 and D_2. The surface displacements are given by the superposition of a series of dislocations of varying slip rate $\dot{s}(z)$:

$$v(x) = \sum_{i=0}^{N} \Delta\dot{s}_i \mathbf{G}(D_1 + i\Delta z), \quad N = \frac{D_2 - D_1}{\Delta z}. \quad (6)$$

where we use $\Delta z = 0.1$ km, the depth resolution of the pre-computed Green functions. $\Delta\dot{s}_i$ is the change in slip rate between two neighboring steps,

$$\Delta\dot{s}_i = \dot{s}(D_1 + (i+1)\Delta z) - \dot{s}(D_1 + i\Delta z).$$

We consider two slip rate profiles with depth: linear and elliptical. In the first case, slip rate increases linearly from zero at depth D_1 to v_0 at D_2 so that $\Delta\dot{s} = v_0/(N+1)$, a constant. The elliptical profile is designed to match more closely the predicted

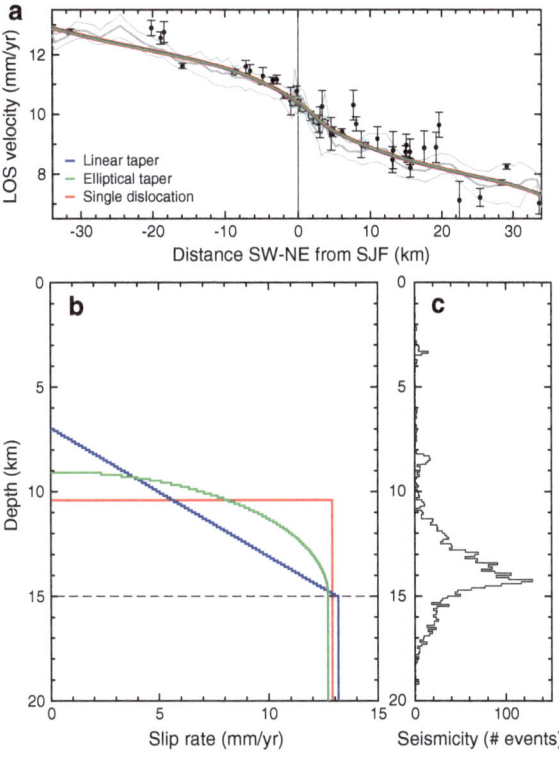

Figure 6

Comparison of three models in a heterogeneous domain. At the surface, the three models are geodetically indistinguishable (**a**), despite different assumptions regarding locking at depth (**b**). For comparison, **c** shows histogram of seismicity occurring within 3 km of the fault and 10 km of the center of the geodetic profile, between 1981 and 2011 (HAUKSSON *et al.* 2012)

profile of slip in numerical models incorporating rate-state friction (e.g., TSE and RICE 1986; LAPUSTA *et al.* 2000), and mimics a nearly constant stressing rate with depth within the transition zone. In this case, $\Delta \dot{s}_i$ is defined by:

$$
\Delta \dot{s}_i = \sqrt{1 - \frac{(D_2 - D_1 - (i+1)\Delta z)^2}{(D_2 - D_1)^2}} - \\
\sqrt{1 - \frac{(D_2 - D_1 - i\Delta z)^2}{(D_2 - D_1)^2}}. \tag{7}
$$

These slip distributions are meant only to illustrate the effect of (unknown) details of slip in the transition zone on the geodetic data and moment accumulation rate. We fix $D_2 = 15$ km in both cases so that, as in the single dislocation model, the models have only two free parameters (v_0 and D_1).

For the linear profile, the best-fitting values are $v_0 = 13.1 \pm 0.8$ mm/year and $D_1 = 7.0 \pm 1.8$ km; for the elliptical profile $v_0 = 12.7 \pm 1.1$ mm/year

and $D_1 = 9.1 \pm 1.6$ km. These results are summarized in Fig. 6. Note that the predicted surface velocities are indistinguishable (Fig. 6a) although both slip rate profiles vs. depth are significantly different from the single dislocation model (Fig. 6b). This is in good agreement with results obtained by KING and WESNOUSKY (2007), who showed that tapered or discrete coseismic slip distributions can produce virtually identical surface deformation.

For each model, we can compare the rate of moment released by creep above 15 km depth to the cumulative seismic moment of earthquakes in the same depth interval. The moment released by creep per unit length along the fault is given by

$$
\frac{\Delta M_{\text{creep}}}{L} = \int_0^{D_2} \mu(z)\, s(z)\, \mathrm{d}z, \tag{8}
$$

where μ is the seismically inferred shear modulus in Fig. 3a, and s is the total modeled slip over a given time period. For the single dislocation model, the moment released by stable creep above 15 km over a 30-year time period would be 6.9×10^{16} N m/km, equivalent to a magnitude 6.0 earthquake over a 20 km-long segment of the fault. For the linear taper of slip with depth, the moment release is 6.3×10^{16} N m/km, and for the elliptical profile it is 6.8×10^{16} N m/km. In comparison, the total moment released by microseismicity recorded within 5 km of the fault, averaged over a 50 km-long segment of the SJF centered at Anza over the same time period was 1.5×10^{16} N m/km, based on the catalog of HAUKSSON et al. (2012). This average includes the high rates of seismic activity in the trifurcation and Hot Springs fault areas to the south and north of the Anza seismic gap, respectively. Even so, the seismic contribution is just 20–25 % of the total moment release required by the slip models, implying that significant frictional heterogeneity is required to support this explanation for the observed strain rate. For comparison, WDOWINSKI (2009) determined a seismic release rate equivalent to <1 mm/year of displacement along the lower 6 km of the central SJF, or 3–4 mm/year in the more seismically active area south of Anza. These rates are similar to our results, which are equivalent to 2–3 mm/year of cumulative displacement in the deep section of the fault.

5. Additional Elastic Modulus Reduction

Alternately, we may consider how large a reduction in shear modulus within a compliant fault zone near the surface could fully account for the strain rate while allowing a locking depth consistent with the depth of seismicity. Because of the computational effort required to evaluate Green functions in a heterogeneous domain, incorporation of the material properties as parameters in the inversion would be prohibitive. Instead, we make use of an analytic solution for the deformation due to a half-elliptical inclusion in an elastic half space (MAHRER 1981). Consider an infinitely long half-elliptical inclusion with depth a, half-width b and shear modulus μ_{in} embedded in a half space with shear modulus μ_0, as shown in Fig. 7a. MAHRER (1981) showed that for the case of simple shear, the strain rate within the ellipse is

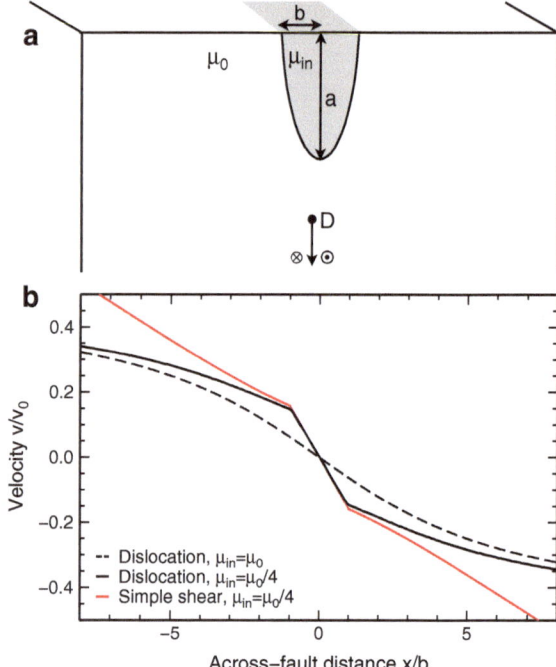

Figure 7

a Geometry of an elliptical inclusion in an elastic half space. **b** Comparison of the analytic solution for an elliptical inclusion under simple shear (*red*) (MAHRER 1981) with the numerical result for a dislocation below the elliptical inclusion (*black*), and a dislocation in a homogeneous elastic half space (*dashed*). Parameters are $a = 3b$, $\mu_{in} = \mu_0/4$, and dislocation locking depth $D = 5b$

$$\dot{\epsilon}_{xy} = \frac{\sigma_0/\mu_0(1 + \phi)}{1 + \phi/r}, \qquad (9)$$

where $\phi = a/b$ and $r = \mu_0/\mu_{in}$. If the locking depth is much greater than the depth of the compliant zone, then the strain rate is nearly constant within the zone and may be approximated by simple shear. The validity of this assumption is demonstrated in Fig. 7b, by comparison of the analytic solution with the numerical result for a buried dislocation beneath the same elliptical inclusion. Although the boundary conditions are different, the strain rate within the inclusion predicted by Eq. (9) is in excellent agreement with the numerical result. Alternative models are possible; JOLIVET et al. (2009) addressed a similar question on the northern SAF using the analytic solution for a dislocation below a rectangular inclusion in a half space. In this case, we feel the elliptical inclusion provides a better approximation to the observed geometry of the fault zone (see Fig. 3a).

Table 2

Required shear modulus ratio for varying compliant fault zone depths, computed using Eq. (10) taking the half-width b = 1.5km and strain enhancement ratio f = 2.4

Compliant zone depth (km)	Aspect ratio φ	μ_0/μ_{in}
3	2	8.0
6	4	3.7
9	6	3.1
12	8	2.9
15	10	2.8
∞	∞	2.4

Relative to a homogeneous half-space, Eq. (9) predicts that the strain rate is enhanced within the compliant zone by a factor $f = (1 + \phi)/(1 + \phi/r)$. Given a measurement of the near-field strain rate enhancement and the dimensions of the inclusion, we can directly solve for the shear modulus ratio:

$$r = \frac{\phi f}{1 + \phi - f}. \qquad (10)$$

In the case of the SJF, the strain rate recorded by the GPS data within the fault zone is 0.60 ± 0.1 μrad/yr. If the fault is locked at 15 km and slipping at a rate of 12 mm/year, the predicted strain rate would be $\epsilon_{xy} = v/\pi D = 0.25$ μrad/yr in a homogeneous half space. The required strain rate enhancement is, therefore, $f = 2.4$, and we take the fault zone half-width $b = 1.5$ km. The remaining unknown parameter is the compliant zone depth a. From Eq. (10) it may be seen that depths <2.1 km, which result in an aspect ratio <1.4, cannot produce the required strain rate even if the shear modulus in the fault zone is zero. We consider a range of larger depths in Table 2; if the zone is infinitely deep, Eq. (10) reduces to the case of a vertical compliant layer (RYBICKI and KASAHARA 1977), in which case a shear modulus reduction by a factor of 2.4 would be required. The tomographic results (Fig. 3a) and theoretical models of damage due to dynamic earthquake ruptures (e.g., KANEKO and FIALKO 2011) do not support a deep low rigidity zone, and for smaller depths the required degree of reduction in shear modulus (a factor of three or more) does not appear to be physically plausible if the material behaves elastically, given the range of properties observed for crustal materials.

6. Discussion

Early geodetic surveys across the San Jacinto fault at Anza suggested an unusually high rate of interseismic strain accumulation across the fault, although these observations were subject to significant uncertainties (LISOWSKI *et al.* 1991). The combination of updated campaign GPS and InSAR data presented here demonstrate unambiguously the presence of a zone of enhanced fault-parallel shear at Anza (Fig. 1c). The maximum shear strain rate determined from GPS is 0.60 ± 0.1 μrad/year across a 3 km-wide zone, asymmetrically offset to the northeast of the fault trace. The InSAR results (MANZO *et al.* 2011) suggest a higher strain rate of 0.79 μrad/year (Fig. 1c). These values are approximately twice the strain rate in the neighboring crust to either side of the fault, giving rise to an unrealistically small predicted locking depth when interpreted with a dislocation model in a homogeneous elastic half space (Fig. 4a). We have considered three potential mechanisms that might explain this behavior: shallow fault creep, a reduced effective elastic modulus within a fault damage zone, or occurrence of steady creep in the lower part of the seismogenic zone.

In the first case, GPS occupations of the Mitchell road alignment array at Anza allow us to rule out shallow, localized fault creep as a contributor to the high gradients in the surface velocity field (Fig. 2). The 20-year timespan of the measurements provides an accuracy of 0.2 mm/year, a significant improvement over earlier results (LOUIE *et al.* 1985). The lack of observable creep stands in contrast with models of earthquake cycle governed by rate- and state-dependent friction laws, which predict that the upper several kilometers of mature strike-slip faults should undergo stable creep in the late interseismic period, even if the material is velocity-weakening (MARONE and SCHOLZ 1988; SCHOLZ 1998; TSE and RICE 1986; LAPUSTA *et al.* 2000; KANEKO *et al.* 2013). One possibility is that the local state of stress is transpressional, so that increased normal stress on the fault may help suppress shallow creep.

Second, the anomalous strain may be attributed to a reduced shear modulus within a compliant fault zone. Zones of this type, extending up to several kilometers from the fault, have been inferred in several cases, based on both geodetic (RYBICKI and KASAHARA 1977; LISOWSKI *et al.* 1991; FIALKO *et al.* 2002; CHEN and FREYMUELLER 2002; FIALKO 2004; HAMIEL and FIALKO 2007; JOLIVET *et al.* 2009; CAKIR *et al.* 2012) and seismic evidence (SPUDICH and OL-SEN 2001; COCHRAN *et al.* 2009; LEWIS *et al.* 2005). Depending on the magnitude of shear modulus reduction and the geometry of the zone, the near-fault strain rate may be significantly increased. A massive low-velocity zone has been imaged along the SJF at Anza by seismic tomography (ALLAM and BEN-ZION 2012). Additionally, the zone of elevated strain rate is asymmetric with respect to the fault (Fig. 1c), consistent with observations of asymmetric damage inferred from geologic observations and LiDAR-derived drainage density maps (DOR *et al.* 2006; WECHSLER *et al.* 2009) and dynamic rupture models which predict asymmetric damage arising from preferred rupture directivity (e.g., BEN-ZION and SHI 2005; AMPUERO and BEN-ZION 2008).

We computed surface velocities due to dislocations in a half-space with elastic properties inferred from seismic tomography (ALLAM and BEN-ZION 2012), and found that the observed shear modulus reduction of 20–40 % does contribute significantly to the near-fault strain rate (Fig. 3), resulting in an increase of the best-fitting locking depth by 35 %, to 10.4 ± 1.3 km (Fig. 4b). However, if the fault is locked to 15 km as expected from the distribution of microseismicity (HAUKSSON *et al.* 2012), the shear modulus would have to be reduced by a factor of 2.4 or more within the compliant fault zone (Table 2) to explain the geodetic data, a contrast that is too high given available seismic observations.

One mechanism that might contribute to the enhanced strain rate is a reduced yield strength within the fault zone, leading to unrecoverable plastic deformation, or distributed slip on a large number of subparallel fault strands. In addition to the seismically observed compliant zone, such yielding could result in a much smaller apparent shear modulus from the geodetic perspective. Some form of inelastic yielding has been suggested as an explanation for the shallow slip deficit observed during large (M ~7) strike-slip earthquakes in southern California as well as other regions (SIMONS

et al. 2002; FIALKO *et al.* 2005; KANEKO and FIALKO 2011). If this yielding persists throughout much of the interseismic period, the accumulation of unrecoverable strain could have significant implications for geologic slip rates, the state of crustal stress, and consequently the seismic hazard. Alternately, yielding might occur only toward the end of the interseismic period when the highest stresses are reached, in which case it would account for a comparatively small amount of the total strain. Addressing this question will require geodetic data of comparable density and accuracy across a number of faults of different maturity and at different points in the seismic cycle, and an improved knowledge of the effects of both coseismic damage and interseismic healing on the material yield strength over time.

Another possibility, proposed by WDOWINSKI (2009), is that the elevated surface strain rate arises from deep interseismic creep extending into the nominally unstable zone above 15 km depth. Such an upward extension of creep is predicted by rate and state frictional models late in the earthquake cycle, if the critical slip-weakening distance is assumed to be much larger than laboratory estimates suggest (e.g., TSE and RICE 1986; LAPUSTA *et al.* 2000). In these models, accelerated aseismic creep propagates into the velocity-weakening layer until the conditions for nucleation of seismic rupture are met (SCHOLZ 1998). In some cases, models with a large slip-weakening distance predict creep within the seismogenic zone at a significant fraction of the interseismic rate (LAPUSTA and RICE 2003). However, SAVAGE (2006) found that the maximum surface strain rate in frictionally controlled models of deep slip should be less than that predicted by a dislocation model that is fully locked within the seismogenic zone, even for amplitudes of creep within the deep part of the seismogenic zone reaching one-third of the total fault slip rate. In such cases, the slip-weakening distance is large enough that the resulting nucleation size required for seismic rupture would prohibit the occurrence of microseismicity on the fault plane, unless there is significant frictional heterogeneity.

The presence of frictional heterogeneity on this segment of the SJF could allow for both active microseismicity and large areas of stable creep, possibly at a rate high enough to impact the surface strain rate. However, it should also lead to qualitatively different microseismic behavior such as repeating earthquakes and well defined spatial patterns of seismicity, such as observed along the SAF near Parkfield (NADEAU et al. 1995; RUBIN et al. 1999; WALDHAUSER et al. 2004). At present, microseismic behavior of this type has not been conclusively observed, although preliminary studies suggest that repeating events may in fact be occurring (Taira and Burgmann, pers. comm.), so this explanation for the elevated strain rate may merit further consideration.

To constrain better the rate and depth distribution of deep creep under this interpretation, we considered two tapered slip rate profiles in addition to the simplified single dislocation model, shown in Fig. 6. The sensitivity of the geodetic data to the details of the slip distribution at the base of the seismogenic zone is nearly zero, although the total moment release rate is well constrained. Thus, all of the models require significant ongoing moment release above the apparent frictional stability transition at ~ 15 km at a rate equivalent to 4–6 times the total moment released as microseismicity within this zone over the past 30 years. The effective aseismic moment released by such creep would be equivalent to a magnitude 6.0 earthquake every 30 years within a 20 km-long segment. Under this interpretation, the best-fitting geodetic locking depth of 10.4 ± 1.3 km reflects an accurate measure of the depth of strain accumulation and potential future moment release that is significantly different from the depth inferred from seismic observations alone.

Finally, we note that uncertainties in the fault geometry and the presence of other active faults may further modify the results. For example, the Earthquake Valley fault located 25 km southwest of the SJF is not typically considered in geodetic models of the region, but recent geologic evidence suggests that it may slip at a long-term rate of 2–3 mm/year (ROCKWELL et al. 2013). If one includes this fault in the above models, the best-fitting SJF locking depth increases to 11.1 ± 0.8 km. This would slightly reduce the potential contributions of inelastic yielding and/or heterogeneous frictional properties discussed above.

7. Conclusions

New geodetic observations confirm a relatively narrow (2–3 km wide) zone with an elevated rate of shear strain across the San Jacinto fault at Anza. We have shown that a reduced elastic shear modulus within the fault zone as inferred by seismic tomography can explain a significant part of the elevated strain rate but cannot fully account for the small locking depths inferred using dislocation models. We ruled out the presence of shallow creep via the re-occupation with GPS of an alignment array installed at Anza in 1990. There are two remaining interpretations of the elevated strain rate: additional elastic or unrecoverable inelastic yielding occurring within the shallow fault zone, or significant stable creep within the deep part of the seismogenic zone. The first explanation, if confirmed, would have significant implications for geologically inferred slip rates and hazard assessment on this and other mature crustal faults, as yielding during the interseismic period would reduce the cumulative coseismic offsets expressed on the fault trace. Such behavior would also affect the state of stress in the crust and the subsequent coseismic release of strain energy. This hypothesis predicts that inelastic deformation and an increased shear strain rate should be geodetically observable in the vicinity of other active faults which are approaching the end of the interseismic period. Alternately, stable creep within the deep part of the fault zone could explain the geodetic strain rate, but requires the presence of significant frictional heterogencity. Such creep, if present, would represent approximately four to six times more moment release than the observed microseismicity, indicating that the deep part of the seismogenic zone is predominantly velocity-strengthening.

Acknowledgments

This work was supported by NSF (grant EAR-0908042) and SCEC. The authors would like to acknowledge Duncan Agnew, Peng Fang and Robert King for their invaluable assistance with the GPS data processing. The original installment of the survey monuments was supported by NSF grant EAR-9104810.

REFERENCES

ALLAM AA, BEN-ZION Y (2012) *Seismic velocity structures in the southern california plate-boundary environment from double-difference tomography.* Geophys J Int 190(2):1181–1196. doi:10.1111/j.1365-246X.2012.05544.x

AMPUERO JP, BEN-ZION Y (2008) *Cracks, pulses and macroscopic asymmetry of dynamic rupture on a bimaterial interface with velocity-weakening friction.* Geophys J Int 173:674–692, doi:10.1111/j.1365-246X.2008.03736.x

BARBOT S, FIALKO Y, SANDWELL D (2009) *Three-dimensional models of elasto-static deformation in heterogeneous media, with applications to the Eastern California Shear Zone.* Geophys J Int 179:500–520

BARBOT S, LAPUSTA N, AVOUAC J-P (2012) *Under the Hood of the Earthquake Machine: Toward Predictive Modeling of the Seismic Cycle.* Science 336:707–710, doi:10.1126/science.1218796

BECKER TW, HARDEBECK JL, ANDERSON G (2005) *Constraints on fault slip rates of the Southern California plate boundary from GPS velocity and stress inversions.* Geophys J Int 160:634–650, doi:10.1111/j.1365-246X.2004.02528.x

BEHR WM, ROOD DH, FLETCHER KE, GUZMAN N, FINKEL R, HANKS TC, HUDNUT KW, KENDRICK KJ, PLATT JP, SHARP WD, J WR, YULE RJ (2010) *Uncertainties in slip-rate estimates for the Mission Creek strand of the southern San Andreas fault at Biskra Palms Oasis, southern California.* Geol Soc Am Bull 122:1360–1377, doi:10.1130/B30020.1

BEN-ZION Y, PENG Z, OKAYA D, SEEBER L, ARMBRUSTER J, OZER N, MICHAEL A, BARIS S, AKTAR M, KUWAHARA Y, ITO H (2003) *A shallow fault-zone structure illuminated by trapped waves in the Karadere-Duzce branch of the North Anatolian Fault, western Turkey.* Geophys J Int 152:699–699, doi:10.1046/j.1365-246X.2003.01870.x

BEN-ZION Y, SHI Z (2005) *Dynamic rupture on a material interface with spontaneous generation of plastic strain in the bulk.* Earth Planet Sci Lett 236:486–496, doi:10.1016/j.epsl.2005.03.025

BLISNIUK K, ROCKWELL T, OWEN LA, OSKIN M, LIPPINCOTT C, CAFFEE MW, DORTCH J (2010) *Late Quaternary slip rate gradient defined using high-resolution topography and* [10]*Be dating of offset landforms on the southern San Jacinto Fault zone, California.* J Geophys Res 115:B08401, doi:10.1029/2009JB006346

BOCK Y, NIKOLAIDIS RM, DE JONGE PJ (2000) *Instantaneous geodetic positioning at medium distances with the Global Positioning System.* J Geophys Res 105:28,223–28,253

BROCHER TM (2005) *Empirical relations between elastic wavespeeds and density in the earth's crust.* Bulletin of the Seismological Society of America 95(6), doi:10.1785/0120050077

CAKIR Z, ERGINTAV S, OZENER H, DOGAN U, AKOGLU AM, MEGHRAOUI M, REILINGER R (2012) *Onset of aseismic creep on major strike-slip faults.* Geology 40:1115–1118, doi:10.1130/G33522.1

CHEN Q, FREYMUELLER J (2002) *Geodetic evidence for a near-fault compliant zone along the San Andreas fault in the San Francisco Bay area.* Bull Seismol Soc Am 92:656–671

CHEN T, LAPUSTA N (2009) *Scaling of small repeating earthquakes explained by interaction of seismic and aseismic slip in a rate and state fault model.* J Geophys Res 114:B01311. doi:10.1029/2008JB005749

CLEVELAND WS (1979) *Robust locally weighted regression and smoothing scatterplots.* J Am Statist Assoc 74:829–836, doi:10.2307/2286407

CLEVELAND WS, DEVLIN SJ (1988) *Locally-weighted regression: An approach to regression analysis by local fitting.* J Am Statist Assoc 83:596–610, doi:10.2307/2289282

COCHRAN ES, LI YG, SHEARER PM, BARBOT S, FIALKO Y, VIDALE JE (2009) *Seismic and geodetic evidence for extensive, long-lived fault damage zones.* Geology 37:315–318, doi:10.1130/G25306A.1

DOR O, ROCKWELL TK, BEN-ZION Y (2006) *Geological observations of damage asymmetry in the structure of the san jacinto, san andreas and punchbowl faults in southern california: A possible indicator for preferred rupture propagation direction.* Pure appl geophys. 163:301–349, doi:10.1007/s00024-005-0023-9

FAY N, HUMPHREYS G (2005) *Fault slip rates, effects of elastic heterogeneity on geodetic data, and the strength of the lower crust in the Salton Trough region, southern California.* J Geophys Res 110:B09401, doi:10.1029/2004JB003548

FIALKO Y (2004) *Probing the mechanical properties of seismically active crust with space geodesy: Study of the co-seismic deformation due to the 1992 M_w7.3 Landers (southern California) earthquake.* J Geophys Res 109:B03307, doi:10.1029/2003JB002756

FIALKO Y (2006) *Interseismic strain accumulation and the earthquake potential on the southern San Andreas fault system.* Nature 441:968–971

FIALKO Y, SANDWELL D, AGNEW D, SIMONS M, SHEARER P, MINSTER B (2002) *Deformation on nearby faults induced by the 1999 Hector Mine earthquake.* Science 297:1858–1862, doi:10.1126/science.1074671

FIALKO Y, SANDWELL D, SIMONS M, ROSEN P (2005) *Three-dimensional deformation caused by the Bam, Iran, earthquake and the origin of shallow slip deficit.* Nature 435:295–299

FUIS GS, SCHEIRER DS, LANGENHEIM VE, KOHLER MD (2012) *A new perspective on the geometry of the san andreas fault in southern california and its relationship to lithospheric structure.* Bull Seism Soc Am 102:236–251

FUKUDA J, JOHNSON KM (2010) *Mixed linearnon-linear inversion of crustal deformation data: Bayesian inference of model, weighting and regularization parameters.* Geophys J Int 181:1441–1458, doi:10.1111/j.1365-246X.2010.04564.x

GOULTY NR, BURFORD RO, ALLEN CR, GILMAN R, JOHNSON CE, KELLER RP (1978) *Large creep events on the Imperial Fault, California.* Bull Seism Soc Am 68:517–521

HAMIEL Y, FIALKO Y (2007) *Structure and mechanical properties of faults in the North Anatolian Fault system from InSAR observations of coseismic deformation due to the 1999 Izmit (Turkey) earthquake.* J Geophys Res 112:B07,412, doi:10.1029/2006JB004777

HAUKSSON E, YANG W, SHEARER PM (2012) *Waveform relocated earthquake catalog for southern california (1981 to 2011).* Bull Seism Soc Am 102(5):2239–2244, doi:10.1785/0120120010

HERRING TA, KING RW, MCCLUSKY SC (2010) Introduction to GAMIT/GLOBK. Massachusetts Institute of Technology, Cambridge, Massachusetts

JOHNSON HO, AGNEW DC, WYATT FK (1994) *Present-day crustal deformation in southern California.* J Geophys Res 99:23951–23974

JOLIVET R, BURGMANN R, HOULIE N (2009) *Geodetic exploration of the elastic properties across and within the northern San Andreas Fault zone.* Earth Planet Sci Lett 288:126–131, doi:10.1016/j.epsl.2009.09.014

KANEKO Y, FIALKO Y (2011) *Shallow slip deficit due to large strike-slip earthquakes in dynamic rupture simulations with elasto-plastic off-fault response.* Geophys J Int 186:1389–1403, doi:10.1111/j.1365-246X.2011.05117.x

KANEKO Y, FIALKO Y, SANDWELL DT, TONG X, FURUYA M (2013) *Interseismic deformation and creep along the central section of the North Anatolian fault (Turkey): InSAR observations and implications for rate-and-state friction properties.* J Geophys Res 118:316–331, doi:10.1029/2012JB009661

KELLER RP, ALLEN CR, GILMAN R, GOULTY NR, HILEMAN JA (1978) *Monitoring slip along major faults in Southern California.* Bull Seism Soc Am 68(4):1187–1190

KING GCP, WESNOUSKY SG (2007) *Scaling of fault parameters for continental strike-slip earthquakes.* Bull Seism Soc Am 97(6):1833–1840, doi:10.1785/0120070048

LAPUSTA N, RICE J, BEN-ZION Y, ZHENG G (2000) *Elastodynamic analysis for slow tectonic loading with spontaneous rupture episodes on faults with rate- and state-dependent friction.* J Geophys Res 105(B10):23,765–23,789, doi:10.1029/2000JB900250

LAPUSTA N, RICE J (2003) *Nucleation and early seismic propagation of small and large events in a crustal earthquake model.* J Geophys Res 108(B4):2205, doi:10.1029/2001JB000793

LEWIS MA, PENG Z, BEN-ZION Y, VERNON FL (2005) *Shallow seismic trapping structure in the san jacinto fault zone near anza, california.* Geophys J Int 162:867–881, doi:10.1111/j.1365-246X.2005.02684.x

LI Y, VERNON FL (2001) *Characterization of the san jacinto fault zone near anza, california, by fault zone trapped waves.* J Geophys Res 106:30,671–30,688, doi:10.1029/2000JB000107

LIN G, SHEARER PM, HAUKSSON E (2007) *Applying a three-dimensional velocity model, waveform cross correlation, and cluster analysis to locate southern California seismicity from 1981 to 2005.* J Geophys Res 112:B12309, doi:10.1029/2007JB004986

LINDSEY EO, FIALKO Y (2013) *Geodetic slip rates in the southern San Andreas fault system: effects of elastic heterogeneity and fault geometry.* J Geophys Res 118, doi:10.1029/2012JB009358

LISOWSKI M, SAVAGE J, PRESCOTT WH (1991) *The velocity field along the San Andreas fault in central and southern California.* J Geophys Res 96:8369–8389

LOHMAN RB, SIMONS M (2005) *Some thoughts on the use of InSAR data to constrain models of surface deformation: Noise structure and data downsampling.* Geochem Geophys Geosyst 6, doi:10.1029/2004GC000841

LOUIE JN, ALLEN CR, JOHNSON DC, HAASE PC, COHN SN (1985) *Fault slip in Southern California.* Bulletin of the Seismological Society of America 75(3):811–833

LUDWIG WJ, NAFE JE, DRAKE CL (1970) Seismic refraction. In: Maxwell AE (ed) The Sea, vol 4, Wiley-Interscience, New York, pp 53–84

LUNDGREN PE, HETLAND A, LIU Z, FIELDING EJ (2009) *Southern San Andreas–San Jacinto fault system slip rates estimated from earthquake cycle models constrained by GPS and interferometric synthetic aperture radar observations.* J Geophys Res 114:B02403, doi:10.1029/2008JB005996

LYONS S, SANDWELL D (2003) *Fault creep along the southern San Andreas from interferometric synthetic aperture radar, permanent scatterers, and stacking.* J Geophys Res 108 doi:10:1029/2002JB001,831

LYONS SN, BOCK Y, SANDWELL DT (2002) *Creep along the Imperial fault, Southern California, from GPS measurements.* J Geophys Res 107, doi:10.1029/2001JB000763

MACKAY DJC (2003) Information Theory, Inference, and Learning Algorithms. Cambridge University Press, Cambridge, UK

MAHRER KD (1981) *Surface deformation in a crustal setting with a long-surfaced non-uniformity.* Tectonophys 76:T1–T11

MANZO M, FIALKO Y, CASU F, PEPE A, LANARI R (2011) *A quantitative assessment of DInSAR measurements of interseismic deformation: the Southern San Andreas Fault case study.* Pure Appl Geophys doi:10.1007/s00024-011-0403-2

MARONE C (1998) *Laboratory-derived friction laws and their application to seismic faulting.* Annu Rev Earth Planet Sci 26:643–696

MARONE C, SCHOLZ CH (1988) *The depth of seismic faulting and the upper transition from stable to unstable slip regimes.* Geophys Res Lett 15:621–624

MEADE, BJ and HAGER, BH (2005) *Block models of crustal motion in southern California constrained by GPS measurements.* J Geophys Res 110, doi:10.1029/2004JB003209

MINSON SE, SIMONS M, BECK JL (2013) *Bayesian inversion for finite fault earthquake source models Itheory and algorithm.* Geophys J Int 133:568–584, doi:10.1093/gji/ggt180

NADEAU R, JOHNSON L (1998) *Seismological studies at Parkfield VI: Moment release rates and estimates of source parameters for small repeating earthquakes.* Bull Seism Soc Am 88:790–814

NADEAU RM, FOXALL W, MCEVILLY TV (1995) *Clustering and periodic recurrence of microearthquakes on the San Andreas fault at Parkfield, California.* Science 267:503–507, doi:10.1126/science.267.5197.503

NEAL RM (2003) *Slice sampling.* Ann Stat 31:705–767

NUR A, MAVKO J (1974) *Postseismic viscoelastic rebound.* Science 183:204–206

OSKIN M, PERG L, BLUMENTRITT D, MUKHOPADHYAY S, IRIONDO A (2007) *Slip rate of the Calico fault: Implications for geologic versus geodetic rate discrepancy in the Eastern California Shear Zone.* J Geophys Res 112:B03,402, doi:10.1029/2006JB004451

PETERSEN MD, WESNOUSKY SG (1994) *Fault slip rates and earthquake histories for active faults in Southern California.* Bull Seism Soc Am 84:1608–1649

PLATT JP, BECKER TW (2010) *Where is the real transform boundary in California?* Geochemistry, Geophysics, Geosystems 11, doi:10.1029/2010GC003060

ROCKWELL T, LOUGHMAN C, MERIFIELD P (1990) *Late quaternary rate of slip along the San-Jacinto fault zone near Anza, Southern California.* J Geophys Res 95:8593–8605

ROCKWELL TK, MAGISTRALE H, HARADEN C, HIRABAYASHI CK (1992) *Emplacement of alignment arrays along the Elsinore and San Jacinto fault zones, Southern California.* USGS Open File Rep pp Grant No. 14–08–0001–G1771

ROCKWELL TK, SEITZ G, DAWSON T, YOUNG J (2006) *The long record of san jacinto fault paleoearthquakes at hog lake: Implications for regional patterns of strain release in the southern san andreas fault system.* Seismol Res Lett 77(2):270

ROCKWELL TK, AKCIZ SO, GORDON E (2013) *Paleoseismology of the Aqua Tibia - Earthquake Valley fault, eastern strand of the Elsinore fault zone.* Seismol Res Lett 84(2):332

RUBIN AM, GILLARD D, GOT JL (1999) *Streaks of microearthquakes along creeping faults.* Nature 400:635–641, doi:10.1038/23196

RYBICKI K, KASAHARA K (1977) *A strike-slip fault in a laterally inhomogeneous medium.* Tectonophysics 42:127–138

RYMER M (2000) *Triggered surface slips in the Coachella Valley area associated with the 1992 Joshua Tree and Landers, California, earthquakes.* Bull Seism Soc Am 90:832–848

RYMER MJ, BOATWRIGHT J, SEEKINS LC, YULE JD, LIU J (2002) *Triggered surface slips in the Salton Trough associated with the*

1999 Hector Mine, California, Earthquake. Bull Seism Soc Am 92:1300–1317

SANDERS CO (1990) *Earthquake depths and the relation to strain accumulation and stress near strike-slip faults in Southern California*. J Geophys Res 95:4571–4762

SANDERS CO, KANAMORI H (1984) *A seismotectonic analysis of the Anza seismic gap, San Jacinto fault zone, Southern California*. J Geophys Res 89:5873–5890

SAVAGE J (1998) *Displacement field for an edge dislocation in a layered half-space*. J Geophys Res 103:2439–2446

SAVAGE J, BURFORD R (1973) *Geodetic determination of relative plate motion in central California*. J Geophys Res 78:832–845

SAVAGE JC (1990) *Equivalent strike-slip earthquake cycles in half-space and lithosphere-asthenosphere earth models*. J Geophys Res 95:4873–4879

SAVAGE JC (2006) *Dislocation pileup as a representation of strain accumulation on a strike-slip fault*. J Geophys Res 111:B04,405, doi:10.1029/2005JB004021

SAVAGE JC, PRESCOTT WH (1978) *Asthenosphere readjustment and the earthquake cycle*. J Geophys Res 83:3369–3376

SCHOLZ CH (1998) *Earthquakes and friction laws*. Nature 391:37–42, doi:10.1038/34097

SHARP R, RYMER M, LIENKAEMPER J (1986) *Surface displacement on the Imperial and Superstition Hills faults triggered by the Westmoreland, California, earthquake of 26 April 1981*. Bull Seism Soc Am 76:949–965

SHEN ZK, KING RW, AGNEW DC, WANG M, HERRING TA, DONG D, FANG P (2011) *A unified analysis of crustal motion in Southern California, 1970–2004: The SCEC crustal motion map*. Journal of Geophysical Research 116(B11):1–19, doi:10.1029/2011JB008549. http://www.agu.org/pubs/crossref/2011/2011JB008549.shtml

SIEH K, WILLIAMS P (1990) *Behavior of the southernmost San Andreas fault during the past 300 years*. J Geophys Res 95:6629–6645

SIMONS M, FIALKO Y, RIVERA L (2002) *Coseismic deformation from the 1999 Mw7.1 Hector Mine, California, earthquake, as inferred from InSAR and GPS observations*. Bull Seism Soc Am 92:1390–1402

SMITH-KONTER BR, SANDWELL DT, SHEARER P (2011) *Locking depths estimated from geodesy and seismology along the san andreas fault system: Implications for seismic moment release*. J Geophys Res 116(B6):1–12, doi:10.1029/2010JB008117. http://www.agu.org/pubs/crossref/2011/2010JB008117.shtml

SPUDICH P, OLSEN K (2001) *Fault zone amplified waves as a possible seismic hazard along the Calaveras fault in Central California*. Geophys Res Lett 28:2533–2536

SUDHAUS H, JONSSON S (2008) *Improved source modelling through combined use of InSAR and GPS under consideration of correlated data errors: application to the June 2000 Kleifarvatn earthquake, Iceland*. Geophys J Int 176:389–404, doi:10.1111/j.1365-246X.2008.03989.x

TAKEUCHI C, FIALKO Y (2012) *Dynamic models of interseismic deformation and stress transfer from plate motion to continental transform faults*. J Geophys Res 117:B05403, doi:10.1029/2011JB009056

TARANTOLA A (2005) Inverse Problem Theory and Methods for Model Parameter Estimation. Society for Industrial and Applied Mathematics, Philadelphia, doi:10.1137/1.9780898717921

THATCHER W, HILEMAN JA, HANKS TC (1975) *Seismic slip distribution along the san jacinto fault zone, southern california and its implications*. Geological Society of America Bulletin 86:1140–1146

TONG X, SANDWELL DT, SMITH-KONTER B (2013) *High-resolution interseismic velocity data along the San Andreas Fault from GPS and InSAR*. J Geophys Res 118, doi:10.1029/2012JB009442

TSE ST, RICE JR (1986) *Crustal earthquake instability in relation to the depth variation of frictional slip properties*. J Geophys Res 91:9452–9472

VAN DER WOERD J, KLINGER Y, SIEH K, TAPPONNIER P, RYERSON F, MERIAUX A (2006) *Long-term slip rate of the southern San Andreas Fault from ^{10}Be-^{26}Al surface exposure dating of an offset alluvial fan*. J Geophys Res 111:B04407, doi:10.1029/2004JB003559

WALDHAUSER F, ELLSWORTH WL, SCHAFF DP, COLE A (2004) *Streaks, multiplets, and holes: High-resolution spatio-temporal behavior of Parkfield seismicity*. Geophys Res Lett 31, doi:10.1029/2004GL020649

WDOWINSKI S (2009) *Deep creep as a cause for the excess seismicity along the San Jacinto fault*. Nature Geoscience 2(12):882–885, doi:10.1038/ngeo684

WECHSLER N, ROCKWELL TK, BEN-ZION Y (2009) *Application of high resolution DEM data to detect rock damage from geomorphic signals along the central San Jacinto Fault*. Geomorphology 113:82–96, doi:10.1016/j.geomorph.2009.06.007

WEI M, SANDWELL D, FIALKO Y (2009) *A silent M4.8 slip event of October 3-6, 2006, on the Superstition Hills fault, Southern California*. J Geophys Res 114:B07402, doi:10.1029/2008JB006135

WEI M, SANDWELL D, SMITH-KONTER B (2010) *Optimal combination of InSAR and GPS for measuring interseismic crustal deformation*. Adv Space Res 46:236–249

WEI M, SANDWELL D, FIALKO Y, BILHAM R (2011) *Slip on faults in the Imperial Valley triggered by the 4 April 2010 Mw 7.2 El Mayor-Cucapah earthquake revealed by InSAR*. Geophys Res Lett 38:L01308, doi:10.1029/2010GL045235

WILLIAMS SDP, BOCK Y, FANG P, JAMASON P, NIKOLAIDIS RM, PRAWIRODIRDJO L, MILLER M, JOHNSON J (2004) *Error analysis of continuous GPS position time series*. J Geophys Res 109:B03412, doi:10.1029/2003JB002741

(Received August 17, 2013, revised November 24, 2013, accepted November 26, 2013, Published online December 22, 2013)

Pure Appl. Geophys. 171 (2014), 2955–2965
© 2014 Springer Basel
DOI 10.1007/s00024-014-0783-1

| Pure and Applied Geophysics

Large Earthquake Hazard of the San Jacinto Fault Zone, CA, from Long Record of Simulated Seismicity Assimilating the Available Instrumental and Paleoseismic Data

G. Zöller[1] and Y. Ben-Zion[2]

Abstract—We investigate spatio-temporal properties of earthquake patterns in the San Jacinto fault zone (SJFZ), California, between Cajon Pass and the Superstition Hill Fault, using a long record of simulated seismicity constrained by available seismological and geological data. The model provides an effective realization of a large segmented strike-slip fault zone in a 3D elastic half-space, with heterogeneous distribution of static friction chosen to represent several clear step-overs at the surface. The simulated synthetic catalog reproduces well the basic statistical features of the instrumental seismicity recorded at the SJFZ area since 1981. The model also produces events larger than those included in the short instrumental record, consistent with paleo-earthquakes documented at sites along the SJFZ for the last 1,400 years. The general agreement between the synthetic and observed data allows us to address with the long-simulated seismicity questions related to large earthquakes and expected seismic hazard. The interaction between $m \geq 7$ events on different sections of the SJFZ is found to be close to random. The hazard associated with $m \geq 7$ events on the SJFZ increases significantly if the long record of simulated seismicity is taken into account. The model simulations indicate that the recent increased number of observed intermediate SJFZ earthquakes is a robust statistical feature heralding the occurrence of $m \geq 7$ earthquakes. The hypocenters of the $m \geq 5$ events in the simulation results move progressively towards the hypocenter of the upcoming $m \geq 7$ earthquake.

Key words: Earthquake dynamics, Earthquake interaction, forecasting, and prediction, Statistical seismology, Seismicity and tectonics.

1. Introduction

The San Jacinto Fault Zone (SJFZ) is part of the San Andreas system in southern California and is located between the Elsinore Fault to the SW and the San Andreas fault to the NE. The total length is around 230 km and the faulting is associated primarily with right-lateral motion, with slip rates in the range 10–20 mm/year (ROCKWELL et al. 1990; FIALKO 2006). The seismicity along the SJFZ is characterized by strong variability along strike. An accurate relocated earthquake catalog from 1 January 1981 until 30 June 2011 includes earthquakes with magnitudes up to $m = 5.4$ (HAUKSSON et al. 2012). The largest recorded earthquake in the study area is the M6.6 event on 9 April 1968 on the Coyote Creek segment. Other historic earthquakes dating back to the nineteenth century have approximately similar size. The relations of WELLS and COPPERSMITH (1994) imply that earthquakes with magnitudes around $m = 7.5$ should be possible if most of the fault takes part in a rupture. The lack of such an event in the instrumental data may simply reflect recurrence intervals larger than the observational period. Other explanations include a general inability of system-size ruptures on the SJFZ due to step-over regions that cannot be jumped. Records of paleo-earthquakes at the Hog Lake and Mystic Lake sites over the last 1,400 years support the occurrence of large earthquakes on the SJFZ (ONDERDONK et al. 2013; ROCKWELL et al. 2014), although the magnitude of these events is not clear. In particular, ONDERDONK et al. (2013) report evidence for seven large events with recurrence times between 160 and 210 years from paleoseismological studies of the Mystic Lake site; some of them fall within quiet periods of the Hog Lake site (ROCKWELL et al. 2006, 2014). The observation that at least three are similar in age between the two sites suggests that the rupture may have jumped the step-over at the San Jacinto Valley. Because the last paleo-earthquake occurred about 200 years ago, the next large event is likely overdue.

---
[1] Institute of Mathematics, University of Potsdam, 14469 Potsdam, Germany. E-mail: zoeller@uni-potsdam.de

[2] Department of Earth Sciences, University of Southern California, Los Angeles, CA 90089-0740, USA. E-mail: benzion@usc.edu

To provide additional understanding on the possibility that the SJFZ is capable of producing large events, and the resulting consequences for the current seismic hazard, we use simulated seismicity based on a model that is adjusted to reproduce key available observations. The model accounts for earthquake-slip, aseismic creep, strong heterogeneities, evolving threshold dynamics, and long-range interactions (BEN-ZION 1996, 2012; ZÖLLER et al. 2005). The model input includes dimensions of the computational grid on which seismicity is simulated, boundary conditions, locations of major stepovers, average local stress drops during brittle instabilities, and creep coefficients that increase with depth and toward the ends of the computational domain. The constitutive parameters are chosen to account for large paleo-earthquakes between (and sometimes across) the major stepovers, and reproduce the space-time-magnitude statistical properties of the instrumental seismicity ($m \leq 5.4$). Model simulation results for several thousands of years allow us to overcome the problem of a short observational period and provide insights on the likely statistical properties of large earthquakes ($m \gg 5.4$).

The general approach of generating synthetic seismicity to overcome problems of sparse data and large statistical uncertainties became popular in recent years (POLLITZ 2011; RICHARDS-DINGER and DIETERICH 2012; ROBINSON et al. 2011; TULLIS et al. 2012). While it is possible to add various ingredients to the model, such as a network of multiple faults and more complicated rheologies, the added degrees of freedom are not necessarily well-constrained. The relatively simple model used in this work can produce a wide range of realistic phenomena (e.g., space-time distributions of slip and hypocenters, different types of frequency–size and temporal event statistics, scaling of source time functions, and more), and reproduce successfully various observed features of seismicity on large fault zones (BEN-ZION et al. 2003; MEHTA et al. 2006; ZÖLLER et al. 2008). The relative simplicity of the model allows us to assimilate efficiently the available information from instrumental and paleoseismic data on the SJFZ, and to derive related statistical estimates of the large-earthquake hazard on the fault.

2. Study Area and Model

The study area with the distribution of earthquakes and other features of interest is shown in Fig. 1a. We consider the portion of the San Jacinto Fault Zone between Cajon Pass and the Superstition Hill Fault, with a length of about 150 km. This choice covers the fault portion where significant paleoseimic data are available and future large earthquakes are generally expected. The investigated region includes two major step-over regions: the first at San Jacinto Valley and the second at Coyote Ridge. The Mystic Lake and Hog Lake sites of paleoseismological studies (ONDERDONK et al. 2013; ROCKWELL et al. 2006, 2014) are also shown.

The model provides an effective realization (BEN-ZION 1996; ZÖLLER et al. 2007) of a large, segmented strike-slip fault zone of 150-km length and 25-km depth, covered by a computational grid with 128×21 cells in a 3D elastic half-space (Fig. 1b). The boundary conditions consist of constant velocity motion $v_{pl} = 14$ mm/year around the fault (FIALKO 2006). It is easy to add to the boundary conditions fluctuations associated with the occurrence of large earthquakes outside the computational grid (BEN-ZION and RICE 1993). However, we assume that such fluctuations have transient effects that are not essential for the large spatio-temporal earthquake patterns that are the focus of this work. The model dynamics are governed by static-kinetic friction and power law creep, as described in BEN-ZION (1996) and ZÖLLER et al. (2005), with heterogeneous distributions of static/kinetic friction and creep properties. The creep rates are assumed to increase toward the border of the computational grid. The idealized boundary conditions at the lateral model edges prevent us from discussing detailed results near the model edges, but this is not expected to significantly affect the large-scale patterns along most of the computational grid.

The changing thickness of the seismogenic crust reflected in the instrumental catalog (HAUKSSON et al. 2012) is modeled by varying the brittle-ductile transition depths from 18 km near Cajon Pass at the NW end of the study area to 10 km near the SE end of the fault. Details on simulating the transition from brittle to ductile deformation are given in BEN-ZION (1996). The major stepovers are accounted for by quenched

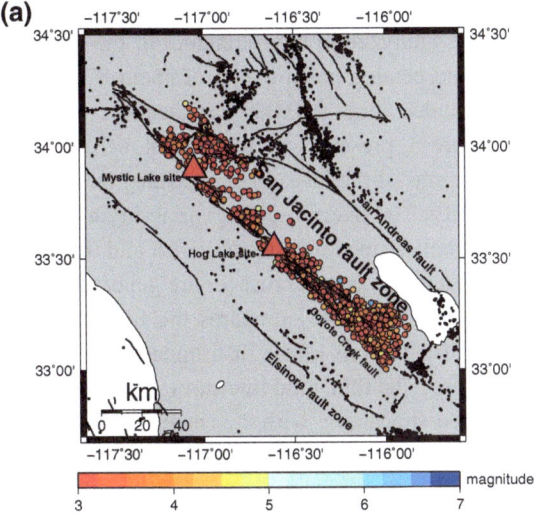

(a)

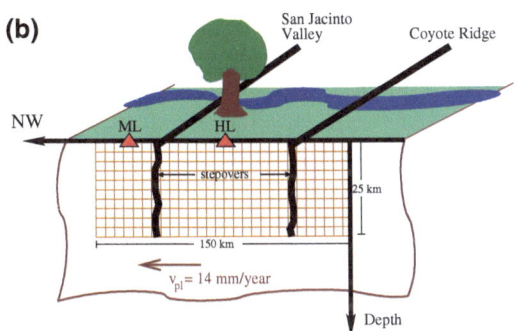

(b)

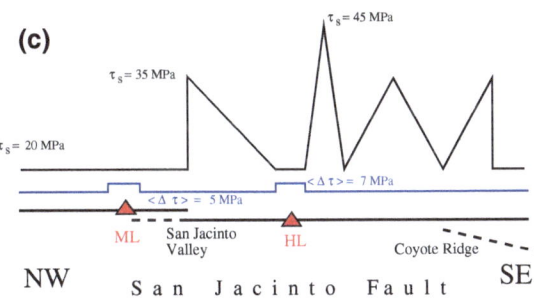

(c)

Figure 1

a Spatial distribution of seismicity in the study area around the San Jacinto fault zone with $m \geq 3$ earthquakes *color-coded* by magnitudes. The *black dots* show earthquakes in the surrounding regions; the Mystic Lake and Hog Lake sites are shown as *red triangles*; **b** sketch of the model for simulated seismicity; **c** dependence of static strength and local stress drop along strike

heterogeneities of the static strength τ_s that should be reached for brittle failure (Fig. 1c). The step-over at San Jacinto valley is a sharp increase followed by a

gradual decrease of $\tau_s(i)$ for a cell i along strike from NW to SE, while the step-over at Coyote Ridge marks the beginning of the complex trifurcation area and is modeled as a series of three spikes. We note that the specific shapes of the heterogeneities have only little influence on the catalog statistics. It is primarily important that the gradient of the static strength is high enough to stop an evolving rupture and thus simulate the step-over; the degree of heterogeneity can be varied further by the number of spikes.

When a computational cell i with shear stress $\tau(i;t)$ at time t reaches the static strength, $\tau(i;t) \geq \tau_s(i)$, the stress drops to a lower arrest stress $\tau(i;t) \rightarrow \tau_a(i)$, and strength is reduced to a dynamic level, $\tau_s(i) \rightarrow \tau_d(i)$, for the duration of the rupture. We add to the hypocenter coordinates a normally distributed noise accounting for location errors. The local reduction to arrest stress $\Delta\tau(i) = \tau_s(i) - \tau_a(i)$ is generally around 5 MPa, and it is increased at the sites of Hog Lake and Mystic Lake to 7 MPa to produce large (paleo) earthquakes. The values $\tau_d(i)$ are determined using a dynamic overshoot coefficient $DOS = (\tau_s(i) - \tau_a(i))/(\tau_s(i) - \tau_d(i)) = 1.25$; for more details see BEN-ZION (1996). The model stress drops are the average values of the differences between the evolving initial and final stresses over the failure areas of the simulated earthquakes.

3. Results

Using the model described in the previous section, we simulate a long earthquake history with events in the magnitude range $m \in [4.01; 7.58]$. The minimum magnitude is determined by the size of the computational cell. Refining the grid can be used to decrease the minimum magnitude at the expense of higher computational effort. The early period of simulated earthquake history is typically characterized by transient behavior, which lasts until the system dynamics stabilize. As an indicator for the dynamical state, we calculate the mean stress on the computational grid for each time when an earthquake occurs. This quantity becomes confined in a particular range as soon as the stable dynamical regime is reached.

95

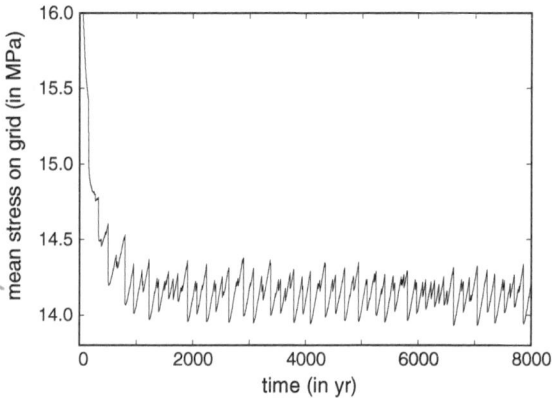

Figure 2

Mean stress on the computational grid for the first 8,000 years of simulated earthquake history. To avoid transient effects, the first 5,000 years are not used for the analysis

Figure 2 shows the stress history in the first 8,000 simulation years, which exhibits after some time typical "stick-slip" behavior with fractal characteristics (BEN-ZION *et al.* 2003; ZÖLLER *et al.* 2008). While visual inspection suggests a transient period of about 1,200 years, we use a more conservative value

and remove the first 5,000 years from the catalog which is analyzed in the remainder of the study. The examined catalog has a length of about 87,000 years and includes about 220,000 earthquakes.

Figure 3 presents a comparison of the spatial distributions of the instrumental catalog and synthetic seismicity. The overall spatial variability along strike (a), the bell-shaped depth profile (b), and the variable thickness of the brittle crust (c) are generally in good agreement. Figure 4a compares the frequency–magnitude distributions of the instrumental and synthetic seismicity. The thin solid line shows an example for a simulated subcatalog with the same time coverage of 30.5 years as the instrumental catalog. The heavy thick line refers to the complete synthetic catalog, which can be considered as a stack of several small catalogs. For both the instrumental and simulated seismicity, the b-value is close to unity. The roll-off at $m \approx 4$ for the synthetic seismicity stems from discretization effects. Large characteristic-type events occur with slightly increased probability compared to the scaling relation up to $m \approx 6.2$. Figure 4b shows the distribution of earthquake stress

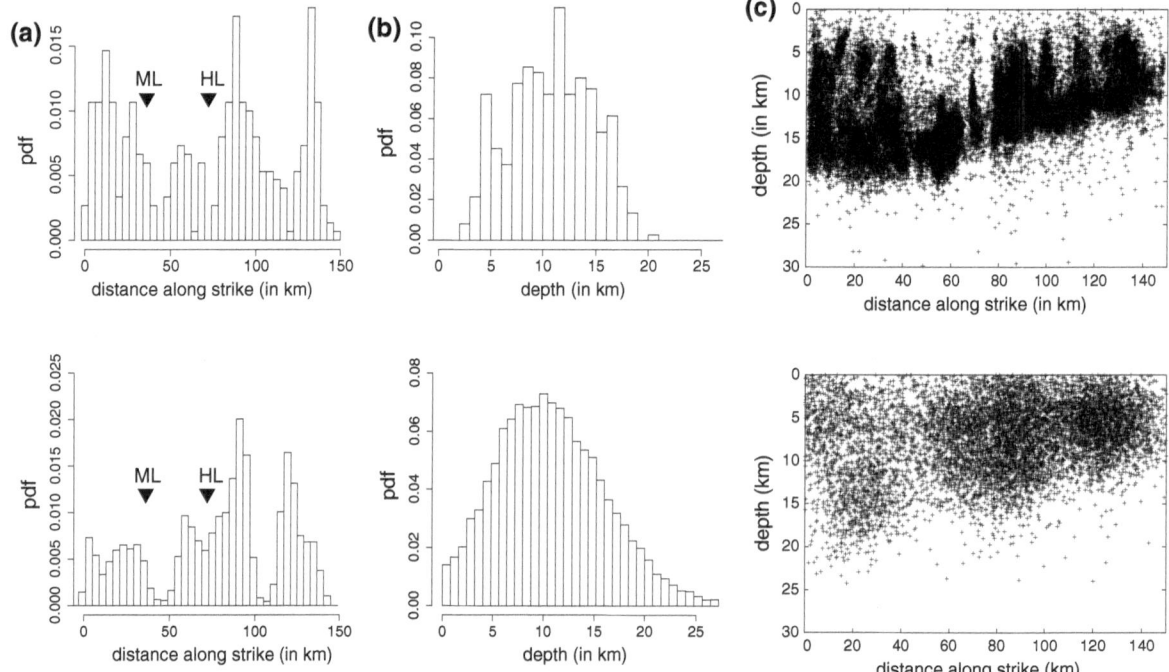

Figure 3

Spatial distributions along-strike (**a**) and with depth (**b**) of the instrumental and synthetic earthquake catalogs. **c** Shows the observed and simulated earthquakes as a function of depth and distance along strike

drops for events with $m \geq 5$. Smaller earthquakes are excluded because, due to the limited grid resolution, they have a smaller range of possible magnitude and stress drop values. In the range not determined by the cell size (Fig. 4b), the model stress drops are in good agreement with the observed lognormal distribution in Fig. 10 of SHEARER *et al.* (2006).

Earthquake recurrence times are presented in Fig. 5. For comparison, the observational catalog is only considered for magnitudes above the lower threshold of the model catalog, namely $m \geq 4.01$, although it is complete down to smaller magnitudes. Therefore, the probability density function (pdf) of recurrence times of the observed data is based on only 35 earthquakes. The figure suggests overall exponential recurrence times for both catalogs, indicating a Poisson process with a coefficient of variation CV ≈ 1. A fit of the exponential distribution with rate λ is given as a dashed line, while a corresponding fit with the Weibull distribution is shown by a dotted line. We note that a Weibull distribution with a shape parameter $k = 1$ is the same as the exponential distribution. Both distributions in Fig. 5 are in overall good agreement with exponential behavior.

The goodness-of-fit of the exponential distribution in Fig. 5 may be evaluated by a Kolmogorov-Smirnov (KS) test. In particular, we applied a one-sample and two-sided KS test using the exponential distribution with the rate estimated from the catalog as a null hypothesis. For the instrumental catalog in Fig. 5a, the p value of the KS test, $p = 0.4581$, is clearly higher than commonly chosen levels of confidence like $\alpha = 0.05$, implying that the exponential distribution cannot be

rejected. In the case of the synthetic catalog (Fig. 5b), the exponential distribution is clearly rejected ($p < 10^{-10}$). This result is, at first glance, surprising. However, Fig. 5b shows a mismatch of data and fit for small recurrence times (<1.2 years). This deviation arises from discretization effects in time and space. The discretization in time is associated with a finite minimum time difference between two earthquakes. The smallest (single-cell) events have only a narrow range of possible magnitude values. In turn, the portion of stress that is transferred by such events has also a small number of preferred values resulting in only a few possible recurrence times. Repeating the KS test for recurrence times ≥ 1.2 years results in $p = 0.1014$, leading again to failure of rejection of the exponential distribution based on commonly used levels of confidence. The exponential function can therefore be considered as a well-fitting distribution for recurrence times of earthquakes with magnitudes >4, if model discretization effects are ignored.

Next, we focus on magnitudes that are rarely, if at all, included in the instrumental catalog. Figure 6a and b show, respectively, recurrence time PDFs for synthetic earthquakes with $m \geq 6$ and $m \geq 7$. For illustration, we fit a Weibull distribution with density $f(t) \sim (t/\mu)^{k-1} \exp\left[-(t/\mu)^k\right]$ to the data (dashed line). The Weibull distribution is used here for two reasons: first, other commonly used distributions like the lognormal or the Brownian Passage Time show poor agreement with the data in this range of recurrence times. Second, if the shape parameter k is unity, the Weibull distribution degenerates to the exponential distribution, indicating a memoryless Poisson

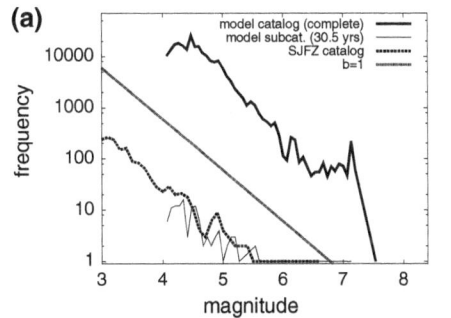
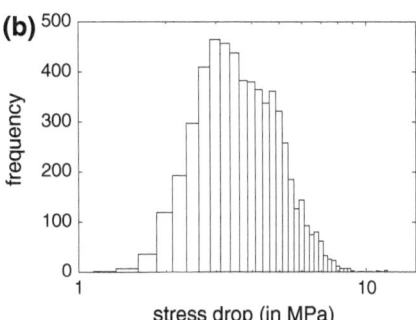

Figure 4

a Frequency–size distributions of observational catalog (*dotted line*), the complete synthetic catalog (*heavy solid line*), and an example for a subcatalog covering the same time interval (30.5 years) as the observational catalog (*thin solid line*) in relation to the Gutenberg-Richter distribution with $b = 1$ (*straight line*); **b** distribution of stress drops for synthetic earthquakes with $m \geq 5$ ($\langle \Delta\sigma \rangle = 4.2$ MPa)

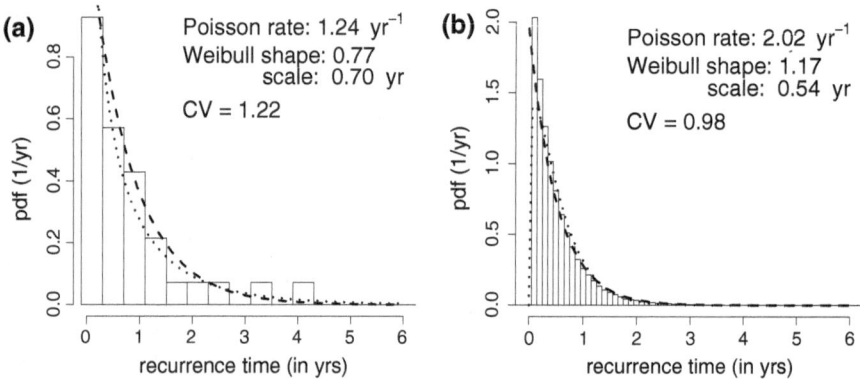

Figure 5

Recurrence time distributions for $m \geq 4.01$ in **a** the instrumental (*left*, 35 events) and **b** the synthetic (*right*, 221.888 events) catalogs. The *dashed lines* indicate fits using exponential distributions (Poisson process) with rate λ and the *dotted lines* indicate fits of the Weibull distribution

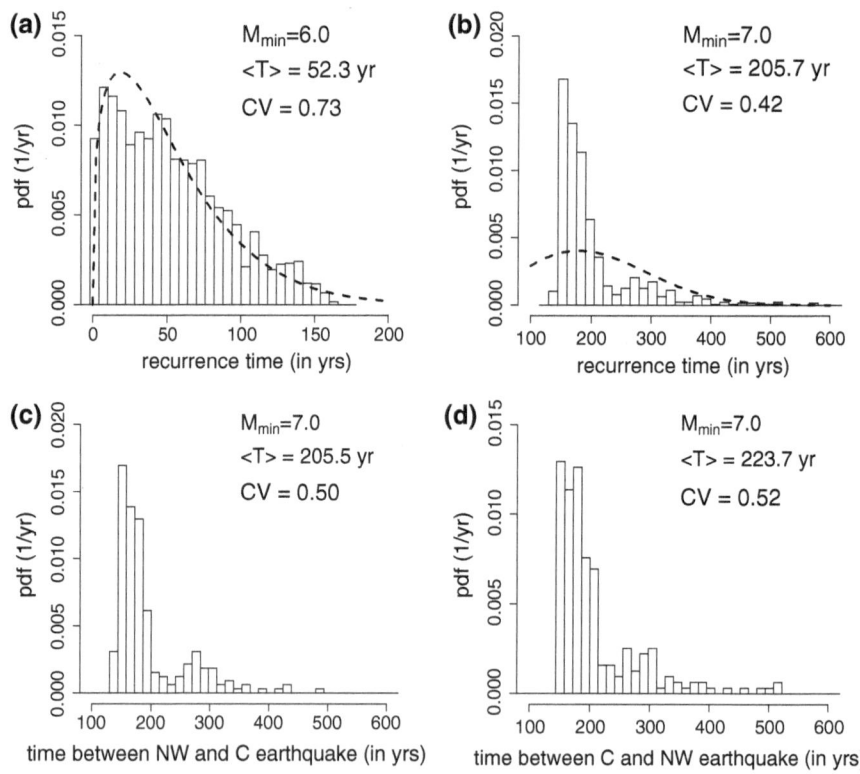

Figure 6

PDF of waiting times for synthetic earthquakes with magnitudes **a** $m \geq 6$ and **b** $m \geq 7.0$. *Dashed lines* are fits of the Weibull distribution to the data. **c**, **d** Show the pdf of waiting times between an earthquake on the NW segment and a subsequent earthquake on the central segment and vice versa. To produce the plots, waiting times from ten catalogs with randomized initial stresses have been stacked

process. Therefore, the value of k might estimate to which degree the recurrence time properties are consistent with a Poisson process. Figure 6a shows gradual deviations from the exponential statistics in

Fig. 5b; the coefficient of variation decreases from 0.98 (Fig. 5b) to 0.78 (Fig. 6a), and the Weibull shape parameter is now $k = 1.29$. The recurrence times for $m \geq 7$ are centered around 150 years with

$k = 2.44$. While the overall fit in Fig. 6b is obviously poor, the tail is roughly consistent with the Weibull distribution. In general, we observe with increasing magnitude cutoff a transition from the exponential distribution to the Weibull (or other non exponential) distribution. This is consistent with the overall random occurrence of small (and moderate) earthquakes, while large earthquakes are characterized by a peaked distribution that allows for quasiperiodic earthquake recurrence. We do not attempt to resolve what is most likely the "true" distribution, because in the following we use directly the simulation results and not the fitted distributions. Commonly used theoretical distributions include the lognormal and the Brownian Passage Time distribution (ZÖLLER et al. 2007, 2008; BEN-ZION 2008).

We note that the small peak at around 300 years is related to the fact that the transition from random intermediate-size earthquakes to large quasi-periodic earthquakes is gradual rather than sharp: due to small differences of the magnitude values of $m \approx 7$ earthquakes, some events with magnitudes slightly smaller than $m = 7$ are not included in this figure, although they are likely part of quasi-periodical large events. This leads to a second mode at about 2×150 years in the recurrence time distribution, which is artificial in the sense that it is affected by the chosen magnitude threshold. The coefficient of variation CV = 0.42 is lower than in Fig. 6a, but is still associated with considerable irregularity. A 90% fraction of the recurrence times has values between 109 and 158 years. To examine the interaction between different sections of the SJFZ, we show in Fig. 6c and d the time intervals between pairs of events with $m \geq 7$ on the central section around the Hog Lake site and the NW section around the Mystic Lake site. To have enough event pairs, results of ten synthetic catalogs with randomized initial stresses have been stacked. We find that Fig. 6c and d are in overall agreement with the PDF in Fig. 6b, suggesting that there is little difference whether subsequent large earthquakes occur on the same or separate segments. The results show no significant evidence that large earthquakes on the NW and the central segments are statistically correlated in time.

The paleoseismological findings of ONDERDONK et al. (2013) indicate seven surface-rupturing earthquakes with recurrence intervals between 159 and 210 years. Even if the 18 paleo-earthquakes within 4,000 years from the Hog Lake site (ROCKWELL et al. 2006, 2014) are included, this number is too small to allow for a statistically robust comparison with the model. In this context, we also note that no precise magnitude information of the paleo-earthquakes are available. However, the ability of the model to produce a broad range of realistic phenomena in general (BEN-ZION 2008; ZÖLLER et al. 2009), and results consistent with observations associated with the SJFZ in particular (Figs. 3, 4, 5) motivates us to use next the synthetic seismicity to estimate the hazard associated with large earthquakes on the SJFZ.

In the following, we quantify the information gain from the paleoseismological data and the numerical model results with respect to the large earthquake hazard. We consider two cases: first, we solely use the instrumental earthquake catalog; second, we include the additional results based on the simulated recurrence time distribution $F(t)$ in Fig. 6b, and the knowledge that the elapsed time t_0 since the last large paleo-event is about 200 years. In this second case, the conditional probability $P(\Delta t | t_0)$ that the next event with $m \geq 7$ will occur in a time interval of length Δt from now, given that the last event occurred $t_0 = 200$ years ago, is given by

$$P(\Delta t | t_0) = \frac{F(t_0 + \Delta t) - F(t_0)}{1 - F(t_0)}. \qquad (1)$$

Results are provided in the second column of Table 1 for three time horizons Δt. For the former case, where only the instrumental catalog is available, Eq. (1) can not be applied, because we do not have the data to estimate the recurrence time distribution $F(t)$ for M7 earthquakes. Therefore, we use the extreme value approach of EPSTEIN and LOMNITZ (1966) based on Poissonian occurrence of earthquakes with intensity λ and magnitudes following the Gutenberg-Richter distribution $F_{GR}(m) = 1 - 10^{-b(m-m_0)}$ with a given b-value. The probability that the largest earthquake in a time interval of length Δt has a magnitude $m \geq 7$ is given in this case by

$$G(\Delta t) = 1 - \exp\left[-\lambda \Delta t (1 - F_{GR}(7))\right], \qquad (2)$$

where the exponential term is the Gumbel distribution that belongs to the family of the Generalized Extreme Value (GEV) distributions. We emphasize

99

Table 1

Probability for an earthquake with m ≥ 7 in the San Jacinto fault zone

Δt (in years)	$P(\Delta t\|t_0)$ model+ observation (%)	$G(\Delta t)$ instr. catalog $m \geq 3$ (%)	$G(\Delta t)$ instr. catalog $m \geq 3$, declustered (%)
10	17.2	1.2	2.0
50	37.9	9.1	9.7
100	65.5	17.4	18.5

First column: future time horizon; second column: calculation with Eq. (1) based on data and model; third column: calculation with Eq. (2) based only on the instrumental catalog; fourth column: the same with the declustered catalog (see text for more information)

that the use of Eq. (2) is in line with traditional probabilistic seismic hazard assessments; in particular, earthquakes in general are assumed to follow a stationary Poisson process characterized by exponentially distributed recurrence times. The corresponding hazard function is time-independent and, therefore, the probabilities for future events do not depend on the past.

In contrast, if the recurrence times follow a Weibull distribution, the hazard will be small immediately after a large earthquake and then increase with evolving time. After a certain time, it will exceed the constant hazard of the Poisson process, given that the next large event has not occurred yet. Whether or not this state is reached after 200 years of "quiescence" in the SJFZ can be elaborated by a comparison of Eq. (2) based on the Poisson process with Eq. (1) based on the recurrence times of the simulated seismicity. For the b value in $F_{GR}(m)$ and the Poisson intensity λ in Eq. (2), we use the following maximum likelihood estimates: $\hat{b} = 1.10$ and $\hat{\lambda} = 12.36\,\text{year}^{-1}$, where the magnitude of completeness is set to $m_0 = 3$ from visual inspection of the scaling regime in the distribution. The corresponding results are shown in the third column of Table 1. Because earthquake clustering may violate the Poisson assumption, we repeat the calculation with a catalog that has been declustered with the method of REASENBERG (1985). Using the standard parameters for California leads to identifying about 90 % of all earthquakes as aftershocks with a new b value estimation of $\hat{b} = 0.64$. Due to this likely

overestimation of the aftershock frequency, we reduce the interaction radius in Reasenberg's method from ten to five crack radii leading to a fraction of 64 % of aftershocks and $\hat{b} = 0.98$. The corresponding results are given in the fourth column of Table 1.

A comparison of both approaches indicates that the hazard increases significantly if not only the instrumental catalog information, but also the model results in combination with the paleoseismological knowledge are taken into account. We recall that 90 % of recurrence times in Fig. 6b are in the range 109 years $\leq T_r \leq 158$ years. Given the good general agreement between the statistical properties of the available data and model catalog, the results strongly suggest that an earthquake with $m \geq 7$ is likely to occur within the next decades; even if it occurred now, the recurrence time would be in the tail of the distribution. For the next 100 years the probability of occurrence is 65.5 %, which is in clear contrast to the small 17.4 % value from the instrumental data or 18.5% from the declustered catalog.

Finally, we note the increased number of $m \sim 5$ earthquakes observed in the study area. While only seven earthquakes with moment magnitude $M_w \geq 5$ have been recorded in the study area during about 50 years since 1 January 1932 (HUTTON et al. 2010), three such events occurred since 2001 along with numerous $m > 4$ earthquakes, including an $M_w = 5.3$ event in 2010 and an $M_w = 4.7$ event on March 11, 2013, most of them in the region southeast of Anza. Although these numbers are not sufficiently high to provide clear statistical evidence for seismicity rate changes in this magnitude range, we can search for this pattern in the synthetic data.

Analyzing stacked synthetic seismicity, including about 800 earthquakes with $m \geq 7$, we count the number of events with $m \geq 5$ preceding each $m \geq 7$ earthquake in a time window of 15 years. The resulting distribution is provided in Fig. 7a. On average, one M7 event is preceded by about 5.6 M5 events within 15 years. For the entire catalog, the average number of M5 earthquakes within a 15-year time window is 4.1 (dashed vertical line in Fig. 7a). The statistical significance of the increased number of M5 earthquakes preceding an M7 event can be evaluated by applying the Student's t test. In particular, we calculate the 99 % confidence interval I_{99} for the true

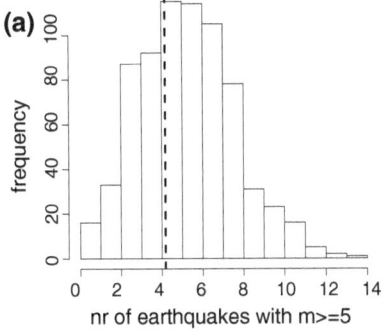
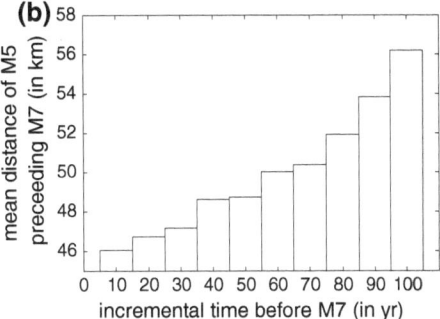

Figure 7

Properties of $m \geq 5$ earthquakes preceding an $m \geq 7$ event within a time window of 15 years. **a** Distribution of the events; the mean number is $n_{\text{before M7}} = 5.8$. The *dashed vertical lines* denotes the mean number of all $m \geq 5$ earthquakes within 15 years in the catalog $n_{\text{catalog}} = 4.1$. **b** Mean spatial distance of the preceding M5 events as a function of time to the M7 earthquake. For instance, the box at time "10 years" denotes that the M5 earthquakes that occurred between 0 and 10 years have an average distance of about 46 km to the upcoming M7 event; the box at time "20 years" refers to a time between 10 and 20 years before the M7 event, etc.

expected value underlying the distribution in Fig. 7a: $I_{99} = [4.9; 5.3]$. Consequently, the mean value of M5 earthquakes within 15 years for the entire catalog, $n_{\text{catalog}} = 4.1$, is clearly smaller than the corresponding mean value before the M7 events. The statistical significance in terms of a one-sample Student's t test is $>99\%$.

The model results can also be used to study spatial correlations of M7 earthquakes and the preceding M5 events. In Fig. 7b we show results for time slices of 10 years before an M7 earthquake. For each time slice, we compute the average distance between the hypocenter of M5 events and the upcoming M7 earthquake. The plot shows a progressive decrease of the distance from about 56 km to about 46 km, with decreasing time to the main event. In particular, the hypocenters of the M5 events migrate on avergage toward the hypocenter of the future M7 event. This behavior is consistent with the emergence of an intermittent spatially correlated stress field before the occurrence of a large event (BEN-ZION *et al*. 2003; ZÖLLER *et al*. 2005). These findings emphasize that the recently observed $m \sim 5$ earthquakes in the SJFZ might be part of the preparation process of an upcoming large earthquake in the region around Anza.

4. Discussion

We analyze observed and simulated seismicity data on the SJFZ, with a focus on properties and

hazards associated with large events. The record of instrumental seismicity is too short, and paleoseismic data are too sparse, to infer on occurrence of large earthquakes rupturing most of the fault. Based on large-scale geometrical properties of the SJFZ (overall dimensions, major stepover regions) and basic physics including frictional failure, dislocation creep, and long-range elastic stress transfer (OKADA 1992; BEN-ZION 1996), we designed a numerical model that duplicates the statistical properties of the observed instrumental and paleoseismic data. A long record of simulated seismicity produced by the model allows us to make statistically-robust statements on the behavior of large earthquakes. The results indicate that there is little interaction between moderate to large earthquakes on the section of the SJFZ around the Hog Lake paleoseismic site and corresponding events around the Mystic Lake site. Nevertheless, occasional earthquakes rupture most of the SJFZ and include the sections around both sites. Combining all the information from the instrumental catalog, paleoseismic data, and generated synthetic seismicity, we find that a large earthquake ($m \geq 7$) is expected with high probability within the next decades, e.g., with 38 % in 50 years and with 66 % in 100 years. This finding is supported by the recently observed growing number of intermediate-size ($m \sim 5$) earthquakes on the SJFZ, which is a robust feature of the preparation process of large earthquakes in the simulated seismicity. The model results suggest that the hypocenter of the large future event

101

is likely to be close (e.g., within 50 km) to the hypocenters of the preceding M5 earthquakes. The combined information provides evidence that a large earthquake is in preparation on the SJFZ, probably in the region around Anza.

Acknowledgments

The manuscript benefitted from the constructive comments of two anonymous reviewers. Figure 1a has been produced with the GMT software of WESSEL and SMITH (1991). GZ acknowledges support from the Potsdam Research Cluster for Georisk Analysis, Environmental Change and Sustainability (PROGRESS). YBZ acknowledges support from the National Science Foundation (grant EAR-0908903).

REFERENCES

BEN-ZION, Y. (1996), Stress, slip, and earthquakes in models of complex single-fault systems incorporating brittle and creep deformations, J. Geophys. Res. 101, 5677–5706.

BEN-ZION, Y. (2008), Collective behavior of earthquakes and faults: Continuum-discrete transitions, evolutionary changes and corresponding dynamic regimes, Rev. Geophys, 46, RG4006, doi:10.1029/2008RG000260.

BEN-ZION, Y. (2012), Episodic tremor and slip on a frictional interface with critical zero weakening in elastic solid, Geophys. J Int. 189, 1159–1168, doi:10.1111/j.1365-246X.2012.05422.x.

BEN-ZION, Y., ENEVA, M., and LIU, Y. (2003), Large Earthquake Cycles And Intermittent Criticality On Heterogeneous Faults Due To Evolving Stress And Seismicity, J. geophys. Res. 108, 2307, doi:10.1029/2002JB002121.

BEN-ZION, Y., and RICE, J. R. (1993), Earthquake Failure Sequences Along a Cellular Fault Zone in a 3D Elastic Solid Containing Asperity and Non-Asperity Regions, J. geophys. Res. 98, 14,109–14,131.

EPSTEIN, B., and LOMNITZ, C. (1966), A model for the occurrence of large earthquakes, Nature 211, 954–956.

FIALKO, Y. (2006), Interseismic strain accumulation and the earthquake potential on the southern San Andreas fault system, Nature 441, 968–971.

HAUKSSON, E., YANG, W., and SHEARER, P. M. (2012), Waveform Relocated Earthquake Catalog for Southern California (1981 to 2011), Bull. seism. Soc. Am. 102, 2239–2244, doi:10.1785/0120120010.

HUTTON, K., WOESSNER, J., and HAUKSSON, E. (2010), Earthquake monitoring in Southern California for Seventy-Seven years (1932-2008), Bull. seism. Soc. Am. 100, 423-446, doi:10.1785/0120090130.

MEHTA, A. P., DAHMEN, K. A., and BEN-ZION, Y. (2006), Universal mean moment rate profiles of earthquake ruptures, Phys. Rev. E., 73, 056104.

OKADA, Y. (1992), Internal deformation due to shear and tensile faults in a half space, Bull. seism. Soc. Am. 82, 1018–1040.

ONDERDONK, N. W., ROCKWELL, T. K., McGILL, S. F., and MARLIYANI, G. (2013), Evidence for seven surface ruptures in the past 1600 years on the Claremont fault at Mystic Lake, northern San Jacinto fault zone, California, Bull. seism. Soc. Am. 103, 519–541, doi:10.1785/0120120060.

POLLITZ, F. F. (2011), Epistemic uncertainty in California-wide synthetic seismicity simulations, Bull. seism. Soc. Am. 101, 2481–2498, doi:10.1785/0120100303.

REASENBERG, P. (1985), Second-order moment of central California seismicity, J. geophys. Res. 90, 5479–5495.

RICHARDS-DINGER, K., and DIETERICH, J. H. (2012), RSQSim Earthquake simulator, Seismol. Res. Lett., 83, 983–990, doi:10.1785/0220120105.

ROBINSON, R., VAN DISSEN, R., and LITCHFIELD, N. (2011), Using synthetic seismicity to evaluate seismic hazard in the Wellington region, New Zealand, Geophys. J Int. 187, 510–528, doi:10.1111/j.1365-246X.2011.05161.x.

ROCKWELL, T. K., T. E. DAWSON, J. YOUNG and GORDON SEITZ (2014), A 21 event, 4,000-year history of surface ruptures in the Anza Seismic Gap, San Jacinto Fault: Implications for long-term earthquake production on a major plate boundary fault, Pure Appl. Geophys., in review.

ROCKWELL, T. K., SEITZ, G., DAWSON, T., and YOUNG, J. (2006), The long record of San Jacinto fault paleoearthquakes at Hog Lake: Implications for regional patterns of strain release in the southern San Andreas fault system, Seismol. Res. Lett. 77, 270.

ROCKWELL, T., LOUGHMAN, C., and MERIFIELD, P. (1990), Late Quaternary rate of slip along the San Jacinto fault zone near Anza, Southern California, J. geophys. Res. 95, 8593–8605.

SHEARER, P. M., PRIETO, G. A., and HAUKSSON, E. (2006), Comprehensive analysis of earthquake source spectra in southern California, J. geophys. Res. 11, B06303, doi:10.10292005JB003979.

TULLIS, T. E., RICHARDS-DINGER, K., BARALL, M., DIETERICH, J. H., FIELD, E. H., HEIEN, E. M., KELLOGG, L. H., POLLITZ, F. F., RUNDLE, J. B., SACHS, M. K., TURCOTTE, D. L., WARD, S. N., and YIKILMAZ, M. B. (2012), Generic Earthquake Simulator, Seismol. Res. Lett. 83, 959–963, doi:10.1785/0220120093.

WELLS, D. L. and COPPERSMITH, K. J. (1994), New empirical relationships among magnitude, rupture length, rupture width, rupture area, and surface displacement, Bull. seism. Soc. Am. 84, 974–1002.

WESSEL, P. and SMITH, W. H. F. (1991), Free software helps map and display data, Eos Trans. AGU 72, 441.

ZÖLLER, G., BEN-ZION, Y., HOLSCHNEIDER, M. and HAINZL, S. (2007), Estimating recurrence times and seismic hazard of large earthquakes on an individual fault, Geophys. J Int. 170, 1300–1310, doi:10.1111/j.1365-246X.2007.03480.x.

ZÖLLER, G., HAINZL, S., BEN-ZION, Y., and HOLSCHNEIDER, M. (2006), Earthquake activity related to seismic cycles in a model for a heterogeneous strike-slip fault, Tectonophysics, 423, 137–145, doi:10.1016/j.tecto.2006.03.007

ZÖLLER, G. HAINZL, S., and HOLSCHNEIDER, M. (2008), Recurrent large earthquakes in a fault region: What can be inferred from small and intermediate events? Bull. seism. Soc. Am. 98, 2641–2651, doi:10.1785/0120080146.

ZÖLLER, G. HAINZL, S., HOLSCHNEIDER, M., and BEN-ZION, Y. (2005), Aftershocks resulting from creeping sections in a heterogeneous *fault*, Geophys. Res. Lett. *32*, L03308, 137–145, doi:10.1029/2004GL021871

ZÖLLER, G., HAINZL, S., BEN-ZION, Y., and HOLSCHNEIDER, M. (2009), *Critical states of seismicity: From models to practical* seismic hazard estimates, *Encyclopedia of Complexity and System Science, R. Meyers (Eds.)* 9, 7853–7872, Springer.

(Received August 29, 2013, revised January 8, 2014, accepted January 21, 2014, Published online February 13, 2014)

Pure Appl. Geophys. 171 (2014), 2967–2991
© 2014 Springer Basel
DOI 10.1007/s00024-014-0853-4

Imaging Faults and Shear Zones Using Receiver Functions

VERA SCHULTE-PELKUM[1,2] and KEVIN H. MAHAN[2]

Abstract—The geometry of faults at seismogenic depths and their continuation into the ductile zone is of interest for a number of applications ranging from earthquake hazard to modes of litho-spheric deformation. Teleseismic passive source imaging of faults and shear zones can be useful particularly where faults are not outlined by local seismicity. Passive seismic signatures of faults may arise from abrupt changes in lithology or foliation orientation in the upper crust, and from mylonitic shear zones at greater depths. Faults and shear zones with less than near-vertical dip lend themselves to detection with teleseismic mode-converted waves (receiver functions) provided that they have either a contrast in isotropic shear velocity (V_s), or a contrast in orientation or strength of anisotropic compressional velocity (V_p). We introduce a detection method for faults and shear zones based on receiver functions. We use synthetic seismograms to demonstrate common features of dipping isotropic interfaces and contrasts in dipping foliation that allows determination of their strike and depth without making further assumptions about the model. We proceed with two applications. We first image a Laramide thrust fault in the western U.S. (the Wind River thrust fault) as a steeply dipping isotropic velocity contrast in the middle crust near the surface trace of the fault; further downdip and across the range, where basin geometry suggests the fault may sole into a subhorizontal shear zone, we identify a candidate shear zone signal from midcrustal depths. The second application is the use of microstructural data from exhumed ductile shear zones in Scotland and in the western Canadian Shield to predict the character of seismic signatures of present-day deep crustal shear zones. Realistic anisotropy in observed shear fabrics generates a signal in receiver functions that is comparable in amplitude to first-order features like the Moho. Observables that can be robustly constrained without significant tradeoffs are foliation strike and the depth of the foliation contrast. We find that an anisotropy of only a few percent in the shear zone is sufficient to generate a strong signal, but that the shear zone width is required to be >2 km for typical frequencies used in receiver function analysis to avoid destructive interference due to the signals from the boundaries of the shear zone.

Key words: Faults, shear zones, receiver functions, fault imaging, snisotropy, crustal deformation.

1. Introduction

Seismically active faults in the upper crust can be traced by the seismicity they generate. More difficult to establish is the subsurface geometry of faults deforming by aseismic creep or those with infrequent seismicity. Below seismogenic depths, faults are assumed to merge into shear zones with ductile deformation (e.g. SIBSON 1977; Fig. 1a) of unclear width. The transition from the seismogenic to ductile portion of crustal faults has received much attention recently after the discovery of low-frequency earthquakes, tectonic tremor, and slow slip events (e.g. BEROZA and IDE 2009; SHELLY *et al.* 2009). The geometry of shear zone continuation is of interest for determining modes of deformation in the crust and lithospheric mantle; for instance, in establishing whether faults that are steep at the surface sole out within the crust (listric fault), or cut down to or even through the Moho (planar fault; e.g. STERN and MCBRIDE 1998; ZHU 2000; WITTLINGER *et al.* 2004; WILSON *et al.* 2004; TITUS *et al.* 2007; PEDRERA *et al.* 2010). The ability to image faults and shear zones in the absence of seismicity on the fault provides useful information for seismic hazard estimation as well as lithosphere dynamics.

Methods for passive source illumination of faults include fault zone head waves (BEN-ZION and MALIN 1991; MCGUIRE and BEN-ZION 2005) and other body waves from nearby microseismicity (e.g. EBERHARDT-PHILLIPS and MICHAEL 1998; MCGUIRE and BEN-ZION 2005; ROECKER *et al.* 2006; THURBER *et al.* 2006; LIN *et al.* 2007; TAPE *et al.* 2009; BULUT *et al.* 2012; ALLAM and BEN-ZION

Report: Topical Volume on Crustal Fault Zones.

[1] Cooperative Institute for Research in Environmental Sciences, University of Colorado Boulder, 2200 Colorado Ave, Boulder, CO 80309-0399, USA. E-mail: vera.schulte@gmail.com
[2] Department of Geological Sciences, University of Colorado Boulder, Boulder, CO 80309-0399, USA. E-mail: kevin.mahan@colorado.edu

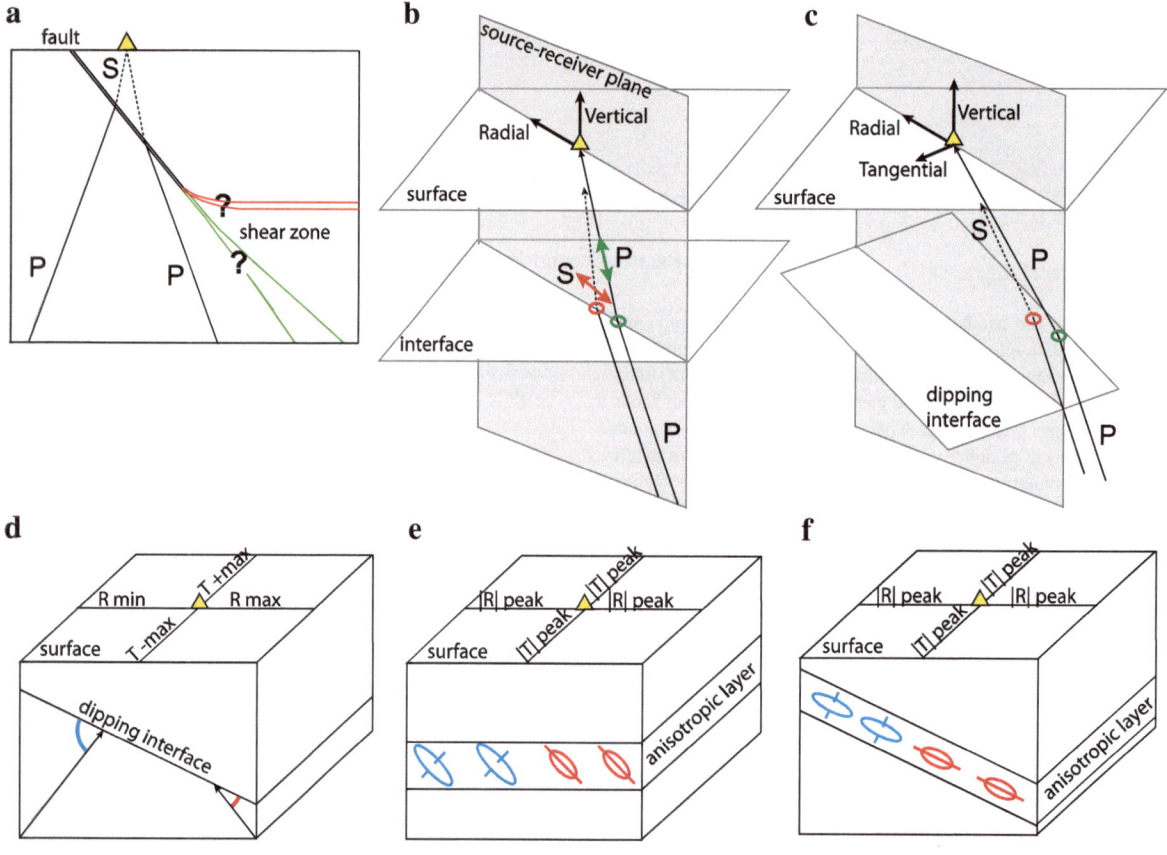

Figure 1

a Conceptual sketch of a fault with a narrow fault zone in the upper crust (*parallel black lines*) and two possible continuation geometries in the deep crust where ductile behavior dominates (widening into a broader shear zone—*green*; soling out into a subhorizontal detachment—*red*). The fault dip is 60° from horizontal. Also sketched are a seismic station (*yellow triangle*) with ray incidence angles for teleseismic P (*solid*) to S (*dotted*) conversions (shown for a ray parameter of 6.15 s/°, an intermediate value for the teleseismic P ray parameter range) assuming constant background V_p and V_s. Fault geometries drawn after SIBSON (1983) and PASSCHIER and TROUW (2005). Teleseismic P incidence angles in the middle crust range from roughly ~15 to 30°, shown in the sketch are 20° (1:1 vertical to horizontal scale). **b** Cartoon illustrating basic receiver function ray and particle motion geometry for a horizontal isotropic converter (interface). *Yellow triangle* is seismic station, *grey plane* is source-receiver plane. *Green circle* and *arrow* indicate P (*solid line*) refraction point and particle motion, *red circle* and *arrow* P to S conversion point and S (*dotted*) particle motion. **c** Same as **b**, but for a dipping interface. **d** Block diagram sketching the amplitude (polarity) behavior seen on radial (R) and tangential (T) component receiver functions as a function of backazimuth. Downdip angle between incident P and interface in indicated *red*, updip in *blue*. **e** Same as **d**, but for a horizontal layer with anisotropy with a plunging symmetry axis. *Blue symbols* indicate slow axis symmetry, *red* fast axis symmetry. |R| and |T| refer to absolute radial and tangential conversion amplitudes; polarity flips occur between the peaks. **f** Same as **e**, but with additional layer dip, here aligned with layer anisotropy

2012; ALLAM *et al.* 2014; ZHEGLOVA *et al.* 2012) as well as ambient noise surface waves (ROUX *et al.* 2011; HILLERS *et al.* 2013). The body wave methods (fault zone head waves and reflections) are limited to settings with nearby seismicity, while the short-period ambient noise surface waves are limited in depth penetration at short wavelengths and lose horizontal resolution at longer wavelengths. Teleseismic body waves are nearly ubiquitous, in contrast to body waves from local seismicity, and

suffer from less of a depth versus lateral resolution tradeoff than surface waves do. Converted and reflected waves in particular are sensitive to velocity and anisotropy contrasts that may delineate a fault or shear zone. Because of their steep incidence angles, teleseismic converted waves provide good illumination of subhorizontal to steeply dipping faults (Fig. 1), but not of subvertical ones since only some raypaths at a given station will pierce a vertical interface.

A popular method for isolating converted waves in teleseismic arrivals is the receiver function technique (PHINNEY 1964; BURDICK and LANGSTON 1977; VINNIK 1977), which takes motion on one component as a proxy for the source wavelet to be removed (deconvolved) from the other components in order to find conversions near the seismic station. For imaging within the crust, P-to-S conversions are ideal (S receiver functions have a lower frequency content resulting in insufficient intracrustal resolution). The most popular application of receiver functions is to detect subhorizontal isotropic shear velocity contrasts such as the Moho. The focus of this paper is to image crustal faults and shear zones via dipping velocity contrasts and contrasts in rock fabric (anisotropy). Figure 1 illustrates some of the cases targeted with our method.

We remind the reader of the fundamental principles of receiver function analysis for the flat-layered, isotropic contrast case, in order to provide a context for the behavior exploited in the more complex settings. In Fig. 1b, an incident teleseismic P wave (illustrated as rays) impinges on an impedance contrast under a station. For direct conversions, the shear velocity contrast has by far the largest effect on conversion amplitude in terms of material parameters (density contrasts become significant for reverberations). At the contrast, some of the incident wave is transmitted as P and some converts to an S wave. The converted S arrival propagates more slowly from the interface to the receiver than the transmitted (direct) P wave. The delay time between direct P and converted S is thus used to measure the depth to the contrast. Converting the delay time to interface depth requires knowledge of P and S velocity between interface and station; errors in shallow velocities can affect the calculated depth for arrivals from deeper structure significantly (e.g. SCHULTE-PELKUM and BEN-ZION 2012). The amplitude of the converted arrival carries information on the strength of the shear velocity contrast at the interface. It is also a function of the angle between the incident P wave and the interface (ZOEPPRITZ 1919; see e.g. AKI and RICHARDS 1980 or SHERIFF and GELDART 1995) and thus varies with epicentral distance to the source even in the horizontal interface case. The conversion amplitude is positive for a slow-over-fast contrast and negative

for a fast-over-slow contrast (the coordinate convention used in this paper is vertical up, radial away from the source, tangential left from positive radial, i.e. left-handed; note that some codes such as ANI-REC by LEVIN and PARK 1998 use a right-handed system).

The seismogram recorded at the station includes complexities of the incident P wave, such as the source time function and any effects incurred by the P wave on the path between the source and the interface, from which it is difficult to separate the converted arrivals. A conversion from a horizontal, isotropic interface is confined to the source-receiver plane (grey plane in Fig. 1b), i.e. to the vertical and radial components; no P or S motion is expected on the tangential component. Because of the steep incident angle of teleseismic waves, the along-ray P particle motion is predominantly recorded on the vertical component and the converted ray-perpendicular S particle motion on the radial component (Fig. 1b). The receiver function technique uses the vertical component seismogram as a proxy for the incident P wave and deconvolves it from the radial component; the process can be thought of as finding matches to the vertical component waveform in the radial component with a delay time between the two. The deconvolution introduces errors in the recovered conversion amplitude (OWENS et al. 1984; AMMON et al. 1990).

The same principles as in the flat-layered case apply for a dipping isotropic interface (Fig. 1c, d). In this case, the direct P wave is bent out of the source-receiver plane for all incident backazimuths except for those in exact up- and down-dip direction, and the converted S wave has an SH component (out of the source-receiver plane) as well as an SV component (within the source-receiver plane). After deconvolution of the vertical from the horizontal components, the bent direct P wave and the converted S wave are seen on both the radial and tangential component receiver function (Fig. 1c).

The dip of the converter leads to azimuthal variations in the receiver function arrival amplitude and delay time (Fig. 1d). Assume the material contrast at the dipping interface is the same as in the horizontal interface in Fig. 1b. Because of the dependence on the angle between the incident wave and the interface (larger amplitude for smaller angle), the conversion from the downdip backazimuth has an increased amplitude (Fig. 1d) and the updip conversion a decreased amplitude compared to a horizontal interface conversion with the same slowness. The tangential component shows no conversion for incidence along the up- and down-dip azimuths and maximum amplitude conversions for incidence perpendicular to the dip, i.e. along strike. The pattern is thus one of maximal variations to the radial component conversion perpendicular to strike and maximal tangential component amplitudes parallel to strike (Fig. 1d). Converted arrivals are also earliest from the updip direction and latest from the downdip direction because of the difference in path length to the interface.

The next case that will be discussed in detail in the paper is that of a subhorizontal shear zone with anisotropy (Fig. 1e). In its simplest form, anisotropy can be approximated as having hexagonal symmetry (also named transverse isotropy, TI), where the symmetry axis can either have fast or slow phase velocity compared to the plane orthogonal to it. Subhorizontal shear can produce anisotropy with a vertical symmetry axis (also named radial anisotropy, or vertical TI–VTI), for instance, due to the alignment of mica fast crystallographic planes with the shear plane, or variable distribution of amphibole fast axis crystallographic orientations within the shear plane (overall slow axis symmetry in both cases). A different end member is a horizontal symmetry axis (named azimuthal anisotropy or horizontal TI–HTI), for instance, due to horizontal shear-parallel preferential alignment of amphibole fast crystallographic axes (overall fast axis symmetry). Shear fabric can contain foliation or lineation at an angle to the shear plane, for instance, after small amounts of shear strain, away from the center of a shear zone, or in the case of S planes, in S–C type composite shear fabric; this case is termed plunging axis or tilted axis anisotropy or TTI, with slow or fast axis symmetry possible.

We illustrate the tilted axis case in Fig. 1e (all three cases are modeled in Sect. 2). The top and bottom of the anisotropic layer each produce a converted S wave. The direct P wave is not bent, so no arrival is seen at zero delay time on the tangential component. For hexagonal symmetry anisotropy with a plunging symmetry axis (Fig. 1e), the converted S wave has the largest radial amplitudes for the up- and downplunge incidence and the largest tangential component amplitudes in the along-strike directions, similar to the dipping isotropic case (Fig. 1d). Amplitude peaks on the same component at opposing backazimuths have opposite polarities. Since the depth to the converting interfaces is constant, there is no delay time variation with backazimuth for constant ray parameters. Splitting of the converted S wave in shear zones of widths considered here is minimal and ignored in this paper. The annotated azimuthal amplitude behavior on the radial and tangential components applies to slow axis as well as fast axis symmetry aligned as shown in Fig. 1e; the symmetry axis tilt is opposite between the two cases, but the strike perpendicular to the symmetry axis can be determined independently without necessitating a choice between fast or slow axis symmetry. In the slow axis symmetry case, the strike is that of the foliation; in the fast axis symmetry case, the determined "strike" is the horizontal line that is perpendicular to the lineation.

The amplitude behavior as a function of backazimuth in the case of a dipping shear zone with aligned anisotropy (Fig. 1f) is again similar to that seen for dipping isotropic interfaces (Fig. 1d) and plunging axis anisotropy (Fig. 1e). In the absence of an isotropic velocity contrast at the boundaries of the anisotropic layer, the direct P wave is not bent (no zero delay time arrival on the tangential component). As in the isotropic dipping interface case (Fig. 1d), the dip of the shear zone leads to differences in path length to the station after conversion resulting in delay time variations with backazimuth. The azimuthal amplitude/polarity behavior is again similar to that in the cases of plunging anisotropy in a horizontal layer (Fig. 1e) and isotropic dip (Fig. 1d).

We model all cases in Fig. 1 in detail below. Inversion of receiver functions for dipping interfaces and anisotropy is typically underdetermined and subject to significant tradeoffs (e.g. SAVAGE 1998), requiring choices such as fast versus slow axis symmetry. We show that the commonalities in amplitude/polarity behavior as a function of backazimuth sketched in Fig. 1 can be exploited in fault and shear zone imaging without a need to resolve all parameters.

We first demonstrate the expected fault-generated signals in synthetic seismograms in Sect. 2. We then investigate two types of settings: (1) a steeply dipping Laramide thrust fault with possible soling out of the fault into a subhorizontal shear zone, where we image the fault as an isotropic dipping contrast in the upper crust and detect a potential subhorizontal detachment in the middle crust (Sect. 3); (2) Two former middle to lower crustal shear zones that are now exhumed in Scotland and in the western Canadian Shield, where we use microstructural characteristics of the shear fabric to predict a measurable seismic signature that could be analogous to that from similar present-day deep crustal shear zones (Sect. 4).

2. Method

2.1. Dipping Isotropic Contrasts

The displacement on a fault can juxtapose different lithologies across the fault. A resulting isotropic shear velocity contrast across a dipping fault leads to characteristic signatures in receiver functions. We show a synthetic example in Fig. 2. The synthetic seismograms were calculated using the raysum code by FREDERIKSEN and BOSTOCK (2000). Unlike in real seismograms, the source time function is a simple pulse with a chosen pulse width and we do not perform a deconvolution; the synthetics are equivalent to receiver functions except for P–P scattering which is a minor effect. On the radial component (used in most receiver function studies to image horizontal interfaces), the amplitude of a converted arrival increases with shallower incidence of the ray relative to the interface. In the case of a dipping interface, the radial component, therefore, shows a characteristic pattern with backazimuth with a maximum downdip and a minimum updip, 180° apart (Fig. 2). The azimuthal variation is superimposed on an azimuthally invariant positive polarity amplitude produced by a horizontal interface with the same velocity contrast.

In the plane-layered isotropic case, coupling is strictly P-SV and particle motion remains in the source-receiver plane. Dipping interfaces in an isotropic model lead to particle motion outside the source-receiver plane, i.e. P-SH coupling (e.g.

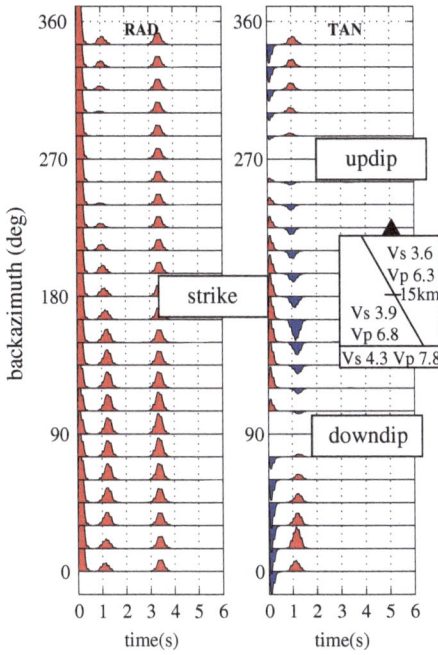

Figure 2

Synthetic seismograms for the model shown in *inset*, with a 60° east-dipping interface with 0.3 km/s V_s contrast. Horizontal axes are delay time in seconds after direct P arrival. Vertical axes are backazimuth looking north (0°), east (90°), etc. from the station. P–S conversions from the dipping interface are seen at $\sim$1–1.5 s delay time. Note the amplitude maximum in the downdip backazimuth on the radial component (*left*) and down- and updip polarity flips on the tangential component (*right*). Conversions from a Moho at 30 km depth (seen at 3.4 s on the radial component) are shown for a comparison of conversion amplitude. The zero delay time arrival on the tangential component is due to out-of-plane ray bending of direct P at the dipping interface. Ray parameter is set to 0.06 s/km for all backazimuths. The amplitude scale is the same for all synthetic and data plots throughout the paper

LANGSTON 1977; SAVAGE 1998; HAYES and FURLONG 2007). The introduction of a dipping interface has two effects: it bends the direct P wave (seen at zero delay time) so that it is seen as an arrival with azimuthal variation not only on the radial, but also the tangential component, and causes an out-of-plane P-to-SH conversion for all incidence azimuths except for incidence from the exact up- and downdip directions. The resulting arrival changes polarity across the up- and downdip nodes on the tangential component (Fig. 2). Other settings that generate P-SH coupling are anisotropy (discussed in Sect. 2.2) and scattering (not explored further here; see ABERS 1998; JONES and PHINNERY 1998).

Finding polarity nodes on the tangential component and corresponding extrema on the radial component, therefore, allows determination of the strike of the dipping interface (from the azimuthal position of the nodes) and the depth to the dipping interface (from the average delay time of the azimuthally varying arrival). The dip direction can be determined from the sense of polarity flip only if there is a priori information on whether the interface constitutes a slow-over-fast or a fast-over-slow contrast. However, the converting interface is closer to the station in the updip backazimuth and farther from the station in the downdip backazimuth. The converted arrival is, therefore, earlier (i.e. has a smaller delay time relative to the direct P arrival) in the updip direction and later in the downdip direction. The conversion amplitude is also a function of the relative angle between the incident P wave and the converting interface, with shallower incidence producing larger amplitude conversions. However, using the arrival amplitude to determine the dip angle is difficult because of a strong tradeoff between dip angle of the interface and velocity contrast across the interface, since larger V_s contrasts also generate larger conversion amplitudes.

2.2. Contrasts in Foliation

Faults may only produce offsets within the same or similar lithologies, with no resulting bulk isotropic velocity contrast across the fault interface. However, the fault offset can produce a contrast in foliation across the fault, in either foliation orientation, strength of the associated anisotropy, or both. Similarly, ductile shear zones may display a contrast in foliation between the shear zone and the material surrounding it. If such a foliation contrast is pierced by teleseismic rays, it can generate high-amplitude converted waves, particularly if the contrast interface or the foliation is at a sharply oblique angle to the ray incidence angle (i.e. if the fault or the foliation are dipping). A few percent of velocity anisotropy are sufficient to generate a signal above normal noise levels and an isotropic velocity contrast is not required. High conversion amplitudes are generated by contrasts in V_p anisotropy, while V_s anisotropy only makes for much smaller conversion amplitudes.

This is opposed to the case of conversions from the isotropic case where V_s contrast controls conversion amplitude and pure V_p contrasts generate no conversion.

We calculate synthetic radial and transverse component seismograms for a model of an isotropic crust with an anisotropic midcrustal layer. The model has no isotropic velocity contrast. The anisotropy has hexagonal symmetry with a slow axis of symmetry, 6 % peak-to-peak anisotropy in V_p and V_s, and a shape factor η of 0.7, all of which are reasonable approximations for deformed crustal rocks (e.g. JI and SALISBURY 1993; GODFREY et al. 2000; RASOLOFOSAON et al. 2000; SHERRINGTON et al. 2004; TATHAM et al. 2008; PORTER et al. 2011; WARD et al. 2012). Slow axis symmetry with a faster orthogonal plane is a valid approximation for crustal anisotropy in many cases, particularly for rock with mica alignment, although deformation fabric in rocks including amphibole (TATHAM et al. 2008; JI et al. 2013) and quartz (WARD et al. 2012) can introduce a fast axis symmetry component. Figure 3 shows synthetic radial and tangential component seismograms for a flat layer with a horizontal slow symmetry axis. The anisotropic layer generates conversions from the isotropic to anisotropic contrasts at the top and bottom of the anisotropic layer that vary in backazimuth with 180° periodicity. A pattern like this would be expected for a layer with vertical foliation.

Foliation within the center of a ductile shear zone tends to align parallel to the shear plane after modest amounts of strain (e.g. PASSCHIER and TROUW 2005). However, foliation at a moderately high angle to the shear plane is possible even after large strain due to either the S-plane in a composite S–C fabric (BERTHÉ et al. 1979; see also LLOYD et al. 2009) and/or an oblique S-dominated fabric that develops at the lower strain margins of the shear zone (RAMSAY and GRAHAM 1970). Figure 4 shows seismograms for the same model as in Fig. 3, except with a plunging (tilted) symmetry axis, as one would observe with dipping foliation. The converted phases now have higher maximum amplitudes and a 360°-periodic component. The tangential component has polarity nodes in the up- and downplunge azimuths, and the radial component has the highest amplitudes in the updip direction of foliation. Higher amplitudes are

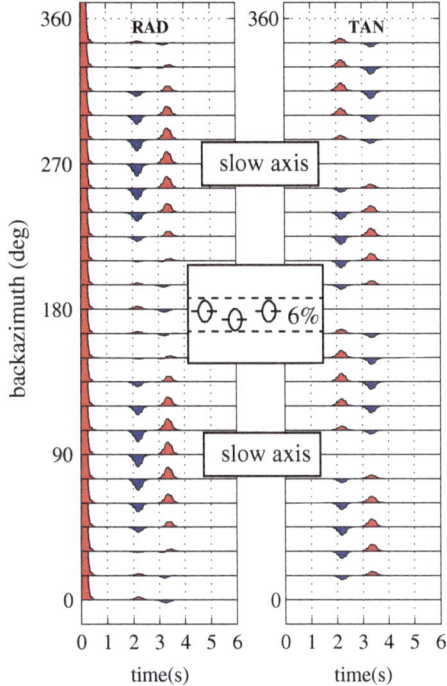

Figure 3

Synthetic seismograms (*left* radial component, *right* tangential component, axes and ray parameter as in Fig. 2) for a model with constant isotropic crustal V_p and V_s, with an anisotropic layer sandwiched within the isotropic halfspace from 20 to 30 km depth (see *inset*). There is no isotropic velocity contrast between the anisotropic layer and the material surrounding it. The anisotropy has a peak-to-peak strength of 6 % in V_p and V_s and has a hexagonal symmetry with shape factor $\eta = 0.7$ (e.g. SHERRINGTON et al. 2004), which are typical values for deformed middle crust. The symmetry axis here is horizontal. Note the 180° periodic conversions from the top (~ 2 s) and bottom of the layer (~ 3.5 s). Direct P is not bent out of plane, so no zero time arrival is visible on the tangential component, although out-of-plane conversions (later arrivals on tangential component) occur when the ray enters and exits the anisotropic layer

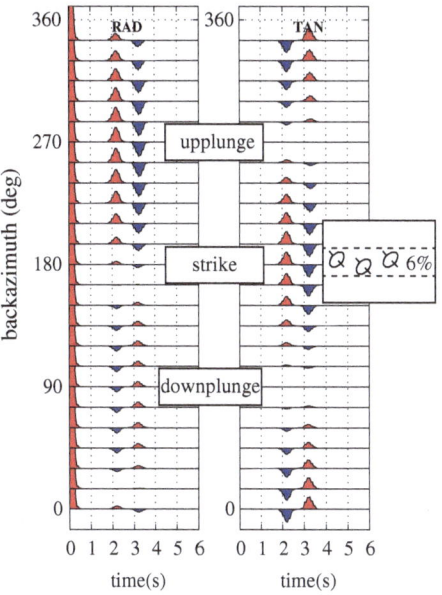

Figure 4

Same as Fig. 3, except the symmetry axis now plunges 45° to the east (backazimuth 90°). Note the dominant 360°-periodic symmetry with nodes in the up- and downplunge orientations on the tangential component and the highest radial amplitudes in the updip orientation, and the increased amplitudes especially in the tangential component compared to the horizontal axis case (Fig. 3)

around 1 (FARRA et al. 1991), while typical crustal rocks range from 0.4 to 0.9 (JI and SALISBURY 1993; GODFREY et al. 2000; SHERRINGTON et al. 2004; PORTER et al. 2011). Larger η result in a signal appearance similar to an intracrustal high-velocity layer, while smaller η mimic a low-velocity layer (Fig. 5). An intermediate value of 0.7 leads to vanishing conversions in this geometry. The synthetic results will be applied to data interpretation in Sect. 3.

3. Application to Laramide Thrust Fault

In this section, we image a moderately steeply dipping thrust fault with an isotropic V_s contrast and test for a seismic signature of an associated mid-/lower crustal subhorizontal shear zone that the fault may sole into. We use the Laramide Wind River thrust fault in Wyoming (Fig. 6) that has previously been imaged by an active source COCORP high-angle reflection experiment (SMITHSON et al. 1979; Fig. 7). Constraints from structural geology (ERSLEV 1986; STEIDTMANN and MIDDLETON 1991) and active

seen when the foliation as well as the shear zone have a steep dip (see shear zone synthetics in Sect. 4).

Signals from subhorizontal foliation are modeled in Fig. 5. We approximate horizontal foliation as slow axis symmetry with a vertical symmetry axis (horizontal fast plane). The resulting signal is constant with backazimuth on the radial component and zero on the tangential component (Fig. 5). The signal strength and polarity strongly depend on the anisotropic shape factor η, a parameter that describes the off-axis velocity dependence of a hexagonal elasticity tensor (BABUSKA and CARA 1991; SHERRINGTON et al. 2004; PORTER 2011). Mantle material has η

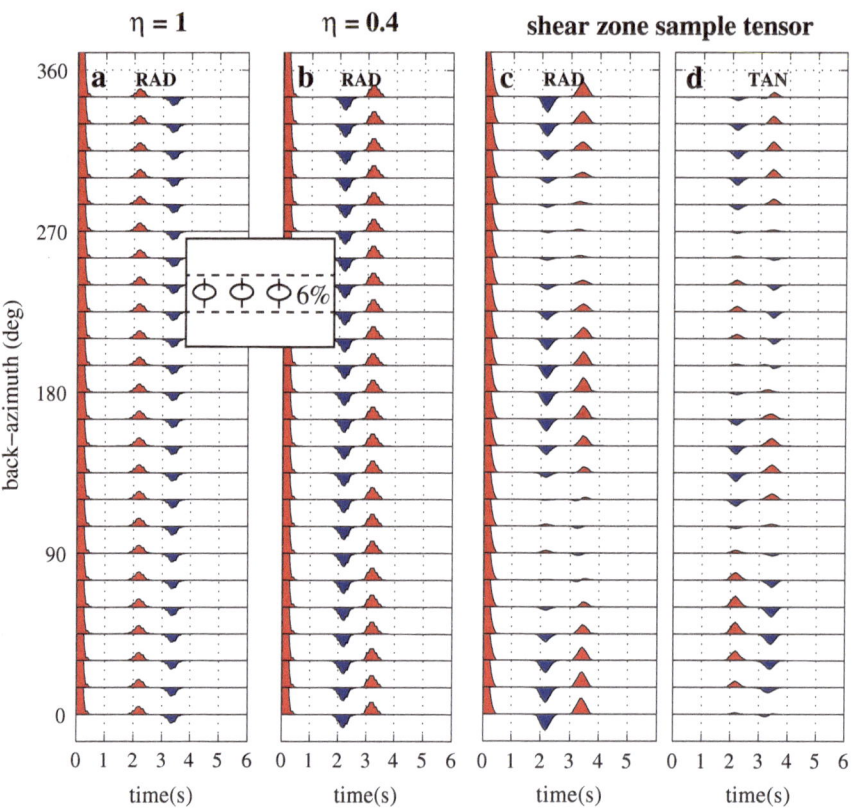

Figure 5

Synthetic seismograms with a sandwiched horizontal anisotropic layer with no isotropic velocity contrast as in Fig. 3, except with vertical symmetry axis (horizontal foliation). **a** Shape factor $\eta = 1$ within the anisotropic layer. The tangential component (not shown) is zero at all azimuths. The signal mimics two isotropic velocity contrasts, with a velocity increase followed by a velocity decrease with depth (a high-velocity layer) although the model has no isotropic contrasts. **b** Radial component for the same model, but with a shape factor of $\eta = 0.4$. The result now mimics a low-velocity layer (the amplitudes are reversed compared to **a**). The shape factor of crustal elasticity tensors in the hexagonal approximation typically ranges from 0.4 to 0.9. The average value of 0.7 used in this model results in near zero conversions on both components (not shown). **c** Radial component for horizontal foliation in the same model using a tensor obtained from a midcrustal shear zone (Grease River shear zone, see Sect. 4). **d** Same as **c**, but tangential component. Using the most deformed tensor from the Upper Badcall shear zone (Sect. 4) with horizontal foliation results in a nearly identical pattern

source seismics suggest 14 vertical km (SMITHSON et al. 1979) of NE-over-SW directed thrust displacement on a fault with up to 48° dip (SMITHSON et al. 1979), which resulted in basement rocks in the hanging wall sitting atop sedimentary rocks in the footwall (Fig. 7), leading to a fast-over-slow isotropic velocity contrast. A distinction between various proposed modes of Laramide lithospheric deformation (pure shear thickening, buckling, rooting into a sub-horizontal detachment, lithospheric faulting—ERSLEV 2005) could be made by imaging the geometry of the fault below ductile depths.

Three broadband stations (J18A, a Transportable Array station; BW06, a U.S. permanent network

station equipped with a STS-2 sensor in a surface vault; and station PD31 from the international monitoring network with a borehole KS-54000 sensor nearly collocated with BW06) sit atop the hanging wall of the fault west of the northern and central portion of the range, while the COCORP line ran through the southern portion (Fig. 6). All three stations are ~10 km inboard from the geologically inferred surface expression of the thrust fault. We calculated P receiver functions (from incident teleseismic P and P_{diff} phases) using the iterative time domain technique of LIGORRIA and AMMON (1999). Details of event selection and processing are described in SCHULTE-PELKUM and MAHAN (2014). All three

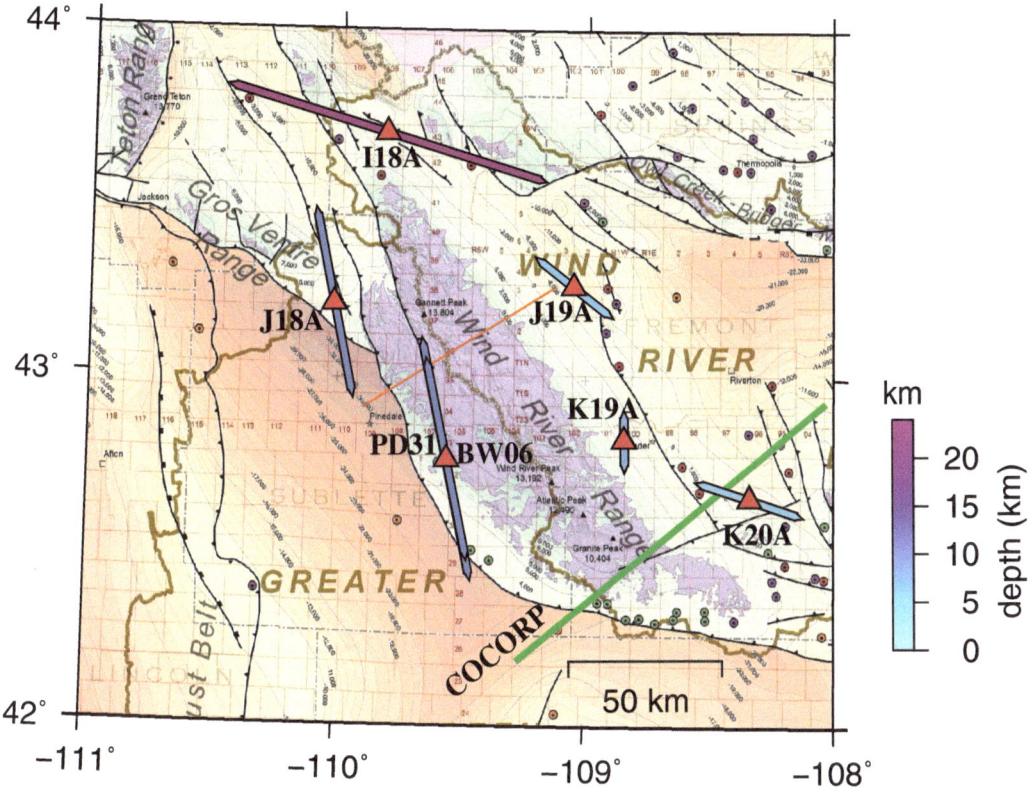

Figure 6

Map of the Wind River range and surroundings with stations (*red triangles*) discussed in text. The background map is a portion of a Precambrian basement map of Wyoming (Lynds 2013, Plate 2; updated from Blackstone 1993), with *purple* colors representing Precambrian outcrops, *green–yellow–brown* colors showing basement elevation above (+) or below (−) mean sea level with contour lines in 1,000 ft contour intervals (multiples of 5,000 ft have heavy contour lines), and faults in *black*. *Colored dots* show well control points for basement depth; *purple dots* are wells drilled to the Precambrian, *green* into or through the Precambrian, all other *colored dots* are wells drilled to younger units. The approximate location of the COCORP reflection line is indicated in *green*, location of a structural cross section (Steidtmann and Middleton 1991) in *orange*, and fault traces in *black*. *Bars* at the stations are from this study and show the strike of the largest amplitude $\sin(\phi)$ periodic arrival observed at each station, representing the strike of the strongest subsurface dipping isotropic contrast or foliation contrast, with the *bar color* showing approximate depth of the converter from the surface (see *color scale*). Stations PD31 and BW06 are collocated

stations atop the thrust fault (PD31, BW06, J18A) show matching azimuthally varying arrivals on the radial and tangential components (Figs. 8, 9, 10) as modeled in Sect. 2. Figure 8a shows azimuthally binned radial and tangential component receiver functions observed at station PD31 (Fig. 6). Note the polarity flips in the prominent tangential component arrival near 1 s. Figure 8b shows radial and tangential component synthetics for a model based on the COCORP results (Smithson *et al.* 1979; Figs. 7, 8c).

On the radial component, the synthetics show the arrival from the dipping basement-over-sediment (fast-over-slow) interface at 1 s, the arrival from the bottom of the sediment wedge near 2 s, and the Moho arrival near 4 s. The tangential component shows the polarity reversed arrival near 1 s from the dipping fault contrast and the corresponding zero delay bent direct P. Corresponding arrivals are seen in the data (Fig. 8a). The position of the polarity flips on both components in the synthetics (Fig. 8b) is controlled by the strike of the dipping fault and matches the observed polarity flip azimuths (Fig. 8a). Using an upper basement velocity of 5.79 km/s (Smithson 1979) and a V_p/V_s ratio of 1.73 (Brocher 2005) to convert the arrival from the dipping interface to depth results in a depth of the fault under the station that matches the active source results (Fig. 7a; Steidtmann and Middleton 1991). Slight variations in arrival times over backazimuth are

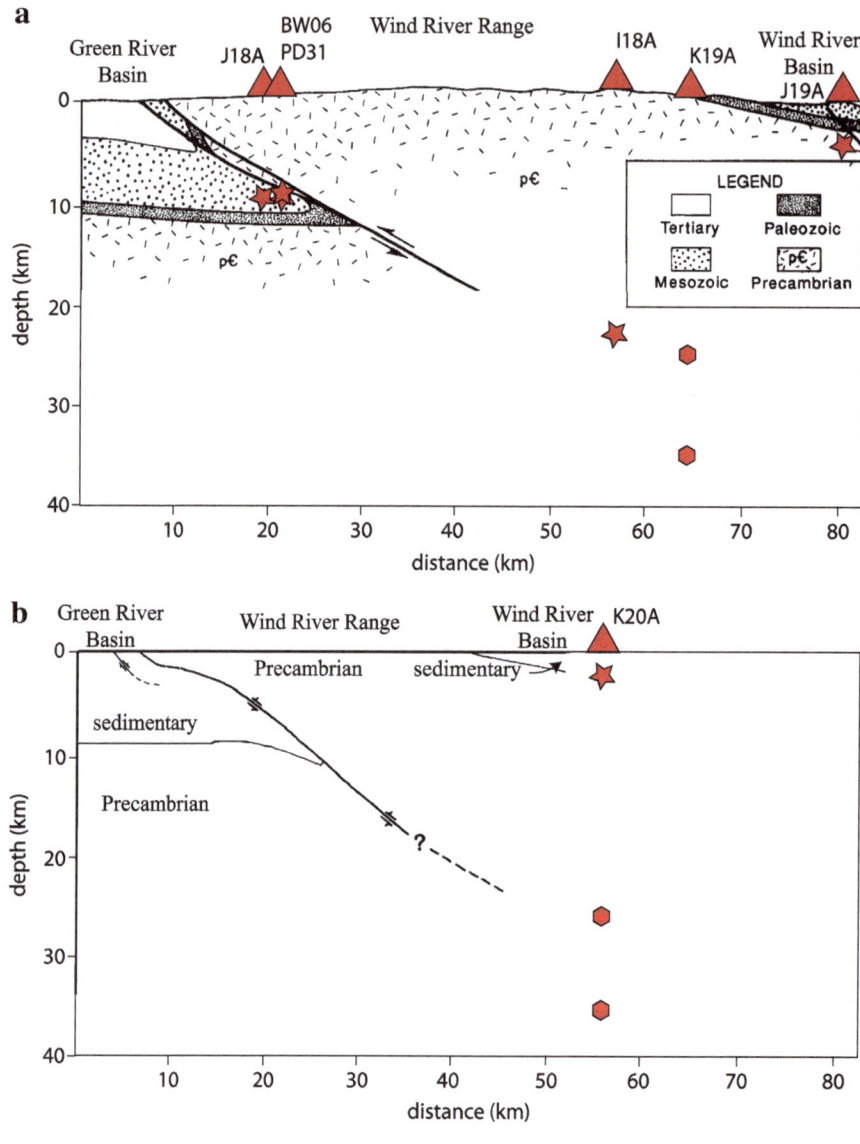

Figure 7

a Structural cross section across the central Wind River range (*orange line* in Fig. 6; Steidtmann and Middleton 1991, Fig. 4d therein; after Stone 1987). **b** Structural cross section across southern Wind River Range (*green line* in Fig. 6) after Smithson *et al.* (1979)(Fig. 17 therein). *Red triangles* are station locations at distances projected perpendicularly onto the Wind River thrust trace (Fig. 6). *Red stars* are depths of imaged dipping structures. Hexagons mark top and bottom of possible subhorizontal shear zone

consistent with the northeastward dip of the contrast (with later arrivals from the downdip direction), although the variation is modest; the azimuthal polarity behavior is a much more obvious and robust observable of the fault conversion than the azimuthal delay time variation. The observations are thus consistent with an isotropic velocity contrast of the basement-over-sediment thrust geometry inferred from the COCORP profile (Fig. 7).

Reading the azimuthally varying signal in the radial component receiver functions is complicated by the presence of azimuthally invariant arrivals from the flat-layered structure. Figure 9 shows a simplified way to pick fault strike. In Fig. 9a, the average radial component receiver function was subtracted from each individual radial component receiver function to isolate the azimuthally varying signal $(R - R_0)$. In Fig. 9b, the tangential component receiver functions

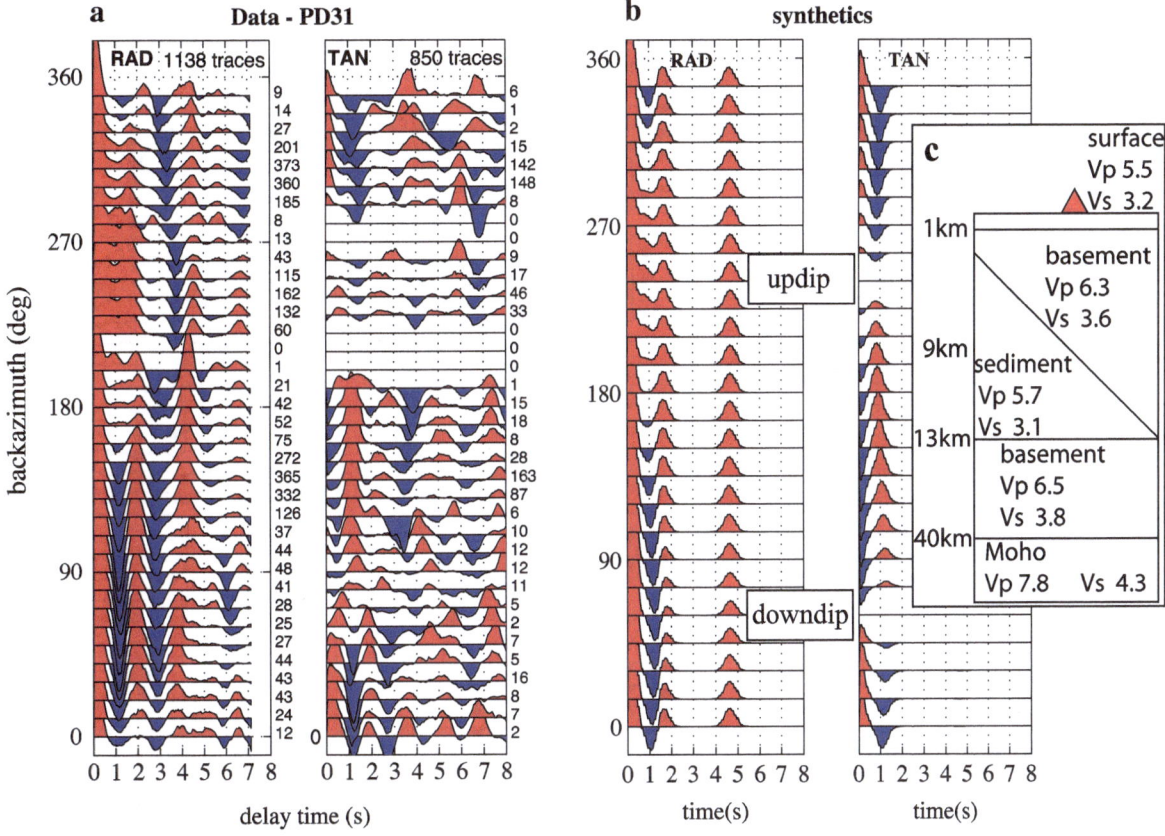

Figure 8

a Azimuthally binned receiver functions for station PD31 stop the Wind River overthrust (Fig. 6); (*left*) radial component average receiver functions, (*right*) tangential. The number on the *right* of each trace indicates how many individual receiver functions contributed to the trace in the azimuth bin. Bins are 10° wide and smoothing was applied by 5° into each neighboring azimuth bin. **b** Synthetic seismograms for a model of the Wind River thrust (see **c**). Amplitude scale is the same in all panels of **a** and **b**, and time scale is corrected to vertical incidence. **c** Sketch of model used to calculate synthetic seismograms in **b**. The fault dips down 44° to ENE and strikes NNW. Moho arrival at 4–5 s in radial sections

were plotted after shifting them in backazimuth by $+90°$ ($T(\phi - 90°)$); note that shifting the receiver functions in backazimuth by $+90°$ is the equivalent of shifting the backazimuth by $-90°$) so that the polarity nodes are now located along the strike of the fault and the signal matches the azimuthal component in the radial receiver functions ($R - R_0$). This treatment allows joint azimuthal binning of the corrected radial and tangential components (Fig. 9c) for more accurate determination of the azimuths of polarity reversal (fault strike). We fit a $\sin(\phi)$ periodic function over backazimuth to the combined receiver function amplitudes at each time step as described in SCHULTE-PELKUM and MAHAN (2014) to determine the azimuthal position of extrema and nodes. The resulting strike of

162° matches the mapped fault strike near PD31 (Fig. 6). Figure 10 shows receiver functions before and after the same treatment for the other two stations atop the overthrust (BW06, Fig. 10a; J18A, Fig. 10b). The result at BW06 matches those at the collocated station PD31 closely (Figs. 6, 7, 9c, 10a); a slight reduction of signal amplitude at BW06 is likely due to the difference in noise conditions between its surface sensor compared to the presumably quieter borehole sensor at PD31. The largest $\sin(\phi)$ periodic arrival at station J18A has a strike (162°; Figs. 6, 10b) very similar to the two stations to the south, also matching the strike of local surface faulting, with an amplitude close to that seen at BW06. Unlike BW06 and PD31, J18A is located on a significant sediment layer

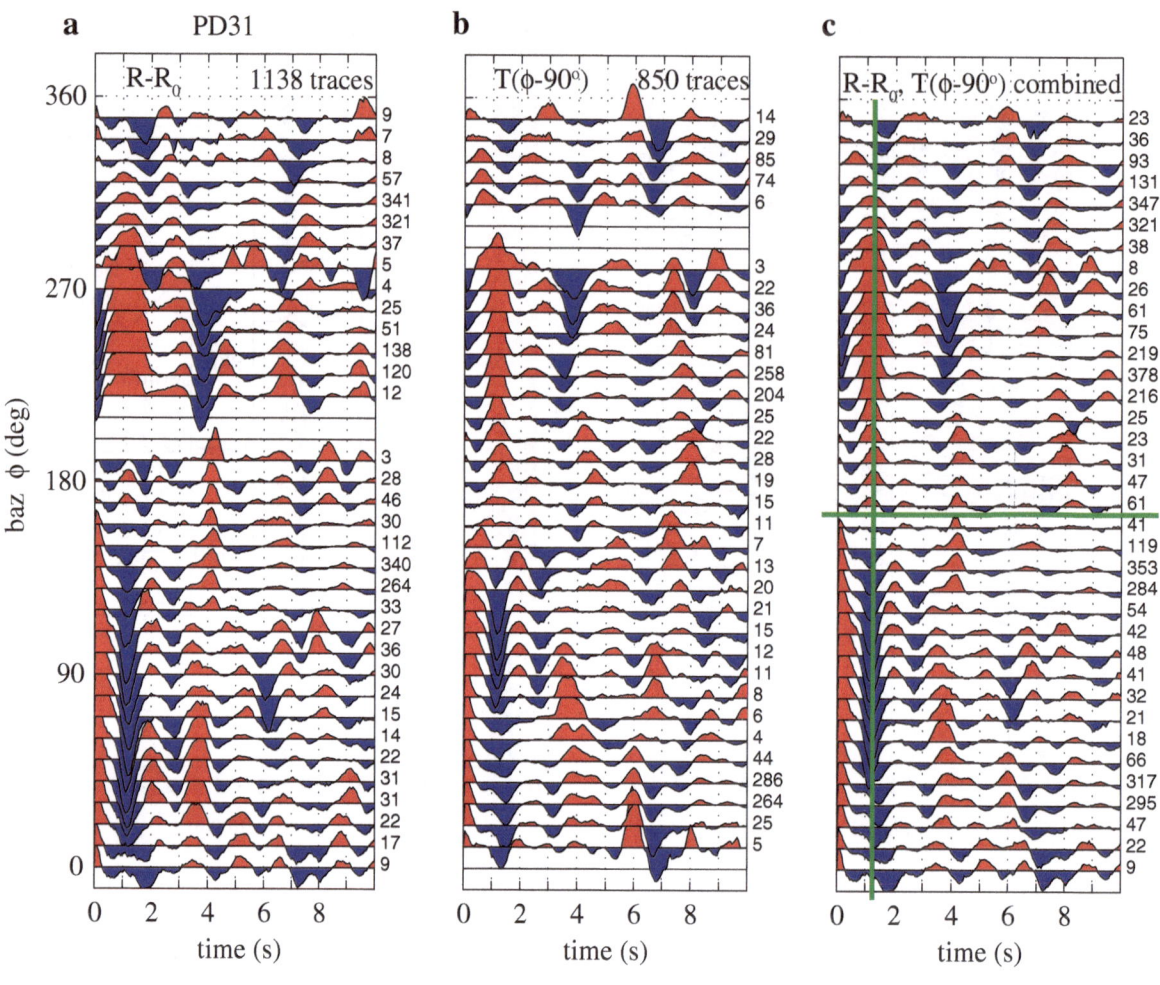

Figure 9

Azimuthally binned receiver functions from the same station (PD31) as in Fig. 9a. **a** Radial component after subtracting the azimuthal average. **b** Tangential component receiver functions after shifting the backazimuth clockwise by 90°. Note the match with **a**. **c** $R - R_0$ from **a** and $T(\phi - 90°)$ from **b**, binned together. Delay time of fault conversion and its strike as determined by finding the maximum amplitude $\sin(\phi)$ periodic fit over backazimuth ϕ shown as *green lines*. The number on the right indicates the number of receiver functions per bin in each panel. Azimuthal binning includes 5° smoothing into each neighboring bin

(Fig. 6). We use a sediment V_p value of 3.36 km/s published by (SMITHSON *et al.* 1979) and a V_p/V_s ratio of 1.9 (BROCHER 2005) to convert the delay time of the dipping converter arrival to depth, with $V_p = 5.79$ km/s and $V_p/V_s = 1.73$ as for the stations discussed above below the sediment-basement interface, and obtain a depth to the fault similar to that at PD31 and BW06 (Fig. 7a).

In order to search for evidence of the thrust soling into a subhorizontal detachment (Fig. 1a) on the inboard eastern side of the Wind River Range, we analyzed data from four TA stations east of the range above the possible location of a detachment (Fig. 6). Since the deeper

Figure 10

Azimuthally binned receiver functions for stations BW06 and J18A atop the Wind River overthrust. **a** All for BW06, from *left* to *right*: Radial receiver functions (Moho at 4–5 s) as in Fig. 8a; tangential receiver functions as in Fig. 8b; $R - R_0$ as in Fig. 9a, with the average radial receiver function that was subtracted from each trace in **a** shown in *green* and *yellow* on *top*; $T(\phi - 90°)$ as in Fig. 9b. **b** Same as in **a**, but for station J18A, Moho appears at 5–6 s on R and $R - R_0$. *Green lines* show strike azimuth and delay time of the fault conversion as determined by solving for the maximum amplitude $\sin(\phi)$ fit over backazimuth; polarity flips at the strike azimuth

portion of the fault and a possible detachment presumably juxtapose basement against basement with only a few kilometers of accumulated strain, a signal, if it

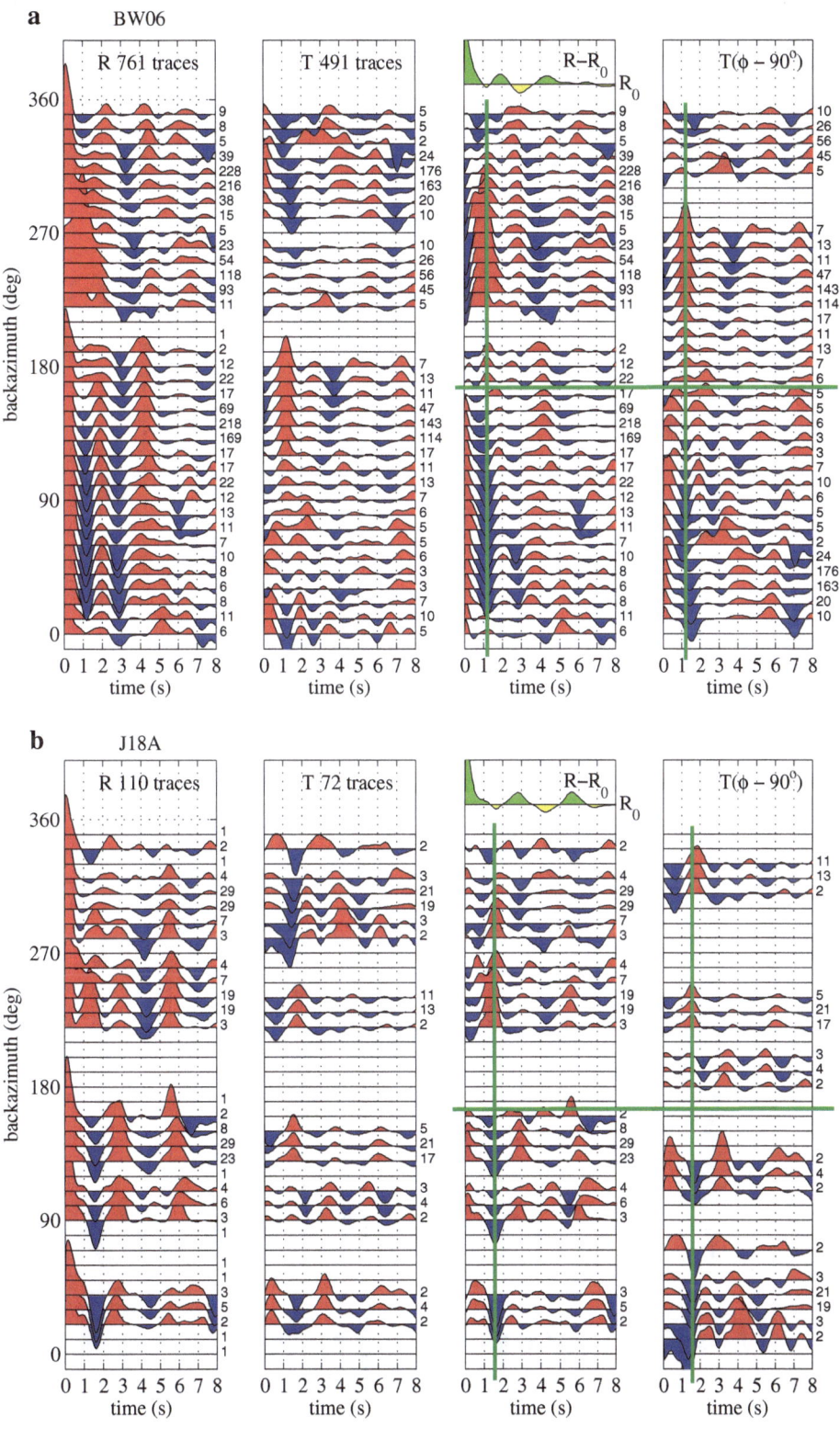

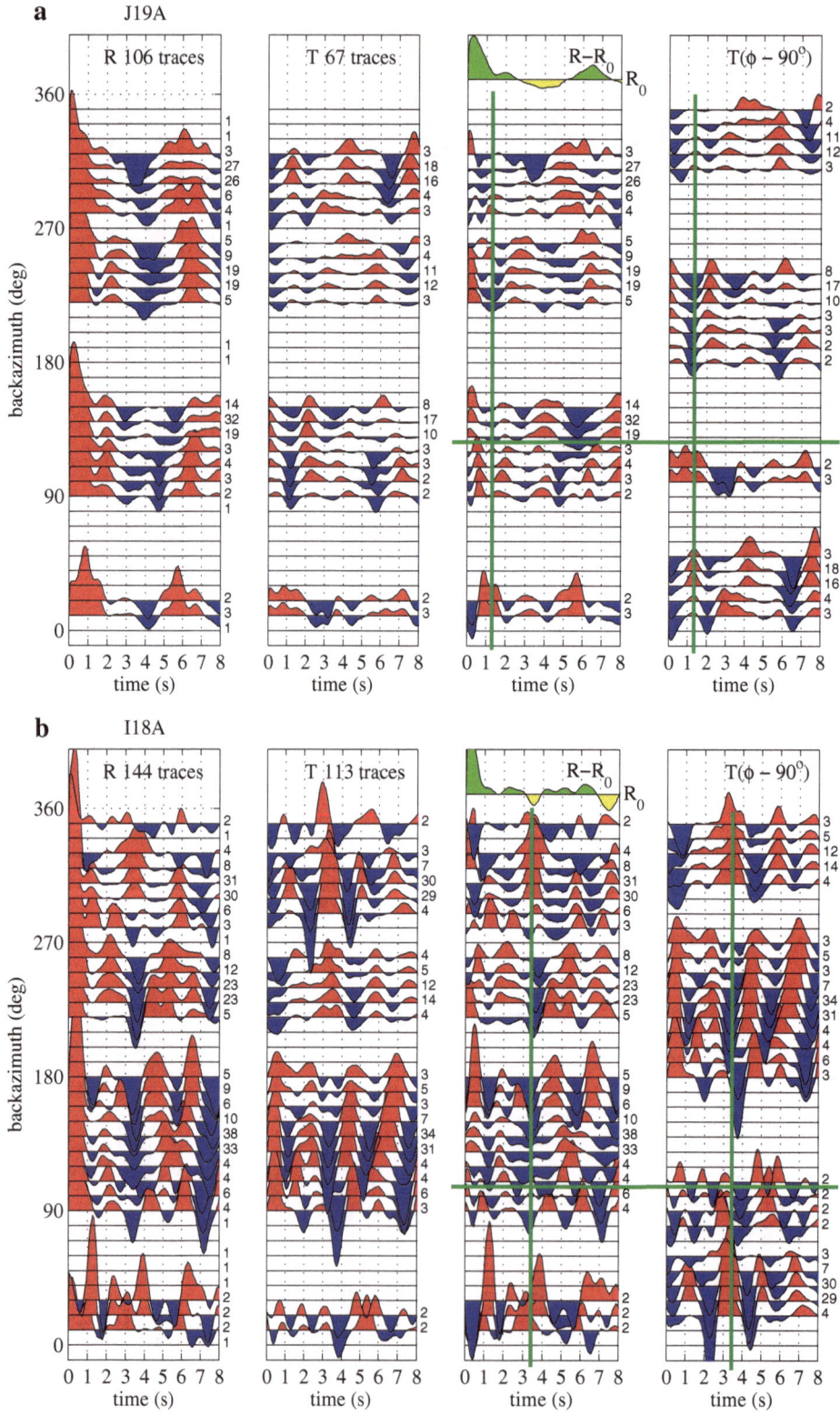

Figure 11
As in Fig. 10, but for stations J19A (**a**; Moho ~6 s) and I18A (**b**; Moho unclear)

exists, would likely have to stem from a foliation contrast, possibly due to shear fabric developed on the detachment which is presumed to be below the brittle-ductile transition (SIBSON 1983). Deep crustal fabric associated with shortening has been observed in the Himalaya (NABELEK et al. 2009; SCHULTE-PELKUM et al. 2005); however, the amount of strain on the Wind River Thrust is much smaller than the hundreds of kilometers on the Main Himalayan Thrust, and Laramide midcrustal deformation likely occurred under lower temperatures (e.g. lowermost crustal residence temperatures in northern Wyoming craton were ~500 °C for the last 1.5 b.y.; BLACKBURN et al. 2011; MAHAN 2006). Neither factor is favorable for the development of a shear zone wide enough to be detected by teleseismic P receiver functions; the signal pulse width is on the order of 0.5–1 s, and a subhorizontal layer should be ~5 km thick to avoid destructive interference between the conversions from its top and bottom. The rays have longer travel paths between the boundaries of steeply dipping shear zones so that 2–3 km can be a sufficient thickness (Sect. 4).

There is no clear signature of a shear zone with high-angle fabric at the TA stations on the east side of the Wind River Range (J19A, Fig. 11a; K19A, Fig. 12a; K20A, Figs. 6, 12b). While the stations show azimuthally varying arrivals, their amplitudes are much smaller than those at the stations west of the range (PD31, BW06, J18A) and the depths are not consistent with a downdip extension of the thrust fault (which is presumably as deep or deeper than at the stations on the west side). A strong azimuthally varying signal is seen at station I18A in the northeast (Figs. 6, 11b). It is unclear whether this signal is related to the Wind River thrust; it may more likely be due to an unrelated younger structure, since the station sits at the boundary between the Wind River Range and the Absaroka range which is originally Laramide age as well, but has been overprinted with Tertiary volcanics. The largest amplitude arrivals with sin(ϕ) azimuthal periodicity at stations J19A (Fig. 11a) and K20A (Fig. 12b) match the strike and depth of the sediment-basement contact (Figs. 6, 7). K19A (Fig. 12a), situated on the northeastern flank of the range just before the basement dips down beneath the Wind River Basin east of the range, shows a smaller arrival with no obvious match to mapped subsurface structure (Fig. 6).

An interesting midcrustal arrival pair is seen on the radial component of station K19A (Fig. 12b). The arrival pair consists of a positive polarity arrival at 2.7 s delay time and a negative polarity arrival at 4.0 s when picked on the average radial component receiver function (green-yellow trace in Fig. 12a). This signal may be caused by horizontal shear-parallel foliation within a shear zone of 5–6 km thickness with isotropic material above and below (Fig. 5). A signal on the tangential component is not generated by this geometry and is not observed at K19A (Fig. 12a). Depths of the radial component arrivals using the same V_p and V_p/V_s values as above are shown in Fig. 7a. The depth range is deeper than the dipping portion of the fault imaged by COCORP and could be consistent in position and depth with a shear zone associated with a subhorizontal detachment (Fig. 7). The positive-over-negative signature would likely require a shape factor on the high end of the range seen in crustal materials (Fig. 5), although a systematic analysis shape factor as a function of composition in deformed crustal rocks has yet to be performed.

An alternative interpretation of the arrival pair is a several-kilometer-thick isotropic, high-velocity, midcrustal layer generating a positive velocity contrast with depth at its top and a negative velocity contrast at its bottom. Such a layer may be generated by a thick mafic sill, or a set of mafic sills, possibly associated with the southern termination of a lower crustal high-velocity layer observed in northern Wyoming and in Montana (termed "7.xx" layer in some publications; e.g. GORMAN et al. 2002; see MAHAN et al. 2012). An isotropic high-velocity layer is unlikely to be generated by shear along a detachment; the basement rocks exposed in the Wind River Range are high-grade metamorphic (amphibolite- to granulite-facies), and shear under midcrustal P–T conditions, especially if hydrous, would likely cause retrogression to amphibolite or lower facies with a concomitant decrease in seismic velocity, rather than an increase as observed in the data.

Multiples from shallow crustal conversions are another mechanism for generating signals in this delay time range. However, station K19A does not show characteristics of sediment-affected receiver functions. Station K20A in the Wind River Basin

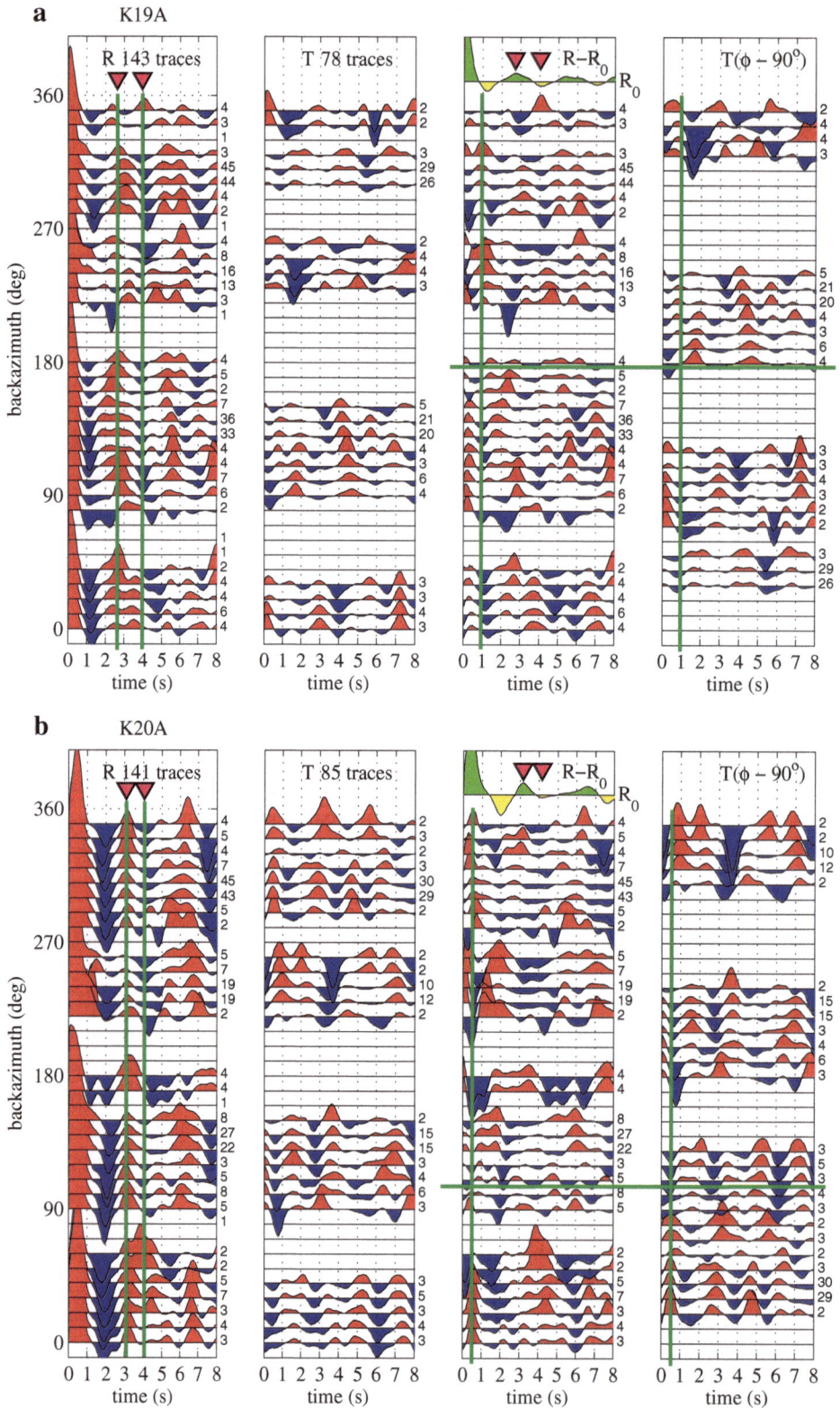

Figure 12

As in Fig. 10, but for stations K19A (**a**; Moho 5–6 s) and K20A (**b**; Moho 6–7 s). Additionally, *magenta triangles* in first and third panel and *green lines* in first panel from the left mark arrivals in the average radial receiver function (*green/yellow*) that may be associated with a subhorizontal shear zone or deep isotropic horizontal contrast

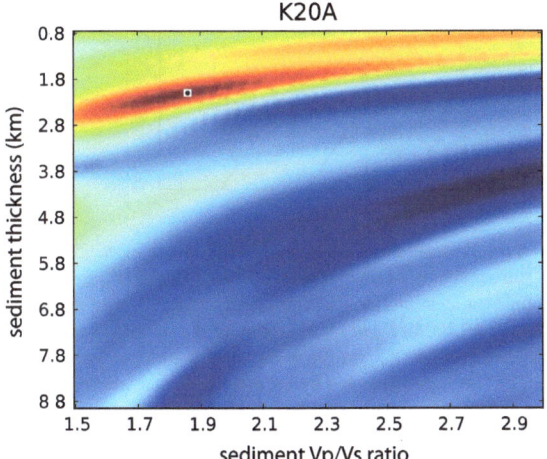

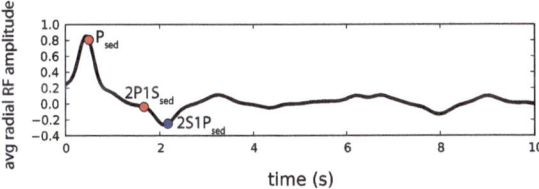

Figure 13

Top high-frequency *H–κ* stack (ZHU and KANAMORI 2000) of sediment reverberations (Yeck *et al.* 2013) to determine sediment thickness at station K20A in the Wind River Basin (Figs. 6, 7, 13). The result matches the mapped value (Fig. 6). *Bottom* average radial component receiver function and predicted arrival time of direct conversion from sediment-basement interface and its multiples using the values determined from **a**, identifying the prominent negative arrival near 2 s as the second reverberation. Figure courtesy of YECK (2013, pers. comm.)

does show a basin signal (delayed and broadened direct P arrival and a negative amplitude second sediment reverberation; Figs. 12, 13) matching mapped basement depth (Fig. 6), but also shows separate arrivals near 3 and 4 s (Fig. 12b) similar to arrivals at comparable delay times at K19A. As for the case of K19A, the midcrustal signal at K20A may be due to a mafic high-velocity layer, or alternatively it may be similar to the radial arrivals in the synthetic example for subhorizontal shear zone foliation (Fig. 5a).

Station J19A has the thickest sediment layer among the stations analyzed here (nearly 5 km, which is more than double the thickness at J18A and K20A), and in this case the sediment reverberations obscure mid to lower crustal arrivals between 2 and 5 s (negative amplitude arrivals in Fig. 11a, leftmost panel). Denser station coverage in the area would aid the interpretation of the midcrustal signal; with the presently available station coverage, we can not draw any conclusions on whether a deep crustal shear zone is visible in the data, although there are several candidate signals in the expected depth range. However, we can exclude the presence of a deep dipping isotropic velocity contrast or deep dipping foliation at K19A and K20A associated with the deep extension of the thrust based on the lack of high-amplitude, sin(φ) periodic late arrivals.

4. Predicted Signals Based on Geological Data from Exhumed Shear Zones

In this section, we attempt to gain ground truth estimates of the seismic signal expected from shear zones in the deep crust. The input data are microstructure-based calculations of seismic wavespeeds in combination with structural maps of rocks in exhumed deep crustal shear zones. We consider two settings, both of which represent amphibolite-facies deformation in the middle to lower crust. The first setting is the Upper Badcall shear zone in NW Scotland (TATHAM *et al.* 2008) and the second is the Grease River shear zone in the Western Canadian Shield, northern Saskatchewan (DUMOND *et al.* 2013).

4.1. Upper Badcall Shear Zone

Figure 14 shows the geometry of the test case based on the Upper Badcall shear zone. A steeply dipping shear zone with foliation parallel to the shear zone boundaries is surrounded by rocks with weakly developed foliation with a different orientation inherited from older processes. The geometry is simplified from that shown by TATHAM *et al.* (2008, their Figs. 1a and b). We use the elasticity tensors of the 'least deformed' and 'most deformed' samples from TATHAM *et al.* (2008, see their Fig. 4).

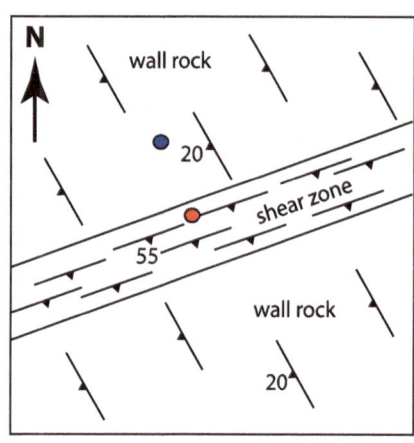
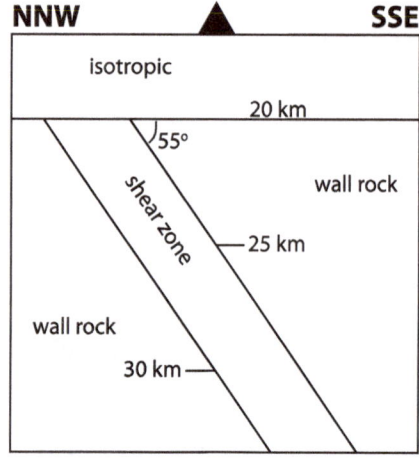

Figure 14

Left map view of a highly simplified version of the structural map of the Upper Badcall shear zone, Scotland (TATHAM *et al.* 2008, their Fig. 1a). The shear zone strikes 70° with steeply SSE-dipping foliation and is surrounded by weaker shallow WSW-dipping fabric striking at a wide angle to the shear zone. Tensors used for seismic modeling are from samples at locations indicated by *blue* (wall rock; 'least deformed' in TATHAM *et al.* 2008) and *red circles* (shear zone; 'most deformed' in TATHAM *et al.* 2008). *Right* Cross section of the simplified shear zone model. While the shear zone mapped by TATHAM *et al.* (2008, their Fig. 1b) has near vertical dip, the dip was set to 55° from horizontal in the model because the 2-D modeling code does not allow layers to pinch out within the region sampled by the rays. Foliation within the shear zone parallels the shear zone dip. The mapped shear zone is only 100 m thick; in order to separate conversions from the *top* and *bottom* of the shear zone in time, it is set to 2.8 km thickness in the model

Approximate sample locations are indicated in Fig. 14 (see exact locations in TATHAM *et al.* 2008, their Fig. 1a; sample 2 is the 'least deformed' or wall rock sample, and sample 7 is the 'most deformed' or shear zone sample). The elasticity tensors were obtained by TATHAM *et al.* (2008) via collecting electron backscatter diffraction (EBSD) data of thin section areas to characterize the deformation fabric and mineral crystallographic orientation in the samples, and using the AniCh5/MTEX software of MAINPRICE (1990) and MAINPRICE and HUMBERT (1994) to calculate bulk properties based on EBSD mapping and single crystal properties.

We calculate synthetic seismograms for a simplified model with an isotropic upper crust and a middle and lower crust consisting of the buried dipping shear zone with the 'most deformed' tensor, sandwiched between two layers with the elastic tensor of the 'least deformed' sample. The tensors are oriented as in the field. The actual shear zone is 100 m wide; we vary shear zone width to determine the influence on the predicted seismic signal. The shear zone dip is somewhat shallower than in the field to accommodate limitations of the modeling code, which does not

handle three-dimensional structure (layers may not pinch out within the area traversed by rays) and to avoid exceptional behavior when rays travel within the fault zone. Maximum V_p anisotropy of the sample used to represent the shear zone is 6.0 %; for the sample used to represent the deep crust outside the shear zone, it is 3.8 %. Both tensors have a large hexagonal symmetry component with some recognizable point maxima due to amphibole alignment (TATHAM *et al.* 2008; velocity hemispheres in their Fig. 4d).

Synthetic seismograms were calculated using a version of FREDERIKSEN and BOSTOCK's (2000) raysum code modified to allow input of a complete elastic tensor. Results are shown in Fig. 15. Conversions appear from the bottom and top of the shear zone, with a weaker conversion from the transition to the isotropic upper crust. Because of the steep dip and the large depth of the modeled shear zone, the variation in delay time (later arrivals from downdip azimuths) is obvious. As is the case of the synthetic anisotropic tensor (Fig. 4), the position of the nodes on the tangential component and the maximum amplitude on the radial component are in the up- and downdip

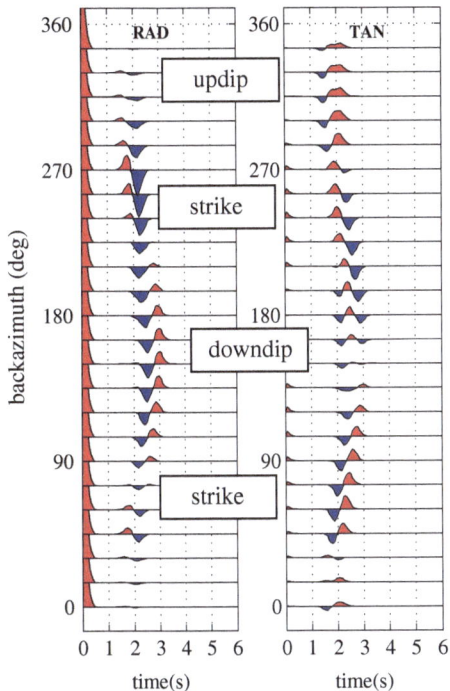

upper frequency limit used in receiver functions can be pushed higher, such that the resolution limit is a function of the site.

Changing the foliation orientation outside the shear zone does not change the conversions sufficiently to remove the destructive interference, as the signal remains dominantly that of the foliation orientation within the shear zone; the signal appearance is controlled by the largest foliation contrast. Putting the strike of the foliation within the shear zone at an angle to the boundaries of the shear zone has minor effects on the observed azimuthal patterns until the angle becomes large, at which point the superposition between the effects of the foliation orientation and the shear zone orientation creates complicated effects. The latter scenario may not be very likely since one would expect the strike of the shear foliation to be related to the strike of the shear zone itself.

Figure 15

Synthetic radial (*left*) and tangential component (*right*) seismograms calculated for the simplified Upper Badcall shear zone model in Fig. 14. The horizontal contrast between isotropic upper crust and weakly foliated wall rock is small and only generates a weak arrival. The prominent arrival pair is the top and bottom side conversion from the dipping shear zone. As in the isotropic (Figs. 2, 8, 9, 10) and synthetic anisotropic (Fig. 4) cases, the polarity flip on the tangential component is aligned with the dip orientation, with polarity in the radial component constant across the dip orientation

4.2. Grease River Shear Zone

The second shear zone example is from the Athabasca granulite terrane in the western Canadian Shield in northern Saskatchewan, Canada. The simplified model (Fig. 16) is based on structural mapping presented by DUMOND et al. (2013, their Fig. 2A). The model is set up as in the Badcall shear zone example, with an isotropic upper crust overlying a dipping shear zone. The Grease River shear zone is much wider than the Upper Badcall shear zone (several kilometers; DUMOND et al. 2013, Fig. 2A therein).

The elasticity tensors used for calculating synthetic seismograms are based on EBSD measurements on a sample taken within the highly deformed portion of the shear zone (Fig. 16). Velocity spheres of the sample are shown in Fig. 17. While the wall rock has higher temperature granulite facies mineralogy, the shear zone itself was synkinematically retrogressed and hydrated (MAHAN 2006) and contains aligned amphibolite facies mineral assemblages. The shear zone anisotropy is dominated by aligned hornblende and biotite with a maximum peak-to-peak V_p anisotropy of 6.7 % (Fig. 17).

Since measurements on samples outside the shear zone are not available yet, we estimate a tensor

orientation of the shear-parallel foliation, and, therefore, constrain foliation and shear zone strike. The foliation within the shear zone dominates the signal; replacing the weak fabric outside the shear zone with isotropic material results in a very similar pattern. Shear zone widths of less than ~2 km lead to destructive interference between the signals from the upper and the lower shear zone boundary because the opposite polarity signals begin to overlap in delay time given the pulse width used here (~1 s, typical for teleseismic receiver functions). When the shear zone narrows to ~0.5 km, only very small amplitude portions of the conversions remain visible. A 100 m wide shear zone such as mapped by TATHAM et al. (2008) would not be detected with teleseismic signal frequencies. Depending on noise conditions, the

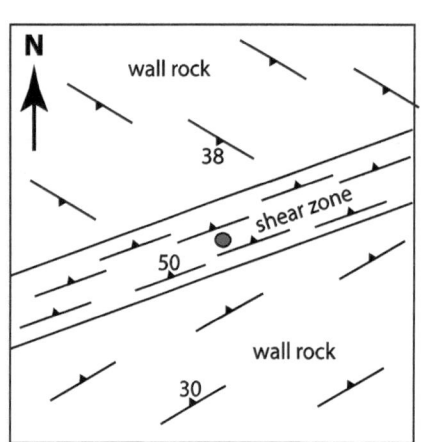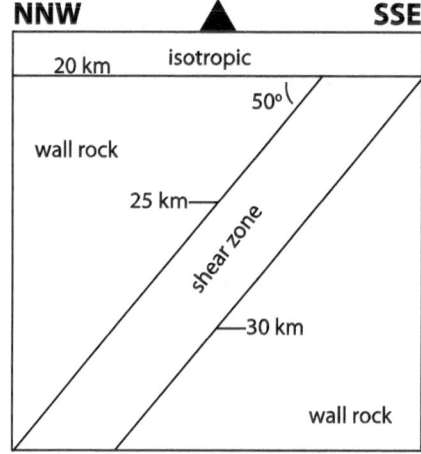

Figure 16

Left map view of a highly simplified version of the structural map of the Grease River shear zone, western Canadian Shield (DUMOND *et al.* 2013, their Fig. 2a). The shear zone strikes 70° with steeply NNW-dipping foliation and is bordered by weaker intermediate-angle SW-dipping fabric to the north, and intermediate-angle NW-dipping fabric to the south. *Right* Cross section of the simplified shear zone model. The original shear zone dip of ~70° mapped by DUMOND *et al.* (2013) was reduced to 50° in the model because of modeling geometry limitations. Foliation within the shear zone parallels the shear zone dip. The mapped shear zone is 2–3 km wide, comparable to the 3.2 km width used for modeling. The sample used for modeling was collected within the shear zone (*red dot*)

calculated from the shear zone sample using the same assemblage, but without the biotite. This is based on the observation that hornblende is stable in the higher grade wallrock, but the stable K-bearing mineral is alkali feldspar instead of mica (WILLIAMS *et al.* 2000; DUMOND *et al.* 2010). The resulting proxy tensor for the wall rock has a maximum 4.1 % V_p anisotropy (Fig. 17).

The synthetic seismograms (Fig. 18) show a very similar pattern to those from the Upper Badcall scenario. The tangential component polarity flips and the highest amplitudes in the radial component occur in the up-/downdip orientation of the foliation and shear zone dip. The higher anisotropy within the shear zone controls the appearance of the signal, while the orientation of the weaker inherited fabric surrounding the shear zone (which is oriented differently between the two sides of the shear zone, in contrast to the Upper Badcall setting) has a very minor influence on the appearance of the conversions. As in the other cases, the strike and depth of the shear zone can be constrained by polarity flips on the tangential component.

5. Discussion and Conclusions

We have demonstrated characteristic signals in receiver functions arising from crustal structure associated with faults and shear zones. While the structural features can be fundamentally different (across-fault isotropic velocity contrast from contrasts in lithology versus foliation changes within similar lithology across a shear zone), the converted waves share the same identifying characteristics of a 360° periodicity in radial and tangential components, with tangential component polarity nodes accompanied by local radial amplitude extrema identifying strike orientation. Parameters that are uniquely constrained are strike and depth (from delay time) of the fault or shear zone foliation. Parameters that trade off in determining the amplitude and sense of polarity of the conversions are isotropic contrast dip, magnitude of isotropic contrast, the sign of isotropic velocity contrast, foliation dip, strength of anisotropy associated with foliation, and shape factor of the anisotropy. None of these latter parameters change the azimuthal position of the tangential nodes and radial extrema.

Grease River - shear zone

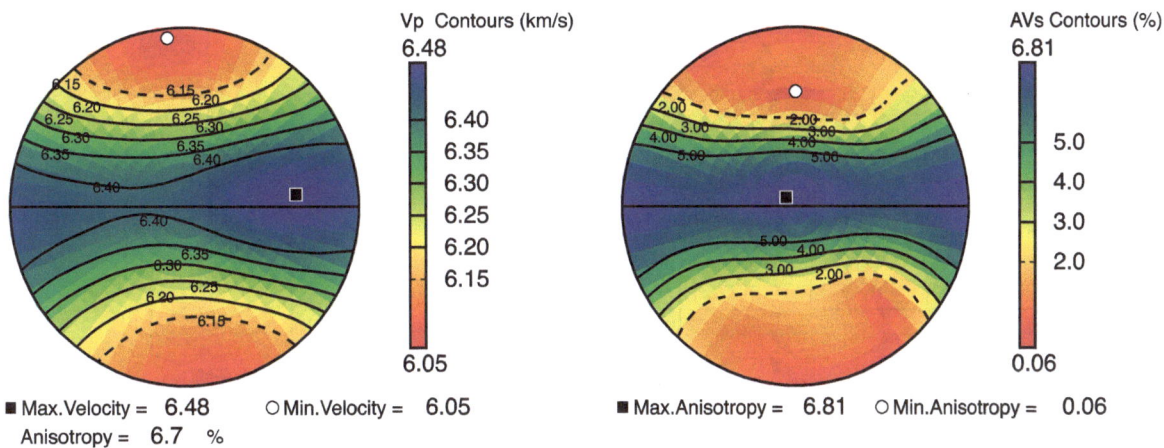

Grease River - wall rock

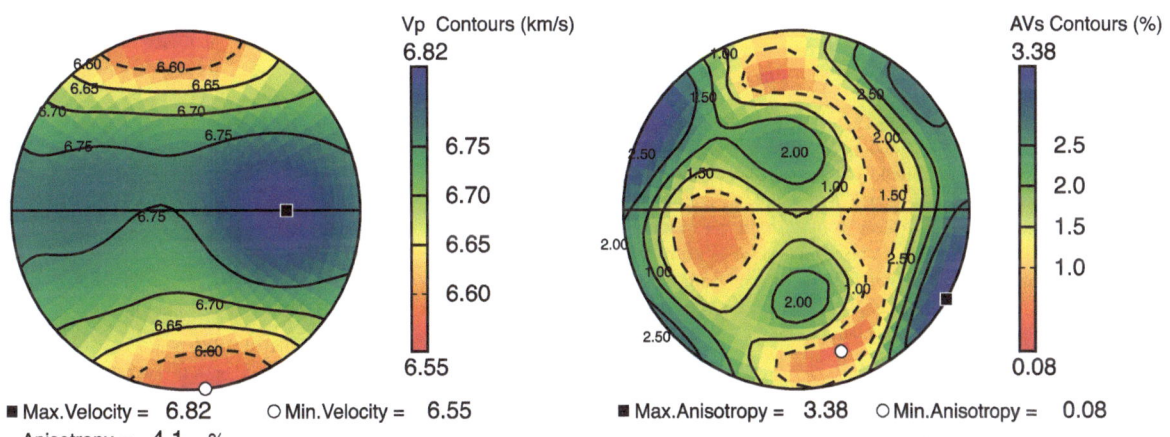

Figure 17

Lower hemisphere velocity plots for V_p (*left*) and ΔV_s (*right*) for elasticity tensors of samples from the Grease River shear zone. The shear plane is horizontal in each plot. The V_p and V_s hemispheres at *top* are for a sample collected within the highly deformed shear zone (*red dot* in Fig. 16). The V_p and V_s hemispheres shown on the *bottom* are for a tensor derived from the scanned shear zone sample after removing the biotite contribution from the elasticity tensor calculation. We use this tensor as a proxy for the wall rock surrounding the shear zone

A narrow shear zone with well developed anisotropy within the shear zone and sharp contrasts to the surrounding medium generates destructive interference between the conversions from either side of the shear zone, such that the signal amplitude is greatly diminished for shear zone widths of less than ∼2 km using typical receiver function pulse lengths of the order of 1 s. Imaging of narrower shear zones may be possible for quiet stations by using higher frequencies. For wider shear zones, an anisotropy

contrast of a few percent is sufficient to generate a first order signal in standard receiver functions (i.e. conversion amplitudes comparable to that from usual receiver function targets like the Moho).

Our method allows mapping of basic structural information on faults and shear zones to depth without assumptions on or inversions for composition, absolute velocities, and details of anisotropic symmetry (LEVIN and PARK 1997, 1998; LEIDIG and ZANDT 2003; VERGNE *et al.* 2003; SHERRINGTON *et al.*

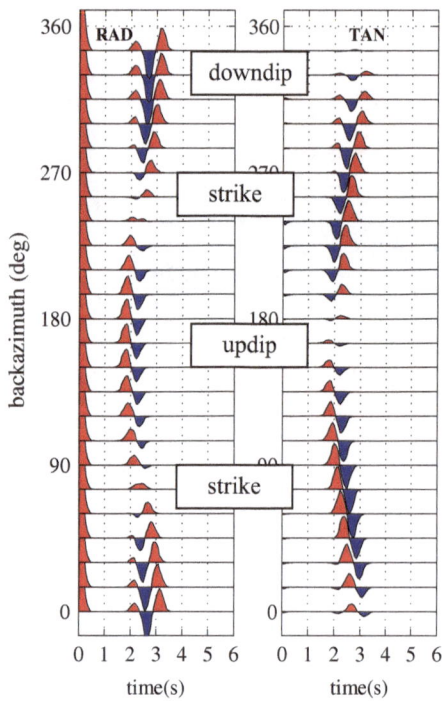

Figure 18

Synthetic radial (*left*) and tangential component (*right*) seismograms calculated for the simplified Grease River shear zone model in Fig. 16. The prominent arrival pair is the top and bottom side conversion from the dipping shear zone. As in the isotropic (Figs. 2, 8, 9, 10) and synthetic anisotropic (Fig. 4) cases, the polarity flip on the tangential component is aligned with the dip orientation, with polarity in the radial component constant across the dip orientation

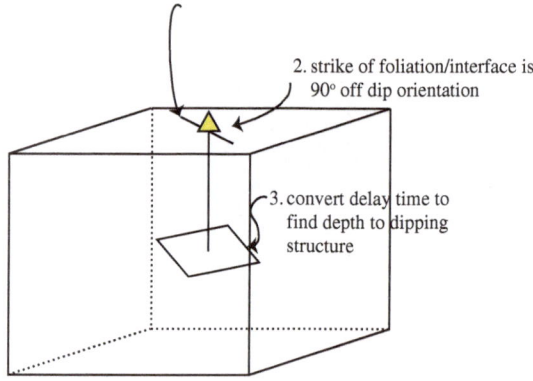

1. find polarity flips in tangential component with large amplitudes (between nodes) in radial component: these are up-/downdip azimuth

2. strike of foliation/interface is 90° off dip orientation

3. convert delay time to find depth to dipping structure

Figure 19

Summary of the technique. Required is a dipping isotropic interface or a contrast in foliation that is pierced by teleseismic P with sufficient azimuthal coverage to determine polarity reversals in the tangential component. In that case, the steps above determine the strike and depth of the underlying structure. The observations are independent of whether the target is a dipping isotropic interface or dipping foliation, and independent of the sense and strength of isotropic velocity contrast and strength or details of anisotropic symmetry, so that strike and depth of fault-associated features can be mapped without having to resolve the attendant tradeoffs

2004; HAYES and FURLONG 2007; OZACAR and ZANDT 2009; PORTER *et al*. 2011; ECKHARDT and RABBEL 2011; LIU and NIU 2012). Figure 19 summarizes the common features in seismograms from dipping foliation contrasts or isotropic contrast interfaces. Imaging can be performed with a single station provided the available teleseismic events cover at least two backazimuthal quadrants and ideally more. The minimum required azimuthal coverage depends on the strike orientation of the structure relative to the event backazimuths. While the method is single-station based in principle, tracing the same structure across several stations is ideal. The common features found between dipping isotropic velocity contrasts and dipping foliation make the technique particularly useful for mapping faults and shear zones since a distinction between the former two features is not required.

Acknowledgments

This material is based upon work supported by the National Science Foundation under Grant Numbers EAR-1251193, EAR-1053291, EAR-0948581, EAR-0750035, and EAR-1252295. Waveforms and metadata for TA, US, and IM networks were accessed via the Incorporated Research Institutions for Seismology (IRIS) Data Management System, specifically the IRIS Data Management Center; the IRIS DMS is funded through the National Science Foundation and specifically the GEO Directorate through the Instrumentation and Facilities Program of the National Science Foundation under Cooperative Agreement EAR-1063471. Data from the TA network were made freely available as part of the EarthScope USArray facility, operated by IRIS and supported by the National Science Foundation, under Cooperative Agreements EAR-0323309, EAR-0323311, EAR-0733069. We thank Will Yeck for performing sediment thickness stacking and providing Fig. 14. Comments by editor Yehuda Ben-Zion and two anonymous reviewers helped improve the manuscript significantly. GMT software was used to prepare some figures (WESSEL and SMITH 1998) and AniCh5/ MTEX was used to calculate elastic properties (MAINPRICE 1990; MAINPRICE and HUMBERT 1994).

REFERENCES

ABERS, G.A. (1998), *Array measurements of phases used in receiver-function calculations: Importance of scattering*, Bull. Seis. Soc. Am. 88, 313–318.

AKI, K. and RICHARDS, P.G. (1980), *Quantitative seismology: Theory and methods*, v. 1: W.H. Freeman and Co.

ALLAM, A.A., BEN-ZION Y. (2012), *Seismic velocity structures in the Southern California plate-boundary environment from double-difference tomography*, Geophys. J. Int., *190*, 1181–1196.

ALLAM, A.A., Y. BEN-ZION and Z. PENG (2014), *Seismic imaging of a bimaterial interface along the Hayward fault, CA, with fault zone headwaves and direct P arrivals*, Pure Appl. Geophys., in press.

AMMON C.J., G.E. RANDALL, and G. ZANDT, (1990), *On the non-uniqueness of receiver function inversions*, J. Geophys. Res., *95*, 15,303–15,318.

BABUSKA, V. and M. CARA, *Seismic Anisotropy in the Earth*, Kluwer Academic Publishers, Dordrecht, 1991.

BEN-ZION, Y. and P. MALIN (1991), *San Andreas fault zone head waves near Parkfield, California*, Science, *251*, 1592–1594.

BEROZA, G., and S. IDE (2009), *Deep tremors and slow quakes*, Science, *324*, 1025–1026.

BERTHÉ, D., P. CHOUKROUNE, and P. JEGOUZO (1979), *Orthogneiss, mylonite and coaxial deformation for granites – example of the South Armorica Shear Zone*, Journ. Struct. Geol. *1*, 31–42.

BLACKBURN, T.J., BOWRING, S., SCHOENE, B., MAHAN, K.H., DUDAS, F. (2011), *U-Pb Thermochronology: creating a temporal record of lithosphere thermal evolution*, Contributions to Mineralogy and Petrology, *162*, 479–500, doi:10.1007/s00410-011-0607-6.

BLACKSTONE, D.L. JR. (1993), *Precambrian basement map of Wyoming – Outcrop and structural configuration*: Geological Survey of Wyoming [Wyoming State Geological Survey] Map Series MS-43, scale 1:1,000,000.

BROCHER, T. (2005), *Empirical relations between elastic wavespeeds and density in the Earth's crust*, Bull. Seis. Soc. Am., *95*, 2081–2092, doi:10.1785/0120050077.

BULUT, F., Y. BEN-ZION and M. BONHOFF (2012), *Evidence for a bimaterial interface along the Mudurnu segment of the North Anatolian Fault Zone from polarization analysis of P waves*, Earth Planet. Sci. Lett., *327–328*, 17–22, doi:10.1016/j.epsl. 2012.02.001.

BURDICK, L. J., LANGSTON, C.A. (1977), *Modeling crustal structure through the use of converted phases in teleseismic body waveforms*, Bulletin of the Seismological Society of America, *67*(3), 677–691.

DUMOND, G., P. GONCALVES, M.L. WILLIAMS, and M.J. JERCINOVIC (2010), *Subhorizontal fabric in exhumed continental lower crust and implications for lower crustal flow: athabasca granulite terrane, western Canadian Shield*, Tectonics, *29*, TC2006.

DUMOND, G., K.H. MAHAN, M.L. WILLIAMS, and M.J. JERCINOVIC (2013), Transpressive uplift and exhumation of continental lower crust revealed by synkinematic monazite reactions, *Lithosphere*, *5*, 507–512.

EBERHART-PHILLIPS, D., and A.J. MICHAEL (1998), *Seismotectonics of the Loma Prieta, California, region determined from three-dimensional V_p, V_p/V_s, and seismicity*, J. Geophys. Res., *103*(B9), 21,099–21,120, doi:10.1029/98JB01984.

ECKHARDT, C. and W. RABBEL (2011), *P-receiver functions of anisotropic continental crust: a hierarchic catalogue of crustal models and azimuthal waveform patterns*, Geophys. J. Int. *187*, 439–479.

ERSLEV, E.A. (1986), *Basement balancing of Rocky Mountain foreland uplifts*, Geology, *14*, 259–262.

ERSLEV, E.A. (2005), *2D Laramide geometries and kinematics of the Rocky Mountains, Western U.S.A.*, Karlstrom, K.E. and Keller, G.R., editors, 2005, The Rocky Mountain Region – An Evolving Lithosphere: Tectonics, Geochemistry, and Geophysics: American Geophysical Union Geophysical Monograph *154*, pp. 7–20.

FARRA, V., VINNIK, L.P., ROMANOWICZ, B., KOSAREV, G.L. and KIND, R. (1991), *Inversion of teleseismic S particle motion for azimuthal anisotropy in the upper mantle: a feasibility study*, Geophys. J. Int., *106*, 421–431.

FREDERIKSEN, A.W. and BOSTOCK, M.G. (2000), *Modelling teleseismic waves in dipping anisotropic structures*. Geophysical Journal International, *141*, 401–412.

GORMAN, A.R., R.M. CLOWES, R.M. ELLISA, T.J. HENSTOCKB, A. LEVANDER, G.D. SPENCE, G.R. KELLER, K.C. MILLER, C.M. SNELSON, M.J.A. BURIANYK, E.R. KANASEWICH, I. ASUDEH, and Z. HAJNAL (2002), *Deep Probe: Imaging the roots of western North America*, Can J Earth Sci, *39*(3), 375–398.

GODFREY, N.J., CHRISTENSEN, N.I., and OKAYA, D.A. (2000), *Anisotropy of schists: Contribution of crustal anisotropy to active source seismic experiments and shear wave splitting observations*: Journal of Geophysical Research, *105*, pp. 27,991–28,007, doi:10.1029/2000JB900286.

HAYES, G.P. and K. FURLONG (2007), *Abrupt changes in crustal structure beneath the Coast Ranges of northern California - developing new techniques in receiver function analysis*, Geophys. J Int., *170*(1), 313–336.

HILLERS, G., Y. BEN-ZION, M. LANDES, M. CAMPILLO (2013), *Interaction of microseisms with crustal heterogeneity: A case study from the San Jacinto fault zone area*, Geochem., Geophys., Geosys., *14*, 2182–2197, doi:10.1002/ggge.20140.

JI, S., T. SHAO, K. MICHIBAYASHI, C. LONG, Q. WANG, Y. KONDO, W. ZHAO, H. WANG, and M.H. SALISBURY (2013), *A new calibration of seismic velocities, anisotropy, fabrics, and elastic moduli of amphibole-rich rocks*, Journ. Geophys. Res., *118*, 4699–4728.

JI, S. and M.H. SALISBURY (1993), *Shear-wave velocities, anisotropy and splitting in high-grade mylonites*, Tectonophysics, *221*, 453–473.

JONES, C.H. and R.A. PHINNEY (1998), *Seismic structure of the lithosphere from teleseismic converted arrivals observed at small arrays in the southern Sierra Nevada and vicinity, California*, J. Geophys. Res. *103*, 10065–10090.

LANGSTON, C.A. (1977), *The effect of planar dipping structure on source and receiver responses for constant ray parameter*, Bull. Seism. Soc. Am. 67, 1029–1050.

LEIDIG, M. and G. ZANDT (2003), *Modeling of highly anisotropic crust and application to the Altiplano-Puna volcanic complex of the central Andes*, Journ. Geophys. Res. 108, 2014.

LEVIN, V. and J. PARK (1997), *Crustal anisotropy in the Ural Mountains foredeep from teleseismic receiver functions*, Geophys. Res. Lett., *24*(11), 1283–1286.

Levin, V., Park, J. (1998), *P-SH* conversions in layered media with hexagonally symmetric anisotropy: A CookBook, *Pure Appl. Geophys. 151*, 669–697.

Ligorria, J.P., Ammon, C.J. (1999), *Iterative deconvolution and receiver-function estimation*, Bulletin of the Seismological Society of America 89, 1395–1400.

Liu, H. and Niu, F. (2012), *Estimating crustal seismic anisotropy with a joint analysis of radial and transverse receiver function data*, Geophys. J. Int. *188*(1), 144–164.

Lin, G., P.M. Shearer, E. Hauksson, and C.H. Thurber (2007), *A three-dimensional crustal seismic velocity model for southern California from a composite event method*, J. Geophys. Res. *112*, B11306.

Lloyd, G.E., R.W.H. Butler, M. Casey, D. Mainprice (2009). *Mica, deformation fabrics and the seismic properties of the continental crust*, Earth and Planetary Science Letters, *288*, 320–328.

Lynds, R.M. (2013), *Geologic storage assessment of carbon dioxide (CO_2) in the Laramide basins of Wyoming*, Technical Memorandum No. 3, Wyoming State Geological Survey, Laramie, Wyoming, accessed at http://www.wsgs.uwyo.edu/Research/Energy/CO2-Storage.aspx, 3/2014.

Mahan, K. (2006), *Retrograde mica in deep crustal granulites: Implications for crustal seismic anisotropy*, Geophys. Res. Lett., *33*(24), L24301, doi:10.1029/2006GL028130.

Mahan, K.H., Schulte-Pelkum, V., Blackburn, T.J., Bowring, S.A., and Dudas, F.O. (2012), *Seismic structure and lithospheric rheology from deep crustal xenoliths, central Montana, USA*. Geochemistry Geophysics Geosystems, *13*, Q10012, doi:10.1029/2012GC004332.

Mainprice, D. (1990), *A Fortran program to calculate seismic anisotropy from the lattice preferred orientation of minerals*, Computers & Geosciences *16*: 385–393.

Mainprice, M. and M. Humbert (1994), *Methods of calculating petrophysical properties from lattice preferred orientation data*, Surveys in Geophysics, *15*(5), 575–592, doi:10.1007/BF00690175.

McGuire, J. and Y. Ben-Zion (2005), *High-resolution imaging of the Bear Valley section of the San Andreas Fault at seismogenic depths with fault-zone head waves and relocated seismicity*, Geophys. J. Int., *163*, 152–164.

Nábelek, J., G. Hetényi, J. Vergne, S. Sapkota, B. Kafle, M. Jiang, H. Su, J. Chen, B.-S. Huang, The Hi-Climb Team (2009), *Underplating in the Himalaya-Tibet Collision Zone Revealed by the Hi-CLIMB Experiment*, Science *325*, 1371.

Owens, T.J., S.R. Taylor, and G. Zandt (1984), *Seismic evidence for an ancient rift beneath the Cumberland Plateau, Tennessee: A detailed analysis of broadband teleseismic P waveforms*, J. Geophys. Res., *89*, 7783–7795.

Ozacar, A. and G. Zandt (2009), *Crustal structure and seismic anisotropy near the San Andreas Fault at Parkfield, California*, Geophys. J. Int. *178*, 1098–1104.

Passchier, C.W., and R. Trouw, *Microtectonics* (Springer, Heidelberg, 2005).

Pedrera, A., F.D. Mancilla, Ruiz-Constan, A., Galindo-Zaldivar, J. et al. (2010), *Crustal-scale transcurrent fault development in a weak-layered crust from an integrated geophysical research: Carboneras Fault Zone, eastern Betic Cordillera, Spain*, Geochem. Geophys. Geosyst. *11*, Q12005.

Phinney, R.A. (1964), *Structure of the Earth's crust from spectral behavior of long-period body waves*. J. Geophys. Res. *69*, 2997–3017.

Porter, R., G. Zandt and N. McQuarrie (2011), *Pervasive lower-crustal seismic anisotropy in Southern California: Evidence for underplated schists and active tectonics*, Lithosphere *3*, 201–220.

Ramsay, J.G. and R.H. Graham (1970), *Strain variation in shear belts*, Can. Journ. of Earth Sciences, *7*(3), 786.

Rasolofosaon, P.N.J., W. Rabbel, S. Siegesmund, and A. Vollbrecht (2000), *Characterization of crack distribution: fabric analysis versus ultrasonic inversion*, Geophys. J. Int. *141*, 413–424.

Roecker, S., C. Thurber, K. Roberts, L. Powell (2006), *Refining the image of the San Andreas Fault near Parkfield, California using a finite difference travel time computation technique*, Tectonophysics *426*, 189–205.

Roux, P., M. Wathelet, and A. Roueff (2011), *The San Andreas Fault revisited through seismic noise and surface wave tomography*, Geophys. Res. Lett., *38*, L13319, doi:10.1029/2011GL047811.

Savage, M.K. (1998), *Lower crustal anisotropy or dipping boundaries? Effects on receiver functions and a case study in New Zealand*, Journal of Geophysical Research, *103*, 15,069–15,087.

Schulte-Pelkum, V. and Y. Ben-Zion (2012), *Apparent vertical Moho offsets under continental strike-slip faults from lithology contrasts in the seismogenic crust*, Bull. Seis. Soc. Am., *102*, 2757–2763.

Schulte-Pelkum, V., G. Monsalve, A. Sheehan, M.R. Pandey, S. Sapkota, R. Bilham and F. Wu (2005), *Imaging the Indian Subcontinent beneath the Himalaya*, Nature, *435*, 1222–1225.

Schulte-Pelkum, V. and K.H. Mahan (2014), *A method for mapping crustal deformation and anisotropy with receiver functions and first results from USArray*, Earth Planet. Sci. Lett., in press, http://dx.doi.org/10.1016/j.epsl.2014.01.050.

Shelly, D.R., W.L. Ellsworth, T. Ryberg, C. Haberland, G.S. Fuis, J. Murphy, R.M. Nadeau, R. Buergmann (2009), *Precise location of San Andreas Fault tremors near Cholame, California using seismometer clusters: Slip on the deep extension of the fault?*, Geophys. Res. Lett. *36*, L01303.

Shen, W., M.H. Ritzwoller, and V. Schulte-Pelkum (2013), *A 3-D model of the crust and uppermost mantle beneath the central and western US by joint inversion of receiver functions and surface wave dispersion*, J. Geophys. Res., doi:10.1029/2012JB009602, *118*, 1–15.

Sheriff, R.E., Geldart, L.P. (1995), 2nd Edition. *Exploration Seismology*. Cambridge University Press.

Sherrington, H.F., Zandt, G., Frederiksen, A. (2004), *Crustal fabric in the Tibetan Plateau based on waveform inversions for seismic anisotropy parameters*, Journal of Geophysical Research *109*, B02312.

Sibson, R.H. (1977), *Fault rocks and fault mechanisms*, J. Geol. Soc. London, *133*, 191–213.

Sibson, R.H. (1983), *Continental fault structure and the shallow earthquake source*, J. Geol. Soc. London, *140*, 741–767.

Smithson, S.B., J.A. Brewer, S. Kaufman, J.E. Oliver, C.A Hurich (1979), *Structure of the Laramide Wind River uplift, Wyoming, from COCORP deep reflection data and from gravity data*, Journ. Geophys. Res., *84*, 5955–5972.

STEIDTMANN, J.R. and MIDDLETON, L.T. (1991), *Fault chronology and uplift history of the southern Wind River Range, Wyoming: Implications for Laramide and post-Laramide deformation in the Rocky Mountain foreland,* Geol. Soc. Am. Bull. *103,* 472–485.

STERN, T.A. and J.H. MCBRIDE (1998), *Seismic exploration of continental strike-slip zones,* Tectonophys. *286,* 63–78.

STONE, D.S. (1987), *Wyoming transect A; Regional geological cross sections: Rocky Mountain Transects,* Littleton, Colorado.

TAPE, C., Q. LIU, A. MAGGI and J. TROMP (2009), *Adjoint tomography of the southern California crust,* Science *325,* 988–992.

TATHAM, D.J., LLOYD, G.E., BUTLER, R.W.H., and CASEY, M. (2008), *Amphibole and lower crustal seismic properties.,* Earth and Planetary Science Letters *267,* 118–128.

THURBER, C., H. ZHANG, F. WALDHAUSER, J. HARDEBECK, A. MICHAEL, D. EBERHARDT-PHILLIPS (2006), *Three-dimensional compressional wavespeed model, earthquake relocations, and focal mechanisms for the Parkfield, California, region,* Bull. Seism. Soc. Am. *96,* S38–S49.

TITUS, SARAH J.; MEDARIS, L. GORDON, JR.; WANG, HERBERT F.; *et al.* (2007), *Continuation of the San Andreas fault system into the upper mantle: Evidence from spinel peridotite xenoliths in the Coyote Lake basalt, central California,* Tectonophys., *1–2,* 1–20.

VERGNE, J., WITTLINGER, G., FARRA, V. and SU, H. (2003), *Evidence for upper crustal anisotropy in the Songpan-Ganze (northeastern Tibet) terrane.* Geophysical Research Letters, *30.*

VINNIK, L.P. (1977), *Detection of waves converted from P to SV in the mantle,* Phys. Earth Planet. Interiors *15,* 39–45.

WARD, D., K. MAHAN and V. SCHULTE-PELKUM (2012), *Roles of quartz and mica in seismic anisotropy of mylonites,* Geophys. J. Int., *190*(2), 1123–1134.

WESSEL, P. and W.H. F. SMITH (1998), *New, improved version of generic mapping tools released,* Eos Trans. Am. Geophys. Union *79*(47), 579.

WILLIAMS, M.L., E.A. MELIS, C.F. KOPF, and S. HANER (2000), *Microstructural tectonometamorphic processes and the development of gneissic layering: a mechanism for metamorphic segregation,* J. metamorphic Geol., *18,* 41–57.

WILSON, C.K., C.H. JONES, P. MOLNAR, A.F. SHEEHAN, and O. BOYD (2004), *Distributed deformation in the lower crust and upper mantle beneath a continental strike-slip fault zone: Marlborough fault system, South Island, New Zealand,* Geology *32,* 837–840.

WITTLINGER, G., J. VERGNE, P. TAPPONIER, V. FARRA, G. POUPINET, M. JIANG, H. SU, G. HERQUEL, A. PAUL (2004), *Teleseismic imaging of subducting lithosphere and Moho offsets beneath western Tibet,* Earth Planet. Science Lett. *221,* 117–130.

YECK, W., A.F. SHEEHAN and V. SCHULTE-PELKUM (2013), *Sequential H-κ stacking to obtain accurate crustal thicknesses beneath sedimentary basins,* Bulletin of the Seismological Society of America, *103,* 2142–2150, doi:10.1785/0120120290.

ZHEGLOVA, P., J.R. MCLAUGHLIN, S.W. ROECKER, J.R. YOON, and D. RENZI (2012), *Imaging quasi-vertical geological faults with earthquake data,* Geophysical Journal International, *189,* 1584–1596.

ZHU, L. and H. KANAMORI (2000), *Moho depth variation in Southern California from teleseismic receiver functions,* Journal of Geophysical Research *105,* 2,969–2,980.

ZHU, L. (2000), *Crustal structure across the San Andreas Fault, southern California from teleseismic converted waves,* Earth Planet. Sci. Lett. *179,* 183–190.

ZOEPPRITZ, KARL (1919), *Erdbebenwellen VII. VIIb. Über Reflexion und Durchgang seismischer Wellen durch Unstetigkeitsflächen. Nachrichten von der Königlichen Gesellschaft der Wissenschaften zu Göttingen,* Mathematisch-physikalische Klasse, 66–84.

(Received October 1, 2013, revised April 15, 2014, accepted April 17, 2014, Published online May 9, 2014)

Pure Appl. Geophys. 171 (2014), 2993–3011
© 2014 Springer Basel
DOI 10.1007/s00024-014-0784-0

Seismic Imaging of a Bimaterial Interface Along the Hayward Fault, CA, with Fault Zone Head Waves and Direct P Arrivals

A. A. ALLAM,[1] Y. BEN-ZION,[1] and Z. PENG[2]

Abstract—We observe fault zone head waves (FZHW) that are generated by and propagate along a roughly 80 km section of the Hayward fault in the San Francisco Bay area. Moveout values between the arrival times of FZHW and direct P waves are used to obtain average P-wave velocity contrasts across different sections of the fault. The results are based on waveforms generated by more than 5,800 earthquakes and recorded at up to 12 stations of the Berkeley digital seismic network (BDSN) and the Northern California seismic network (NCSN). Robust identification of FZHW requires the combination of multiple techniques due to the diverse instrumentation of the BDSN and NCSN. For single-component short-period instruments, FZHW are identified by examining sets of waveforms from both sides of the fault, and finding on one (the slow) side emergent reversed-polarity arrivals before the direct P waves. For three-component broadband and strong-motion instruments, the FZHW are identified with polarization analysis that detects early arrivals from the fault direction before the regular body waves which have polarizations along the source-receiver backazimuth. The results indicate average velocity contrasts of 3–8 % along the Hayward fault, with the southwest side having faster P wave velocities in agreement with tomographic images. A systematic moveout between the FZHW and direct P waves for about a 80 km long fault section suggests a single continuous interface in the seismogenic zone over that distance. We observe some complexities near the junction with the Calaveras fault in the SE-most portion and near the city of Oakland. Regions giving rise to variable FZHW arrival times can be correlated to first order with the presence of lithological complexity such as slivers of high-velocity metamorphic serpentinized rocks and relatively distributed seismicity. The seismic velocity contrast and geological complexity have important implications for earthquake and rupture dynamics of the Hayward fault, including a statistically preferred propagation direction of earthquake ruptures to the SE.

Key words: Fault zone head waves, Hayward fault, Bimaterial interface, Seismic imaging of faults, Moveout analysis, Particle motion analysis.

1. Introduction

Major faults with large cumulative slip have prominent bimaterial interfaces which juxtapose rock bodies with different elastic properties. Geological studies show that the active slip zone in large fault structures is often localized along the overall lithology contrast and/or the boundary of a damaged fault zone layer (e.g., SIBSON 2003; SENGOR et al. 2005; DOR et al. 2008). The existence of a bimaterial fault interface can significantly modify various aspects of earthquake physics and related phenomena (BEN-ZION and ANDREWS 1998). These include generation of predominately unilateral or strongly asymmetric ruptures with a preferred propagation direction controlled by the velocity contrast, sense of loading, and sub-shear vs. super-shear rupture speed (e.g., WEERTMAN 1980, 2002; SHI and BEN-ZION 2006; LENGLINE and GOT 2011), stress evolution towards a tensional regime with little frictional heat and possible transient fault opening (e.g. ANDREWS and BEN-ZION 1997; BEN-ZION and HUANG 2002; DALGUER and DAY 2009), generation of very high slip velocities behind the propagating rupture front (e.g., BEN-ZION 2001; AMPUERO and BEN-ZION 2008), asymmetric distributions of aftershocks on the fault (e.g., RUBIN and GILLARD 2000; ZALIAPIN and BEN-ZION 2011), asymmetric off-fault rock damage (e.g., DOR et al. 2006; WECHSLER et al. 2009; MITCHELL et al. 2011) and asymmetric geodetic fields across the fault (e.g., LE PICHON et al. 2005; WDOWINSKI et al. 2007). The directivity effects associated with predominately unilateral bimaterial ruptures can produce strongly asymmetric ground motion, amplifying the motion by a factor of 5 or more in the direction of rupture propagation (e.g., BEN-ZION 2003; OLSEN et al. 2006; BRIETZKE et al. 2009). Detecting and imaging

[1] Department of Earth Sciences, University of Southern California, Los Angeles, CA 90089-0740, USA. E-mail: aallam@usc.edu
[2] School of Earth and Atmospheric Sciences, Georgia Institute of Technology, Atlanta, GA 30332, USA.

properties of bimaterial faults, especially in fault sections near highly populated urban areas, can therefore lead to improved estimates of seismic hazard associated with the faults.

Earthquakes near a fault bimaterial interface generate fault zone head waves (FZHW) that propagate along the interface with the velocity and motion polarity of the faster block, and are radiated from there to the block with slower seismic velocity (BEN-ZION 1989, 1990). The FZHW are the first arriving phases at locations on the slower block with normal distance to the fault less than a critical value that depends on the velocity contrast and propagation distance along the fault (see Sect. 3). FZHW provide the most direct and highest resolution information on across-fault contrasts of rock types at seismogenic depth. A mere detection of FZHW indicates the existence of a bimaterial interface and the sense of velocity contrast across the fault. Analysis of recorded FZHW and the later arriving direct body waves gives basic characterization of the spatial extent along-strike and with depth, and degree of velocity contrast, of the bimaterial fault interface (e.g., BEN-ZION and MALIN 1991; HOUGH et al. 1994; MCGUIRE and BEN-ZION 2005; ZHAO et al. 2010). Such information can provide fundamental input on the internal structure, expected dynamics, and estimates of seismic shaking hazard associated with the fault. Also, since bimaterial interfaces change the distribution of arrival times and first motion polarities near the fault, ignoring their existence can introduce errors and biases into derivations of earthquake locations and focal mechanisms (e.g. MCNALLY and MCEVILLY 1977; OPPENHEIMER et al. 1988; BEN-ZION and MALIN 1991).

In the following sections we provide results associated with detecting and measuring properties of FZHW and direct P arrivals in seismic waveforms from stations near the Hayward fault, a major strike-slip fault on the east side of the San Francisco Bay. Depending upon the type of data available at each station, we use particle motion analysis (BULUT et al. 2012) or compare waveforms across the fault and search for emergent arrivals on the slow side with coherent moveout from the following direct P waves (e.g., BEN-ZION and MALIN 1991; LEWIS et al. 2007; ZHAO et al. 2010). The measured arrival times of FZHW and direct P arrivals are used to estimate the average velocity contrast across different fault sections and to identify locations that produce waveform complexities.

2. Regional and Geologic Setting

The tectonic motion between the Pacific and North American plates in the San Francisco Bay area is taken up by a series of splayed right-lateral faults. With roughly 100 km of cumulative slip, the Hayward fault has been a major component of this plate boundary system since the late Pliocene era (MCLAUGHLIN et al. 1996; GRAYMER et al. 2002; LANGENHEIM et al. 2010). The Hayward fault is about a 120 km long segment of the San Andreas system that bisects the densely-populated cities of Berkeley and Oakland, among others (Fig. 1a). From geodetic measurements, the Hayward fault is estimated to accommodate 7–9 mm/year of slip, which is approximately 15–20 % of the total plate motion (D'ALESSIO et al. 2005; SCHMIDT et al. 2005; EVANS et al. 2012). Though some of this slip is alleviated aseismically due to a high creep rate along the Hayward (BILHAM and WHITEHEAD 1997; BÜRGMANN et al. 2000; SIMPSON et al. 2001; LIENKAEMPER et al. 2012), paleoseismic studies indicate that the Hayward fault frequently ruptures in $M_w > 6.7$ earthquakes with a 161 ± 65 years recurrence interval (LIENKAEMPER et al. 2010); the most recent such event occurred in 1868 (LAWSON 1908; YU and SEGALL 1996). Based on this information, the Hayward is regarded as the most hazardous Bay Area fault, with a 31 % probability to produce a $M_w > 6.7$ event within the next 30 years (FIELD et al. 2009). Considering the dense population (~ 7 million) and potential for devastating co-seismic site effects in the region (e.g., KNUDSEN and WENTWORTH 2000; OLSEN et al. 2008), it is clear that the Hayward fault poses a severe seismic risk to the San Francisco Bay area.

Because of its large total displacement, the Hayward fault juxtaposes several different lithologies with contrasting elastic properties. In general, the western block is composed of the Franciscan Metamorphic complex beneath a shallow (<500 m) cover of Quaternary sediments, while the eastern block

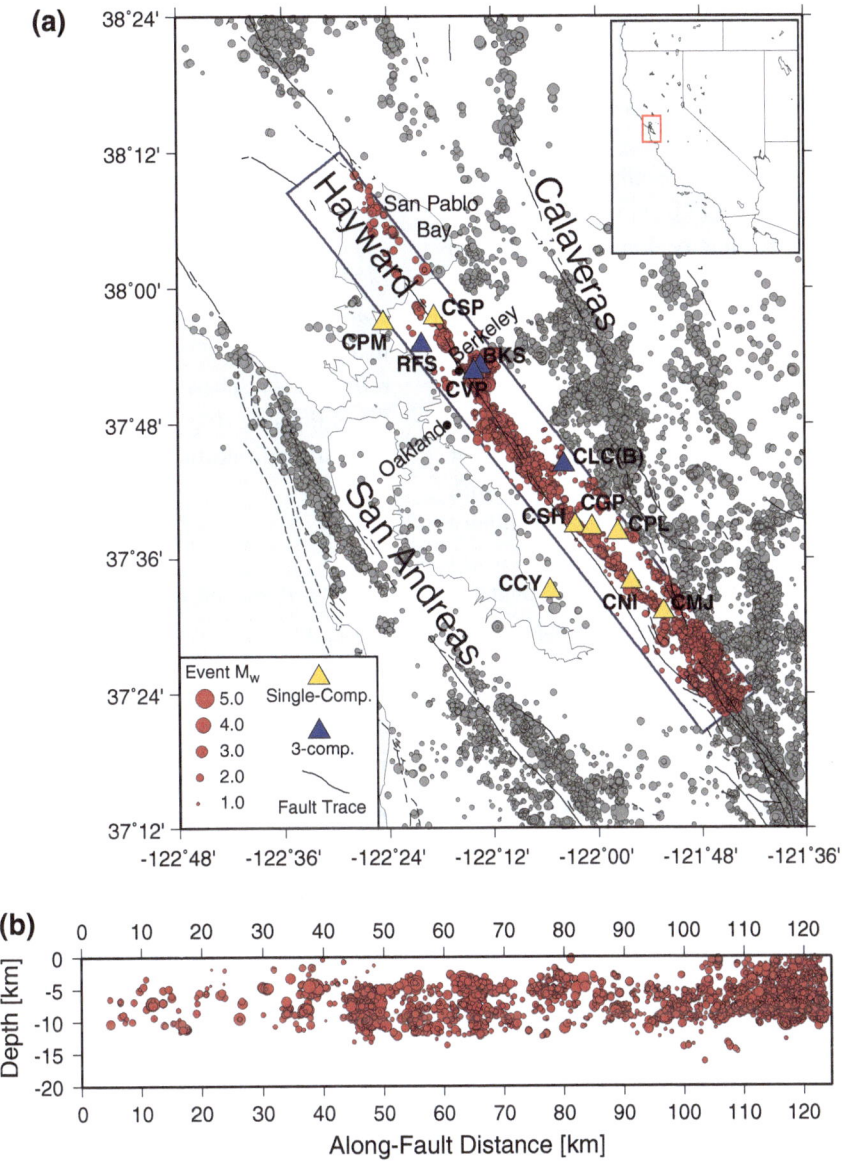

Figure 1

a Location map for the Hayward fault zone region (*inside long rectangle*) with 5,834 M_w > 1.0 earthquakes from the NCEDC catalog (*red circles*), regional seismicity (*grey circles*) and surface traces of large faults (*black lines*). NCSN stations are either single-component geophones (*gold triangles*) or multi-component stations featuring a variety of instruments (*blue triangles*) and all three components of motion. **b** Cross-section view along the Hayward fault showing the depth distribution of seismicity (*red circles*). All of the seismicity inside the *long rectangle* (**a**) is collapsed onto a single vertical plane

contains rocks of the Coast Range Ophiolite and Great Valley sedimentary sequence (GRAYMER *et al.* 2005). In the northernmost portion of the fault, where it intersects San Pablo Bay, Franciscan mélange is in contact at the surface with Cretaceous sandstones. Near the city of Oakland, a complicated zone of serpentinized igneous rocks is in contact at the surface with fine-grained sedimentary units. From Oakland southwards to the junction with the Calaveras fault, metasandstones are in contact at the surface with shales, sandstones, and conglomerates. In all of these cases, western metamorphic rocks have higher seismic velocities than the adjacent sedimentary units on the eastern side, as shown by laboratory

(BROCHER 2008) and tomographic (ZHANG and THUR-
BER 2003; MANAKER et al. 2005; HARDEBECK et al.
2007) studies.

Though a considerable proportion of the total slip is
released by aseismic creep (BILHAM and WHITEHEAD
1997), the Hayward fault is seismically active along its
entire length to a depth of about 15 km (Fig. 1b). There
are significant gaps in seismicity along the fault which
have been interpreted as locked patches (WALDHAUSER
and ELLSWORTH 2000; SCHMIDT et al. 2005) with specific geological features (GRAYMER et al. 2005). The
fault is also more seismically active to the south, where
relocated seismicity (SCHAFF and WALDHAUSER 2005;
WALDHAUSER and SCHAFF 2008) indicates a slight
northeastern dip of the main fault near the junction with
the Calaveras fault. In a fault-normal sense, seismicity
is broadly distributed in the upper 5 km, but is much
narrower at depth. This pattern is generally thought to
be consistent with a localized slip zone at depth (e.g.,

CHESTER and CHESTER 1998; BEN-ZION and SAMMIS
2003; FROST et al. 2009).

3. Fault Zone Head Waves

Fault zone head waves are seismic phases that
spend a large portion of their propagation paths
refracting along a bimaterial fault interface (BEN-ZION
1989, 1990). They are analogous to refracted waves
in a horizontally layered medium such as the mantle-
refracted Pn phase. The FZHW arrive before the
direct P waves (Fig. 2) for stations on the slow side of
the fault with normal distance to the interface $x < x_c$
given by

$$x_c = r \cdot \tan\left[\cos^{-1}(\alpha_2/\alpha_1)\right], \tag{1}$$

where r is the propagation distance along the fault
and α_2 and α_1 are the average P wave velocities of the

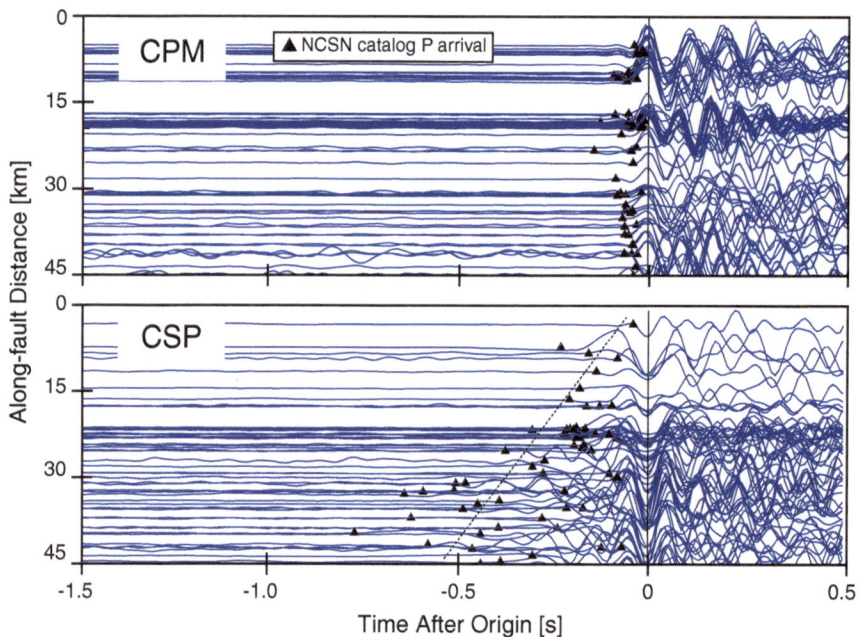

Figure 2
Example sets of seismograms from two stations on opposite sides of the Hayward fault. Station CSP is on the slow NE side at about 3 km
from the fault, while station CPM is on the fast SW side about 10 km from the fault. *Vertical black lines* are the first significant peak within
the time window, *black triangles* are catalog P arrival picks from the NCEDC, and the inclined *dashed line* is the inferred moveout of the early
arrivals relative to the peak for station CSP. At station CPM on the faster side of the fault, the first large peak is consistently impulsive with
considerably larger amplitude than the preceding noise. In contrast, seismograms from the station CSP feature emergent early arrivals before
the first early peak which appear to have a coherent moveout with along-fault propagation distance. First arrivals at CSP generally have the
opposite polarity to that expected for the right-lateral strike-slip focal mechanisms which dominate the earthquake catalog in this region.
Though the NCEDC picking algorithms identify both of these phases as P wave arrivals, there is clearly a different character between the two
stations

slower and faster blocks, respectively (BEN-ZION 1989). The head waves have emergent waveforms that generally increase in amplitude up to the arrivals of the more impulsive body waves, which truncate the tail of the head waves with reversed polarity larger amplitude motion (BEN-ZION 1989, 1990).

For a simple bimaterial interface between two different quarter spaces, the differential arrival time (Δt) between the first arriving FZHW and the following direct P wave grows with the distance traveled along the fault (r) as

$$\Delta t \sim r \left(\frac{1}{\alpha_2} - \frac{1}{\alpha_1} \right) \sim r \left(\frac{\Delta \alpha}{\alpha^2} \right), \qquad (2)$$

where α and $\Delta \alpha$ denote the average and differential P-wave velocities, respectively (BEN-ZION and MALIN 1991). Assuming an average velocity for a given fault zone section, Eq. (2) can be used together with measured moveout data to derive a corresponding estimate of the average velocity contrast across the fault section. BEN-ZION et al. (1992) and LEWIS et al. (2007) constrained the depth variations of the velocity contrasts across sections of the San Andreas Fault with joint inversion of FZHW and P arrival times using simple ray tracing expressions for a structure consisting of a two different horizontally-layered quarter spaces. More recently, BENNINGTON et al. (2013) incorporated head waves into double-difference tomography inversions and obtained images with improved sharpness and degree of across-fault velocity contrast, along with better localization of seismicity, at the Parkfield section of the San Andreas Fault. In this paper we limit the analysis to derivations of average velocity contrasts at different fault sections of the type done by BEN-ZION and MALIN (1991), MCGUIRE and BEN-ZION (2005), and others.

Because the FZHW is radiated from the fault interface, in contrast to the direct P wave that is radiated from the hypocenter location, the horizontal motion polarity provides an important feature that can be used to distinguish between and measure the arrivals of the two phases (BULUT 2012). The accuracy of detecting the head wave arrival is limited by its emergent character and the existence of ambient noise. The accuracy of detecting the direct P wave arrival with horizontal

motion polarity at near-fault stations is limited by the fact that a vertical strike-slip fault is a nodal plane of the horizontal motion.

4. Analysis Procedure and Results

Due to the highly variable data quality of the seismic network along the Hayward fault, we combine multiple techniques in order to distinguish and pick arrival times of head wave and direct P phases. For 3-component seismic stations, we apply a horizontal particle motion polarization analysis in short moving time windows and search for changes of polarization direction from random (noise) to approximately fault normal (head wave) and epicenteral (P wave) directions. For single-component (vertical geophone) stations, we compare first arrivals for each event recorded at a variety of stations on both sides of the fault and search for systematic emergent slow-side first motions that have the same polarity as first arrivals (direct P waves) on the faster side of the fault. In both cases, the arrival times of identified head and direct P waves are picked and assigned a quality rating based on the clarity of separation of the two phases, consistency of polarity with right-lateral focal mechanisms, signal-to-noise-ratio, and impulsiveness of the direct P arrival. A series of consistency checks are then applied to the entire dataset, reducing by about 15 % the final number of phase picks. The following subsections provide additional details on each analysis technique, along with descriptions of the data and results.

4.1. Data and Pre-processing

We examined 5,834 earthquakes (Fig. 1) recorded at up to 28 stations of the Berkeley digital seismic network (BDSN) and the Northern California seismic network (NCSN). Event information was taken from the double-difference relocated catalog of WALDHAUSER and SCHAFF (2008), which spans 1984–2009. Waveforms and routine phase arrival picks were obtained from the Northern California Earthquake Data Center (NCEDC). The employed events range in size from M_w 1.0 to M_w 4.0, and have a maximum fault-normal distance of 8 km and a maximum depth of 16.2 km. The instruments include 15 single-component L4 velocity geophones, six K2 Episensor accelerometers, four Guralp 40T broadband seismometers, and three Wilcoxon borehole accelerometers. Of the 28 stations examined, eight are sufficiently close to the fault on the slow (NE) side to record head wave first arrivals (Eq. 1).

For each record, we remove the mean and apply a 2-pass (acausal) 2-pole 1 Hz high-pass Butterworth filter in order to remove long-period noise. Since the same pre-processing is applied to all data, it does not affect our key conclusions on the existence and extent of a major bimaterial interface in the structure of the Hayward fault, based on comparisons of waveforms at stations across the fault. In the appendix we demonstrate that the pre-processing has minor effects on the target analysis signals, limited to producing some uncertainties on the precise properties of the early head waves. However, this is inevitable given the emergent character of FZHW and has little effect on our main results. We integrate the resulting seismograms numerically using cumulative trapezoidal integration to produce displacement records.

Figure 2 shows example sets of seismograms at two stations on the opposite sides of the fault that illustrate the significant differences in the character of seismograms recorded at the different crustal blocks. Station CSP is on the slow NE side about 3 km from the fault, while station CPM is on the fast SW side about 10 km from the fault. At both stations, the vertical-component velocity seismograms are aligned on the first large peak (vertical black lines) within 2 s of the NCEDC catalog picked arrival time (black triangles). On the faster crustal block, the first large peak is consistently impulsive with considerably larger amplitude than the preceding portion of the seismograms, which appears to consist of incoherent random noise. Subsequent peaks in the seismograms also line up for waveforms originating at similar along-fault distances. The NCEDC catalog picks are also roughly aligned, with no significant moveout with along-fault propagation distance. In contrast, the seismograms from the slower side of the fault include emergent early arrivals before the first impulsive peak, which appear to have a coherent moveout with along-fault propagation distance and are candidate head waves. The NCEDC picking algorithms identify these early emergent phases as P-wave arrivals, but

they clearly differ in character from the arrivals at the station CPM on the other side of the fault. In the subsequent sections, we analyze the seismograms from both sides of the fault in much greater detail.

4.2. *Particle Motion Polarization*

Following the method of BULUT *et al.* (2012), we identify head waves in data observed on two horizontal components using particle motion analysis summarized briefly below. As mentioned, the head waves travel along and are radiated from the fault bimaterial interface. They are the first arriving seismic phases at stations on the slower side closer to the fault than x_c of Eq. (1), and are expected to exhibit polarity change from random-like motion charactering the preceding noise to polarization that turns toward the fault-normal azimuth (Fig. 3). The particle motion of the direct P wave is along the source-receiver propagation path, so the transition from the FZHW to direct P waves is expected to be associated with a change of polarity toward the source-receiver backazimuth. To track changes of polarization directions, we follow BULUT *et al.* (2012) and calculate polarizations in narrow time windows

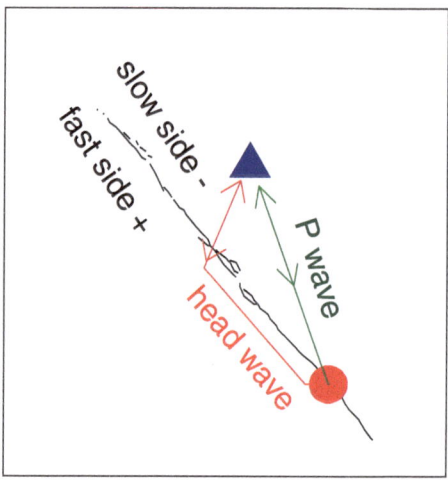

Figure 3
Schematic diagram showing the difference in horizontal polarization of the direct P wave compared to the first-arriving head wave. For a slow-side station near the fault, the direct P arrives polarized directly along the source receiver backazimuth. The preceding head wave, on the other hand, has polarization more normal to the strike of the fault

of the recorded horizontal displacement seismograms with the algorithm of JURKEVICS (1988).

If we combine all three components of motion in a single time window of length N into an $N \times 3$ matrix X, the covariance matrix can be computed as

$$S = \frac{XX^T}{N} = \begin{bmatrix} S_{zz} & S_{zn} & S_{ze} \\ S_{zn} & S_{nn} & S_{ne} \\ S_{ze} & S_{ne} & S_{ee} \end{bmatrix}. \qquad (3)$$

The eigenvalues $(\lambda_1, \lambda_2, \lambda_3)$ and eigenvectors (u_1, u_2, u_3) of S give, respectively, the amplitudes and directions of the axes of the polarization ellipse. The polarization of incoming P waves is assumed to be the horizontal orientation of rectilinear motion corresponding to the largest eigenvector λ_1 given by

$$Az = \tan^{-1}\left(\frac{u_{21}}{u_{31}}\right). \qquad (4)$$

For a regular direct P wave Az should be along the source-receiver backazimuth, whereas for FZHW it should be oriented towards the fault interface. Additionally, the degree of polarization can be calculated as

$$q = 1.0 - ((\lambda_2 + \lambda_3)/2\lambda_1), \qquad (5)$$

which will be 1.0 when particle motion is purely rectilinear as in an isolated body wave, and 0 for a perfectly uncorrelated motion.

In practice, all of the eigenvalues are nonzero due to correlated seismic noise and scattering. Nonetheless, the head and direct P waves can be identified tentatively by comparing the largest eigenvalues λ_1 of consecutive time windows: head waves will have larger λ_1 than the noise, and P waves will have larger λ_1 than head waves. The arrivals of the head and direct P waves can be substantiated further and picked by identifying transitions of particle motion from approximately random toward the fault-normal and from there toward the backazimuth direction. By using a moving time window, the ratio of eigenvalues can be maximized to accurately pick each phase arrival.

For the employed data in this study, we found that it is useful to examine horizontal particle motion using a moving window with a length of ten samples, which corresponds to a time window of 0.1 or 0.2 s depending on the instrument's sampling rate. To

begin, we identify the initial phase arrival by finding the first (after the origin) time window in which the largest eigenvalue is at least five times that of the background noise. Though this is an ad hoc choice, it is likely a conservative one; a ratio of 3 is probably sufficient to distinguish phase arrivals from noise. Second, we determine if this arrival is a head wave or direct P wave based upon the azimuth Az and degree q of the polarization calculated using Eqs. (4) and (5). If these parameters are inconsistent with the expectations for a direct P arrival, and are consistent with those for FZHW, we identify the first phase tentatively as a head wave and continue the moving window analysis to maximize the eigenvalue ratio of the next phase arrival. In general, this second arrival will be polarized along the backazimuth and will have a higher eigenvalue ratio. A consistency of eigenvalue ratios and polarization directions gives us confidence that the second phase is a direct P wave, and that the preceding arrival is a head wave.

Figure 4a, b illustrate the discussed procedure with example results for two separate events with significantly different epicentral distances recorded each by a single station (BKS) on the slow NE side of the fault. In general, the difference in amplitude, and, thus, in the largest eigenvalue, between the FZHW and direct P wave decreases with increasing along-fault propagation distance. For comparison, Fig. 4c provides example particle motion results for a station (RFS) on the fast SW side of the fault. The particle motion analysis shows in this case only an impulsive P wave arrival polarized along the backazimuth with a large λ_1 and a large q. The stations on the slower side of the fault have generally less impulsive arrivals and a lower signal-to-noise ratio. Nevertheless, FZHW and P-wave arrivals are clearly identified and picked using the above analysis procedure at three stations (blue triangles in Fig. 1) on the NE side of the fault.

4.3. Vertical Component Analysis

For stations which only record a single (vertical) component of motion, we apply a waveform comparison method (e.g., BEN-ZION and MALIN 1991; LEWIS et al. 2007; ZHAO et al. 2010). We begin by selecting events with $M_w > 1.5$, because smaller

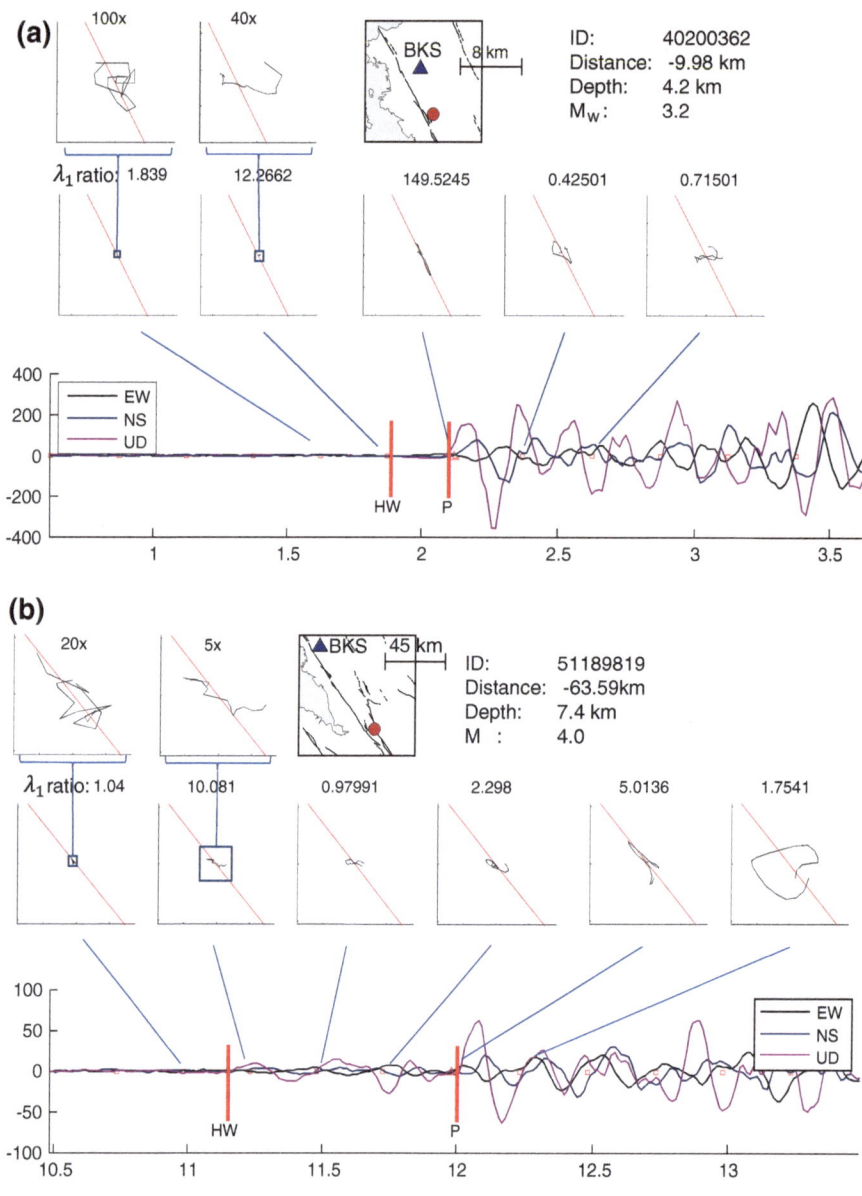

Figure 4

a Horizontal particle motion at station BKS for an event ~10 km along-fault showing an emergent head wave arrival followed by an impulsive backazimuth-polarized direct P wave. For each window, the ratio of the largest eigenvalue (for the particle motion covariance matrix in Eq. 1) in that window to the largest eigenvalue in the previous time window is shown. Though the head wave is relatively low amplitude compared to the P wave, it is much higher amplitude than the preceding noise. Additionally, the head wave is not polarized along the backazimuth, as is the later direct P. **b** Similar to **a**, but for an event much further away (~65 km along-fault) recorded at the same station. The head wave emerges from the noise with a high eigenvalue ratio, followed by an impulsive higher amplitude P wave with a strong backazimuth polarization. The observation of head waves at such large distances indicates a consistent fault zone interface with contrasting velocity. **c** Similar to **a** and **b**, but for a medium distance event (~15 km along-fault) recorded at fast-side station RFS. An impulsive backazimuth-polarized P wave has no significant preceding phase

events tend not to produce long-range coherent phases. Though smaller events in principle could be used, the generally low signal-to-noise ratio in the instrument response bands of the NCSN single-component stations precludes reliable detection of an emergent phase before the P wave. This leaves

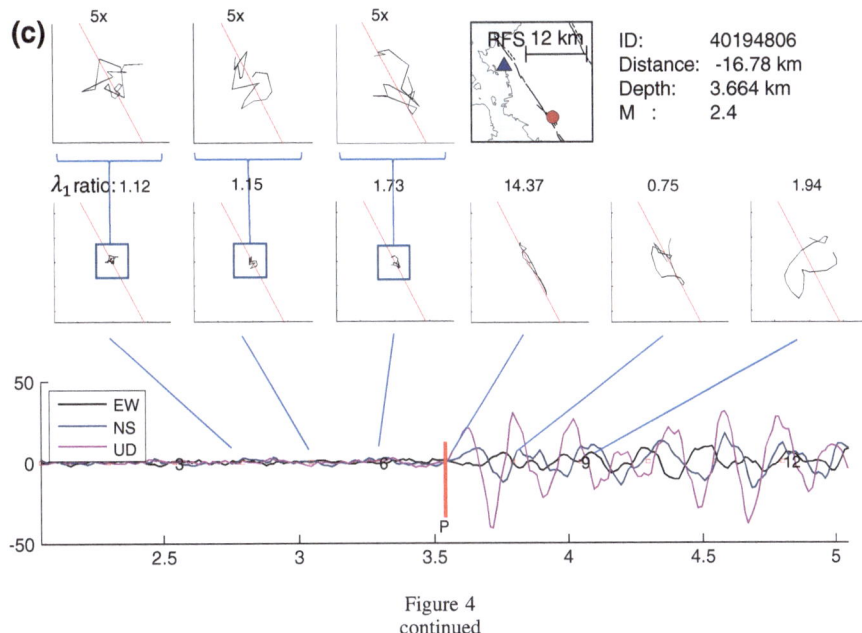

Figure 4
continued

1,808 candidate events, each recorded by an average of twelve stations. Of these events, approximately half are well-recorded by any given station. The remaining events (red circles in Fig. 1) provide good spatial coverage all along the Hayward fault.

After event selection, we compare all of the available seismograms for each event individually, arranging the waveforms by along-fault propagation distance and coloring them according to the side of the fault on which they are recorded (Fig. 5a). Head waves are identified as emergent arrivals with the polarity of the fast-side direct P wave. Direct P waves are identified as relatively impulsive high-amplitude arrivals. Although it is difficult to quantify the uncertainty of the picks we estimate that it is less than 0.1 s; the picking error is certainly much less than the separation time between head wave and direct P arrivals.

Additionally, we make the simplifying assumption that the motion polarity should be consistent with right-lateral focal mechanisms; most events along the Hayward fault are associated with near-vertical strike-slip mechanisms aligned with the seismicity trend (HARDEBECK et al. 2007). However, there is some complexity in focal mechanisms at both the southern junction with the Calaveras fault and the northern stepover to the Rodgers Creek fault zone,

which may explain some of the complexity in observed waveforms.

Next, we assign quality ratings of "good", "fair", or "poor" based on the clarity of separation and relative amplitude of the two phases, consistency of polarity with right-lateral focal mechanisms, signal-to-noise-ratio, and impulsiveness of the direct P arrival (Fig. 5b). Good quality measurements are characterized by a high signal-to-noise ratio, with an emergent first-motion followed by an impulsive higher amplitude arrival. Fair quality ratings are generally assigned where there is not clear separation between the two phases. Poor quality measurements mostly have a low signal-to-noise ratio and lack separation. In the following sections, we exclude 'poor' arrivals from further consideration.

4.4. Consistency Checks

After assembling a dataset of FZHW and direct P arrivals, we re-examine each individual pick in the context of each station. First, we create stack-plots of all the picks aligned on the peak of the direct P arrival (Fig. 6). For reference, we plot the original NCEDC catalog P arrival time, the new P pick, and the head wave pick. We then evaluate each pick separately, using both velocity and displacement seismograms to

139

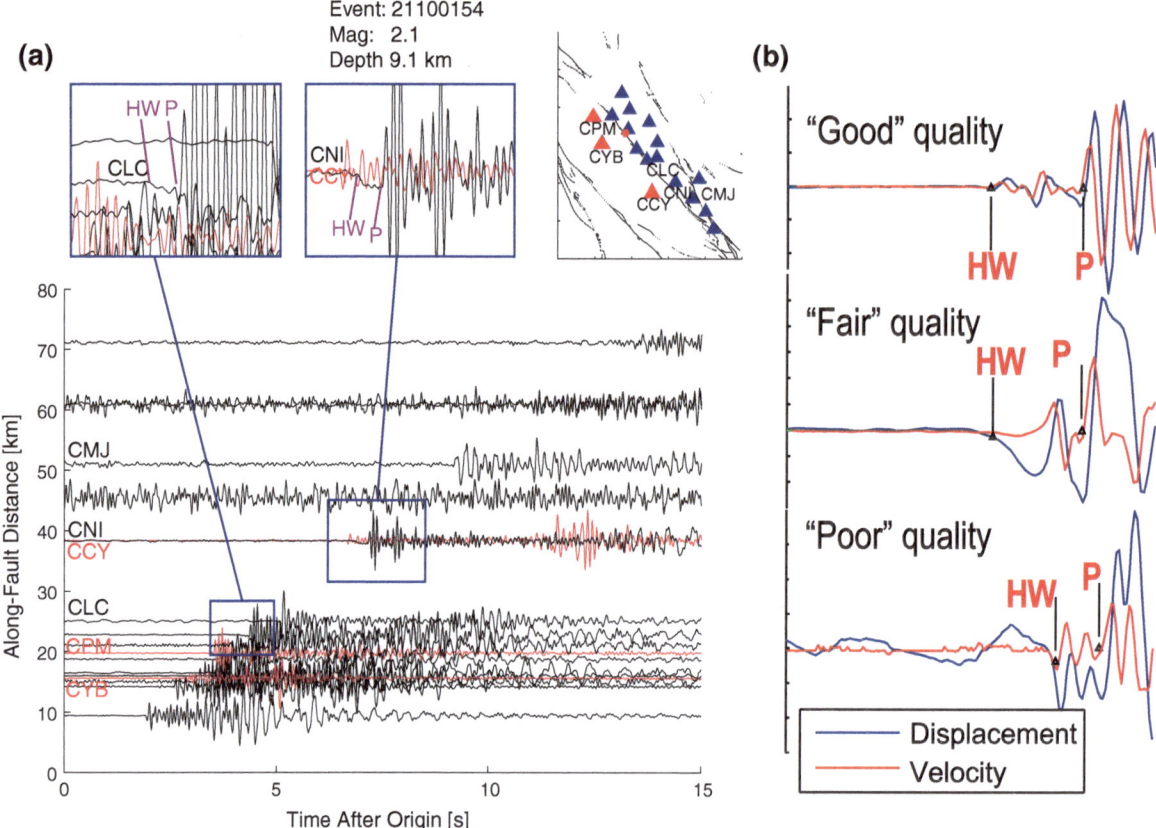

Figure 5

a Example head waves for single-component stations picked by comparing waveforms from many stations for a single event from both the fast (*red line*) and slow (*black line*) sides of the fault. Stations CLC and CNI both show emergent first arrivals which have downward first-motion polarity (indicated in the *zoomed-in* views), which is opposite to the expected polarity produced by the common regional right-lateral focal mechanisms. Stations CCY and CPM feature more impulsive arrivals with the correct first motion. **b** Examples of 'good', 'fair', and 'poor' quality measurements made on single-component station CMJ. The good quality measurement features low-amplitude first arrivals followed by a very clear impulsive phase in both velocity and displacement. The fair measurement is more emergent, with the head wave only being clearly observed in displacement, and the P wave only clearly observed in velocity. The poor measurement features very high amplitude noise in displacement, with no obvious difference in character of P and head wave except motion polarity

adjust slightly the head wave or P pick for about 10 % of the phases, and removing any obviously erroneous picks.

Next, we fit a least-squares regression line separately to the P wave and head wave arrival times at each station, and consider picks with a residual higher than an ad hoc value of 10 % to be outliers. This step is applied mainly to remove misidentification problems from the picked arrival times, which occur when the head waves and P waves are mistaken for each other. In fitting the regression line, we make no *a priori* assumptions; the line is not constrained to have a positive slope or to pass through the origin. All eight stations on the

slow side of the Hayward fault show positive moveouts (Fig. 6) that are discussed further below. We calculate a residual for each phase pick (picked arrival time minus regression prediction) and remove all picks for which the residual is greater than 10 % of the predicted arrival time (residual/prediction >0.1). This process reduces the total amount of data by ∼15 % at each station.

4.5. *Estimating Velocity Contrast*

Once the final catalog of FZHW and direct P arrivals is assembled, the moveout of delay time Δt between the two phases at each station with

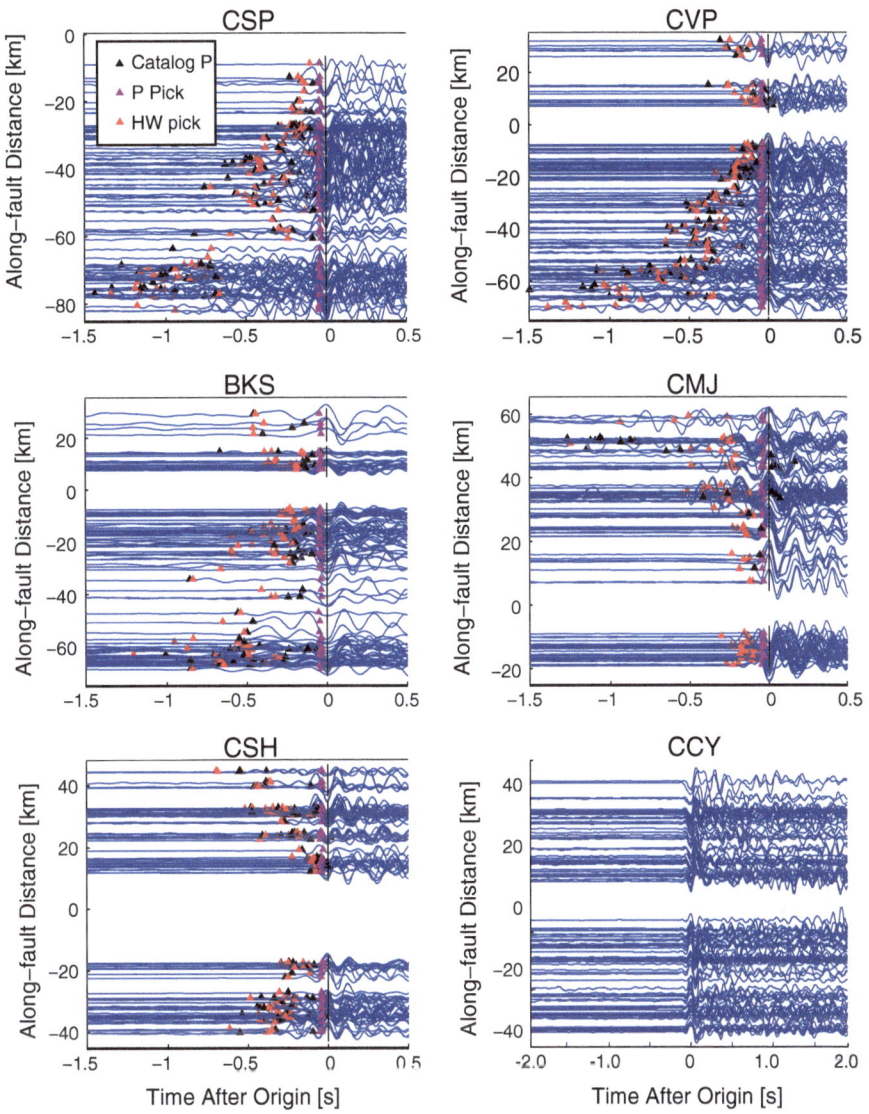

Figure 6
Stack plots showing velocity seismograms for five slow-side stations and one fast-side station for comparison arranged by along-fault event distance. The traces are all aligned on the peak of the picked P wave arrival (*black vertical line*). NCSN catalog P arrival times (*black triangles*), our P picks (*purple triangles*), and our head wave picks (*red triangles*) are shown for each trace. Stations CVP, BKS, and CCY are 3-component with picks generated from particle motion analysis. Fast-side station CCY features impulsive P wave arrivals with no preceding phase. The five slow-side stations show a clear moveout with along-fault propagation distance

propagation distance along the fault can be used (Eq. 2) to estimate the average velocity contrast across the bimaterial interface sampled by the data. Using the measured arrival times of the direct P waves, we estimate that the average P wave velocity for the examined area around the Hayward fault is ~4.85 km/s. This value is consistent with previous velocity estimates for the creeping section of the San

Andreas Fault south of Hollister (MᴄGᴜɪʀᴇ and Bᴇɴ-Zɪᴏɴ 2005; Lᴇᴡɪꜱ *et al.* 2007), and it varies less than 5 % from station to station. The variations are most likely due to the distribution of phase propagation distances at each station; a station with a longer average propagation distance will record arrivals that travelled through deeper, and thus faster, crustal material.

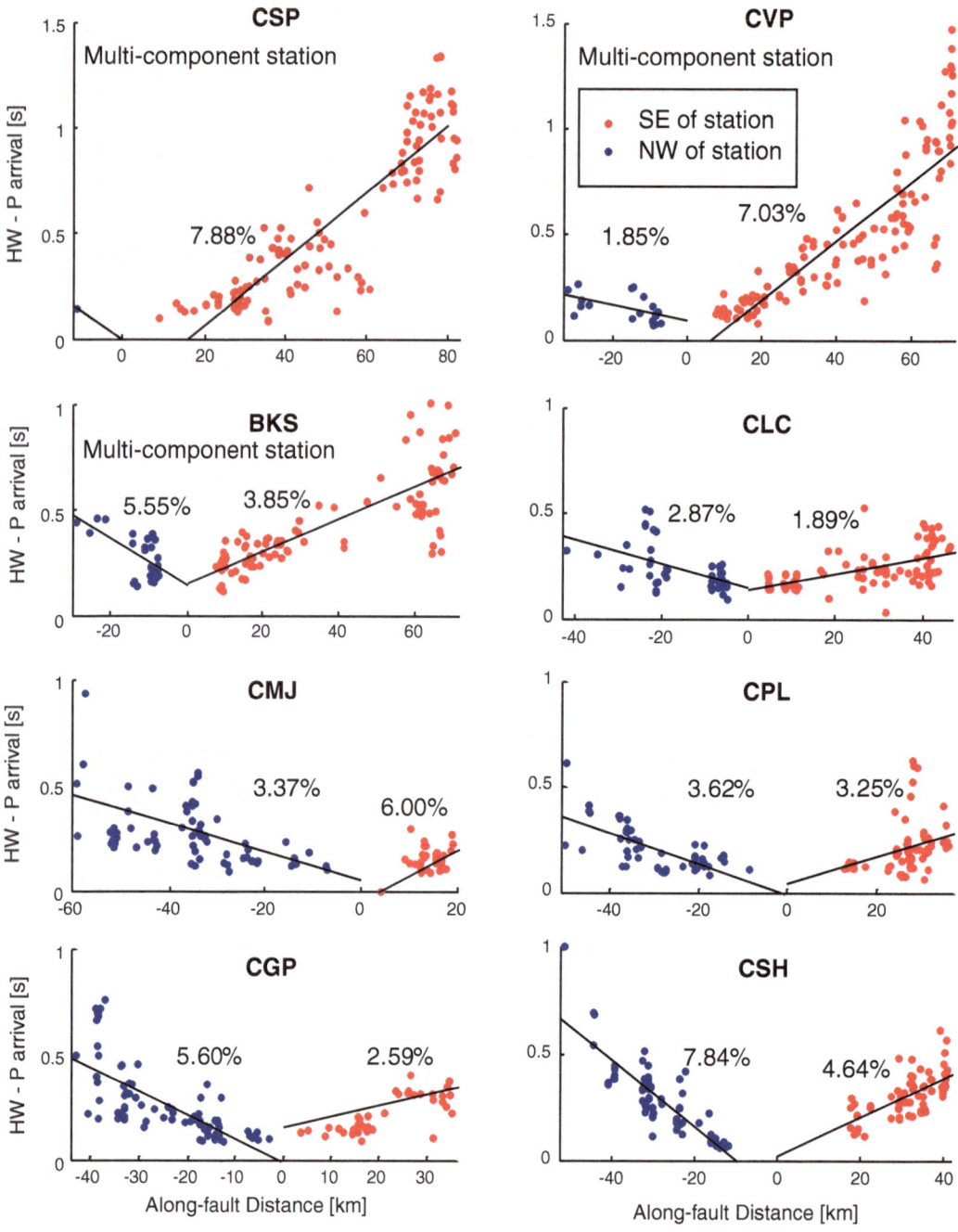

Figure 7
Differential travel time (HW−P arrival times) shown for eight slow-side stations at which head waves were observed. Distance zero is the station location; events to the northwest are shown as *blue circles*, and events to the southeast are shown as *red circles*. The differential time increases as a function of along-fault distance on average according to the best-fit least-squares regression lines (*black*). Based on the slope of the best-fit line, the average velocity contrast in either along-fault direction is calculated for each station according to Eq. (2). Stations with all three components of motion, which were picked using the particle motion analysis (Sect. 4.2), are indicated

Figure 7 displays the measured time delays between head and direct P waves, and least-squares moveout lines, at different along-strike directions of

each of the eight used stations on the slow side of the Hayward fault. Given the average nature of Eq. (2), the estimated velocity contrasts in Fig. 7 are based on the average P wave velocity (~4.85 km/s) for the entire study area. Station CSP, located at the northern end of the fault zone, has the highest contrast at almost 8 %. Station CSH, at the southern end of the fault, has a similarly high contrast. The lowest contrast of ~2 % is observed at station CLC, a 3-component station. Other stations feature a range of measured values.

A remarkable aspect of our results is the coherent propagation of FZHW over large distances along the Hayward fault. For station CSP, located furthest to the NW, head wave arrivals are detected as much as 1.4 s before the direct P wave from events as far away as 80 km, near the junction with the Calaveras fault. Station CSP has three components, so head wave arrivals were picked using particle motion. The consistent moveout observed at CSP is interrupted only at along-fault distances of −60 to −50 km. Station CVP shows a similar pattern in the moveout of the observed head wave signals, with a maximum Δt of 1.5 s at a distance of -70 km. Stations BKS and CMJ, more centrally located along the fault, show much greater scatter in the observed moveout profiles, though it should be noted that BKS is a three-component station with head wave arrivals picked using particle motion. Station CMJ, while showing scatter in head wave arrival times, has quite consistent P and post-P waveforms. Station CSH features clear head wave arrivals with consistent moveout in both directions along the fault.

5. Discussion

Tomographic images of the Hayward fault structure based on body waves and volumetric cells show clear contrasts of rock bodies across the fault (ZHANG and THURBER 2003; MANAKER et al. 2005; HARDEBECK et al. 2007). However, these studies do not distinguish between a sharp bimaterial interface, which may serve as an "attractor" of earthquake ruptures and modify considerably their properties (e.g., BEN-ZION and ANDREWS 1998; BRIETZKE and BEN-ZION 2006), and a smeared lithology contrast with little effect on earthquake dynamics. To distinguish between these possibilities, we conduct thorough analyses of horizontal particle motion and detailed

comparisons of waveforms to identify head wave arrivals at stations near the Hayward fault. The FZHW are only observed on the NE side of the fault, and are found to propagate over a large range (~80 km) of along-fault distances. This indicates the existence of a persistent bimaterial interface along most or the entire seismogenic zone of the fault. In all, we obtain 1,448 head wave and 1,538 direct P arrivals on the slow side of the fault, as well as 413 direct P arrivals at three stations on the fast side. Results for stations CSP, CVP, BKS, CMJ, CSH, and CCY are shown in Fig. 6 as station-centric stack plots of vertical displacement seismograms centered on the P wave arrival. At station CCY, on the SW (fast) side of the fault, P waves are the first impulsive arrivals with a high signal-to-noise ratio and consistent first-motion polarity. Similar results characterize the waveforms at station CPM on the SW side (Fig. 2). In contrast, the stations on the NE (slow) side record coherent emergent FZHW with opposite polarity and near fault-normal polarization that precede the direct P arrivals. A consistent moveout with distance is observed at all of the slow-side stations, although some scatter is present in the data.

The estimated velocity contrasts (Fig. 7) are within the range of values obtained for the San Andreas Fault near Parkfield (BEN-ZION et al. 1992; ZHAO et al. 2010) and south of Hollister (McGUIRE and BEN-ZION 2005; LEWIS et al. 2007), the Calaveras fault (ZHAO and PENG 2008), and the 1944 rupture zone and Mudurnu section of the North Anatolian fault (OZAKIN et al. 2012; BULUT et al. 2012). The highest velocity contrasts are measured at stations on either end of the fault zone, indicating that the bimaterial interface is extensive with depth and persistent laterally. The SW side of the fault is 3–8 % faster than the NE side, which is consistent with regional geology (GRAYMER et al. 2005) and tomographic imaging (e.g., HARDE-BECK et al. 2007). A large-scale and sharp bimaterial interface is also consistent with the fact that the Hayward fault is a major component of the plate boundary (D'ALESSIO et al. 2005; SCHMIDT et al. 2005; EVANS et al. 2012) and has been so for the last 12 Ma (McLAUGHLIN et al. 1996; GRAYMER et al. 2002; LAN-GENHEIM et al. 2010).

The spatial variations of the imaged velocity contrast can be assessed from the measured delay

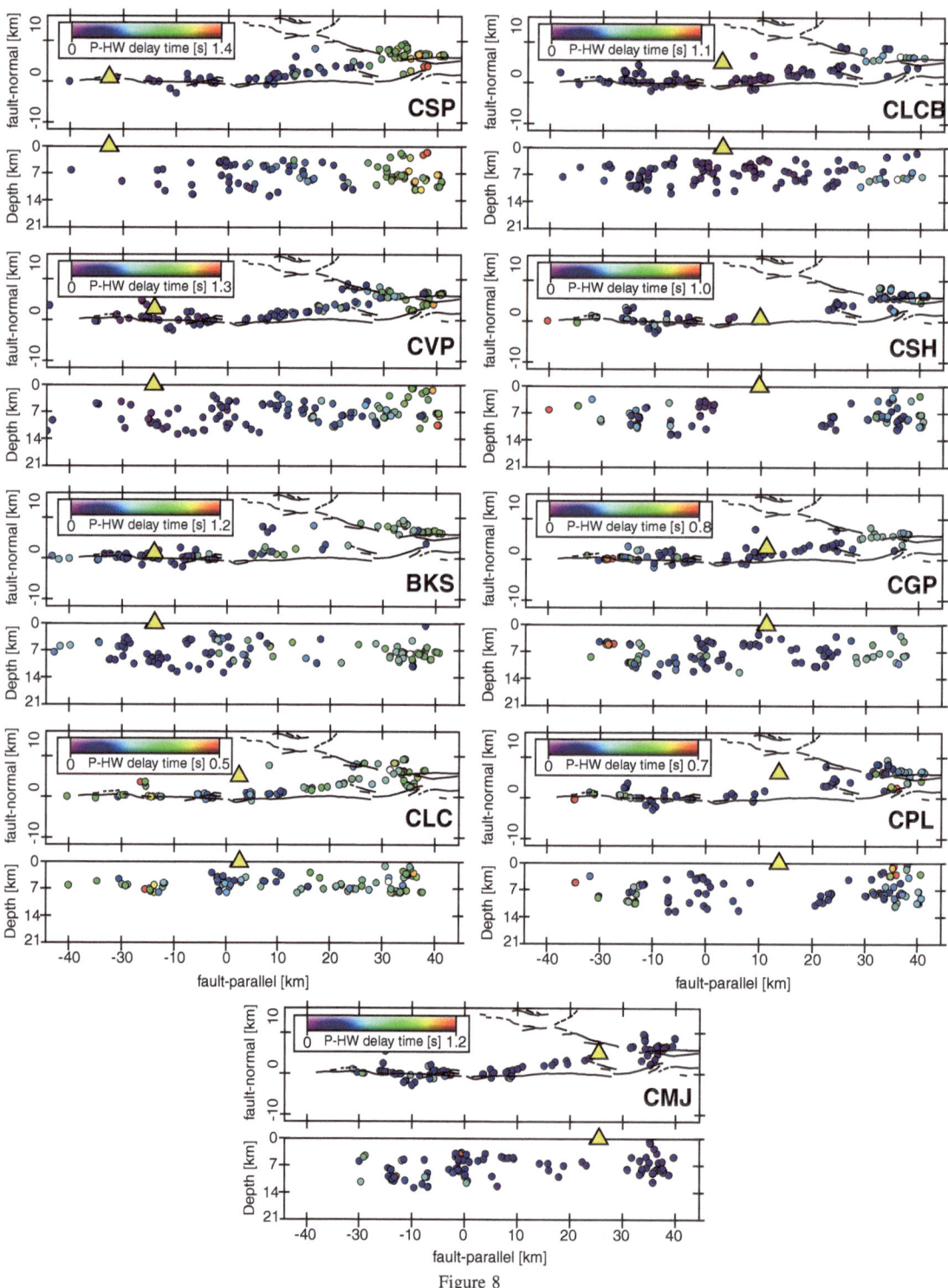

Figure 8

Map-views and cross-sections for nine slow-side stations with differential travel times (head wave minus direct P) plotted with the indicated *color scale* at hypocentral locations. In general, the delay increases with propagation distance, indicating a persistent ~80 km-long across-fault contrast. Complexity in delay times observed for events near the Calaveras junction (35 km along-fault), the City of Oakland (−10 km), and the San Pablo Bay (−30 km) correlate overall with geological and geometrical fault zone complexities. The largest delay times are observed at CSP for two very shallow events with hypocenters very close to the surface trace of the Hayward fault

times that are used for the average moveout results. Figure 8 shows for nine stations the observed delay times (head wave minus direct P arrival time) in map-view and vertical cross-sections at the source locations. In general, the delay increases with increasing along-fault propagation distance, which indicates a continuous interface with persistent velocity contrast along the entire ∼80 km section examined. The observed delay times exhibit scatter in specific locations for multiple stations. For virtually all the stations, events near the Calaveras junction produce a wide range of delay times with only minor changes in event depth and fault-normal distance. The largest delay times of this study, observed at station CSP, originate in this region from two very shallow events that are very close to the surface trace of the Hayward fault. These same events, and others nearby, also produce very large delay times at stations CVP, BKS, CLC, and CPL. Other areas with large scatter in delay times are near the city of Oakland (−10 km in the fault-parallel coordinate) and near the San Pablo Bay (−30 km). Complicated delay time patterns for events near Oakland are observed for short propagation distances at BKS and long-distances at CMJ. Variable delay times for events near San Pablo Bay (stations CGP, CSH, CPL) possibly reflect a small-scale local reversal of contrast polarity (ZHANG and THURBER 2003; GRAYMER et al. 2005). In general, delay time decreases with increasing hypocentral depth, suggesting that the velocity contrast decreases with depth, though there is large variation in this pattern.

The complexities in Δt for events originating near Oakland and near the Calaveras junction correlate with regional geology. WALDHAUSER and ELLSWORTH (2002) used seismicity patterns to define locked fault patches at depth, including one near Oakland. GRAYMER et al. (2005) correlated the same segment with a sliver of complicated metamorphic rocks, including serpentinite. The scatter in Δt of our results (Figs. 7, 8) could also be due to heterogeneity in the source region. Delay times for events originating at the Calaveras junction may be complicated by a reversal in the polarity of the velocity contrast; the NE side of the Calaveras fault is likely higher velocity based on the geology (GRAYMER et al. 2005) and tomographic imaging (MANAKER et al. 2005). In

addition, junctions of faults often produce a higher concentration of damaged rock (FINZI et al. 2009; ALLAM and BEN-ZION 2012; XU and BEN-ZION 2013) that can lead to waveform complexities and local reversals of velocity contrast.

The average 3–8 % contrast in velocity (NE slow) across the Hayward fault indicates a statistically preferred SE propagation direction for typical sub-shear ruptures and related directivity effects (e.g., BEN-ZION and ANDREWS 1998; AMPUERO and BEN-ZION 2008; BRIETZKE et al. 2009; XU et al. 2012). This may be viewed as good news in terms of seismic risk, as the denser population centers and more hazardous site effects (e.g., KNUDSEN and WENTWORTH 2000) are located towards the northern portion of the Hayward fault. For the less common super-shear ruptures, the preferred propagation direction is toward the NW (e.g., WEERTMAN 2002; SHI and BEN-ZION 2006). Some of the scatter in delay times may be produced by distributed seismicity at regions of complexity, involving events at variable fault-normal distances from the central bimaterial fault. This may be reduced by projecting all the events to the central fault surface. Such work and additional analyses of the head and direct P waves are left for a future study.

Acknowledgments

We thank the scientists involved in recording and archiving the data used in this work, Fatih Bulut for providing a core code for particle motion analysis, and Kathleen Ritterbush and Zheqiang Shi for useful discussions. The paper benefited from useful comments by two anonymous referees and Editor Antonio Rovelli. The study was supported by the National Science Foundation (Grant EAR-1315340).

Appendix: Filtering Effects on the Target Signals

Here we examine the consequences of the employed filtering scheme on the target signals of this study. There are tradeoffs in signal phase, amplitude, and polarity between causal and acausal filters. Because head waves are emergent, first-arriving, and low in amplitude compared to body waves, some care must be

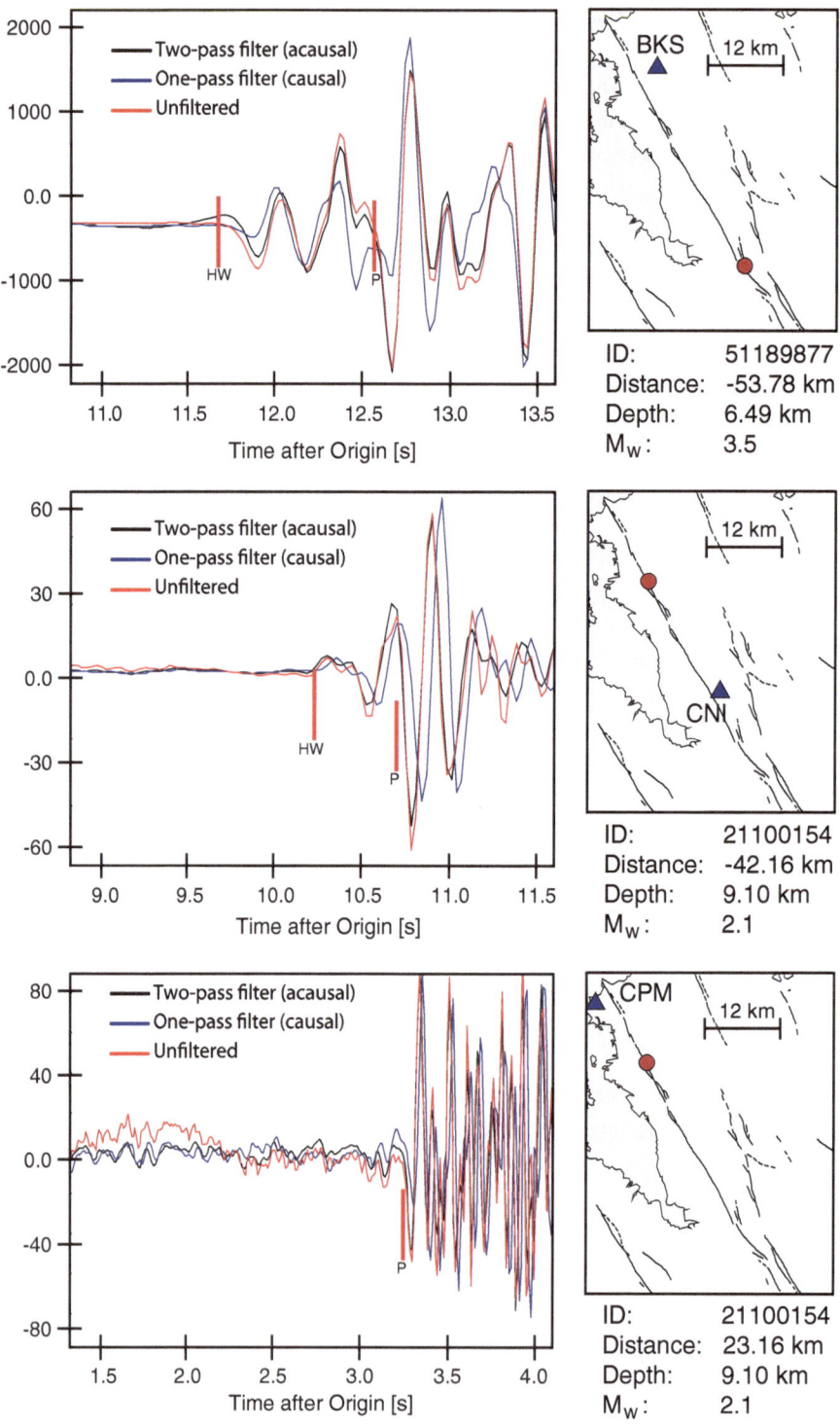

Figure 9
Examples of filtering effects on velocity seismograms at stations BKS (*top*), CNI (*middle*) and CPM (*bottom*). Event and station locations are shown on the right. The raw data (*red lines*) were filtered with a one-pass causal (*blue line*) or 2-pass acausal (*black line*) filter, both of which are 2-pole butterworth highpass filters with a corner frequency 1 Hz

taken in filter design. In general terms, acausal filters can produce small oscillations preceding high-amplitude arrivals (e.g., P waves). Causal filters lack such acausal effects, but produce a phase shift.

Figure 9 shows examples of single-pass (causal) and 2-pass (acausal) filters applied to three stations at different epicentral distances. Both filter types are 2-pole high-pass butterworth filters with 1 Hz corner frequency. For the record at station BKS on the slow NE side of the fault (top) both filters are somewhat problematic; the causal filter's negative phase delay would lead to an overestimation of the P wave arrival time, while the acausal filter changes the polarity and decreases the arrival time of the head wave. At station CNI closer to the fault on the slow side (middle), the acausal filter performs well, introducing no significant artifacts, while the causal filter would lead to a slight overestimation of the headwave arrival time. At station CPM on the fast SW side of the fault (bottom), both filters perform adequately, though a slight phase delay exists in the causally-filtered trace.

Based on these and other waveform comparisons, we choose to apply a 2-pass filter for the general analysis, as the causal filter affects the arrival picks of both the head and P waves. However, we recognize that acausal effects can sometime flip the head wave polarity or lead to a slight underestimation of its arrival time. Though these artifacts can potentially bias the head wave pick, they have only a small effect and do not change the overall consistent moveout observed at all slow-side stations (Fig. 6). Given the approximate character of the estimated (average) velocity contrast values in this work, the applied filter has minor effects on the results.

REFERENCES

ALLAM, A.A. and BEN-ZION, Y., 2012, *Seismic velocity structures in the Southern California plate-boundary environment from double-difference tomography*, Geophys. J. Int., *190*, 1181–1196, doi:10.1111/j.1365-246X.2012.05544.x.

AMPUERO, J.-P. and BEN-ZION, Y., 2008, *Cracks, pulses and macroscopic asymmetry of dynamic rupture on a bimaterial interface with velocity-weakening friction*, Geophys. J. Int., *173*, 674–692, doi:10.1111/j.1365-246X.2008.03736.x.

ANDREWS, D.J., and BEN-ZION, Y., 1997, *Wrinkle-like slip pulse on a fault between different materials*, J. Geophys. Res., *102*(B1), 553–571.

BENNINGTON, N.L., THURBER, C., PENG, Z., ZHANG, H., and ZHAO, P., 2013, *Incorporating fault zone head wave and direct wave secondary arrival times into seismic tomography: Application at Parkfield, California*, J. Geophys. Res., *118*, 1008–1014.

BEN-ZION, Y., 1989, *The response of two joined quarter spaces to SH line sources located at the material discontinuity interface*, Geophys. J. Int., *98*, 213–222.

BEN-ZION, Y., 1990, *The response of two half spaces to point dislocations at the material interface*, Geophys. J. Int., *101*, 507–528.

BEN-ZION, Y., 2001, *Dynamic Rupture in Recent Models of Earthquake Faults*, J. Mech. Phys. Solids, *49*, 2209–2244.

BEN-ZION,, Y., 2003, Appendix 2, Key Formulas in Earthquake Seismology, in *International Handbook of Earthquake and Engineering Seismology*, eds. W. HK Lee, H. Kanamori, P. C. Jennings, and C. Kisslinger, *Part B*, 1857–1875, Academic Press.

BEN-ZION, Y. and ANDREWS, D.J., 1998, *Properties and implications of dynamic rupture along a material interface*, Bull. Seismol. Soc. Am., *88*(4), 1085–1094.

BEN-ZION, Y., KATZ, S., and LEARY, P., 1992, *Joint inversion of fault zone head waves and direct P arrivals for crustal structure near major faults*, J. Geophys. Res., *97*, 1943–1951.

BEN-ZION, Y. and MALIN, P., 1991, *San Andreas fault zone head waves near Parkfield, California*, Science, *251*, 1592–1594.

BEN-ZION, Y. and SAMMIS, C.G., 2003, *Characterization of Fault Zones*, Pure Appl. Geophys., *160*, 677–715.

BEN-ZION, Y., and HUANG, Y., 2002, *Dynamic rupture on an interface between a compliant fault zone layer and a stiffer surrounding solid*, J. Geophys. Res., *107*(B2), Art. No. 2042, doi:10.1029/2001JB000254.

BILHAM, R., and WHITEHEAD, S., 1997, *Subsurface creep on the Hayward fault, Fremont, California*. Geophys. Res. Lett., *24*(11), 1307–1310.

BRIETZKE, G. B. and Y. BEN-ZION, 2006. *Examining tendencies of in-plane rupture to migrate to material interfaces*, Geophys. J. Int., *167*, 807–819, doi:10.1111/j.1365-246X.2006.03137.x.

BRIETZKE, G.B., COCHARD, A. and IGEL, H., 2009, *Importance of biomaterial interfaces for earthquake dynamics and strong ground motion*, Geophys. J. Int., *178*(2), 921–938.

BROCHER, T. M., 2008, *Compressional and shear-wave velocity versus depth relations for common rock types in northern California*, Bull. Seism. Soc. Am., *98*(2), 950 968.

BULUT, F., Y. BEN-ZION and M. BONHOFF, 2012. *Evidence for a bimaterial interface along the Mudurnu segment of the North Anatolian Fault Zone from polarization analysis of P waves*, Earth Planet. Sci. Lett., *327–328*, 17–22, doi:10.1016/j.epsl.2012.02.001.

BÜRGMANN, R., SCHMIDT, D., NADEAU, R.M., D'ALESSIO, M., FIELDING, E., MANAKER, D., and MURRAY, M.H., 2000, *Earthquake potential along the northern Hayward fault, California*, Science, *289*(5482), 1178–1182.

CHESTER, F.M., and CHESTER, J.S., 1998, *Ultracataclasite structure and friction processes of the Punchbowl fault, San Andreas system, California*, Tectonophysics, *295*(1), 199–221.

D'ALESSIO, M.A., JOHANSON, I.A., BÜRGMANN, R., SCHMIDT, D.A., and MURRAY, M.H., 2005, *Slicing up the San Francisco Bay Area: Block kinematics and fault slip rates from GPS-derived surface velocities*, J. Geopys. Res., *110*(B6), B06403.

DALGUER, L.A., and DAY, S.M., 2009, *Asymmetric rupture of large aspect-ratio faults at bimaterial interface in 3D*, Geophys. Res. Lett., *36*(23).

DOR, O., ROCKWELL, T.K., and BEN-ZION, Y., 2006, *Geological observations of damage asymmetry in the structure of the San*

Jacinto, San Andreas and Punchbowl Faults in Southern California: a possible indicator for preferred rupture propagation direction. Pure Appl. Geophys., *163*(2), 301–349, doi:10.1007/s00024-005-0023-9.

DOR, O., YILDIRIM, C., ROCKWELL, T.K., BEN-ZION, Y., EMRE, O., SISK, M., DUMAN, T. Y., 2008. *Geologic and geomorphologic asymmetry across the rupture zones of the 1943 and 1944 earthquakes on the North Anatolian Fault: possible signals for preferred earthquake propagation direction*, Geophys. J. Int., *173*, 483–504, doi:10.1111/j.1365-246X.2008.03709.x.

EVANS, E.L., LOVELESS, J.P., and MEADE, B.J., 2012, *Geodetic constraints on San Francisco Bay Area fault slip rates and potential seismogenic asperities on the partially creeping Hayward fault*, J. Geophys. Res., *117*(B3), B03410.

FIELD, E. H., DAWSON, T. E., FELZER, K. R., FRANKEL, A. D., GUPTA, V., JORDAN, T. H., and WILLS, C. J., 2009, *Uniform California earthquake rupture forecast, version 2 (UCERF 2)*, Bull. Seism. Soc. Am., *99*(4), 2053–2107.

FINZI, Y., E.H. HEARN, Y. BEN-ZION and V. LYAKHOVSKY, 2009, *Structural properties and deformation patterns of evolving strike-slip faults: Numerical simulations incorporating damage rheology*, Pure Appl. Geophys., *166*, 1537–1573, doi:10.1007/s00024-009-0522-1.

FROST, E., DOLAN, J., SAMMIS, C., HACKER, B., COLE, J., and RATSCHBACHER, L., 2009, *Progressive strain localization in a major strike-slip fault exhumed from midseismogenic depths: Structural observations from the Salzach-Ennstal-Mariazell-Puchberg fault system, Austria*, J. Geophys. Res., *114*(B4), B04406.

GRAYMER, R.W., PONCE, D.A., JACHENS, R.C., SIMPSON, R.W., PHELPS, G.A., and WENTWORTH, C.M., 2005, *Three-dimensional geologic map of the Hayward fault, northern California: Correlation of rock units with variations in seismicity, creep rate, and fault dip*, Geology, *33*(6), 521–524.

GRAYMER, R.W., SARNA-WOJCICKI, A.M., WALKER, J.P., MCLAUGHLIN, R.J., and FLECK, R.J., 2002, *Controls on timing and amount of right-lateral offset on the East Bay fault system, San Francisco Bay region, California*, Geol. Soc. Am. Bull., *114*(12), 1471–1479.

HARDEBECK, J.L., MICHAEL, A.J., and BROCHER, T.M., 2007, *Seismic velocity structure and seismotectonics of the eastern San Francisco Bay region, California*, Bull. Seismol. Soc. Am., *97*(3), 826–842.

HOUGH, S. E., Y. BEN-ZION and P. LEARY, 1994, *Fault zone waves observed at the southern Joshua Tree earthquake rupture zone*, Bull. Seism. Soc. Am., *84*, 761–767.

JURKEVICS, A., 1988, *Polarization analysis of three-component array data*, Bull. Seismol. Soc. Am., *78*(5), 1725–1743.

KNUDSEN, K.L., and WENTWORTH, C.M., 2000, Preliminary maps of Quaternary deposits and liquefaction susceptibility, nine-county San Francisco Bay Region, California: a digital database, US Geological Survey.

LANGENHEIM, V.E., GRAYMER, R.W., JACHENS, R.C., MCLAUGHLIN, R.J., WAGNER, D.L., and SWEETKIND, D.S., 2010, *Geophysical framework of the northern San Francisco Bay region, California*, Geosphere, *6*(5), 594–620.

LAWSON, A.C., 1908, The California Earthquake of April 18, 1906: Report of the State Earthquake Investigation Commission, in Two Volumes and Atlas (No. 87), Carnegie institution of Washington.

LE PICHON, A., HERRY, P., MIALLE, P., VERGOZ, J., BRACHET, N., GARCÉS, M., and CERANNA, L., 2005, *Infrasound associated with 2004–2005 large Sumatra earthquakes and tsunami*, Geophys. Res. Lett., *32*(19), L19802.

LENGLINE, O., and GOT, J.-L., 2011, *Rupture Directivity of Micro-Earthquake Sequences near Parkfield, California*, Geophys. Res. Lett., *38*, L08310, doi:10.1029/2011GL047303.

LEWIS, M.A, BEN-ZION, Y., and MCGUIRE, J., 2007, *Imaging the deep structure of the San Andreas Fault south of Hollister with joint analysis of fault-zone head and direct P arrivals*, Geophys. J. Int., *169*, 1028–1042, doi:10.1111/j.1365-246X.2006.03319.x.

LIENKAEMPER, J.J., MCFARLAND, F.S., SIMPSON, R.W., BILHAM, R.G., PONCE, D.A., BOATWRIGHT, J.J., and CASKEY, S.J., 2012, *Long-term creep rates on the hayward fault: evidence for controls on the size and frequency of large earthquakes*, Bull. Seismol. Soc. Am., *102*(1), 31–41.

LIENKAEMPER, J.J., WILLIAMS, P.L., and GUILDERSON, T.P., 2010, *Evidence for a twelfth large earthquake on the southern Hayward fault in the past 1900 years*, Bull. Seismol. Soc. Am., *100*(5A), 2024–2034.

MANAKER, D.M., MICHAEL, A.J., and BÜRGMANN, R., 2005, *Subsurface structure and kinematics of the Calaveras–Hayward fault stepover from three-dimensional VP and seismicity, San Francisco Bay region, California*, Bull. Seismol. Soc. Am., *95*(2), 446–470.

MCGUIRE, J. and BEN-ZION, Y., 2005, *High-resolution imaging of the Bear Valley section of the San Andreas Fault at seismogenic depths with fault-zone head waves and relocated seismicity*, Geophys. J. Int., *163*, 152–164, doi:10.1111/j.1365-246X.2005.02703.x.

MCLAUGHLIN, R.J., SLITER, W.V., SORG, D.H., RUSSELL, P.C., and SARNA-WOJCICKI, A.M., 1996, *Large-scale right-slip displacement on the east San Francisco Bay region fault system, California: implications for location of late Miocene to Pliocene Pacific plate boundary*, Tectonics, *15*(1), 1–18.

MCNALLY, K.C., and MCEVILLY, T.,1977, *Velocity contrast across the San Andreas fault in central California: small-scale variations from P-wave nodal plane distortion*, Bull. Seism. Soc. Am., *67*, 1565–1576.

MITCHELL, T.M., BEN-ZION, Y., and SHIMAMOTO, T., 2011, *Pulverized fault rocks and damage asymmetry along the Arima-Takatsuki Tectonic Line, Japan*. Earth Planet. Sci. Lett., *308*, 284–297, doi:10.1016/j.epsl.2011.04.023.

OLSEN, A.H., AAGAARD, B.T., and HEATON, T.H., 2008, *Long-period building response to earthquakes in the San Francisco Bay Area*. Bull. Seismol. Soc. Am., *98*(2), 1047–1065.

OLSEN, K.B., DAY, S.M., MINSTER, J.B., CUI, Y., CHOURASIA, A., FAERMAN, M., and JORDAN, T., 2006, *Strong shaking in Los Angeles expected from southern San Andreas earthquake*, Geophys. Res. Lett., *33*(7), L07305.

OPPENHEIMER, D.H., REASENBERG, P.A., and SIMPSON, R.W., 1988, *Fault plane solutions for the 1984 Morgan Hill, California, earthquake sequence: evidence for the state of stress on the Calaveras fault*, J. Geophys. Res., *93*, 9007–9026.

OZAKIN, Y., BEN-ZION, Y., AKTAR, M., KARABULUT, H., and PENG, Z., 2012, *Velocity contrast across the 1944 rupture zone of the North Anatolian fault east of Ismetpasa from analysis of teleseismic arrivals*, Geophys. Res. Lett., *39*, L08307, doi:10.1029/2012GL051426.

RUBIN, A. and GILLARD, D., 2000, *Aftershock asymmetry/rupture directivity among central San Andreas fault microearthquakes*, J. Geophys. Res., *105*(B8), 19095–19109.

SCHAFF, D. P., and WALDHAUSER, F., 2005, *Waveform cross-correlation-based differential travel-time measurements at the*

Northern California Seismic Network, Bull. Seismol. Soc. Am., *95*(6), 2446–2461.

SCHMIDT, D.A., BÜRGMANN, R., NADEAU, R.M., and D'ALESSIO, M., 2005, *Distribution of aseismic slip rate on the Hayward fault inferred from seismic and geodetic data,* J. Geophys. Res., *110*(B8), B08406.

SENGÖR, A.M.C., TÜYSÜZ, O., IMREN, C., SAKINÇ, M., EYIDOGAN, H., GÖRÜR, N., and RANGIN, C., 2005, *The North Anatolian fault: a new look,* Annu. Rev. Earth Planet. Sci., *33*, 37–112.

SHI, Z. and Y. BEN-ZION, 2006. *Dynamic rupture on a bimaterial interface governed by slip-weakening friction,* Geophys. J. Int., *165*, 469–484, doi:10.1111/j.1365-246X.2006.02853.x.

SIBSON, R.H., 2003, *Thickness of the seismic slip zone,* Bull. Seismol. Soc. Am., *93*(3), 1169–1178.

SIMPSON, R. W., LIENKAEMPER, J.J., and GALEHOUSE, J.S., 2001, *Variations in creep rate along the Hayward Fault, California, interpreted as changes in depth of creep,* Geophys. Res. Lett, *28*(11), 2269–2272.

WALDHAUSER, F. and SCHAFF, D.P., 2008, *Large-scale relocation of two decades of Northern California seismicity using cross-correlation and double-difference methods,* J. Geophys. Res., *113*, B08311, doi:10.1029/2007JB005479.

WALDHAUSER, F., and ELLSWORTH, W.L., 2000, A *Double-Difference Earthquake Location Algorithm: Method and Application to the Northern Hayward Fault, California,* Bull. Seismol. Soc. Am, *90*, 6, 1353–1368.

WALDHAUSER, F., and ELLSWORTH, W.L., 2002, *Fault structure and mechanics of the Hayward Fault, California, from double-difference earthquake locations,* J. Geophys. Res., *107*(B3), 2054.

WEERTMAN, J., 1980. *Unstable slippage across a fault that separates elastic media of different elastic constants,* J. Geophys. Res., *85*, 1455–1461.

WEERTMAN, J., 2002. *Subsonic type earthquake dislocation moving at approximately* $\sqrt{2}$ × *shear wave velocity on interface between half spaces of slightly different elastic constants,* Geophys. Res. Lett., *29*(10), doi:10.1029/2001GL013916.

WDOWINSKI, S., SMITH, B., BOCK, Y., and SANDWELL, D., 2007, *Diffuse intersesimic deformation across the Pacific-North America plate boundary,* Geology, *35*, 311–314.

WECHSLER, N., T. K. ROCKWELL and Y. BEN-ZION, 2009, *Application of high resolution DEM data to detect rock damage from geomorphic signals along the central San Jacinto Fault,* Geomorphology, *113*, 82–96, doi:10.1016/j.geomorph.2009.06.007.

XU, S. and BEN-ZION, Y., 2013, *Numerical and theoretical analyses of in-plane dynamic rupture on a frictional interface and off-fault yielding patterns at different scales,* Geophys. J. Int., *193*, 304–320, doi:10.1093/gji/ggs105.

XU, S., BEN-ZION, Y. and AMPUERO, J.-P., 2012, *Properties of Inelastic Yielding Zones Generated by In-plane Dynamic Ruptures: I. Model description and basic results,* Geophys. J. Int., *191*, 1325–1342, doi:10.1111/j.1365-246X.2012.05679.x.

YU, E., and SEGALL, P., 1996, *Slip in the 1868 Hayward earthquake from the analysis of historical triangulation data.* Journal of Geophysical Research, *101*(B7), 16101–16.

ZALIAPIN, I. and BEN-ZION, Y., 2011, *Asymmetric distribution of aftershocks on large faults in California,* Geophys. J. Int., *185*, 1288–1304, doi:10.1111/j.1365-246X.2011.04995.x.

ZHANG, H., and THURBER, C.H., 2003, *Double-difference tomography: The method and its application to the Hayward fault, California,* Bull. Seism. Soc. Am, *93*, 1875–1889.

ZHAO, P., and PENG, Z., 2008, *Velocity contrast along the Calaveras fault from analysis of fault zone head waves generated by repeating earthquakes,* Geophys. Res. Lett., *35*(1), L01303.

ZHAO, P., PENG., Z., SHI, Z., LEWIS, M., and BEN-ZION, Y., 2010, *Variations of the Velocity Contrast and Rupture Properties of M6 Earthquakes along the Parkfield Section of the San Andreas Fault,* Geophys. J. Int., *180*, 765–780, doi:10.1111/j.1365-246X. 2009.04436.x.

(Received August 8, 2013, revised January 3, 2014, accepted January 21, 2014, Published online February 12, 2014)

Reprinted from the journal

Pure Appl. Geophys. 171 (2014), 3013–3022
© 2014 Springer Basel
DOI 10.1007/s00024-014-0806-y

Joint Inversion of Body-Wave Arrival Times and Surface-Wave Dispersion for Three-Dimensional Seismic Structure Around SAFOD

HAIJIANG ZHANG,[1] MONICA MACEIRA,[2] PHILIPPE ROUX,[3] and CLIFFORD THURBER[4]

Abstract—We incorporate body-wave arrival time and surface-wave dispersion data into a joint inversion for three-dimensional P-wave and S-wave velocity structure of the crust surrounding the site of the San Andreas Fault Observatory at Depth. The contributions of the two data types to the inversion are controlled by the relative weighting of the respective equations. We find that the trade-off between fitting the two data types, controlled by the weighting, defines a clear optimal solution. Varying the weighting away from the optimal point leads to sharp increases in misfit for one data type with only modest reduction in misfit for the other data type. All the acceptable solutions yield structures with similar primary features, but the smaller-scale features change substantially. When there is a lower relative weight on the surface-wave data, it appears that the solution over-fits the body-wave data, leading to a relatively rough V_s model, whereas for the optimal weighting, we obtain a relatively smooth model that is able to fit both the body-wave and surface-wave observations adequately.

1. Introduction

The crust around the San Andreas Fault Observatory at Depth (SAFOD) has been the subject of many geophysical studies aimed at characterizing in detail the fault zone structure and elucidating the lithologies and physical properties of the surrounding rocks. Seismic methods in particular have revealed the complex two-dimensional (2D) and three-dimensional (3D) structure of the crustal volume around SAFOD (LEES and MALIN, 1990; MICHELINI and MCEVILLY, 1991; EBERHART-PHILLIPS and MICHAEL, 1993; THURBER et al., 2003, 2004; HOLE et al., 2006; ROECKER et al., 2006; BLEIBINHAUS et al., 2007; ZHANG et al., 2005, 2009; BENNINGTON et al., 2008), and the strong velocity reduction in the fault damage zone (LI et al., 1990, 1997, 2004; BEN ZION and MALIN, 1991; KORNEEV et al., 2003; LI and MALIN, 2008; LEWIS and BEN ZION, 2010; WU et al., 2010). Important additional insights have been obtained from magnetotelluric (MT) studies (UNSWORTH et al., 1997, 2000; UNSWORTH and BEDROSIAN, 2004; BECKEN et al., 2008), and some studies have carried out different types of joint geophysical inversion (ROECKER et al., 2004; BENNINGTON et al. in revision).

Here, we experiment with a different type of joint inversion, using body-wave arrival time and surface-wave dispersion data to image the P-wave and S-wave velocity structure of the upper crust surrounding SAFOD. The two data types have complementary strengths—the body-wave data have good resolution at depth, albeit only where there are crossing rays between sources and receivers, whereas the short-period surface waves have very good near-surface resolution and are not dependent on the earthquake source distribution because they are derived from ambient noise. The body-wave data are from local earthquakes and explosions, comprising the dataset analyzed by ZHANG et al. (2009). The surface-wave data are for Love waves from ambient noise correlations, and are from ROUX et al. (2011). We examine how the S-wave model varies as we vary the relative weighting of the fit to the two data sets and in comparison to the previous separate inversion results, and assess whether the "optimal" model, based on the weight corresponding to the corner of a

---
[1] Laboratory of Seismology and Earth's Interior, School of Earth and Space Sciences, University of Science and Technology of China, 96 Jinzhai Road, Hefei 230026, Anhui, China. E-mail: zhang11@ustc.edu.cn
[2] Earth and Environmental Sciences, Los Alamos National Laboratory, Los Alamos, New Mexico 87545, USA.
[3] ISTerre, CNRS, IRD, Université Joseph Fourier, Saint-Martin-d'Hères, France.
[4] Department of Geoscience, University of Wisconsin-Madison, Madison, Wisconsin 53706, USA.

trade-off curve, indeed appears to be optimal or not. We note that due to the indirect coupling of the S-wave and P-wave models only through the hypocenter parameters, the P-wave models obtained with the different weights are virtually indistinguishable from each other and from the original model of ZHANG et al. (2009).

2. Datasets and Processing

Figure 1 shows a map of the events and stations from which body-wave arrival-time and surface-wave dispersion data were obtained (ZHANG et al., 2009). The main source of data is a temporary seismic array known as the Parkfield Area Seismic Observatory (PASO) (THURBER et al., 2003, 2004). All of the PASO array stations had three-component sensors with the majority of them being broadband. We also include data from (1) the UC-Berkeley High Resolution Seismic Network (HRSN), (2) fault-zone guided wave field projects in 1994 (LI et al., 1997) and 2004 (LI and MALIN, 2008), (3) the USGS Central California Seismic Network, and (4) USGS temporary stations. Earthquake and explosion data were also obtained from borehole seismic strings deployed in the SAFOD Pilot Hole in July 2002 (CHAVARRIA et al., 2003) and in the SAFOD main borehole in May

2005 by Paulsson Geophysical Services, Inc., as well as from borehole geophones deployed periodically in the SAFOD borehole between December 2004 and November 2006 by the SAFOD project. Sources of explosion data include a ~5 km long high-resolution reflection/refraction line in 1998 (HOLE et al., 2001; CATCHINGS et al., 2002), a ~50 km long reflection/refraction line in 2003 (HOLE et al., 2006), and PASO "calibration" shots in 2002, 2003, 2004, and 2006. In total, there are 1,563 events including 574 earthquakes, 836 shots (explosions with known location and origin times), and 153 blasts (explosions with known location and uncertain origin times). There are about 90,800 arrival times (72 % P and 28 % S) included in the inversion. For the 574 earthquakes, differential times (approximately 489,000 P and 364,000 S) from event pairs observed at common stations were also calculated using a waveform cross-correlation package, BCSEIS (DU et al., 2004).

The continuous data for the ambient noise analysis were obtained from 30 broadband stations of the PASO array (green triangles in Fig. 1). Since the local ambient seismic noise is dominated by the microseism noise excitation around 0.15 Hz (TANIMOTO et al., 2006), the signals were pre-whitened between 0.15 and 0.35 Hz to allow an extension of the group velocity analysis to higher frequencies. For such directive noise (coming from the Pacific Ocean),

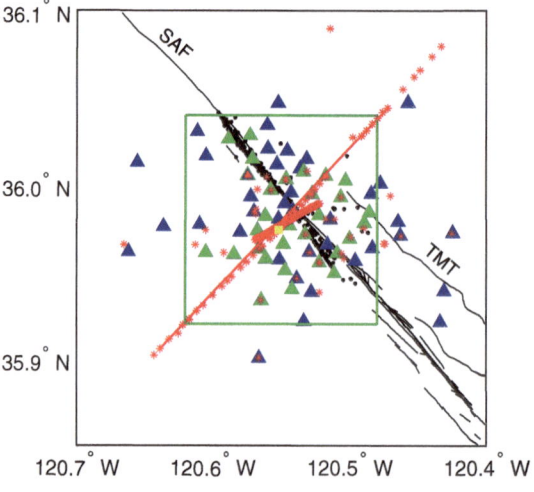

Figure 1

Map of seismicity (black dots), seismic stations (triangles), shots (red stars), and faults (black lines). The green box indicates the model volume shown in Fig. 6. Green triangles are the stations used for the surface-wave analysis. The yellow square indicates the SAFOD site. SAF San Andreas Fault, TMT Table Mountain Thrust

the concept of passive seismic-noise tomography was explored on three-component sensors (ROUX *et al.*, 2011). An optimal rotation algorithm (ORA) was applied to the nine-component correlation tensor measured from each pair of three-component seismometers among the array, that forced each station pair to re-align in the noise direction, a necessary condition to extract unbiased travel-times from passive seismic processing (ROUX, 2009; ROUEFF *et al.*, 2009). Taking advantage of the short distances between the sensors, only a relatively short time period of data (15 days) was needed to obtain adequate correlation results.

Note that no near-field contributions of the surface-wave's tensor were retrieved from the noise correlation, as the dominant noise source was in the far field. This confirmed that noise correlation on long time records simply behaves as a time-domain interferometer that magnifies phase coherence between station pairs in the frequency bandwidth of interest. As a consequence, the small size of the seismic network is both an advantage and a disadvantage regarding tomography inversion: an advantage as the coherence is high for many station pairs, which makes travel time measurements very accurate; a disadvantage since travel times extracted from the noise-correlation tensor are close to zero, which could make residual uncertainty of great importance in velocity measurement errors. In practice, an average signal-to-noise ratio of 60 was obtained at the peak maximum of the 15-day-averaged correlation function for each station pair. With such a high value of the signal-to-noise ratio, the accuracy of the travel-time measurement could, in theory, be close to infinity and, at least in practice, sufficiently high to provide reliable estimations.

Finally, after the rotation was performed, an optimal surface-wave tensor is obtained from which Rayleigh and Love waves were separately extracted for tomography inversion [see Fig. 2 and further details about the Love-wave inversion in ROUX *et al.* (2011)]. The choice of Love waves for the inversion was motivated by (1) their high sensitivity to shallow geophysical structure and (2) the residual pollution of P-waves on the vertical and radial components (ROUX *et al.*, 2005) of the correlation tensor that may bias the dispersion curve of Rayleigh waves. The tomography

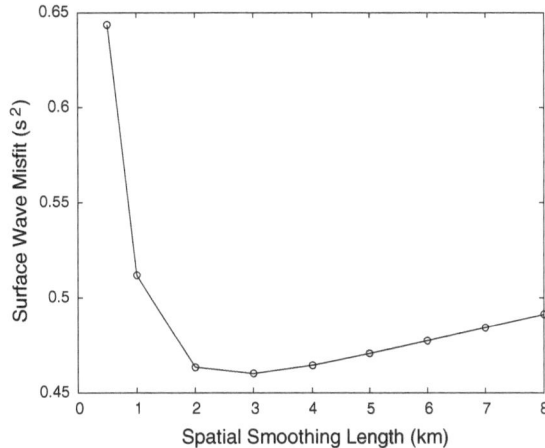

Figure 2

Example misfit *curve*, for the 0.3 Hz band, for the determination of the optimal spatial smoothing length for the Love wave group velocity maps of ROUX *et al.* (2011) that are used in our joint inversion. The plot indicates the minimum falls between 2.0 and 3.0 km, so a value of 2.5 km was adopted

inversion resulted from the combination of 360 selected dispersion curves for Love waves. Straight rays were assumed as propagation paths and a 2.5-km spatial smoothing was applied. The latter was based on the point of minimum misfit as a function of spatial smoothing length (Fig. 2). This shows that the surface-wave group velocity maps were not heavily smoothed. Examples of group velocity maps from ROUX *et al.* (2011) are shown in Fig. 3. The northeast-southwest velocity gradient across the San Andreas Fault (SAF) is clearly visible in the ROUX *et al.* (2011) V_s model, as well as the low-velocity region down to ~ 2.5 km below the surface between the SAFOD borehole and the SAF surface trace.

3. Joint Inversion

First, we describe the basic elements of the (separate) body-wave and surface-wave inversions that are incorporated into our joint inversion and then we explain how the joint inversion is set up.

For this work, we make use of the regional-scale version of the double-difference (DD) tomography algorithm tomoDD (ZHANG and THURBER, 2003, 2006). DD tomography is a generalization of DD location (WALDHAUSER and ELLSWORTH, 2000), simultaneously solving for the 3D velocity structure and seismic event

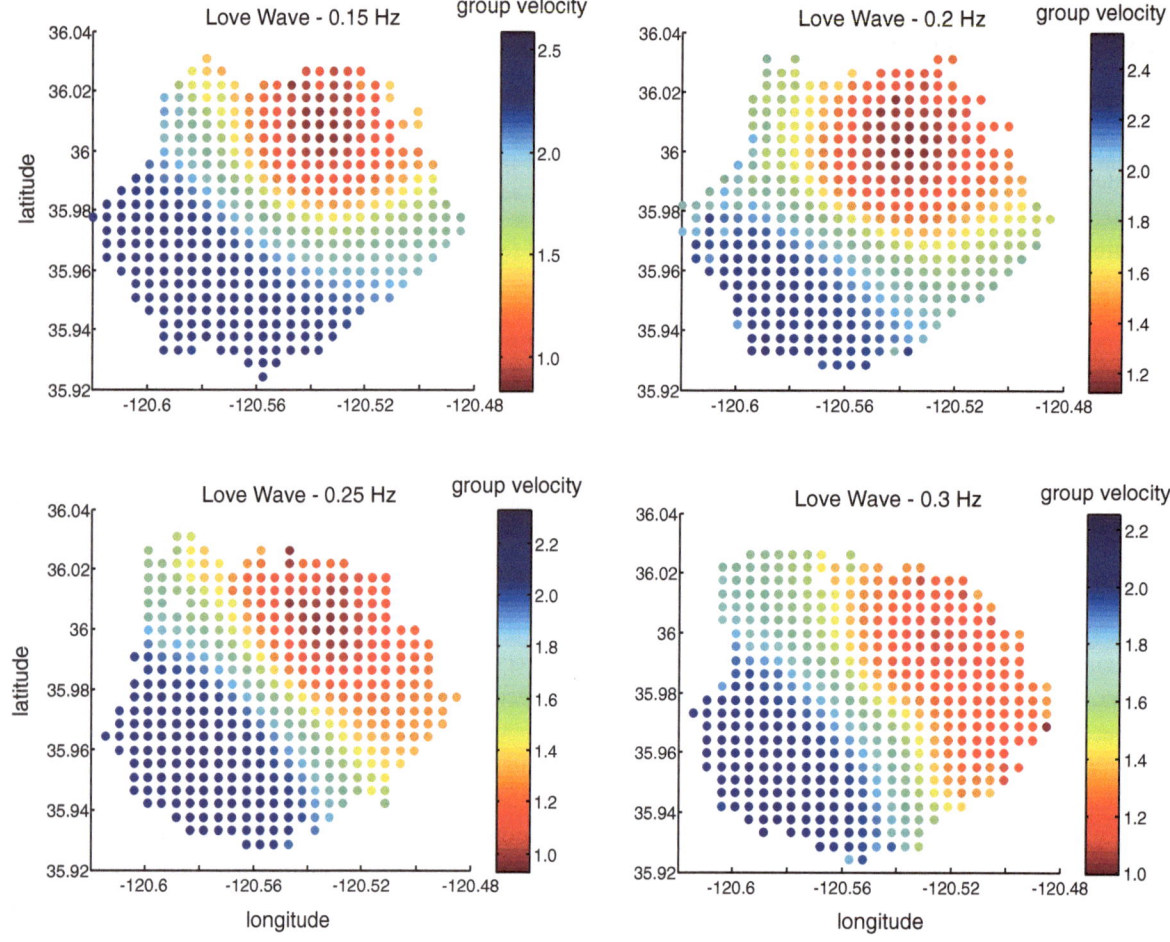

Figure 3
Example Love wave group velocity maps from Roux *et al.* (2011) used in our joint inversion. Velocities are in m/s

locations. The P-wave and S-wave velocity models are inverted for as separate models, although they are mathematically coupled through the hypocenter parameters. DD tomography uses a combination of absolute and more accurate differential arrival times and hierarchically determines the velocity structure from larger scale to smaller scale. LSQR (Paige and Saunders, 1982) is used to invert for model perturbations, and regularization of the inversion is accomplished with a first-difference smoothing operator. The regional-scale version, tomoFDD (Zhang *et al.*, 2004), employs a finite-difference solver [either Podvin and Lecomte (1991), or Hole and Zelt (1995), the latter based on Vidale (1988)] to calculate travel times in a spherical Earth geometry that is converted to a Cartesian system (so-called "sphere in a box").

The surface-wave inversion code that is integrated into the joint inversion algorithm is from Maceira and Ammon (2009) and follows Julià *et al.* (2000). The propagator matrix solver in the algorithm DISPER80 (Saito, 1988) is used for the forward calculation of dispersion curves from layered velocity models. Theoretically, the dispersion curve is a function of shear wave velocity, compressional wave velocity, and density of the media (Bucher and Smith, 1971). However, the sensitivity to P-wave velocity and density is significantly smaller than the sensitivity to S-wave velocity (Takeuchi and Saito, 1972; Aki and Richards, 1980; Bache *et al.*, 1978; Tanimoto, 1991). Therefore, only shear velocity variations are considered for modeling the Love wave dispersion observations.

To accomplish the joint inversion, the body-wave and surface-wave equations are combined in a single system with weighting factors controlling the relative contributions of the two data types to the solution for model perturbations. We note that, in general, the body-wave equations involve the P-wave and S-wave velocity structure and the source locations and/or origin times (for earthquakes and explosions with unknown origin time), whereas the surface-wave equations only involve the S-wave velocity. Thus, the joint inversion equations can be expressed as

$$
\begin{bmatrix}
\mu_1 \mathbf{G}_H^{T_p} & \mu_1 \mathbf{G}_{V_p}^{T_p} & 0 \\
\mu_2 \mathbf{G}_H^{T_s} & 0 & \mu_2 \mathbf{G}_{V_s}^{T_s} \\
0 & 0 & \mu_3 \mathbf{G}_{V_s}^{SW} \\
0 & w_p \mathbf{L}_{V_p} & 0 \\
0 & 0 & w_s \mathbf{L}_{V_s} \\
\lambda_H \mathbf{I} & 0 & 0 \\
0 & \lambda_p \mathbf{I} & 0 \\
0 & 0 & \lambda_s \mathbf{I}
\end{bmatrix}
\begin{bmatrix}
\Delta \mathbf{H} \\
\Delta \mathbf{m}_p \\
\Delta \mathbf{m}_s
\end{bmatrix}
=
\begin{bmatrix}
\mu_1 \mathbf{d}^{T_p} \\
\mu_2 \mathbf{d}^{T_s} \\
\mu_3 \mathbf{d}^{SW} \\
0 \\
0 \\
0 \\
0 \\
0
\end{bmatrix},
$$

(1)

where $\mathbf{G}_H^{T_p}$, $\mathbf{G}_H^{T_s}$, $\mathbf{G}_{V_p}^{T_p}$ and $\mathbf{G}_{V_s}^{T_s}$ are the sensitivity matrices of first P-arrival and S-arrival times with respect to hypocenter parameters, V_p, and V_s, respectively; $\mathbf{G}_{V_s}^{SW}$ is the sensitivity matrix of surface wave dispersion data with respect to V_s; $\mathbf{L}_{V_p}$ and $\mathbf{L}_{V_s}$ are the first-order smoothing matrices for the V_p and V_s models with weights of w_p and w_s, respectively; λ_H, λ_p, and λ_s are the damping parameters for hypocenter parameters, V_p, and V_s model parameters, respectively; $\Delta \mathbf{H}$, $\Delta \mathbf{m}_p$, and $\Delta \mathbf{m}_s$ are perturbations to hypocenter parameters, V_p, and V_s model parameters, respectively; and μ_1, μ_2 and μ_3 are relative weights for body wave P-arrival and S-arrival times and surface wave dispersion data, respectively. For simplicity in our study we set data weights μ_1 and μ_2 for P and S arrival time data to be equal. The damping parameters, smoothing weight parameters, and data weighting parameters are chosen by a combination of requiring a reasonable condition number (mainly controlled by the damping) and a trade-off analysis (examining the data weighting). We retain the same smoothing weight that was found to be optimal by ZHANG et al. (2009). LSQR (PAIGE and SAUNDERS, 1982) is used to solve the system of Eq. (1).

We note that there is a difference in the way the velocity structure is parameterized in the underlying body-wave versus surface-wave components of the joint inversion. For body-wave arrival times, the velocity structure is represented by the value at grid nodes, and interpolation is used to obtain the velocity value at any point. Thus, velocity is treated as a continuous function of position. In contrast, for the surface waves, velocities are defined in latitude-longitude cells and vertical layers such that the velocity is constant within each cell. To accommodate two different model parameterizations, we put the node in the center of each cell. For this study, the grid intervals are 0.0048° in latitude and 0.0044° in longitude, and vary from 0.5 to 3 km in depth (at $Z =$ -0.5, 0, 0.5, 1.0, 2.0, 4.0, 7.0, and 10.0 km).

4. Trade-off Curve and Inversion Results

To explore specifically the effect of the joint inversion on the structural model, we keep all data and parameters constant except for the weighting of the body-wave equations, which we vary over several orders of magnitude, with the surface-wave weight kept fixed at 1. We note that the damping and smoothing we apply to stabilize LSQR and regularize the inversion is the same as that used by ZHANG et al. (2009) in their body-wave only tomography study. We find that for weights below a value of 1 (i.e., equal weighting of body-wave and surface-wave equations), the misfit to both the body-wave and surface-wave data ceases to change, and above 100, the body-wave misfit remains constant. Thus, we consider the trade-off in data fit between body waves and surface waves in the range 1–100 for the body-wave weight (Fig. 4; Table 1). What we find is that there is a strong, smooth variation in data fit as we move from a weight of 1 to a weight of 100, with a "knee" in the curve near a weight of 20. Near this point, a small increase in the body-wave weight leads to a large increase in surface-wave misfit with little decrease in body-wave misfit. Conversely, a small decrease in the body-wave weight leads to a large increase in body-wave misfit with little decrease in surface-wave misfit. Thus, on this basis we tentatively identify a body-wave weight of 20 as optimal.

An independent way to estimate the proper relative weighting of the body-wave and surface-wave

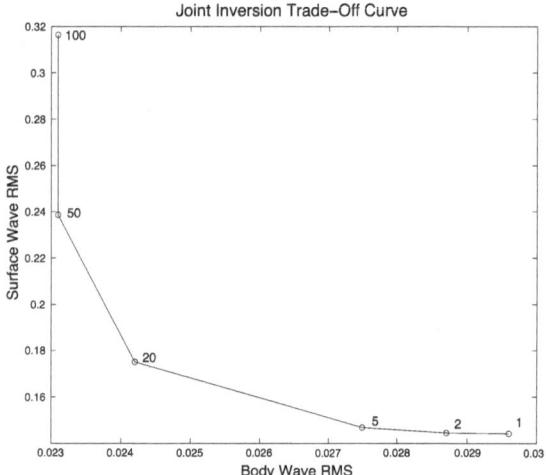

Figure 4
Plot of surface-wave data misfit (in km/s) versus body-wave data misfit (in s) in the joint inversion as a function of the body-wave weight (the surface-wave weight was kept fixed at 1). A body-wave weight of 20 is adopted as optimal

Table 1

Variance reduction for body-wave and surface-wave data as a function of the body-wave weight

Body-wave weight	Body-wave variance reduction (%)	Surface-wave variance reduction (%)
1	77.5	74.0
2	77.7	73.9
5	78.4	73.0
20	79.7	61.6
50	80.4	28.7
100	81.1	1.15

data is in terms of their relative data uncertainties. For the body-wave data, the average pick uncertainty is on the order of 0.02 s, based on the pick quality as estimated directly from the waveform data, which agrees well with the level of RMS misfit we obtain in the trade-off analysis (Fig. 3). Similarly, the surface-wave data uncertainty estimate of 10 % (ROUX et al., 2011) is consistent with the RMS misfit on the order of 0.2 km/s, given that the model velocities are mainly in the range of 1.5–3 km/s. The ratio of 10 between these two uncertainties is of the same order as the optimal weighting ratio of 20–1 from the trade-off curve in Fig. 4. This is analogous to the stochastic inverse (e.g., AKI et al., 1977), for which the ratio of the a priori data variance to model variance is used as the damping factor in the least squares inversion.

Next, we compare the inversion model results for the nominal optimal body-wave weight (Fig. 5c) to those for low (Fig. 5a), moderately low (Fig. 5b), moderately high (Fig. 5d), and high (Fig. 5e) body-wave weights. Horizontal slices through the V_s models for the various cases are shown in Fig. 5 from 500 m above to 2 km below sea level, where the changes are greatest. Moving from the optimal (Fig. 5c) to the moderately low (Fig. 5b) and low (Fig. 5a) body-wave weight cases, the primary features remain present but the model amplitude variations are slightly reduced and smaller-scale features, especially those near the SAF, shrink or disappear. This indicates the lateral smoothing effect of the surface-wave data. Moving from the optimal (Fig. 5c) to the moderately high (Fig. 5d) and high (Fig. 5e) body-wave weight cases, the primary features still remain, but there is a substantial increase in the number and amplitude of the smaller-scale features. Given that the body-wave data misfit is reduced by only 4 % at the cost of a ~ 35–80 % increase in the surface-wave data misfit, we interpret the higher-weight cases to indicate overfitting the noise in the body-wave data.

ROUX et al. (2011) highlighted a "U-shaped" feature containing a low-velocity zone in their ambient noise model. We compare an iso-surface at the same V_s value of 2.5 km/s in our new model (Fig. 6a) to their original result (Fig. 6b). This steep-sided wedge of slow material corresponds to the sedimentary packages penetrated by the main SA-FOD borehole. The joint inversion results reveal a greater depth and along-fault extent of this feature than the previous ambient noise only model.

We also note an area of strongly varying group velocity with frequency in the southeast part of our study area, near 35.97°, −120.50°, northeast of the SAF (Fig. 3). Here, group velocity drops from ~ 2 km/s at 0.15 and 0.2 Hz to ~ 1.2 km/s at 0.3 Hz. This is an area that has been observed to be anomalous in some previous studies. ZHANG et al. (2009) found relatively high V_p here extending from ~ 2 to 7 km depth from body-wave tomography, and ZHANG et al. (2007) found the same area to be highly anisotropic (~ 4 %) from a tomographic inversion of shear wave splitting delays. Thus, it is not surprising to find it to be anomalous in terms of Love wave group velocity behavior as well.

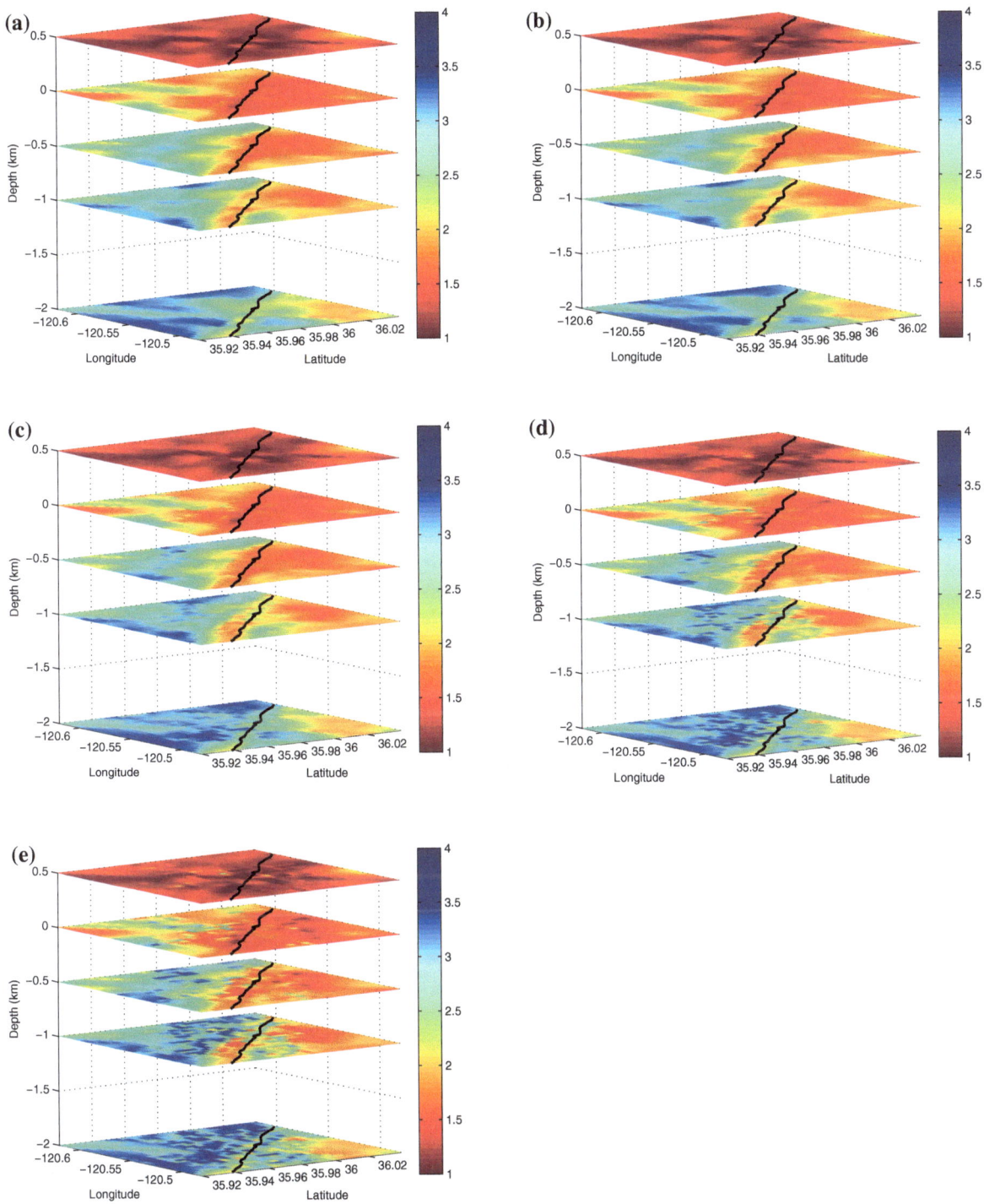

Figure 5

Comparison of velocity model results for different body-wave weights, showing *horizontal slices* at the depths of the shallow *grid nodes* (−0.5, 0, 0.5, 1, and 2 km depth relative to mean sea level). The weights used for the models that are shown: (a) 2, (b) 5, (c) 20, (d) 50, (e) 100. The *black line* is the SAF trace. The view is from the southeast

(a)

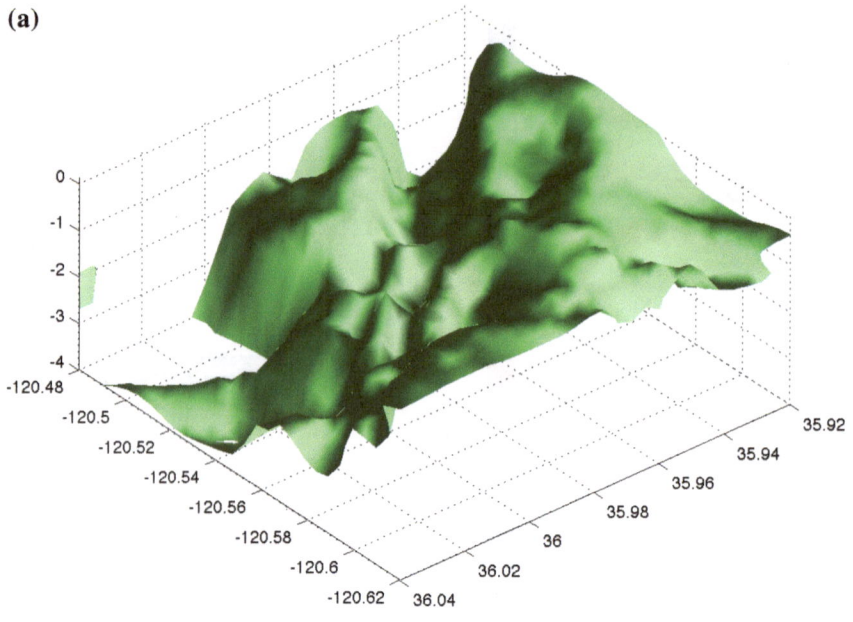

(b)

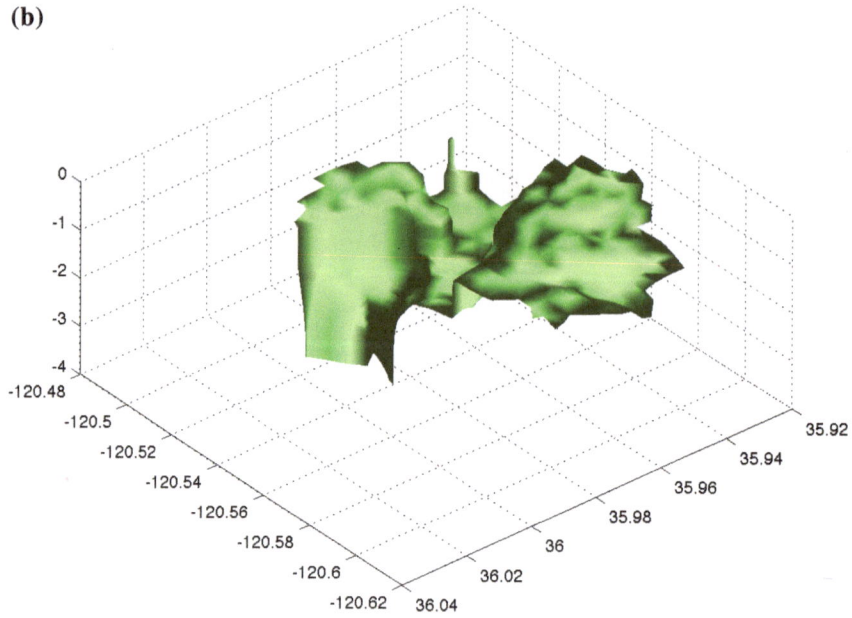

Figure 6
Iso-surface view of the 2.5 km/s V_s surface for (**a**) the weight of 20 model in Fig. 5 compared to (**b**) the same view from Roux *et al.* (2011). The view is from the northwest. Notice the deeper expression and longer extent of the "U-shaped" feature in the joint inversion model in (**a**)

5. Discussion and Conclusions

The joint inversion algorithm expressed by Eq. (1) has advantages compared to separate body-wave arrival time inversion and surface-wave dispersion inversion. Compared to the separate arrival-time inversion, the joint inversion V_s model is more strongly constrained by the combination of body-wave data and surface-wave data. Even though traditional surface-wave inversion typically only derives a V_s model, an updated V_p model is required at each iteration in order to compute the surface wave responses. In traditional surface-wave inversions, this is generally accomplished by assuming a constant V_p/V_s ratio provided by the a priori model. In comparison, the joint inversion determines a more reliable V_p model from P-wave arrival times. As a result, the surface-wave responses are more accurate and the surface-wave data can be better used to invert for the V_s model.

The joint inversion results do not make significant changes to the main features found previously in the crust surrounding SAFOD. Rather, the joint inversion suppresses oscillatory features in the V_s model that do not make a significant change in the fit to the body-wave data. Adding the surface-wave constraints to the body-wave inversion leads to a simpler (smoother) V_s model that is able to fit both the body-wave and surface-wave data adequately.

Acknowledgments

We thank Yehuda Ben-Zion and Antonio Rovelli for organizing the 40th Workshop of the International School of Geophysics on "Properties and Processes of Crustal Fault Zones" in Erice, Sicily, which motivated the present work. We are grateful to two anonymous reviewers for their constructive comments, which we hope have led to substantial improvement of the manuscript. This research presented here was partly supported by the Chinese government's executive program for exploring the deep interior beneath the Chinese continent (SinoProbe-02), Natural Science Foundation of China under Grant No. 41274055, and Fundamental Research Funds for the Central Universities (WK2080000053). This research was also supported by DE-NA0001523 from the US Department of Energy.

References

AKI, K., A. CHRISTOFFERSSON, and E.S. HUSEBYE (1977). *Determination of the three-dimensional structure of the lithosphere*, J. Geophys. Res., 82, 277–296.

AKI, K., and P. G. RICHARDS (1980). Quantitative Seismology: Theory and Methods, W. H. Freeman, San Francisco, CA, USA.

BACHE, T. C., W. L. RODI, and D. G. HARKRIDER (1978). *Crustal structures inferred from Rayleigh-wave signatures of NTS explosions*, Bull. Seismol. Soc. Am., 68, 1399–1413.

BECKEN, M., O. RITTER, S. K. PARK, P. A. BEDROSIAN, U. WECKMANN, and M. WEBER (2008), *A deep crustal fluid channel into the San Andreas Fault system near Parkfield, California*, Geophys. J. Int. 173, 718–732.

BENNINGTON, N., C. THURBER, and S. ROECKER (2008), *Three-dimensional seismic attenuation structure around the SAFOD site, Parkfield, California*, Bull. Seismol. Soc. Am. 98, 2934–2947.

BENNINGTON, N., H. ZHANG, C. H. Thurber, and PAUL A. BEDROSIAN (in revision), *Joint inversion of seismic and magnetotelluric data in the Parkfield region of California using the normalized cross-gradient constraint*, Pure App. Geophys.

BEN-ZION, Y., and P. MALIN (1991), San Andreas fault zone head waves near Parkfield, California. Science 251, 1592–1594.

BLEIBINHAUS, F., J. A. HOLE, T. RYBERG, and G. S. FUIS (2007), *Structure of the California Coast Ranges and San Andreas Fault at SAFOD from seismic waveform inversion and reflection imaging*, J. Geophys. Res. 112, B06315, doi:10.1029/2006JB004611.

BUCHER, R. L., and R. B. SMITH (1971). Crustal structure of the eastern Basin and Range Province and the northern Colorado Plateau from phase velocities of Rayleigh waves, in *The Structure and Physical Properties of the Earth's Crust*, Geophys. Monogr. Ser., vol. 14, edited by J. G. Heacock, pp. 59–70, AGU, Washington, D.C.

CATCHINGS, R. D., M. J. RYMER, M. R. GOLDMAN, J. A. HOLE, R. HUGGINS and C. LIPPUS (2002), *High-resolution seismic velocities and shallow structure of the San Andreas Fault Zone at Middle Mountain, Parkfield, California*, Bull. Seismol. Soc. Am. 92, 2493–2503.

CHAVARRIA, J. A., P. MALIN, R. D. CATCHINGS, and E. SHALEV (2003), *A look inside the San Andreas fault at Parkfield through vertical seismic profiling*, Science 302, 1746–1748.

DU, W.-X., C. H. THURBER, and D. EBERHART-PHILLIPS (2004), *Earthquake relocation using cross-correlation time delay estimates verified with the bispectrum method*, Bull. Seismol. Soc. Am. 94, 856–866.

EBERHART-PHILLIPS, D., and A. J. MICHAEL (1993), *Three-dimensional velocity structure, seismicity, and fault structure in the Parkfield region, central CA*, J. Geophys. Res. 98, 15,737–15,758.

HOLE, J. A., R. D. CATCHINGS, K. C. St. CLAIR, M. J. RYMER, D. A. OKAYA, and B. J. CARNEY (2001), *Steep-dip imaging of the shallow San Andreas fault near Parkfield*, Science 294, 1513–1515.

HOLE, J., T. RYBERG, G. FUIS, F. BLEIBINHAUS, and A. SHARMA (2006), *Structure of the San Andreas fault zone at SAFOD from a seismic refraction survey*, Geophys. Res. Lett. 33, L07312.

HOLE, J. A., and ZELT, B. C. (1995), *3-D finite-difference reflection traveltimes*, Geophys. J. Int. 121, 427–434.

JULIÀ, J., C. J. AMMON, R. B. HERRMANN, and A. M. CORREIG (2000), *Joint inversion of receiver function and surface wave dispersion observations*, Geophys. J. Int. 143, 99–112.

KORNEEV, V. A., R. M. NADEAU, and T. V. MCEVILLY (2003), *Seismological studies at Parkfield IX: Fault-zone imaging using guided wave attenuation*, Bull. Seismol. Soc. Am. 93, 1415–1426.

LEES, J. M., and P. E. MALIN (1990), *Tomographic images of P wave velocity variation at Parkfield, California*, J. Geophys. Res. 95, 21,793–21,804.

LEWIS, M. A., and Y. BEN ZION (2010), *Diversity of fault zone damage and trapping structures in the Parkfield section of the San Andreas Fault from comprehensive analysis of near fault seismograms*, Geophys. J. Int. 183, 1579–1595.

LI, Y.-G., W. L. ELLSWORTH, C. H. THURBER, P. MALIN, and K. AKI (1997), *Fault zone guided waves from explosions in the San Andreas fault at Parkfield and Cienega Valley, California*, Bull. Seism. Soc. Am. *87*, 210–221.

LI, Y.-G., and P.E. MALIN (2008), *San Andreas Fault damage at SAFOD viewed with fault-guided waves*, Geophys. Res. Lett. *35*, L08304, doi:10.1029/2007GL032924.

LI, Y. G., P. C. LEARY, K. AKI, and P. E. MALIN (1990), *Seismic trapped modes in Oroville and San Andreas fault zones*, Science *249*, 763–766.

LI, Y.-G., J. E. VIDALE, and E. S. COCHRAN (2004), *Low-velocity damaged structure of the San Andreas Fault at Parkfield from fault zone trapped waves*, Geophys. Res. Lett. *31*, L12S06, doi:10.1029/2003GL019044.

MICHELINI, A., and T.V. MCEVILLY (1991), *Seismological studies at Parkfield: I, Simultaneous inversion for velocity structure and hypocenters using cubic B-splines parameterization*, Bull. Seism. Soc. Am. *81*, 524–552.

MACEIRA, M. and C. J. AMMON (2009). *Joint inversion of surface wave velocity and gravity observations and its application to Central Asian basins shear velocity structure*, J. Geophys. Res. *114*, B02314, doi:10.1029/2007JB005157.

PAIGE, C. C., and M. A. SAUNDERS (1982), *LSQR: An algorithm for sparse linear equations and sparse least squares*, ACM Trans. Math. Software *8*, 43–71.

PODVIN, P. and I. LECOMTE (1991), *Finite difference computation of traveltimes in very contrasted velocity models: a massively parallel approach and its associated tools.*, Geophys. J. Int. *105*, 271–284.

ROECKER, S., C. THURBER, and D. MCPHEE (2004), *Joint inversion of gravity and arrival time data from Parkfield: New constraints on structure and hypocenter locations near the SAFOD drill site*, Geophys. Res. Lett. *31*, L12S04.

ROECKER, S., C. THURBER, K. ROBERTS, and L. POWELL (2006), *Refining the image of the San Andreas Fault near Parkfield, California using a finite difference travel time computation technique*, Tectonophysics *426*, doi:10.1016/j.tecto.2006.02.026.

ROUX, P., K.G. SABRA, P. GERSTOFT and W.A. KUPERMAN (2005), *P-waves from cross-correlation of seismic ambient noise*, Geophys. Res. Lett. *32*, L19303.

ROUEFF A., P. ROUX and P. RÉFRÉGIER (2009), *Wave separation in ambient seismic noise using intrinsic coherence and polarization filtering*, Signal Process. *89*, 410–421.

ROUX, P. (2009), *Passive seismic imaging with directive ambient noise: Application to surface waves on the San Andreas Fault (SAF) in Parkfield*, Geophys. J. Int. *179*, 367–373.

ROUX, P., A. ROUEFF and M. WATHELET (2011), *The San Andreas Fault revisited through seismic noise and surface-wave tomography*, Geophys. Res. Lett., *38*, L13319.

SAITO, M. (1988), DISPER80: A subroutine package for the calculation of seismic normal mode solutions, in D. J. Doornbos (ed.), *Seismological Algorithms: Computational Methods and Computer Programs*, Academic Press, New York, pp. 293–319.

TAKEUCHI, H., and M. SAITO (1972), Seismic surface waves, in *Methods of Computational Physics*, edited by B. A. Bolt, pp. 217–295, Academic, New York.

TANIMOTO, T. (1991), *Waveform inversion for three-dimensional density and S-wave structure*, J. Geophys. Res., *96*, 8167–8189.

TANIMOTO, T., S. ISHIMARU, and C. ALVIZURI (2006), Seasonality of particle motion of microseisms, *Geophys. J. Int. 166*, 253–266, doi:10.1111/j.1365-246X.2006.02931.x.

Thurber, C., S. Roecker, K. Roberts, M. Gold, L. Powell, and K. Rittger (2003), Earthquake locations and three-dimensional fault zone structure along the creeping section of the San Andreas fault near Parkfield, CA: Preparing for SAFOD, *Geophys. Res. Lett. 30*.

THURBER, C., S. ROECKER, H. ZHANG, S. BAHER, and W. ELLSWORTH (2004), Fine-scale structure of the San Andreas fault zone and location of the SAFOD target earthquakes, *Geophys. Res. Lett. 31*, L12S02.

UNSWORTH M, and P. BEDROSIAN (2004) Electrical resistivity at the SAFOD site from magnetotelluric exploration, *Geophys. Res. Lett. 31*, L12S05, doi:10.1029/2003GL019405.

UNSWORTH M, P. BEDROSIAN, M. EISEL, G. EGBERT, and W. SIRIPUNVARAPORN (2000) Along strike variations in the electrical structure of the San Andreas Fault at Parkfield, California, *Geophys. Res. Lett. 27*, 3021–3024.

UNSWORTH, M. J., P. E. MALIN, G. D. EGBERT, and J. R. BOOKER (1997), Internal structure of the San Andreas fault zone at Parkfield, California, *Geology 25*, 359–362.

VIDALE, J. (1988), Finite-difference calculation of travel times, *Bull. Seis. Soc. Am. 78*, 2062–2076.

WALDHAUSER, F., and W. L. ELLSWORTH (2000). A double-difference earthquake location algorithm: method and application to the northern Hayward Fault, California, *Bull. Seism. Soc. Am. 90*, 1353–1368.

WU, J., J. A. HOLE, and J. A. SNOKE (2010), Fault zone structure at depth from differential dispersion of seismic guided waves: evidence for a deep waveguide on the San Andreas Fault, *Geophys. J. Int. 182*, 343–354.

ZHANG, H., and C. H. THURBER (2003), Double-difference tomography: The method and its application to the Hayward Fault, California, *Bull. Seism. Soc. Am. 93*, 1875–1889.

ZHANG, H., and C. THURBER (2005). Adaptive mesh seismic tomography based on tetrahedral and Voronoi diagrams: Application to Parkfield, California, *J. Geophys. Res. 110*, B04303.

ZHANG, H., and C. THURBER (2006). Development and applications of double-difference tomography, *Pure App. Geophys. 163*, 373–403, doi:10.1007/s00024-005-0021-y.

ZHANG, H., Y. LIU, C. THURBER, and S. ROECKER (2007), *Three-dimensional shear-wave splitting tomography in the Parkfield, California Region*, Geophys. Res. Lett. *34*, L24308. doi:10.1029/2007GL03195.

ZHANG, H., C. THURBER, and P. BEDROSIAN (2009), Joint inversion for V_p, V_s, and V_p/V_s at SAFOD, Parkfield, California, *Geochem. Geophys. Geosyst. 10*, Q11002, doi:10.1029/2009GC002709.

ZHANG, H., C. THURBER, D. SHELLY, S. IDE, G. BEROZA, and A. HASEGAWA (2004), High-resolution subducting slab structure beneath Northern Honshu, Japan, revealed by double-difference tomography, *Geology 32*, 361–364.

(Received September 14, 2013, revised February 17, 2014, accepted February 19, 2014, Published online March 29, 2014)

Pure Appl. Geophys. 171 (2014), 3023–3043
© 2013 Springer (outside the USA)
DOI 10.1007/s00024-013-0748-9

What Do Data Used to Develop Ground-Motion Prediction Equations Tell Us About Motions Near Faults?

DAVID M. BOORE[1]

Abstract—A large database of ground motions from shallow earthquakes occurring in active tectonic regions around the world, recently developed in the Pacific Earthquake Engineering Center's NGA-West2 project, has been used to investigate what such a database can say about the properties and processes of crustal fault zones. There are a relatively small number of near-rupture records, implying that few recordings in the database are within crustal fault zones, but the records that do exist emphasize the complexity of ground-motion amplitudes and polarization close to individual faults. On average over the whole data set, however, the scaling of ground motions with magnitude at a fixed distance, and the distance dependence of the ground motions, seem to be largely consistent with simple seismological models of source scaling, path propagation effects, and local site amplification. The data show that ground motions close to large faults, as measured by elastic response spectra, tend to saturate and become essentially constant for short periods. This saturation seems to be primarily a geometrical effect, due to the increasing size of the rupture surface with magnitude, and not due to a breakdown in self similarity.

Key words: GMPEs, ground-motion prediction equations, magnitude scaling, fault zone, engineering seismology.

1. Introduction

This paper is based on an invited talk with a similar title given at the 40th Workshop of the International School of Geophysics on Properties and Processes of Crustal Fault Zones in Erice, Sicily, Italy, May 18–24, 2013. That talk provided some information on what global compilations of recorded ground motions might say about the subject of the workshop. The global database used in this article is from a recent multi-year effort (the Pacific Earthquake Engineering Research Center's NGA-West2

project: BOZORGNIA *et al.* 2014) to derive ground-motion prediction equations (GMPEs) for shallow earthquakes in active tectonic regions. This article concentrates on the ground-motion data, rather than the equations developed from those data. After discussing the NGA-West2 database, I show examples of motions within and close to the fault damage zones (FDZs) for several specific faults. [FDZ is a commonly used term to describe the zone of intensely fractured rock surrounding the narrow core of a fault in which the primary fault slip occurs (e.g., CAINE *et al.* 1996); the width of a FDZ can vary, but it is often on the order of 100 to 200 m (e.g., BEN-ZION and SAMMIS, 2003)]. This is followed by a comparison of the magnitude scaling of ground motions from recordings in the NGA-West2 database with simulations from a standard seismological model. For a detailed discussion of the information that can be obtained from a rich region-specific database, see the paper by KURZON *et al.* (2014) in this volume.

2. The PEER NGA-West2 Database

The Pacific Earthquake Engineering Center (PEER) NGA-West2 database, developed by ANCHETA *et al.* (2013, 2014), contains 21,336 three-component recordings from 599 shallow crustal earthquakes in active tectonic regions around the world. Great care was taken in developing the database: the recordings were processed in a uniform and consistent manner to provide high-quality seismic intensity measures and metadata, such as source and site properties. The metadata were evaluated by several teams of researchers to ensure consistency in view of the different regions and methods used to obtain the metadata by various researchers.

[1] US Geological Survey, Menlo Park, CA, USA. E-mail: boore@usgs.gov

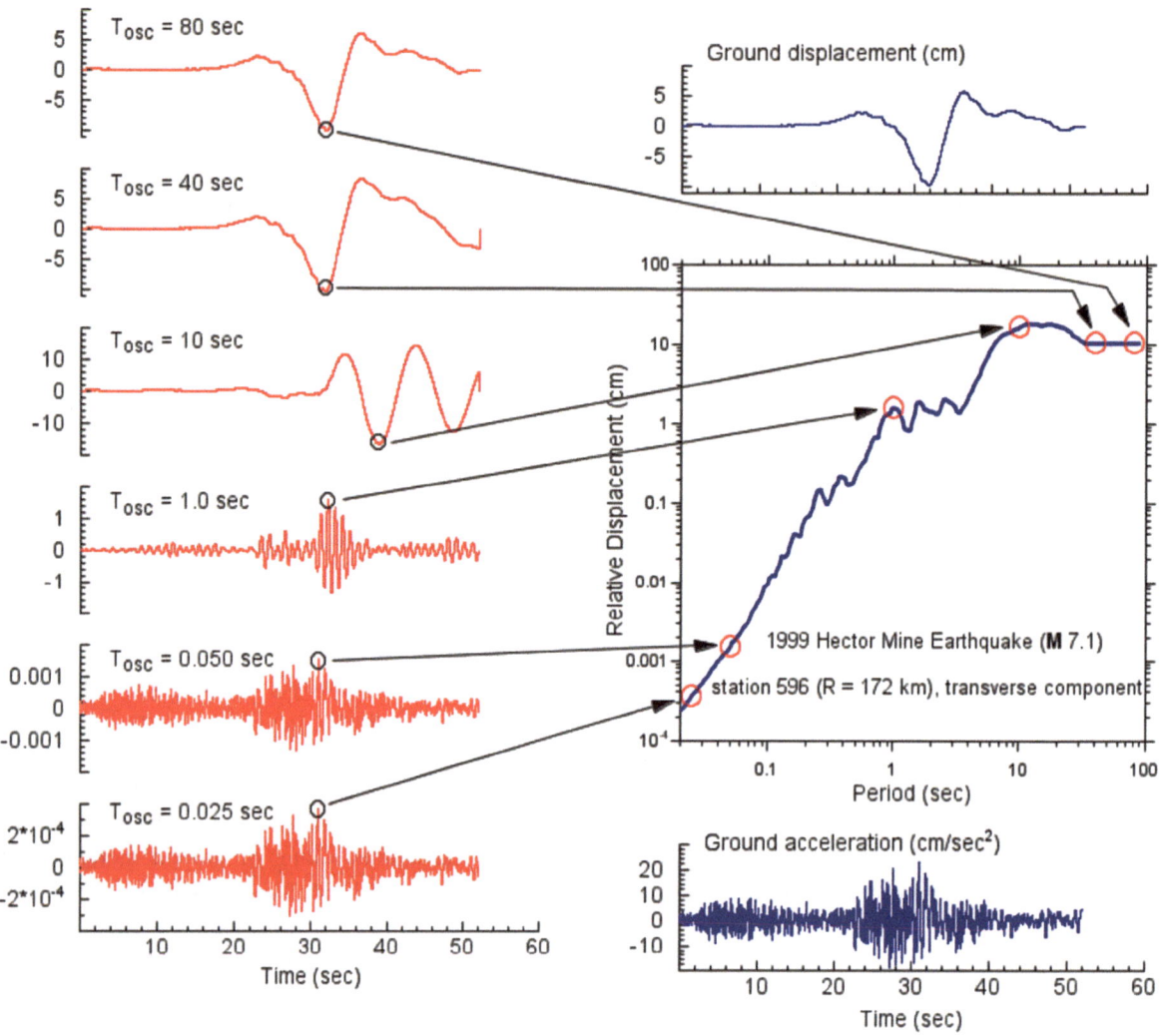

Figure 1

Steps in constructing a response spectrum for an actual recording. The trace in the *lower right-hand* corner is the ground acceleration, and the traces on the *left-hand side* are the displacement response time-series for simple oscillators with different natural periods of vibration (T_{OSC}). The maximum displacement of each oscillator is plotted against its natural period to construct the relative displacement response spectrum. The lowest two time series on the *left* of the figure show that the waveforms of short-period oscillators are very similar to the input acceleration time series, whereas for long-period oscillators, the oscillator response equals that of the ground displacement (shown in the *upper right-hand corner*). **M** is moment magnitude and R is the distance to the fault rupture. (Modified from Fig. 7 in BOMMER and BOORE 2005)

The ground-motion intensity measure used for the NGA-West2 database is 5 %-damped pseudo-absolute response spectral acceleration (PSA). Because many readers of this journal may not be familiar with PSA, I show the basis for its development in Fig. 1. In this case, the example illustrates the construction of the relative displacement response spectrum *SD*, but PSA $= (2\pi/T_{OSC})^2$ SD, where T_{OSC} is the period of the oscillator. In essence, a response spectrum is the peak response of a series of damped, single-degree-of-freedom harmonic oscillators, with periods ranging from very short to very long values (typically 0.01 to 20 s), to a single input acceleration time series. PSA for a very short-period oscillator equals the peak acceleration of the input time series, whereas SD for a very long-period oscillator equals the peak

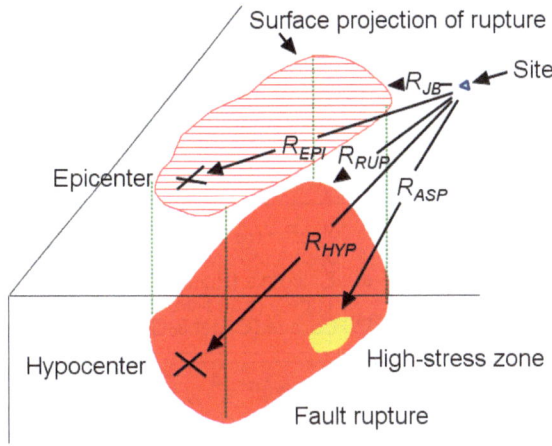

Figure 2

Some distance measures. The most commonly used measures in modern GMPEs are R_{RUP}, the closest distance to the rupture surface, and R_{JB}, the closest horizontal distance to the surface projection of the rupture surface ("JB" for Joyner and Boore, who introduced this measure in JOYNER and BOORE 1981). $R_{JB} = 0.0$ for sites over the fault

displacement obtained from double integration of the input acceleration time series (these asymptotic properties are shown in Fig. 1). Response spectra are useful descriptors of ground-motion intensity because buildings are often well-represented as single-degree-of-freedom oscillators (with fundamental mode resonant periods given approximately by $T_{OSC} = 0.1N$, where N is the number of stories), and thus a single response spectrum plot can be used to estimate the peak forces in buildings with a wide range of resonant periods subjected to the shaking for a particular recording.

The most common metadata used in developing GMPEs are measures of distance, magnitude, and site geology. In the NGA-West2 database, the magnitude measure is moment magnitude **M** (HANKS and KANAMORI 1979). The two main distance measures used in the NGA-West2 project are R_{RUP} and R_{JB}, defined in Fig. 2 (along with a number of other possible measures of distance from a site to a fault rupture surface). The site geology is characterized in the NGA-West2 project by the time-weighted average of the shear-wave velocity from the surface to 30 m (V_{S30}). While it has been argued that such a

velocity may not be representative of the shear-wave velocities at deeper depths, which can affect longer period motions, BOORE *et al.* (2011) show that there is a good correlation of V_{S30} and the shear-wave velocity averaged to depths significantly greater than 30 m (Fig. 3).

The NGA-West2 database contains *PSA* for periods from 0.01 to 20 s. The magnitude-distance distribution of the PSA are shown in Fig. 4 for T_{OSC} of 1.0 and 10.0 s, with the data differentiated by earthquake source mechanism. It is clear from Fig. 4 that there are many fewer data for the long-period oscillator (and in fact, the fall-off in available data begins at a period of about 1.0 s, as shown in BOORE *et al.* 2014); this is an inevitable consequence of the signal-to-noise characteristics of ground motions recorded on accelerographs (which provide the bulk of the data for the larger earthquakes).

The uniformly processed data and carefully evaluated metadata in the NGA-West2 database can be used in several ways that are relevant to the subject of processes and properties of crustal fault zones. Near-fault data are commonly used in inversions for the rupture process of individual earthquakes, and specific recordings can be used to look for amplitude variations and polarization complexities that might be indicative of fault zone properties, some examples of which are given in the next section. The GMPEs developed from the database are a convenient summary of the overall magnitude and distance behavior of a very large number of ground-motion recordings, and as such, they are useful in assessing the magnitude scaling of ground motion, which is intimately related to the source processes of earthquakes. This is discussed in a later section.

3. Ground Motions near Faults: Some Examples of Complexity

Although there is not a metadata field in the database for stations within the FDZ, some idea of the number of such sites is given by the distance metric R_{JB}. Table 1 gives the number of three-component

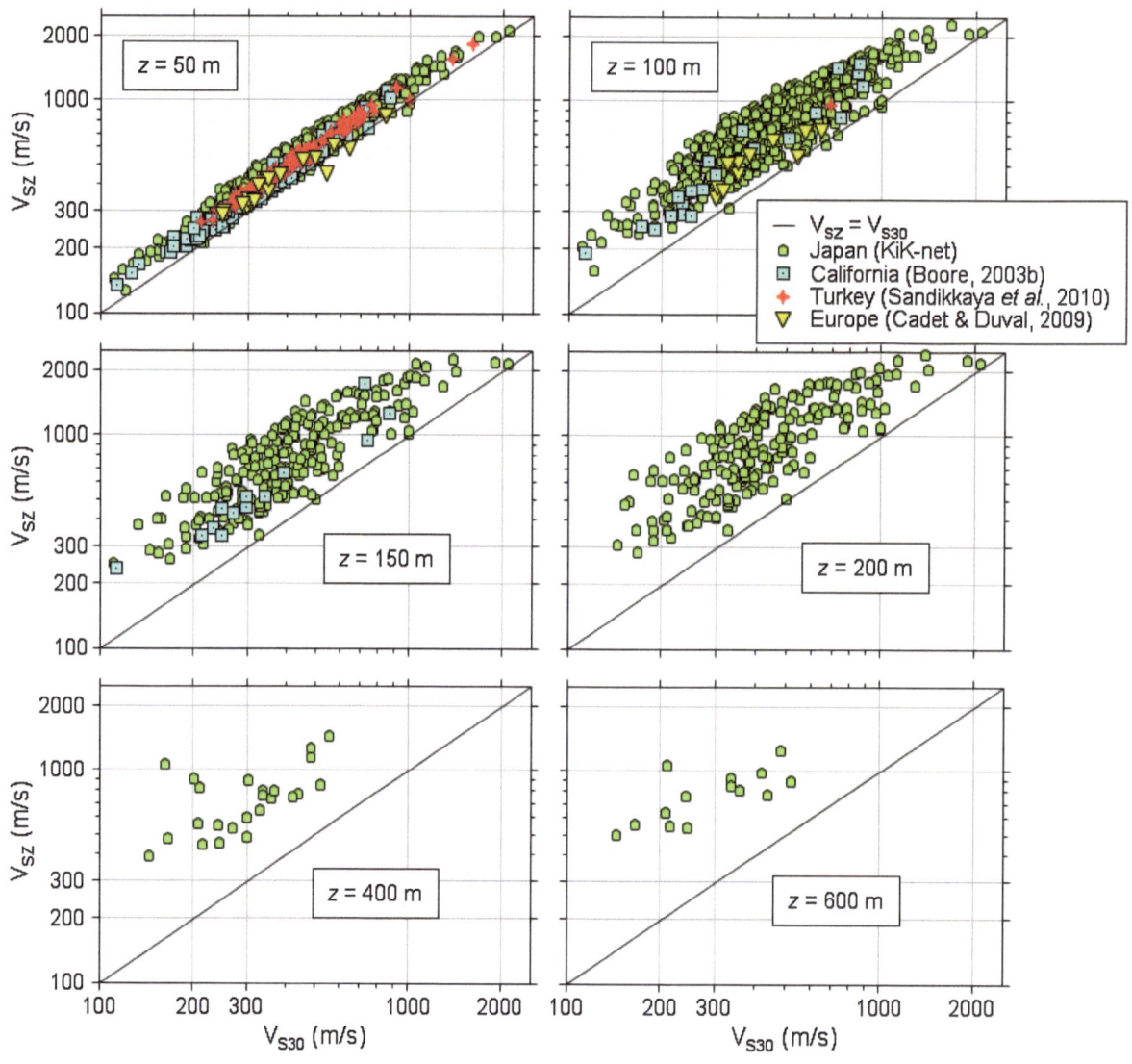

Figure 3
Scatterplot of V_{S30} and V_{SZ} from shear-wave velocity profiles for six averaging depths z (only the profiles for KiK-net stations had profiles to the three greatest values of z). (Modified from Fig. 10 in BOORE 2011, which contains formal correlation coefficients for each graph; these range from 0.98 for $z = 50$ m to 0.79 for $z = 600$ m)

records in the NGA-West2 database for various ranges of R_{JB}, for a number of large and well-recorded earthquakes. While a station with $R_{JB} = 0.0$ km does not imply that the station is within the FDZ (it could be above the zone, for a dipping fault), and R_{JB} can be large even if the station is in the FDZ (if the station is far beyond the horizontal extent of faulting for a specific earthquake), the table does give some idea of the relative scarcity of data within FDZs. Based on the entries in the table, there appear to be relatively

few data in the NGA-West2 database from sites within the fault damage zone (FDZ) of earthquakes.

3.1. Variability of Ground Motions in and near FDZs: The 2004 Parkfield Earthquake

One earthquake with a relatively large number of recordings in or near the FDZ is the **M** 6.0, 2004 Parkfield, strikeslip event that occurred along the San

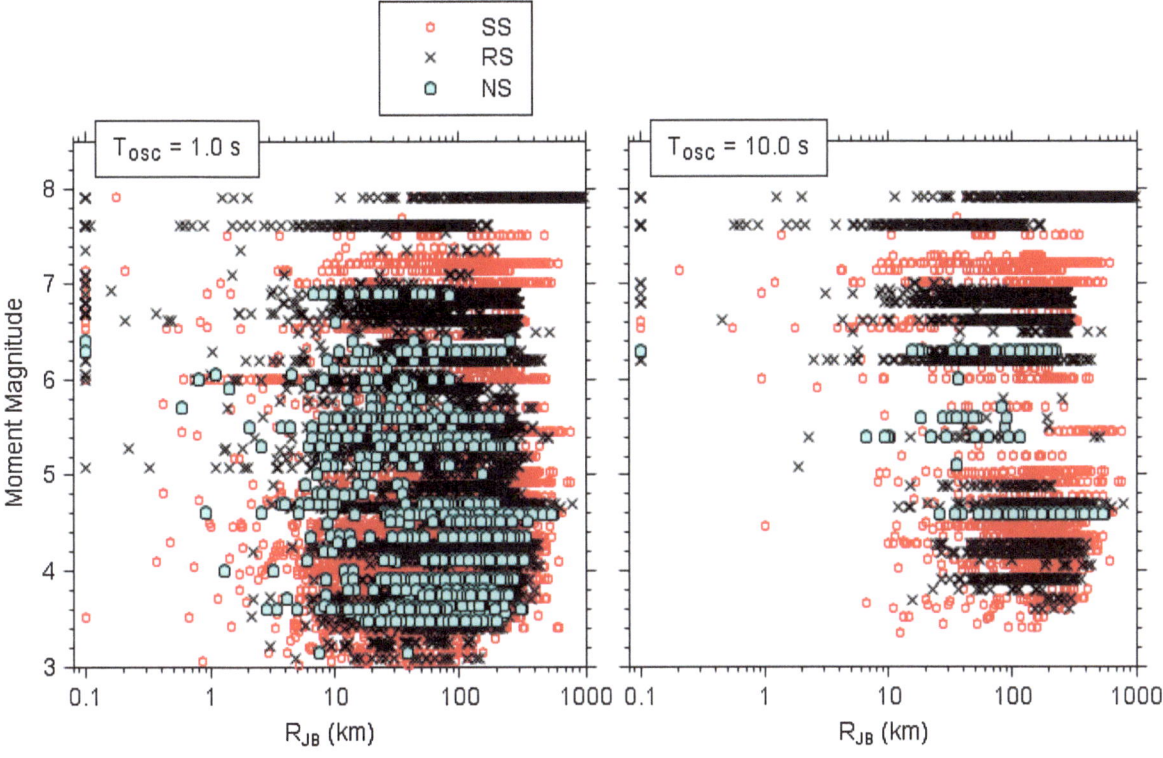

Figure 4

Magnitude-distance distribution of data from the PEER NGA-West2 database, differentiated by fault type (*SS* StrikeSlip, *NS* NormalSlip, *RS* ReverseSlip). The distributions are shown for two oscillator periods, 1.0 and 10.0 s

Table 1

Numbers of records in PEER NGA-West2 database for three near-source distance ranges for selected events

Event	Type	M	$R_{JB} < 2$ km	$R_{JB} < 5$ km	$R_{JB} < 10$ km
Kocaeli	SS	7.6	2	3	4
Chi–Chi	RS	7.5	18	23	42
Duzce	SS	7.1	2	7	9
Denali	SS	7.9	1	1	1
Parkfield	SS	6.0	19	41	63
Wenchuan	RS	7.9	5	6	6

Note that being close to a fault is not necessarily the same as being in the fault zone, particularly for non-vertical faults. It is also true that a station can be in the fault damage zone (FDZ), and yet R_{JB} could be large. The database does not have a field indicating whether or not a station was in the FDZ

Andreas fault. To date, this earthquake has more recordings at near-fault distances than any other earthquake in the world, and many studies have investigated the variations in seismic velocity in and near the FDZ (e.g., EBERHART-PHILLIPS and MICHAEL

1993; THURBER *et al.* 2006). Figure 5 shows many of the stations that recorded the event, along with horizontal ground-motion displacement particle motions (not shown are the GEOS stations of BORCHERDT *et al.* 2006). The spatial similarity of the particle motions of horizontal ground displacement is obvious, with motions close to the surface traces of the fault being polarized in the direction normal to the fault, while those stations farther away are polarized in the fault-parallel direction (BORCHERDT *et al.* 2006; SHAKAL *et al.* 2006). These characteristics are expected from considerations of the radiation pattern of strikeslip earthquakes, as well as the propagation of the rupture to the northwest from the epicenter.

Although there is an overall uniformity to the polarizations, there are substantial variations spatially in the amplitudes of the motions, as shown by detailed maps of PSA for $T_{OSC} = 0.1$ s and $T_{OSC} = 2.0$ s (Figs. 6, 7; Fig. 7 includes data from the GEOS stations). The maps in Fig. 6 are for recordings at the UPSAR array (FLETCHER *et al.* 1992, 2006), a dense

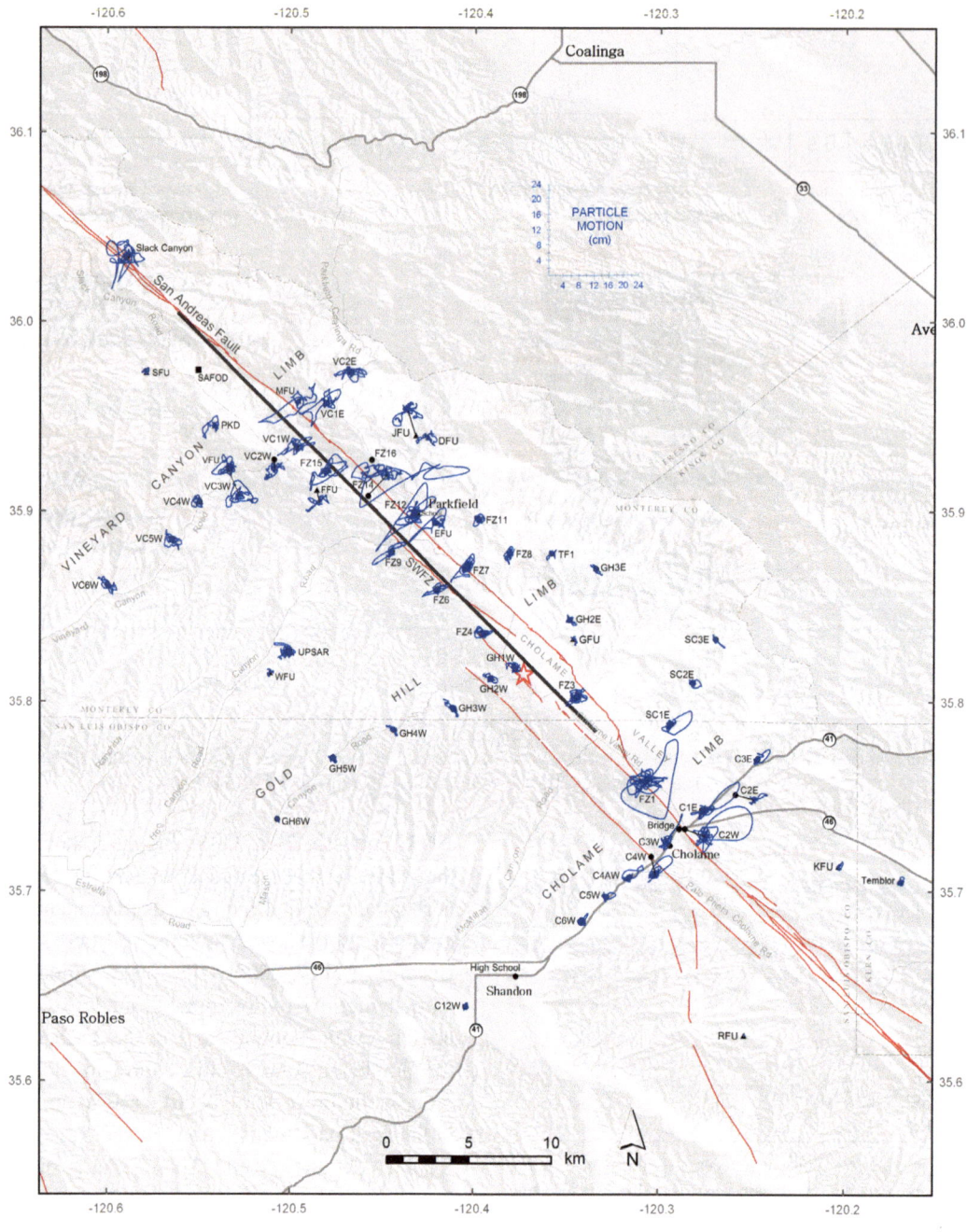

Figure 5

Particle motions from horizontal components of ground displacement time series, derived from double integration of the acceleration time series recordings of the 2004 Parkfield **M** 6 earthquake. The *red lines* are traces of the San Andreas fault, and the *black line* is a simplified fault model used by SHAKAL *et al.* (2006) to determine distances to the recording stations. (Fig. 14 in SHAKAL *et al.* 2006)

array of instruments on ridge tops, underlain by similar geologic conditions (the location of the array is shown at the left center of Fig. 5 and by the green rectangles in Fig. 7). Because the wavelengths of the waves controlling the spectral response amplitudes generally increase with period, I expect there to be less variability for long-period PSA (Fig. 6a) than for short-period PSA (Fig. 6b). Scrutiny of Fig. 6 shows

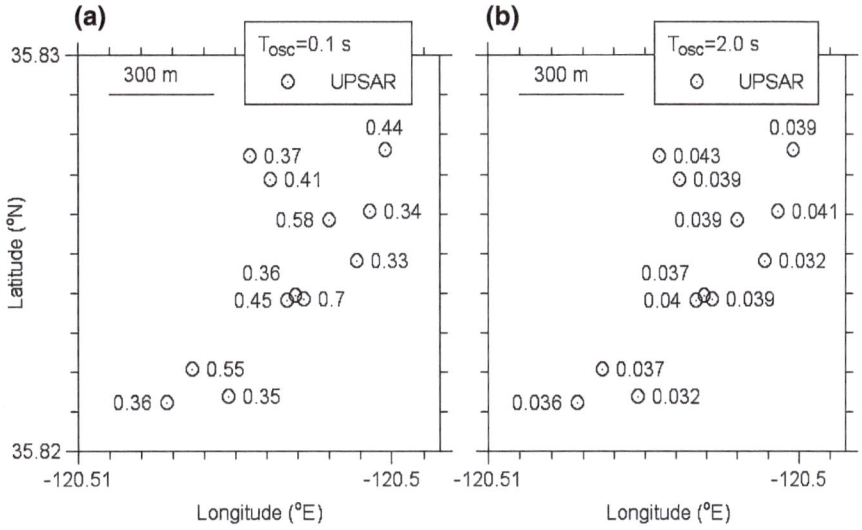

Figure 6

Values of PSA, in g, at UPSAR stations from recordings of the 2004 Parkfield mainshock, for $T_{OSC} = 0.1$ s *(left graph)* and $T_{OSC} = 2.0$ s *(right graph)*

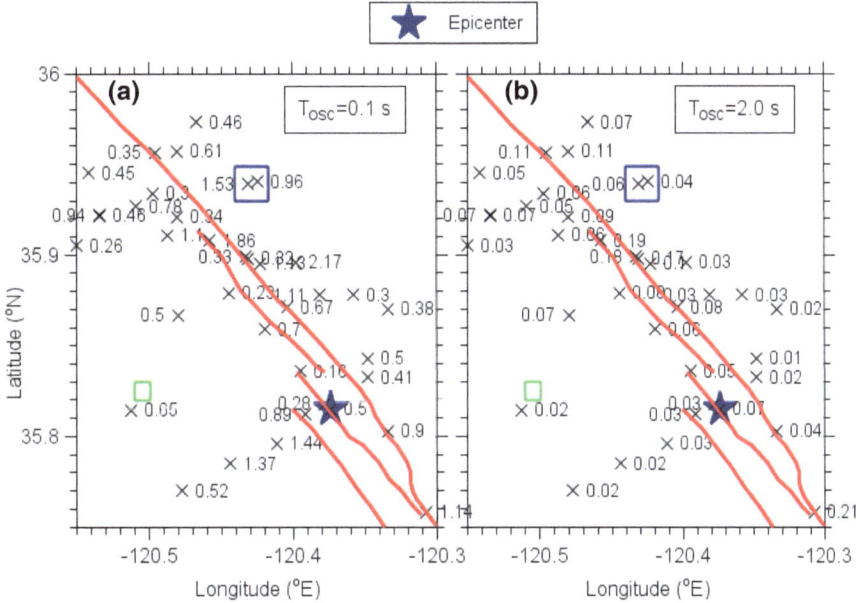

Figure 7

Values of *PSA*, in g, at stations that recorded the 2004 Parkfield mainshock, for $T_{OSC} = 0.1$ s *(left graph)* and $T_{OSC} = 2.0$ s *(right graph)*. Surface fault strands of the San Andreas fault *(red)* from SHAKAL *et al.* (2006). The time series from the stations within the *blue rectangles* are shown in a later figure; the UPSAR array is within the *green rectangles*; the ground-motion values are not shown because they overlap too much to be resolvable; see the previous figure

this generally to be the case. The spatial variability of the motions at the UPSAR array might correspond to the minimum expected for any closely spaced sites, given that the sites are located on geologically similar materials and are removed from any complications due to the fault zone. WANG *et al.* (2006) have studied the variability of the Parkfield mainshock motions recorded at UPSAR in detail.

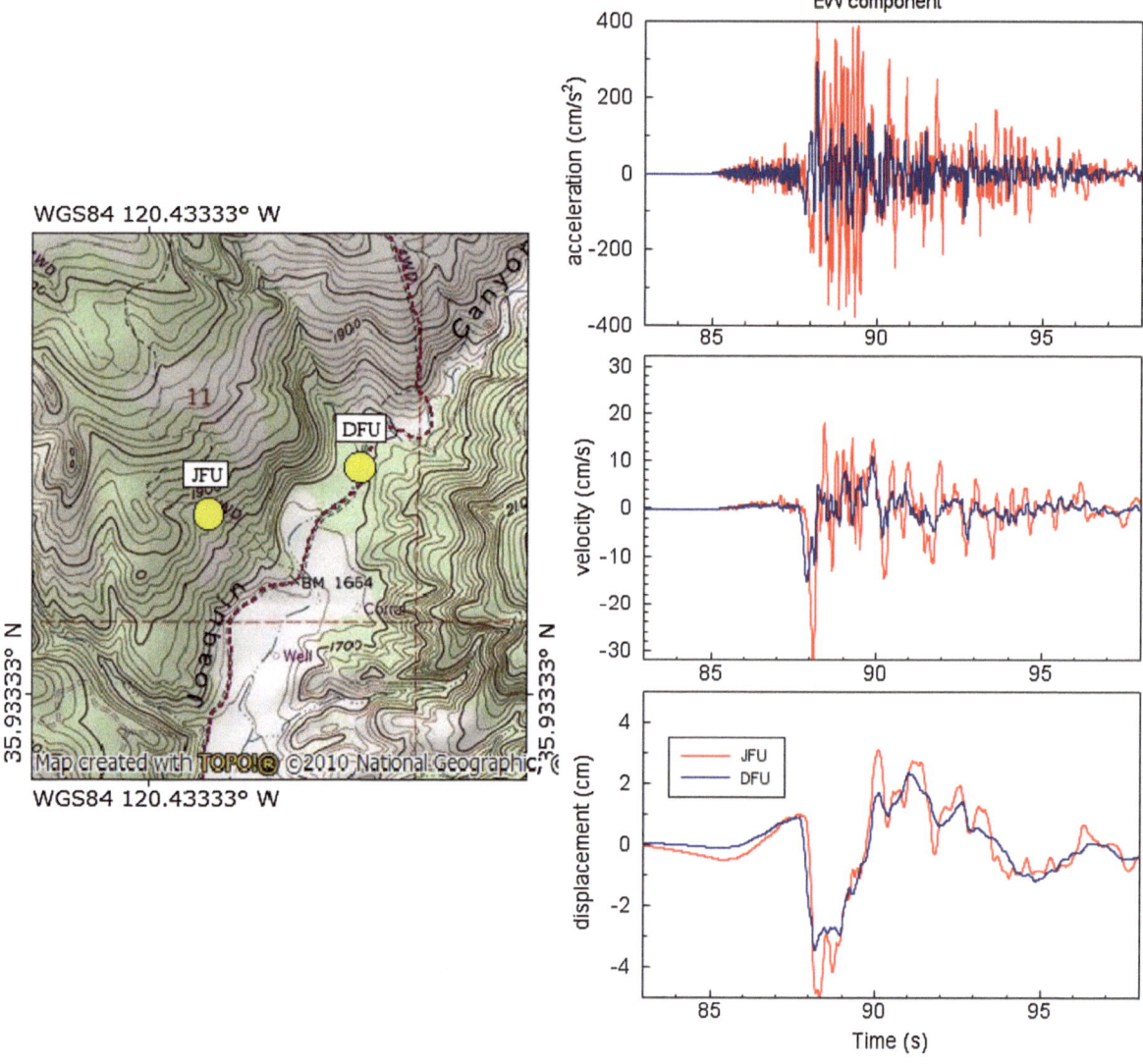

Figure 8
Two nearby acceleration recordings of the 2004 Parkfield mainshock and the velocity and displacement time series derived from the recordings, showing frequency-dependent spatial variability. The distance between DFU and JFU is 590 m. [DFU and JFU are station codes from the GEOS web page (see Sect. 7); they correspond to GEOS stations Donna Lee and Joaquin Canyon, respectively)

The spatial variability of stations within and near the FDZ is shown in Fig. 7. While there can be significant variation in the longer period motions, in general, there is more variation for short-period PSA than for long-period PSA. Unraveling the sources of this variability can be difficult, as there is significant complexity in the geologic properties along the fault, with velocity heterogeneities of varying amounts across the fault (e.g., THURBER 2006) and along-strike variations in the properties of the FDZ (LEWIS and BEN-ZION 2010). In addition, the speed of rupture propagation and amount of fault slip were heterogeneous along the fault (e.g., FLETCHER 2006; LIU *et al.* 2006; ZHAO *et al.* 2010). I make no attempt to unravel these effects, and only point out the amount of variability that exists, leaving it to others to explain the causes of these effects (e.g., see CULTRERA *et al.* 2003; PISCHIUTTA *et al.* 2013; KURZON *et al.* 2014, this volume, for a discussion of variability near other faults).

One example of site-to-site variability that might have different causes is shown in Fig. 8. This shows the recorded accelerations in one horizontal direction and the velocity and displacement time series obtained from the recorded accelerations, for two stations only 590 m apart. As the map shows, station JFU is on a hillside, while station DFU is on the edge of a narrow valley. Thus, there are differences in topography and presumably, underlying geology (DFU probably being underlain by a thin layer of alluvium), both of which can cause differences in ground motion. The stations are outside of the FDZ

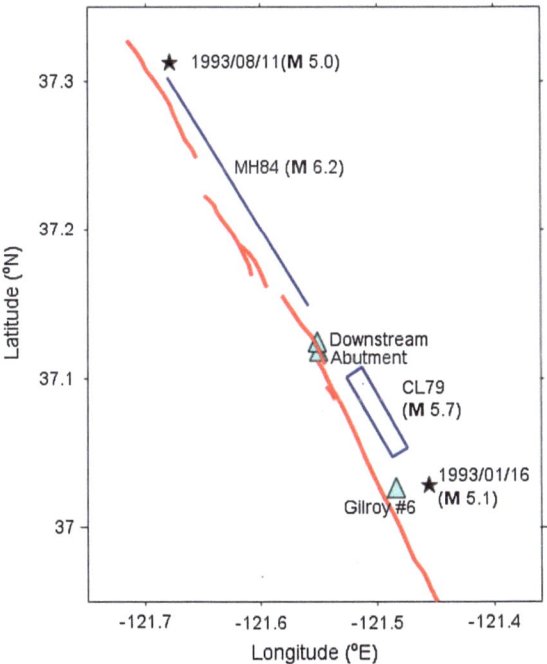

Figure 9
Source and recording station locations along the Calaveras fault in central California. The blue line is the approximate surface trace of the 1984 Morgan Hill earthquake (MH84), and the blue rectangle is the approximate surface projection of the dipping 1979 Coyote Lake earthquake (CL79). The *stars* are epicenters of two smaller earthquakes. Shown are the locations of the Coyote Lake Dam abutment and downstream stations, as well as the Gilroy array station #6. The surface fault strands (*red*) are from figures in BEROZA and SPUDICH (1988), LIU and HELMBERGER (1983)

(they correspond to the stations enclosed in blue rectangles in Fig. 7). The displacements are more similar than the accelerations, both in terms of the amplitudes and the waveforms. This is expected, as the displacements are controlled by motions with lower frequencies, and thus longer wavelengths, than either the velocities or accelerations, and therefore are less sensitive to spatial variations in topography or underlying geology than are the higher-frequency ground velocities or accelerations (e.g., HANKS 1975).

3.2. Complexity in the Polarization of Motions within FDZs: Recordings near the Calaveras Fault

While the previous section illustrated the spatial variability of ground-motion amplitudes at various frequencies, this section compares the polarization of ground displacements for stations within and close to the FDZ of the Calaveras fault in central California. The locations and magnitudes of the earthquakes and station locations, as well as mapped surface strands of the fault, are shown in Fig. 9. In this section, I focus attention on two stations clearly within the FDZ: the abutment and downstream stations at Coyote Lake Dam (see BOORE *et al.* 2004). The downstream station was installed after the 1984 Morgan Hill earthquake. A third station (Gilroy #6) is located over the FDZ of the Calaveras fault, but as suggested in Fig. 10, due to the dip of the fault the station, it is likely not within the FDZ. Furthermore, note the significant difference in V_{S30} (shown at the top of Fig. 10) for the downstream station and Gilroy #6; the latter has a higher V_{S30}, consistent with the location of the station on a ridge and less consistent with a location within the FDZ.

The first example of polarization complexity is shown in Fig. 11. Here, the portion of the ground displacement near the peak motion at both the abutment and downstream stations have similar amplitudes and polarization for the 1993 **M** 5.1 earthquake. The motion is polarized approximately southwest-northeast (and normal to the fault strike), as expected from the radiation pattern of a strikeslip fault located to the southeast of the station. In distinct contrast is the almost fault-parallel polarization of the motion from the 1979 **M** 5.7 earthquake, which had an extended rupture surface that was closer to the

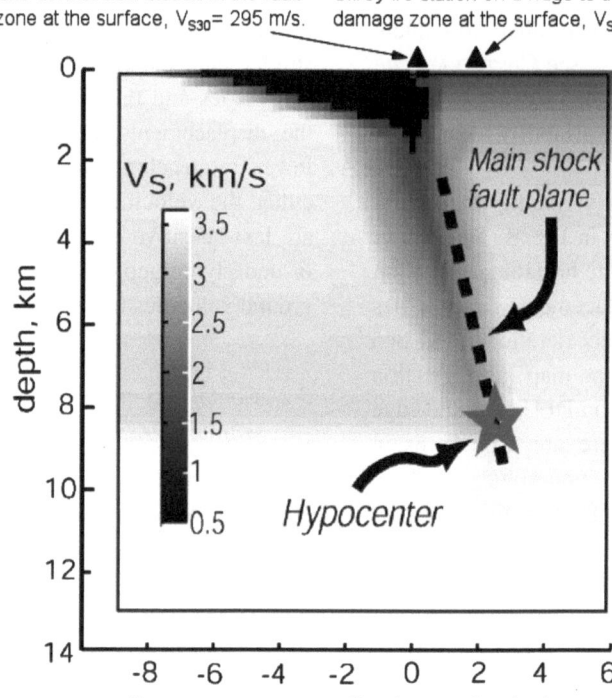

Coyote Lake downstream station in the fault damage zone at the surface, V_{S30}= 295 m/s.

Gilroy #6 station on a ridge to the east of the fault damage zone at the surface, V_{S30}=663 m/s.

Figure 10
An approximate cross section through the Calaveras fault in the vicinity of Gilroy station #6. The locations of the two stations are approximate; V_{S30} values from the PEER NGA-West2 database (see Sect. 7). (Modified from SPUDICH and OLSEN 2001)

station than was the 1993 earthquake. Note that the downstream station was not installed until 1984, so there are no downstream records for the 1979 event.

The second example is for earthquakes to the northwest of the Coyote Lake Dam station. Again I show motions from a moderate earthquake (1984 **M** 6.2) and a smaller event (1993 **M** 5.0). The smaller event is again farther from the station than the larger event (Fig. 12). As in the previous example, the polarization of the small event is close to being fault normal, unlike that of the larger event, although the motions are not as linearly polarized as they were in the previous example.

The final example compares the particle motions for the same earthquake (1984 Morgan Hill) recorded at the stations CLD abutment and Gilroy #6 (Fig. 13). The polarization at Gilroy #6 for the strong portion of the ground displacement is inconsistent with that at the station within the FDZ, even though the azimuths from the stations to the source are about the same.

I have no quantitative explanations for the observed differences of the polarization of the strong portion of the ground displacement, but offer them as clear examples of the complexity of polarization properties of motions within FDZs. The azimuths from the events to the stations are about the same for each of the three examples shown here, but the distances from the stations to the sources of the strongest shaking probably differ. Thus, details of the interaction of the waves and the geometry of the FDZ apparently can have a large effect on the polarization of the ground motion.

4. Distance and Magnitude Dependence of Ground Motions from a Global Database

The previous sections discussed detailed properties of ground motions in and near FDZ for a few specific earthquakes, but those sections do not deal with the

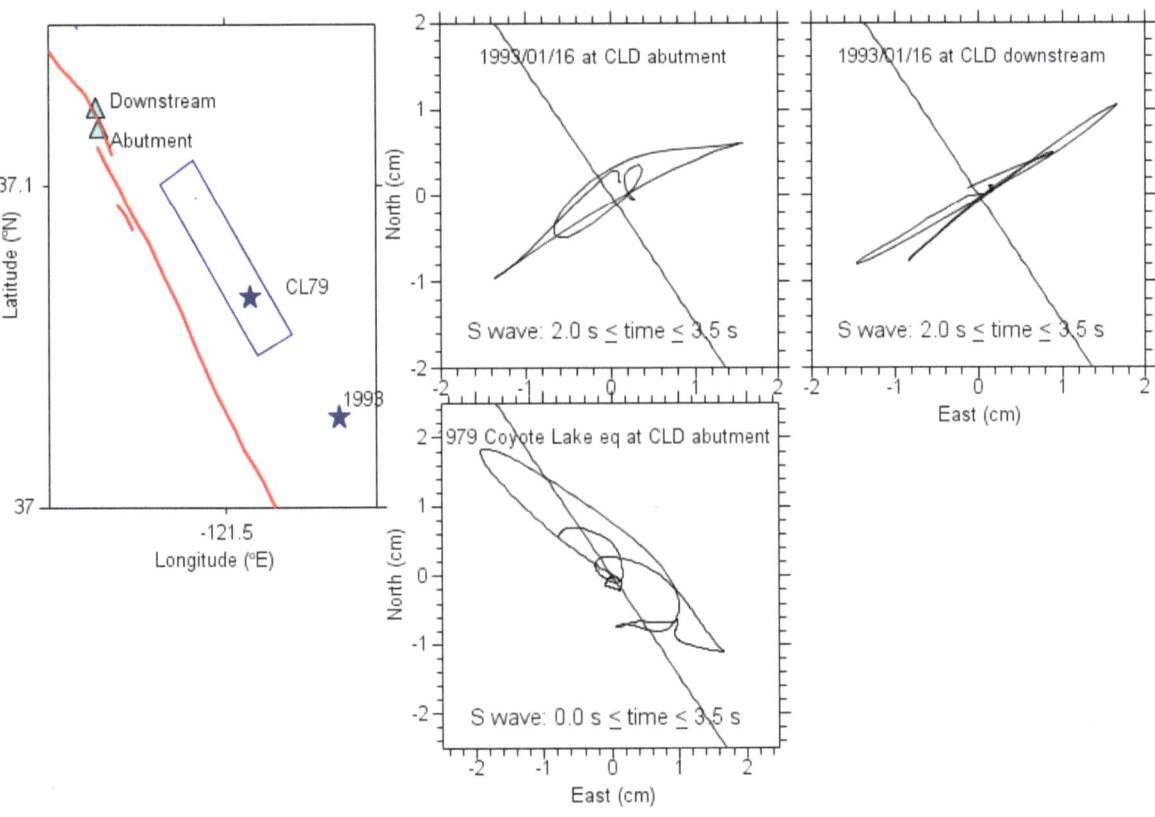

Figure 11
Location map and hodograms of the strongest portion of the horizontal ground displacements recorded from the 16 January 1993 and 1979 earthquakes at Coyote Lake Dam (CLD) abutment (*middle column of graphs*) and the 1993 earthquake at CLD downstream (*right graph*). The *stars* are epicenters

larger question of how ground motion scales on average with magnitude. This scaling is fundamentally controlled by source processes, and thus the observed scaling contains information that is relevant to the subject of this volume. Insight into the average scaling is best obtained from the large database assembled in the NGA-West2 project. To provide an overview of the magnitude and distance dependence of the global data, Fig. 14 shows PSA values for four periods plotted against distance, with magnitude bins indicated by symbols of different color. The data are from strikeslip earthquakes, adjusted to a common V_{S30} value of 760 m/s using the site response equations of SEYHAN and STEWART (2014). This figure shows a number of robust features related to magnitude and distance scaling of ground motions for a wide range of magnitudes and distances, without assuming any functional

forms for this dependence (aside from the V_{S30} adjustment). The main features shown by the data are:

- There is significant scatter in the data. The scatter is larger for small earthquakes and generally increases with distance (at least to distances of about 200 km). In spite of the scatter, however, there are systematic distance and magnitude trends in the data, as discussed in the next items.
- For a single magnitude and for all periods, the motions tend to saturate for large earthquakes; that is, they approach a constant value, as the distance from the fault rupture to the observation point decreases. This can only be concluded definitively for large magnitudes for which the rupture approaches the ground surface, and therefore the distance measure used in Fig. 14 can approach 0.0.

171

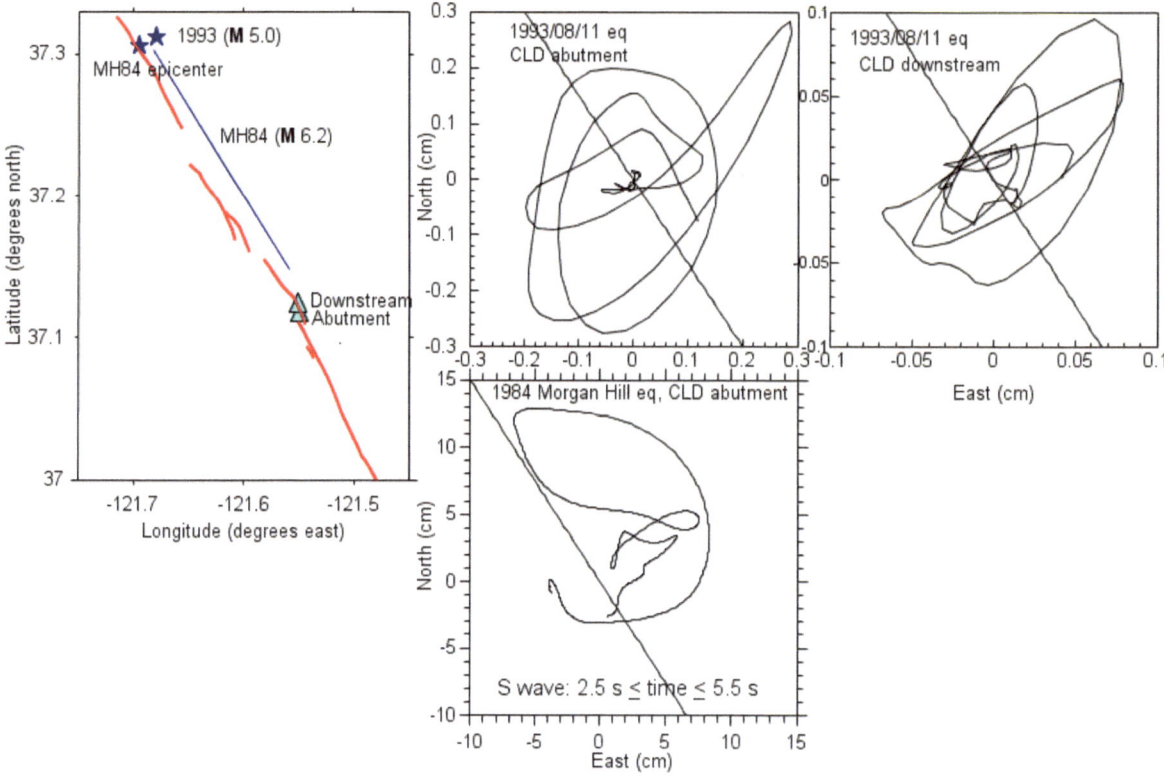

Figure 12

Location map and hodograms of the strongest portion of the horizontal ground displacements recorded from the 11 August 1993 and 1984 earthquakes at Coyote Lake Dam (CLD) abutment (*middle column of graphs*) and the 1993 earthquake at CLD downstream (*right graph*). The *stars* are epicenters

Smaller earthquakes do not reach the surface, and therefore surface observations cannot be used to infer whether or not the motions near the rupture surfaces of small earthquakes saturate.

- At any fixed distance the ground motion increases with magnitude in a nonlinear fashion, with a tendency to saturate for large magnitudes, particularly for shorter period motions. The overall magnitude scaling increases with increasing period, but it is smaller at short distances than at longer distances. For short periods and close distances, there appears to be almost complete saturation for the motions from large earthquakes.

- For a given period and magnitude the median ground motions decay with distance; this decay shows curvature at greater distances on the log–log plot used in Fig. 14. This decay can be parameterized as $\exp(-\alpha R_{\mathrm{RUP}})/R_{\mathrm{RUP}}^{\beta}$, where the terms in

the numerator and denominator are similar to the decay from a point source due to anelastic attenuation and geometrical spreading, respectively. In log–log plots, the anelastic attenuation produces the curvature at greater distances, and the geometrical spreading produces the linear decay at closer distances. Careful inspection of Fig. 14 shows that the apparent geometrical spreading decreases as magnitude increases.

5. Magnitude Scaling of Ground Motions from a Global Database and from Simulations

Although the distance dependence of the ground motion and its inherent variability are interesting

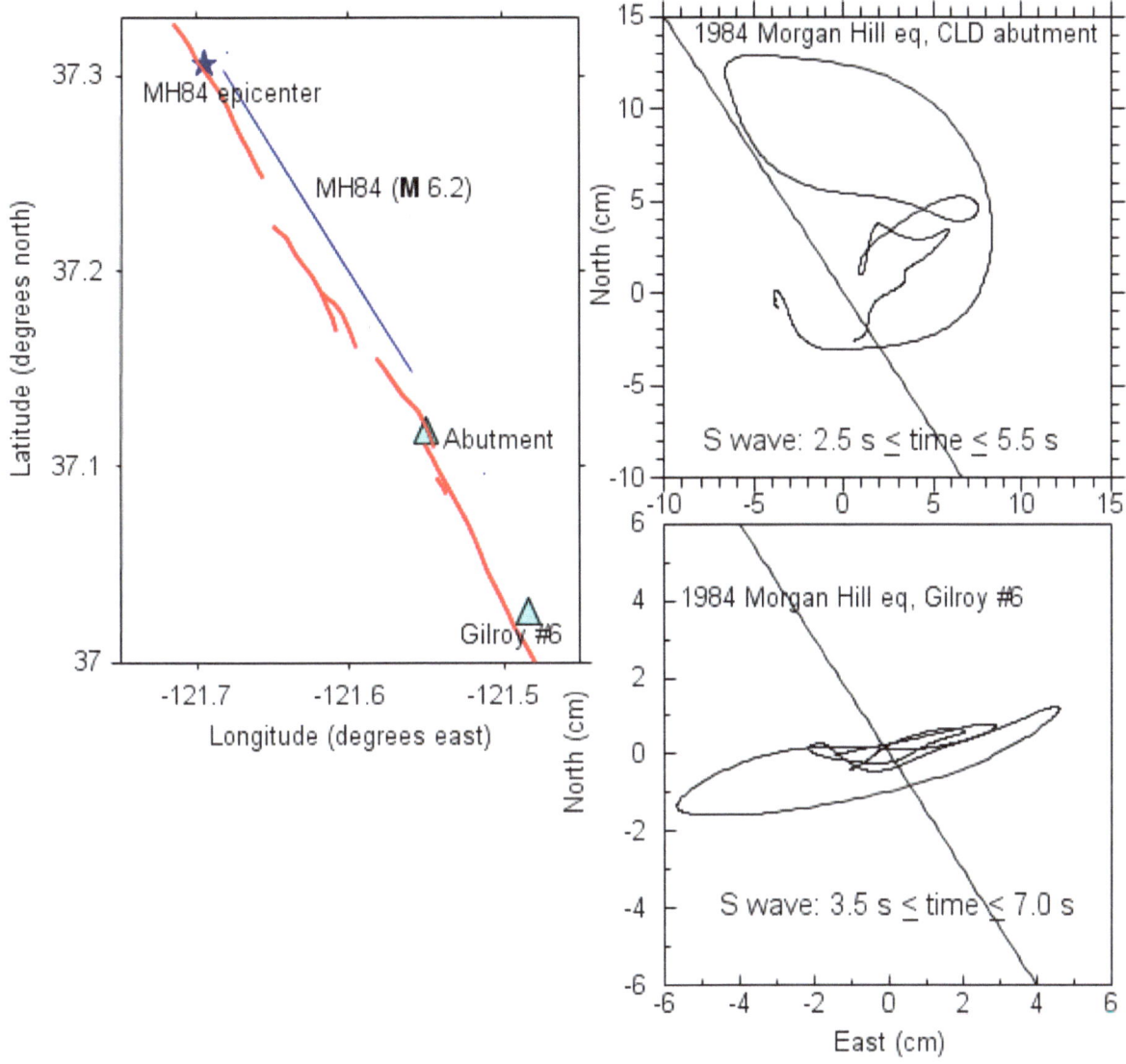

Figure 13
Location map and hodograms of the strongest portion of the horizontal ground displacements recorded from the 1984 earthquake at Coyote Lake Dam (CLD) abutment and Gilroy #6. The *stars* are epicenters

from a seismological perspective and are of critical importance to earthquake engineers, the scaling of the motions with magnitude, particularly at close distances, is more relevant for understanding the properties and processes of crustal fault zones. For that reason, magnitude scaling is discussed in this section. I show the scaling from the data at distances within 4 km of the surface projection of the rupture, as well as at moderate distances (50 to 100 km). I also compare these data to simulations using a standard seismological model based on a point-source representation of the earthquake rupture. While a judicious choice of distance goes a long ways toward accounting for finite-fault effects in the point-source

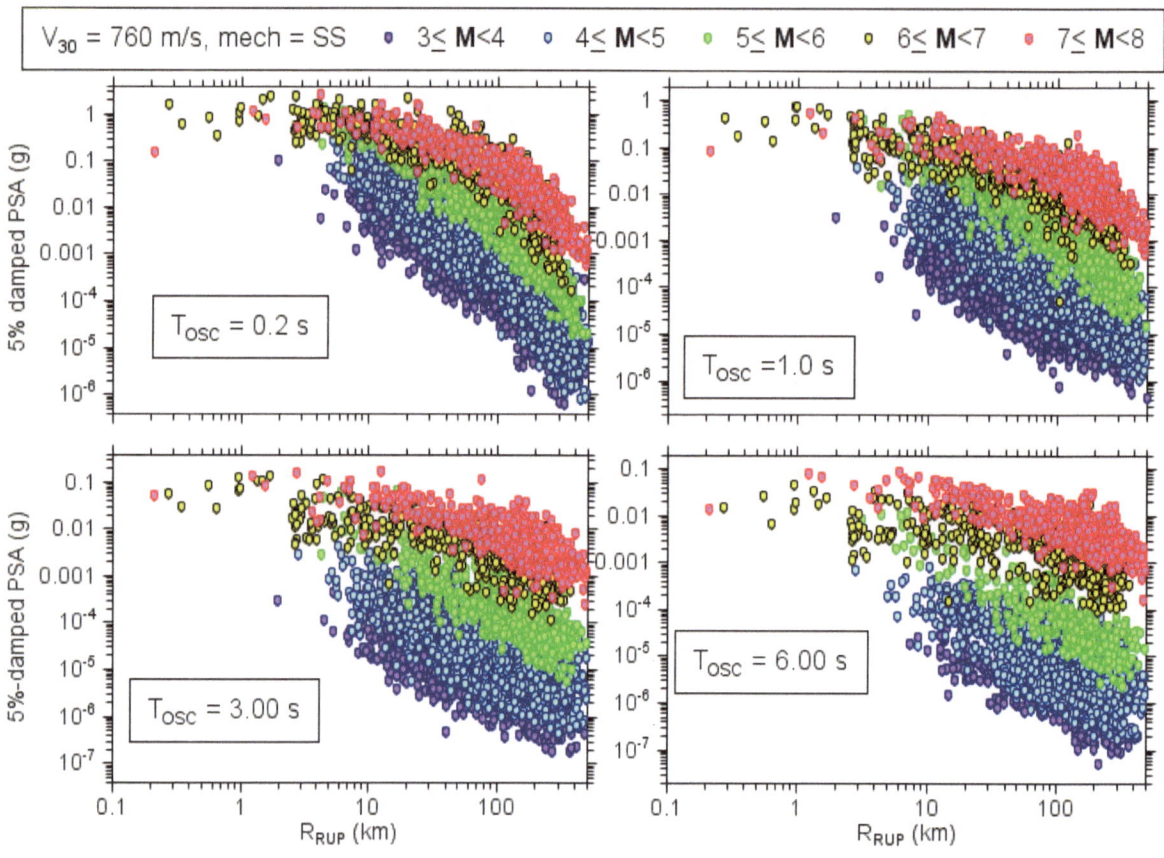

Figure 14
PSA at four periods for strikeslip, mainshock earthquakes, from the PEER NGA-West2 database. All amplitudes adjusted to $V_{S30} = 760$ m/s using the equations of SEYHAN and STEWART (2014). Each point is an observation of ground motion; the graphs for $T_{OSC} = 0.2$ s and $T_{OSC} = 6.0$ s contain 11,307 and 3,342 data points, respectively. (From BOORE et al. 2014)

model, I also compare observed and simulated magnitude scaling at distances near 70 km, where the finiteness of the source is less important.

5.1. Stochastic Modeling of Ground Motion

The simulations of ground motions shown here are based on the stochastic method, first introduced by HANKS and MCGUIRE (1981) and developed by others (see BOORE 2003a, for a review of the method). The basis for the method is shown in Fig. 15. Radiated energy described by the spectra in the top graph is assumed to be distributed randomly over a duration given by the addition of the source duration and a distant-dependent duration that captures the effect of wave propagation and scattering of energy.

The key to the success of the model lies in defining the Fourier acceleration spectrum of ground motion as a function of the predictor variables. The spectra are generally quite simple in shape (Fig. 15, top), with the complexity in the ground motion coming from the assumption that the motion is distributed randomly over a specified duration. The spectrum of ground acceleration is usually based on the multiplication of three functions of frequency, representing the contributions from the source, the propagation path, and the site. These functions generally vary smoothly with frequency. This is not a requirement in the stochastic method, but is a consequence of the explicit choice to model overall behavior of motion rather than the specific motion for a particular earthquake and site. The simplest source spectra are

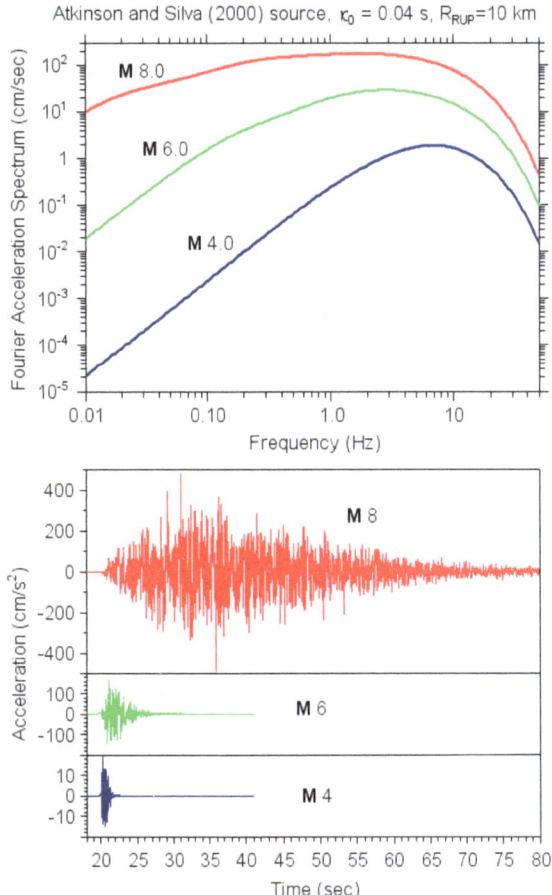

Atkinson and Silva (2000) source, $\kappa_0 = 0.04$ s, $R_{RUP}=10$ km

Figure 15

Basis of the stochastic method. The radiated energy described by the spectra in the *upper part* of the figure is assumed to be distributed randomly over durations equal to the inverse of frequencies related to the transition between the increasing and flat spectral amplitudes. Each time series is one realization of the random process for the actual spectrum shown (note the different ordinate scale for the **M** 4 time series compared to those for the **M** 6 and **M** 8 time series). Various peak ground-motion parameters (such as response spectra, instrument response, and velocity and acceleration) can be obtained by averaging the parameters computed from each member of a suite of acceleration time series or more simply by using random vibration theory, working directly with the spectra. (Modified from BOORE 2003a)

based on the single-corner frequency model discussed by AKI (1967) and BRUNE (1970, 1971). In this model, the spectrum increases with frequency below a corner frequency related to the seismic moment; the low-frequency part of the spectrum scales as seismic moment. Above the corner frequency, the spectrum is flat, with an amplitude related to the seismic moment and to a parameter having the dimensions of stress.

The propagation path is usually represented by a geometrical spreading and apparent anelastic attenuation function (as given the fourth bullet item in Sect. 4). The site amplification is generally given as a monotonic function of frequency that excludes attenuation, but includes amplifications from the source to the ground surface (e.g., BOORE 2013). The attenuation is included through a filter having the form $\exp(-\pi \kappa_0 f)$, where κ_0 is a parameter introduced by ANDERSON and HOUGH (1984); it generally has a value near 0.04 s for rock sites in active tectonic areas and 0.005 s for very hard sites in stable continental regions.

The simulations in this paper use the parameters of ATKINSON and SILVA (2000, hereafter "AS00") for coastal California; the Fourier acceleration spectra in the top graph of Fig. 15 are for this model (which has two low-frequency corner frequencies for the larger earthquakes). AS00 spectra were developed primarily by fitting ground-motion observations from California earthquakes; the propagation-path parameters were taken from RAOOF *et al.'s* (1999) analysis of recordings in southern California, and the site amplification used the BOORE and JOYNER (1997) generic rock amplifications (see BOORE 2013, for some comments related to this amplification).

The stochastic method used here, as implemented in the SMSIM suite of programs (see Sect. 7), assumes that the source is a point in space, which at first glance seems a poor assumption at near-fault distances. The method has been extended to finite-fault simulations (e.g., the EXSIM program of MOTAZEDIAN and ATKINSON 2005, and the modification EXSIM_DMB by BOORE 2009) by breaking the fault into a number of subfaults and adding together the properly time-lagged acceleration time series computed using the point-source stochastic method for each subfault. This, however, requires a specific fault and station geometry, which is awkward when computing simulated motions to be compared to averages of motions for the global database. For this reason, it is desirable to use point-source simulations. The limitations of the point-source simulations can be circumvented to some extent by modifying the distances used in the simulations. For a number of years, the most common distance used in the simulations was R_{RUP}, the closest distance to the

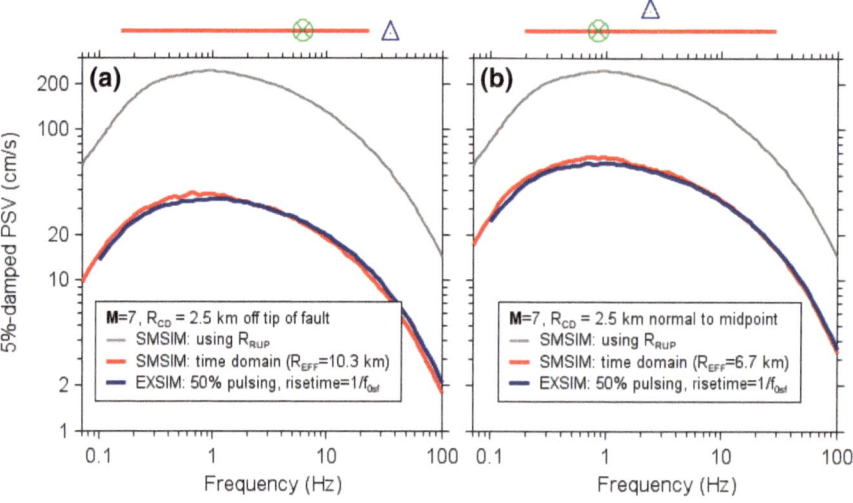

Figure 16

PSV for **M** 7 and R_{RUP} = 2.5 km. The simulation used a finite-fault model (EXSIM) and a point-source model (SMSIM). The station was located off the tip of the fault (*left graph*) and normal to the midpoint of the fault (*right graph*). The fault-station diagrams above each graph are map views. Two distances were used for the SMSIM calculations: R_{RUP} and R_{EFF}. Effective hypocenters at a distance R_{EFF} are shown by the *circles* with crosses through them for the two fault-station scenarios (only the distance R_{EFF} was used in the SMSIM calculations; the positions of the effective hypocenters plotted on the fault are only to provide a visual appreciation for the differences in size between R_{RUP} and R_{EFF}). (Modified from figures in BOORE 2009)

rupture surface, in keeping with the distance used in many empirically based GMPEs. This is not the best measure of distance, however, as most of the motion at a site will arrive from locations farther than the closest distance (e.g., AS00; TORO 2002; SCHERBAUM *et al.* 2006; BOORE 2009). This is shown in Fig. 16, which compares point-source and extended- source simulations. When R_{RUP} is used in the point-source simulations (SMSIM), the computed motions are much larger than from the finite-source (EXSIM) simulations. Agreement between the SMSIM and EXSIM simulations is obtained when an effective distance R_{EFF} similar to a root-mean-square distance to the fault is used in the SMSIM simulations (BOORE 2009). The point of Fig. 16 is to show that the point-source simulations with properly chosen distances can give motions in close agreement to those from the extended-source simulations. For purposes of comparing with the magnitude scaling from the global database, however, it is awkward to use R_{EFF}, as it requires a specific fault-station geometry. Based on extended-source simulations for many source-station geometries, AS00 propose a modification to R_{RUP} that can be used without specifying a source-site

geometry; their modification is used in the simulations shown in this article.

5.2. Observed and Simulated Magnitude Scaling

The observed and simulated magnitude scaling for two distance ranges are shown in Figs. 17, 18, and 19 for PSA at oscillator periods of 0.2, 1.0, and 4.0 s, respectively. The observations are from strikeslip faults and have been adjusted to V_{S30} = 620 m/s (the value for the crustal amplifications used in the simulations; the V_{S30} of 760 m/s used in Fig. 14 is a standard reference velocity used in GMPEs and in building codes). The parameters used in the simulations are those of AS00, which are guided by data from California, so some degree of agreement with the observations should be expected (although the NGA-West2 database contains much more California data than were available to AS00). No attempt has been made to modify the AS00 model used in the stochastic method to match the data. Simulations are shown with and without the AS00 distance modification. At near-fault distances there are large differences in the near-fault simulations, and the

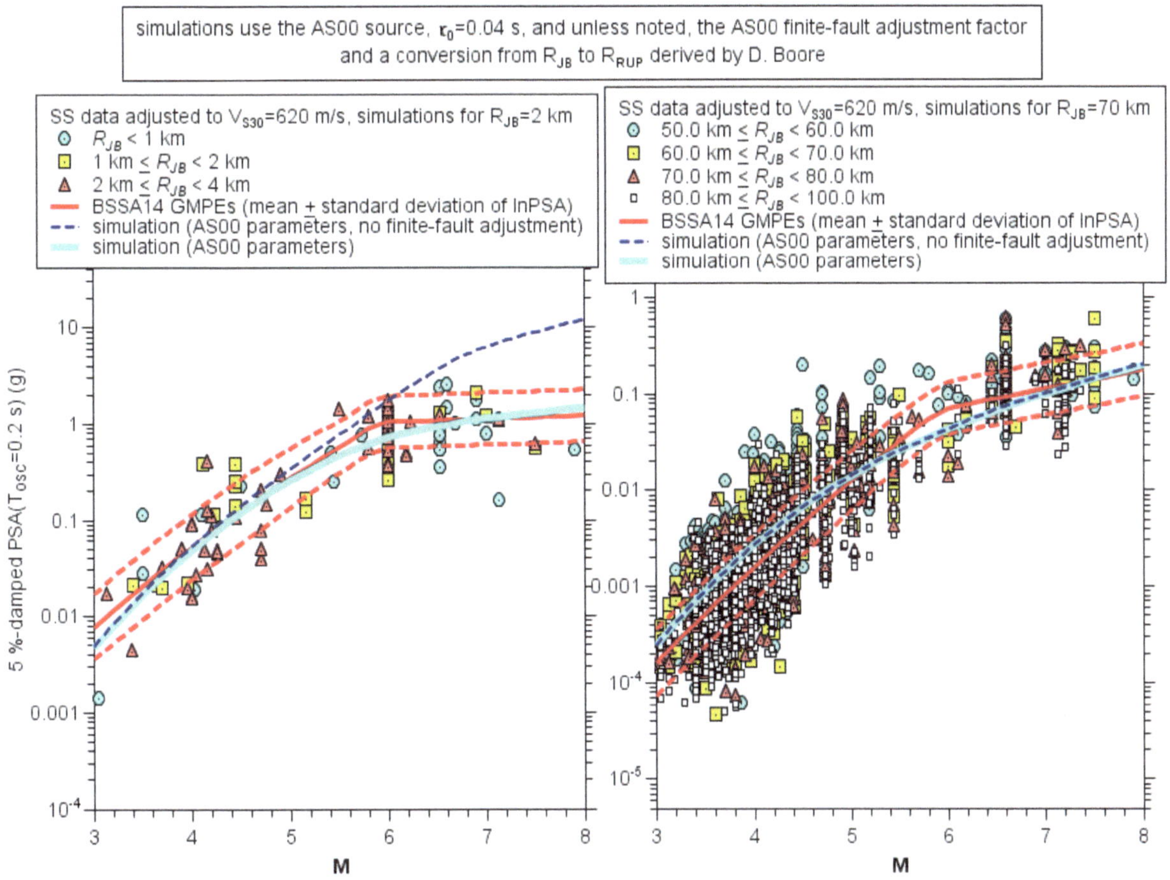

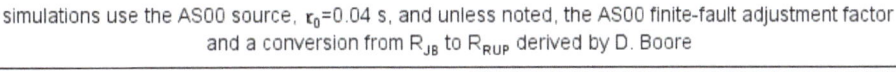

Figure 17

Magnitude scaling for $T_{OSC} = 0.2$ s PSA near-fault (*left graph*) and intermediate-fault (*right graph*) distances, for strikeslip (SS) earthquakes. The *symbols* are data from the NGA-West2 database, adjusted to $V_{S30} = 620$ m/s. The *red curves* (mean and ± one standard deviation of ln PSA) are from the BOORE *et al.* (2014, "BSSA14") GMPEs, and the *cyan* and *dashed blue curves* are from stochastic-method simulations, using the ATKINSON and SILVA (2000, "AS00") model parameters, without (*dashed blue*) and with (*cyan*) the AS00 finite-fault adjustments

simulations using the distance modification are in much better agreement with the data; this again highlights the importance of a modification to the distance used in the point-source simulations at near-source distances. Also shown are motions from the BOORE *et al.* (2014, hereafter "BSSA14") GMPEs (this is the only place in this article where motions from GMPEs are shown); the kink in the BSSA14 magnitude scaling is a result of the simple functional form assumed in developing the GMPEs, although the data do suggest a slope change in the magnitude dependence over a fairly small range of magnitude. The observed magnitude scaling shows strong saturation at close distances for short periods (this has

been noted previously by others—e.g., YAMADA *et al.* 2009, for PGA); much of this saturation is modeled by the simulations when the distance modification is used. The overall good agreement between observed and simulated motions suggests that the magnitude scaling of ground motions over a very wide range of periods and magnitudes is largely explained by a simple model of the seismic source, without the need for additional complexities in the source radiation or a breakdown of self similarity, a conclusion also reached by BALTAY and HANKS (2014). Of course, the success of the simple source, propagation-path, and site-response model may not hold for recordings of individual earthquakes, as presumably much of the

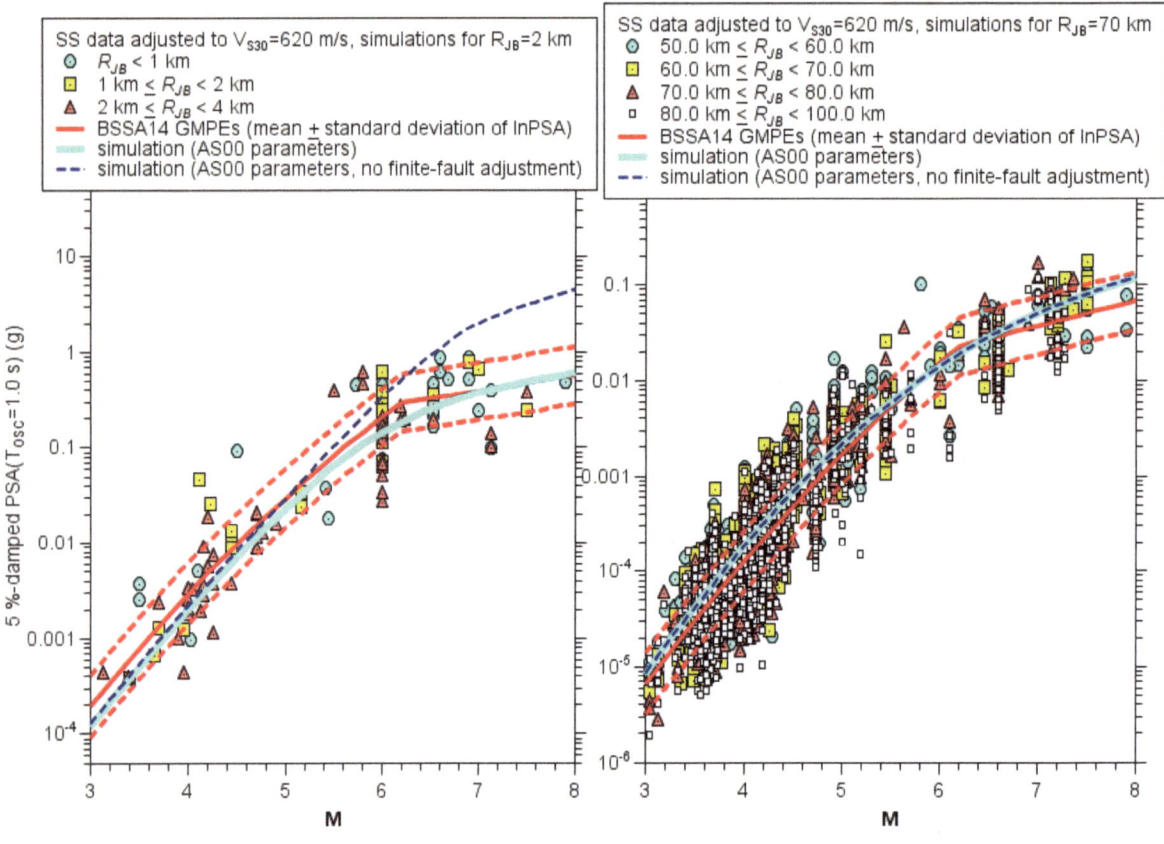

Figure 18

Magnitude scaling for $T_{OSC} = 1.0$ s PSA near-fault (*left graph*) and intermediate-fault (*right graph*) distances, for strikeslip (SS) earthquakes. The *symbols* are data from the NGA-West2 database, adjusted to $V_{S30} = 620$ m/s. The *red curves* (mean and ± one standard deviation of ln PSA) are from the Boore *et al.* (2014, "BSSA14") GMPEs, and the *cyan and dashed blue curves* are from stochastic-method simulations, using the Atkinson and Silva (2000, "AS00") model parameters, without (*dashed blue*) and with (*cyan*) the AS00 finite-fault adjustments

large inherent variability in ground motions shown in Fig. 14 is due to earthquake-dependent and station-dependent complexities in source, path, and site properties. One advantage of using the median ground motions from the large global database is that these complexities average out, thus revealing the overall scaling of motions with magnitude.

6. Conclusions

The global database used in this article is a valuable resource for ground motions from shallow earthquakes in active tectonic regions. Unfortunately,

there is not a field in the database that is a direct indicator of the relation between the station location and the fault damage zone (FDZ) for a given earthquake. The closest distances from a station to the fault rupture and to the projection of the rupture surface to the Earth's surface are included in the database, however, and although imperfect indicators of the station location with respect to the FDZ, these distance measures lead to the conclusion that there are few records in the NGA-West2 database obtained within FDZs. In spite of this, the global database was carefully constructed, and thus can provide reliable data and metadata for investigation of those few earthquakes with recordings within and close to the

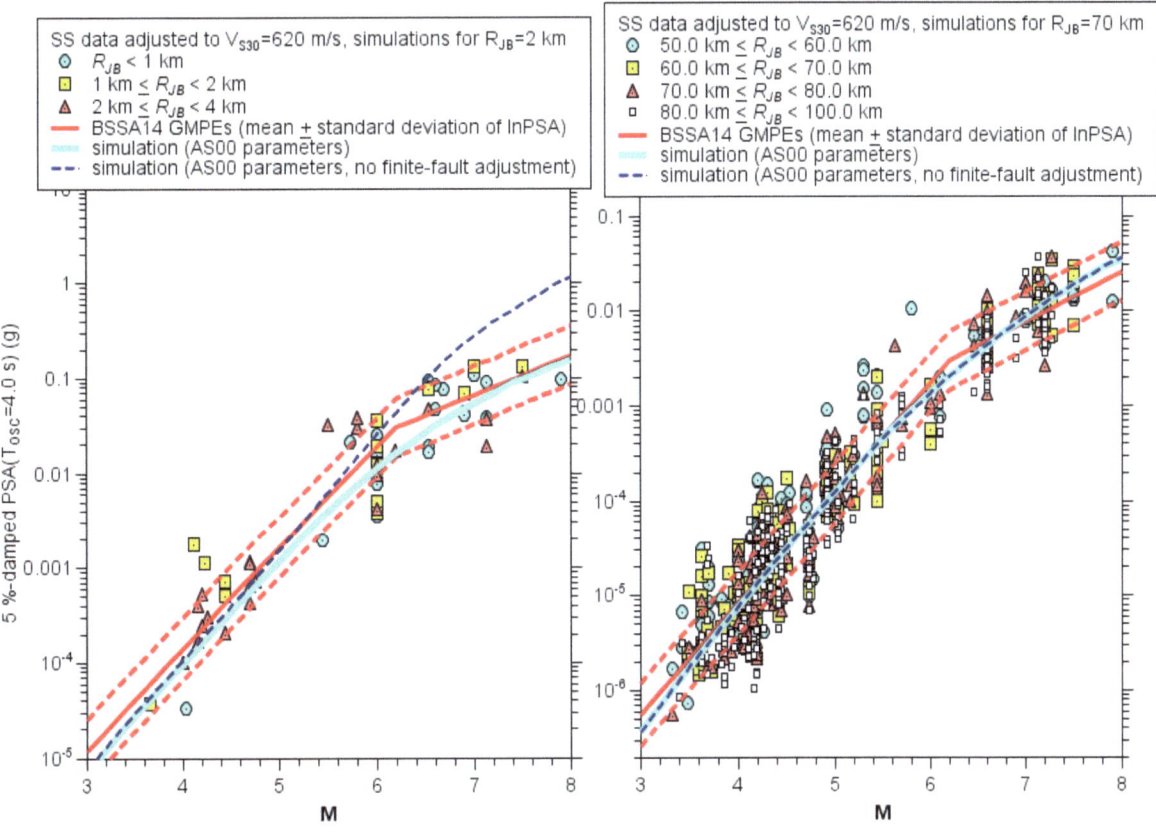

Figure 19

Magnitude scaling for $T_{OSC} = 4.0$ s PSA near-fault (*left graph*) and intermediate-fault (*right graph*) distances, for strikeslip (SS) earthquakes. The *symbols* are data from the NGA-West2 database, adjusted to $V_{S30} = 620$ m/s. The *red curves* (mean and $\pm$ one standard deviation of ln PSA) are from the Boore et al. (2014, "BSSA14") GMPEs, and the *cyan* and *dashed blue curves* are from stochastic-method simulations, using the Atkinson and Silva (2000, "AS00") model parameters, without (*dashed blue*) and with (*cyan*) the AS00 finite-fault adjustments

FDZ. Several examples of these motions are discussed in this article, the results emphasizing the complexity of the ground motions, both in terms of spatial variation of the amplitudes of the motion and variability in the direction of ground-motion polarization, which seem to depend on details of the source-to-station propagation path. The NGA-West2 database also provides some insight into the overall scaling of motions with magnitude for different periods of ground motion. The data show that the magnitude-to-magnitude increase of motions at a given distance becomes smaller as magnitude increases, with short-period motions at near-fault distances attaining almost constant values (i.e., complete saturation) for large magnitudes. The

magnitude scaling is in good agreement with simple models of the source and path effects, showing that the magnitude saturation is largely a geometrical effect of the increasing fault size with magnitude, and not due to a fundamental change in the stress release along faults as earthquakes increase in size.

7. Data Sources

Most of the data used in this article came from the PEER NGA-West2 database, available from http://peer.berkeley.edu/ngawest/databases/ (last accessed 28 Oct 2013). Time series came from http://peer.berkeley.edu/peer_ground_motion_database/ (last accessed 28 Oct 2013), http://nsmp.wr.usgs.gov/ GEOS/PRK/parkfield.html (last accessed 28 Oct 2013) for GEOS data, and from sources discussed in the papers from which the figures were taken. The SMSIM software used to compute the magnitude scaling is available from http://www.daveboore.com/ software_online.html (last accessed 28 Oct 2013).

Acknowledgments

I thank Yehuda Ben-Zion and Antonio Rovelli for the invitation to attend the 40th Workshop of the International School of Geophysics on PROPERTIES AND PROCESSES OF CRUSTAL FAULT ZONES in Erice, Sicily, Italy, May 18–24, 2013, and for the encouragement to prepare a paper based on my talk. I also thank Joe Fletcher, Tony Shakal, and Paul Spudich for providing figures, Chris Dietel for corrected station coordinates for the GEOS stations, and Annemarie Baltay, Carola Di Alessandro, Tom Hanks, and an anonymous person for constructive reviews that led to significant improvements in the article.

REFERENCES

AKI, K. (1967). *Scaling law of seismic spectrum*, J. Geophys. Res. *72*, 1217–1231.

ANCHETA, T.D., R.B. DARRAGH, J.P. STEWART, E. SEYHAN, W.J. SILVA, B.S.-J. CHIOU, K.E. WOODDELL, R.W. GRAVES, A.R. KOTTKE, D.M. BOORE, T. KISHIDA, and J.L. DONAHUE (2013). PEER NGA-West2 Database, *PEER Report 2013/03*, Pacific Earthquake Engineering Research Center, University of California, Berkeley, 170 pp.

ANCHETA, T.D., R.B. DARRAGH, J.P. STEWART, E. SEYHAN, W.J. SILVA, B.S.-J. CHIOU, K. E. WOODDELL, R. W. GRAVES, A. R. KOTTKE, D. M. BOORE, T. KISHIDA, and J.L. DONAHUE (2014). *NGA-West2 database*, Earthquake Spectra *30*, (in review).

ANDERSON, J.G. and S.E. HOUGH (1984). *A model for the shape of the Fourier amplitude spectrum of acceleration at high frequencies*, Bull. Seismol. Soc. Am. *74*, 1969–1993.

ATKINSON, G.M. and W. SILVA (2000). *Stochastic modeling of California ground motions*, Bull. Seismol. Soc. Am. *90*, 255–274.

BALTAY, A.S. and T.C. HANKS (2014). *Understanding the magnitude dependence of PGA and PGV in NGA-West2 data*, Bull. Seismol. Soc. Am. *104*, (in review).

BEN-ZION, Y. and C.G. SAMMIS (2003). *Characterization of fault zones*, Pure appl. Geophys. *160*, 677–715.

BEROZA, G.C. and P. SPUDICH (1988). *Linearized inversion for fault rupture behavior: Application to the 1984 Morgan Hill, California, earthquake*, J. Geophys. Res. *93*, 6275–6296.

BOMMER, J.J. and D.M. BOORE (2005). Engineering Geology: Seismology, in *Encyclopaedia of Geology*, R. C. Selley, L. Robin M. Cocks, and Ian R. Plimer (Editors), Elsevier Ltd., doi:10. 1016/B0-12-369396-9/90020-0, 499–515.

BOORE, D.M. (2003a). *Prediction of ground motion using the stochastic method*, Pure and Applied Geophysics *160*, 635–676.

BOORE, D.M. (2003b). A compendium of P- and S-wave velocities from surface-to-borehole logging: Summary and reanalysis of previously published data and analysis of unpublished data, *U. S. Geological Survey Open-File Report 03-191*, 13 pp. Available from http://geopubs.wr.usgs.gov/open-file/of03-191/.

BOORE, D.M. (2009). *Comparing stochastic point-source and finite-source ground-motion simulations: SMSIM and EXSIM*, Bull. Seismol. Soc. Am. *99*, 3202–3216.

BOORE, D.M. (2013). *The uses and limitations of the square-root impedance method for computing site amplification*, Bull. Seismol. Soc. Am. *103*, 2356–2368.

BOORE, D.M. and W.B. JOYNER (1997). *Site amplifications for generic rock sites*, Bull. Seismol. Soc. Am. *87*, 327–341.

BOORE, D.M., V.M. GRAIZER, A.F. SHAKAL, and J.C. TINSLEY (2004). *A study of possible ground-motion amplification at the Coyote Lake Dam, California*, Bull. Seismol. Soc. Am. *94*, 1327–1342.

BOORE, D.M., E.M. THOMPSON, and H. CADET (2011). *Regional correlations of $V_{s\,30}$ and velocities averaged over depths less than and greater than 30 m*, Bull. Seismol. Soc. Am. *101*, 3046–3059.

BOORE, D.M., J.P. STEWART, E. SEYHAN, and G.M. ATKINSON (2014). *NGA-West 2 equations for predicting PGA, PGV, and 5%-Damped PSA for shallow crustal earthquakes*, Earthquake Spectra *30*, (in press).

BORCHERDT, R.D., M.J.S. JOHNSTON, G. GLASSMOYER, and C. DIETEL (2006). *Recordings of the Parkfield 2004 earthquake on the GEOS array: implications for earthquake precursors, fault rupture, and coseismic strain changes*, Bull. Seismol. Soc. Am. *96*, S73–S89.

BOZORGNIA, Y., N.A. ABRAHAMSON, L. AL ATIK, T.D. ANCHETA, G.M. ATKINSON, J.W. BAKER, A. BALTAY, D.M. BOORE, K.W. CAMPBELL, B.S.-J. CHIOU, R. DARRAGH, S. DAY, J. DONAHUE, R.W. GRAVES, N. GREGOR, T. HANKS, I.M. IDRISS, R. KAMAI, T. KISHIDA, A. KOTTKE, S.A. MAHIN, S. REZAEIAN, B. ROWSHANDEL, E. SEYHAN, S. SHAHI, T. SHANTZ, W. SILVA, P. SPUDICH, J.P. STEWART, J. WATSON-LAMPREY, K. WOODDELL, and R. YOUNGS (2014). *NGA-West2 Research Project*, Earthquake Spectra *30*, (in review).

BRUNE, J.N. (1970). *Tectonic stress and the spectra of seismic shear waves from earthquakes*, J. Geophys. Res. *75*, 4997–5009.

BRUNE, J.N. (1971). *Correction*, J. Geophys. Res. *76*, 5002.

CADET, H. and A.-M. DUVAL (2009). *A shear wave velocity study based on the KiK-net borehole data: A short note*, Seismol. Res. Letters *80*, 440–445.

CAINE, S., J.P. EVANS, and C.B. FORSTER (1996). *Fault zone architecture and permeability structure*, Geology *24*, 1025–1028.

CULTRERA, G., A. ROVELLI, G. MELE, R. AZZARA, A. CASERTA, and F. MARRA (2003). *Azimuth-dependent amplification of weak and strong ground motions within a fault zone (Nocera Umbra, central Italy)*, J. Geophys. Res. *108*, 2156, doi:10.1029/ 2002JB001929, 14 pp.

EBERHART-PHILLIPS, D., and A. J. MICHAEL (1993). *Three-dimensional velocity structure, seismicity, and fault structure in the Parkfield region, central California*, J. Geophys. Res. *98*, 15,737–15,758.

FLETCHER, J.B., L.M. BAKER, P. SPUDICH, P. GOLDSTEIN, J.D. SIMS, and M. HELLWEG (1992). *The USGS Parkfield, California, dense seismograph array: UPSAR*, Bull. Seismol. Soc. Am. *82*, 1041–1070.

FLETCHER, J. B., P. SPUDICH, and L. M. BAKER (2006). *Rupture propagation of the 2004 Parkfield, California, earthquake from observations at the UPSAR array*, Bull. Seismol. Soc. Am. *96*, S129–S142.

HANKS, T.C. (1975). *Strong ground motion of the San Fernando, California, earthquake: ground displacements*, Bull. Seismol. Soc. Am. *65*, 193–225.

HANKS, T.C. and H. KANAMORI (1979). *A moment magnitude scale*, J. Geophys. Res. *84*, 2348–2350.

HANKS, T.C. and R.K. MCGUIRE (1981). *The character of high-frequency strong ground motion*, Bull. Seismol. Soc. Am. *71*, 2071–2095.

JOYNER, W.B. and D.M. BOORE (1981). *Peak horizontal acceleration and velocity from strong-motion records including records from the 1979 Imperial Valley, California, earthquake*, Bull. Seismol. Soc. Am. *71*, 2011—2038.

KURZON, I., F.L.VERNON, Y. BEN-ZION, and G. ATKINSON (2014). *Ground motion prediction equations in the San Jacinto Fault Zone –Significant effects of rupture directivity and fault zone amplification*, Pure and Applied Geophysics *171*, (in review).

LEWIS, M.A. and Y. BEN-ZION (2010). *Diversity of fault zone damage and trapping structures in the Parkfield section of the San Andreas Fault from comprehensive analysis of near fault seismograms*, Geophys. J. Int. *183*, 1579–1595.

LIU, H.-L. and D.V. HELMBERGER (1983). *The near-source ground motion of the 6 August 1979 Coyote Lake, California, earthquake*, Bull. Seismol. Soc. Am. *73*, 201–218.

LIU, P., S. CUSTODIO, and R. J. ARCHULETA (2006). *Kinematic inversion of the 2004 M 6.0 Parkfield earthquake including an approximation to site effects*, Bull. Seismol. Soc. Am. *96*, S143–S158.

MOTAZEDIAN, D. and G.M. ATKINSON (2005). *Stochastic finite-fault modeling based on a dynamic corner frequency*, Bull. Seismol. Soc. Am. *95*, 995–1010.

PISCHIUTTA, M., A. ROVELLI, F. SALVINI, G. DI GIULIO, and Y. BEN-ZION (2013). *Directional resonance variations across the Pernicana Fault, Mt Etna, in relation to brittle deformation fields*, Geophys. J. Int., doi:10.1093/gji/ggt031, 11 pp.

RAOOF, M. R.B. HERRMANN, and L. MALAGNINI (1999). *Attenuation and excitation of three-component ground motion in Southern California*, Bull. Seismol. Soc. Am. *89*, 888–902.

SANDIKKAYA, M.A., M.T. YILMAZ, B.S. BAKIR, and Ö. YILMAZ (2010). *Site classification of Turkish national strong-motion stations*, J. Seismol.*14*, 543–563.

SCHERBAUM, F., F. COTTON, and H. STAEDTKE (2006).*The estimation of minimum-misfit stochastic models from empirical ground-motion prediction equations*, Bull. Seismol. Soc. Am. *96*, 427–445, doi:10.1785/0120050015.

SEYHAN, E. and J.P. STEWART (2014). *Semi-empirical nonlinear site amplification from NGA-West 2 data and simulations*, Earthquake Spectra *30*, (in review).

SHAKAL, A., H. HADDADI, V. GRAIZER, K. LIN, and M. HUANG (2006). *Some key features of the strong-motion data from the M 6.0 Parkfield, California, earthquake of 28 September 2004*, Bull. Seismol. Soc. Am. *96*, S90–S118.

SPUDICH, P. and K.B. OLSEN (2001). *Fault zone amplified waves as a possible seismic hazard along the Calaveras fault in central California*, Geophys. Res. Letters *28*, 2533–2536.

THURBER, C., H. ZHANG, F. WALDHAUSER, J. HARDEBECK, A. MICHAEL, and D. EBERHART-PHILLIPS (2006). *Three-dimensional compressional wavespeed model, earthquake relocations, and focal mechanisms for the Parkfield, California, region*, Bull. Seismol. Soc. Am. *96*, S38–S49.

TORO, G.R. (2002). Modification of the TORO et al. (1997) attenuation relations for large magnitudes and short distances: Risk Engineering, Inc. report, http://www.riskeng.com/PDF/atten_toro_extended.pdf.

WANG, G.-Q., G.-Q. TANG, C. R. JACKSON, X.-Y. ZHOU, and Q.-L. LIN (2006). *Strong ground motions observed at the UPSAR during the 2003 San Simeon (M 6.5) and 2004 Parkfield (M 6.0), California, earthquakes*, Bull. Seismol. Soc. Am. *96*, S159–S182.

YAMADA, M., A.H. OLSEN, and T.H. HEATON (2009). *Statistical features of short-period and long-period near-source ground motions*, Bull. Seismol. Soc. Am. *99*, 3264–3274.

ZHAO, P., Z. PENG, Z. SHI, M.A. LEWIS, and Y. BEN-ZION (2010). *Variations of the velocity contrast and rupture properties of M6 earthquakes along the Parkfield section of the San Andreas fault*, Geophys. J. Int. *180*, 765–780.

(Received August 20, 2013, accepted November 22, 2013, Published online December 15, 2013)

Reprinted from the journal

Pure Appl. Geophys. 171 (2014), 3045–3081
© 2014 Springer Basel
DOI 10.1007/s00024-014-0855-2

Ground Motion Prediction Equations in the San Jacinto Fault Zone: Significant Effects of Rupture Directivity and Fault Zone Amplification

I. Kurzon,[1,4] F. L. Vernon,[1] Y. Ben-Zion,[2] and G. Atkinson[3]

Abstract—We present a new set of Ground Motion Prediction Equations (GMPEs) for horizontal Peak Ground Acceleration, Peak Ground Velocity, and 5 % damped pseudo-spectral acceleration (PSA), developed for the San Jacinto Fault Zone (SJFZ) area. Besides using these equations to quantify seismic shaking in the area, the results allow us to examine the physics and local properties controlling the observed ground motions. The analyzed dataset includes $\sim$30,000 observations from $\sim$800 events spanning a magnitude range of $1.5 < M < 6.0$ and recorded by up to 140 stations at epicentral distances ranging from essentially zero to 150 km. The local GMPE is developed for the SJFZ by applying classical regression techniques with predictive variables that include first distance and magnitude, and then site characteristics, rupture directivity, and fault zone amplification. The significance of these effects is determined by measuring the uncertainty-reduction of the GMPE due to each factor. The results show that, in contrast to many regional studies, traditional site characteristic has a relatively minor effect on peak amplitudes in our study area. However, rupture directivity is a significant factor controlling the amplitudes of ground motion even for small events. The dense seismic network and newly developed directivity tool enable us to extract efficiently directivity effects with statistical significance, using the ground-motion dataset during the regression analysis process. The obtained rupture directivities are consistent with the main focal mechanism orientations and surface trace orientations, known from other studies, and predictions for bimaterial ruptures in the trifurcation area of the SJFZ. Fault zone amplification is a second important factor, showing strong impact on the peak ground motion values, with increasing role for the lower frequency range (<10 Hz) examined in the 5 % damped PSA values. We also observe signatures of large amplitude-variances, which indicate additional source-related control on the distribution of amplitudes (besides rupture directivity) for aftershocks close in time and location to the M_L 5.1 earthquake of March 2013. Using the full set of records we present the most complete set of GMPEs for the SJFZ area, including a higher-amplitude prediction for regions in the direction of rupture.

Key words: Local GMPEs, San Jacinto fault zone, directivity effects, fault zone amplification, regression analysis, engineering seismology.

1. Introduction

Ground Motion Prediction Equations (GMPEs) provide a fundamental tool for analysis of seismic hazard, by describing the amplitude and attenuation of peak ground motion values from the events to the recording stations. These equations are typically developed by regression of the peak ground motion values to first-order variables such as magnitude and distance, and possibly several other predictive variables. In the past decade there has been extensive work in this field, such as by DOUGLAS (2001, 2003) and the Next Generation Attenuation project (NGA; ABRAHAMSON *et al.* 2008), in which five different groups have developed sets of GMPEs for shallow events in active tectonic regions (ABRAHAMSON and SILVA 2008; BOORE and ATKINSON 2008; CAMPBELL and BOZORGNIA 2008; CHIOU and YOUNGS 2008; IDRISS 2008) (Note: the NGA project has been updated over the last few years, with NGA-2 GMPEs to be published in Earthquake Spectra in 2014.). These GMPEs were developed on a global or regional scale, combining data from different tectonic settings to enable a larger database to be compiled. However, these large scales (global or regional) tend to generalize and mask some of the local attributes affecting ground motion in specific areas. Moreover, the focus of such GMPEs has tended to be restricted to the upper magnitude range ($M > 5$), and they do not always scale well over a broader magnitude range. Recent studies have extended the lower magnitude limit down to M 3 (e.g.,

[1] University of California San Diego, La Jolla, CA, USA.
[2] University of Southern California, Los Angeles, CA, USA.
[3] University of Western Ontario, London, ON, Canada.
[4] Geological Survey of Israel, Jerusalem, Israel. E-mail: ittaik@gsi.gov.il

Boore *et al.* 2013, and other GMPEs developed as part of the NGA 2 project: http://peer.berkeley.edu/publications/peer_reports_complete.html) and M 2 (Cua and Heaton 2008), but have retained the global-scale nature of the GMPEs.

In this study, we develop GMPEs based on data from a single fault system, focusing on a complex region of the San Jacinto Fault Zone (SJFZ) in southern California. The change from a global or regional scale down to a local scale might reduce the role of some effects (such as tectonic settings) on ground-motion variability, while facilitating the identification of other effects not seen in larger scales, some of which might be unique to the study area. We use a rich database of ground motions in the SJFZ to examine: (a) the main factors controlling amplitudes on a local scale, besides magnitude and distance; (b) if these factors can shed light on some of the unique characteristics of earthquake physics in the study area and properties of the fault zone; (c) the differences in the variance of local GMPEs in comparison to that for regional or global GMPEs; and (d) what information we can gather about the fault zone and the source by comparing the variance and/or residuals of the GMPEs from different datasets within the same local fault zone (for example, different aftershock sequences).

The SJFZ provides a great natural laboratory for examining some of the above questions for several reasons: (a) it is a highly-active seismic fault, producing a significant number of well-recorded small events ranging up to magnitude 4 along with several larger events; (b) it is densely covered by several seismic networks with a recent significant increase in stations due to a SJFZ experiment initiated in 2010 (providing also abundant recordings within the immediate vicinity of the fault zone); and (c) it encompasses a complex system of faults, with relatively large clusters of seismicity not necessarily aligned with the fault traces or converging to distinct surfaces, and with heterogeneous focal mechanisms of events (Hartse *et al.* 1994; Bailey *et al.* 2010). In this study we investigate how the complexity and other characteristics of the fault zone and earthquake processes influence recorded ground motions. By analyzing a high-fidelity ground-motion dataset we uncover some of the more dominant amplitude-controlling factors in the region, clarify their statistical signatures in the data, and apply them to generate fault zone-specific GMPEs for the SJFZ.

2. Geological Settings

The SJFZ is a major component of the strike-slip fault system that accommodates the plate boundary motion in Southern California, branching from the San Andreas Fault at Cajon Pass, crossing the Peninsular Ranges Batholith, and terminating in the desert of the Imperial Valley. The SJFZ has had about 24 km of slip since the latest Pliocene to early Pleistocene (Sharp 1967; Rockwell *et al.* 1990; Dorsey and Roering 2006), and estimated slip rates that vary along strike between 8 and 20 mm/year (Rockwell 2003; Kendrick *et al.* 2002; Fay and Humphreys 2005; Fialko 2006). In this study, we analyze data from four major segments of the SJFZ (Fig. 1) comprising together the central part of the fault: (1) Clark—Anza (CA), (2) Coyote Creek (CC), (3) Buck Ridge (BR), and (4) Hot Springs (HS). The seismicity in the past 30 years (Hutton *et al.* 2008; Hauksson *et al.* 2012) along these segments has been clustered in two main regions (Fig. 1): the Hot Springs cluster at a depth range of 15–22 km, and the trifurcation (TR) cluster at a depth range of 7–17 km; between these two clusters there is a distinct gap (Fig. 1 top-right). The Clark—Anza and the Coyote Creek segments are known to have sustained moderate to large earthquakes based on historical and paleoseismic data (e.g. Salisbury *et al.* 2012; Onderdonk *et al.* 2013, and references therein).

3. Dataset and Data Processing

We analyze ground motions associated with 800 events in the magnitude range of $1.5 \leq M \leq 6$ in a 45×125 km rectangular area around the SJFZ (Fig. 1). The event database consists of the seismicity in the region from February 2010 to May 2012, and three moderate earthquakes (M_L 5.6 [M_W 5.2] June 2005, M_L 5.9 [M_W 5.4] July 2010 and M_L 5.1 [M_W 4.7] March 2013) along with their aftershock sequences. The station distribution includes several networks operating in this region (see Table 1) within

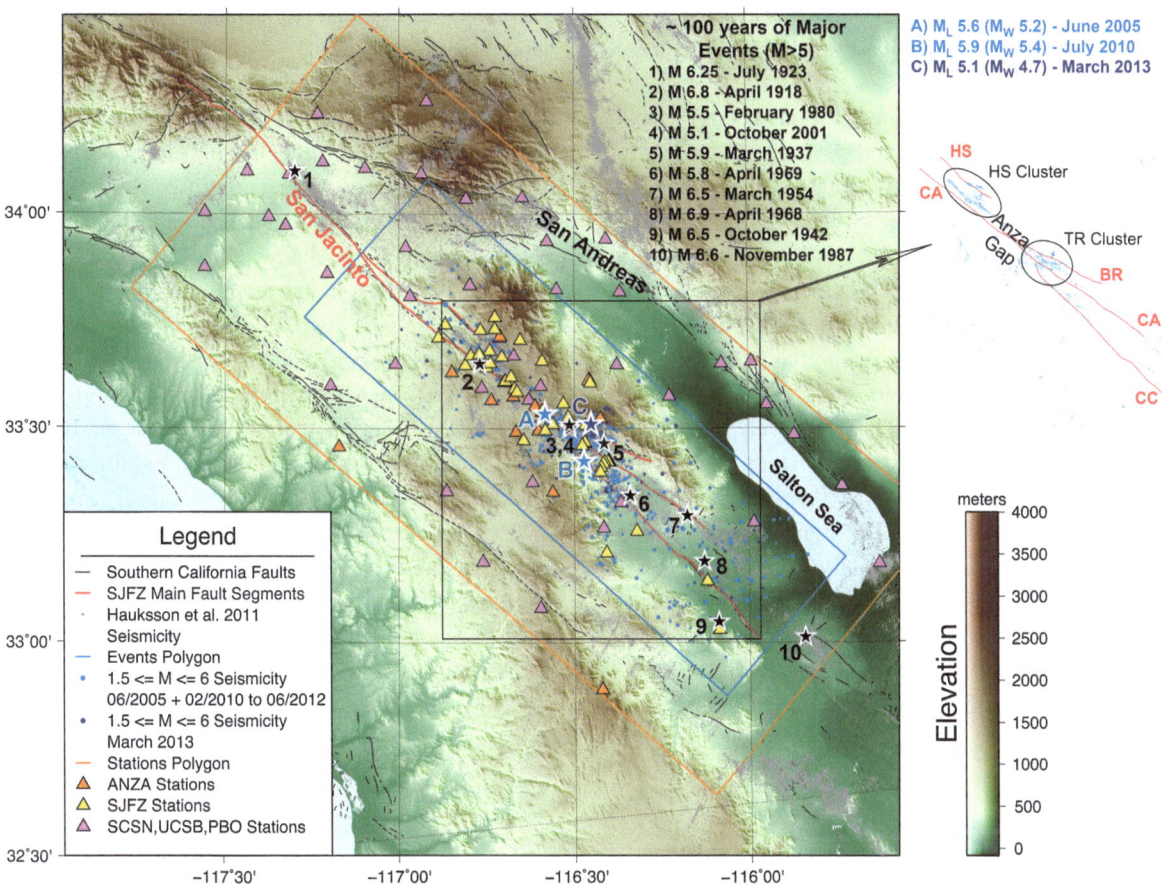

Figure 1

Location map of the SJFZ, showing also the stations and events distribution considered in this study. The map indicates four of the main fault segments (*CC* Coyote Creek, *CA* Clark Anza, *BR* Buck Ridge, and *HS* Hot Springs), the seismicity pattern in the past 30 years according to HAUKSSON *et al.* (2012), and the major events: $M > 6$ in the past 100 years, and $M > 5$ in the past 30 years (VERNON 1989; SALISBURY *et al.* 2012), seen as *black-white stars* referred to by the event list on the upper right side of the map. The events used for this study are located within the blue event rectangular, and include the three main events (*blue-white stars*) and aftershock sequences of M_L 5.6 (M_W 5.2) in June 2005, M_L 5.9 (M_W 5.4) in July 2010, and M_L 5.1 (M_W 4.7) in March 2013. The stations are located within the *orange rectangle* and include several networks, including the new SJFZ project. Altogether, there are 800 stations recorded on ~200 instruments in ~140 seismic stations.

The side map emphasizes the four major fault segments, and the two main seismic clusters, with the Anza gap between them

Table 1

The network-station distribution included in this study

Code	Network	$N_{Stations}$	Description
AZ	ANZA	12	The ANZA network was installed in 1982 and is operated by UCSD. It has 18 stations, from which 12 are within the study area
YN	SJFZ	75	PASSCAL temporary stations for the NSF Continental Dynamics (NSF SJFZ CD) project titled: Collaborative Research: Structural Architecture and Evolutionary Plate-boundary Processes along the San Jacinto Fault Zone
CI	SCSN	40	Southern California Seismic Network operated by Caltech
PB	PBO	10	Plate Boundary Observatory operated by UNAVCO
SB	UCSB	5	Stations operated by University of California, Santa Barbara

a 90 × 275 km rectangle around the event epicenters, allowing in most cases good focal-sphere coverage of the data.

The Peak Ground Acceleration (PGA), Peak Ground Velocity (PGV), and 5 % damped pseudo-spectral acceleration (PSA) datasets are generated using acceleration and velocity waveforms filtered using a 1–30 Hz fourth order band-pass Butterworth filter (with four poles at each corner frequency). The PGA values are mainly used to reveal the factors controlling ground motions, while the PGV and PSA values provide the complete set of GMPEs required for engineering purposes. The relatively high band pass filter of 1–30 Hz used in this study reflects the high frequency content of the data, especially for events in the $M < 3$ range. The signals for the $M > 3$ events were chosen preferentially from the acceleration channels due to clipping of the velocity recordings at some of the close-range stations, while the signals for the $M < 3$ events were preferentially taken from the velocity channels due to their higher

sensitivity and ability to capture lower magnitude data.

Figure 2 displays the magnitude—distance distribution of the recordings and the depth distribution of the events used in the PGA analysis. It is shown in two different colors, representing two different datasets: blue for June 2005 and February 2010 to May 2012, and orange for March 2013. These two datasets reflect two phases in the evolution of the GMPEs presented here (see Sect. 4). The first phase was done before the aftershock sequence of March 2013, generating a set of GMPEs from 650 events and 20,000 Peak Ground Motion records. In the second phase, the March 2013 dataset was merged (addition of 150 events and 15,000 records), providing a new set of GMPEs. There are two main reasons for keeping the results in these two separated phases: (a) there are some insights that were observed within the Phase 1 dataset, which were masked by the Phase 2 complete dataset, and (b) it demonstrates better the limitations of GMPEs in predicting future motions. These issues

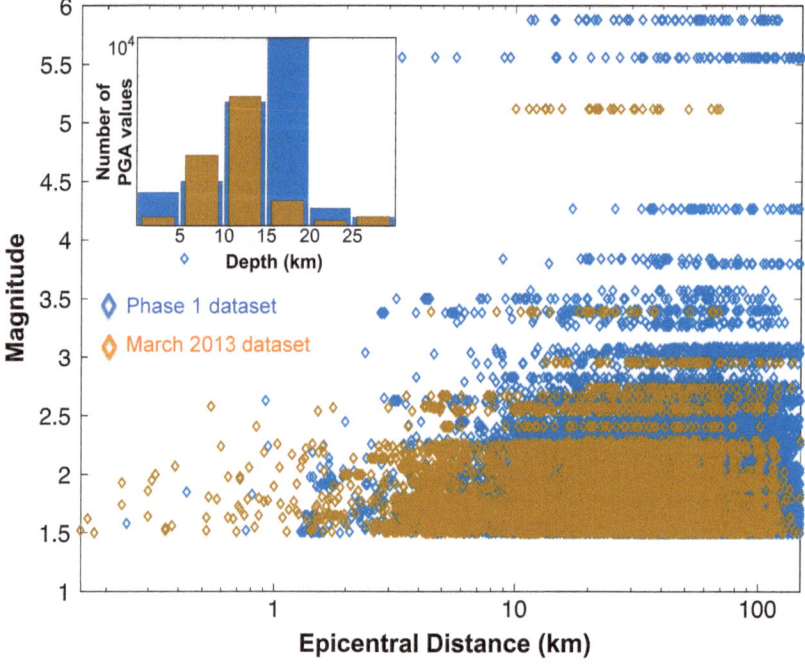

Figure 2

Distance—Magnitude Distribution of the records. Distances are presented by the epicentral distance, and magnitudes are in M_L. The complete dataset is split into two subsets: one for the first phase of this study (*blue*) and the second for the additional records of the M_L 5.1—March 2013 (*orange*). Some of the gaps of the Phase 1 distribution are filled by the new data, especially in the close epicentral distances. The depth histogram is showing a shallower distribution of depth records than the one seen in the Phase 1 dataset are further discussed later in the paper. Note that for the purpose of the GMPEs we use the M_L magnitude.

Peak values of ground motion were picked automatically using a signal to noise ratio (SNR) algorithm that was tuned to capture the main signal and peak value in each waveform. The tuning was done manually by viewing a significant subset (~ 35 %, $\sim 2{,}500$ records) of the actual peak values of the June 2005 and July 2010 aftershock sequences. A total of 35,000 peak ground records were collected for each one of the datasets (PGA, PGV, and for each period of the PSA), from a total of ~ 200 instruments at ~ 140 stations (some stations have a seismometer and an accelerometer).

4. Data Analysis Procedure

We present two different sets of GMPEs. The first set (Phase 1) was generated based on the dataset of June 2005 and February 2010 to May 2012. The second set of GMPEs (Phase 2) is based on the merging of the M_L 5.1 March 2013 aftershock sequence, which shows very different characteristics from what was predicted by the Phase 1—GMPEs, and emphasizes additional aspects in the generation of GMPEs.

Phase 1—June 2005 + February 2010 to May 2012.

We divide our search for the GMPEs and the factors controlling ground motion values in the SJFZ into several stages. First we consider the two most dominant factors, magnitude and distance, and solve for them using the basic formulation provided by BOORE and ATKINSON (2008):

$$Ln(Y_1) = e_1 + e_2(M - M_h) + e_3(M - M_h)^2$$
$$+ [c_1 + c_2(M - M_{ref})] \cdot Ln\left[\sqrt{R_e^2 + h_A^2} \middle/ R_{ref}\right] \quad (1)$$
$$+ c_3\left[\sqrt{R_e^2 + h_A^2} - R_{ref}\right].$$

Here, M is the local magnitude, R_e is the epicentral distance, $e_1, e_2, e_3, c_1, c_2, c_3$ and h_A are regression coefficients, and M_h, M_{ref} and R_{ref} are constants. M_h, M_{ref} and R_{ref} can be chosen largely for convenience, as long as they are within the range of magnitude and distance of the data (see BOORE and ATKINSON 2008,

for further discussion on this issue). We set $M_h = 2.5$, $R_{ref} = 1$ km and $M_{ref} = 3$. The variable Y_1 gives predicted values based on Eq. (1) referred to in this study as the 'basic regression' or 'basic GMPE'.

We can use the hypocentral distance and regress only for six coefficients, or use the epicentral distance and regress for an additional coefficient, the apparent depth h_A. The difference between the two options is minor, but the latter is somewhat advantageous as it avoids the addition of depth uncertainties into the total uncertainty, resulting in slightly a smaller uncertainty (~ 0.5 %). By using the epicentral distances, h_A can be estimated from the maximum amplitude as the epicentral distance approaches zero; deeper apparent depths indicate smaller amplitudes. This is the option chosen in Eq. (1), where e_1, e_2 and e_3 are magnitude related, c_1, c_2 and c_3, are distance related, and h_A is the apparent depth of all records.

In the second stage, we examine additional factors that may affect ground motion peak values using a combination of two main tools. Plotting the observed records together with the basic GMPE, color-coded according to a tentative additional factor, provides a first estimation of the relevance of the examined factor. To quantify that factor we also apply the method used by THOMPSON et al. (2011) to estimate site response:

$$Ln(Y_2) = b_1 + b_2 Ln(X) + b_3 Ln(Y_1), \quad (2)$$

where the right side has a linear relation between the examined new factor X, the basic GMPE value Y_1 and the new predicted values Y_2. This equation is an indication of the linearity of $Ln(X)$ with respect to the new GMPE Y_2. Note that when $b_3 \Rightarrow 1$ Eq. (2) approaches a linear relation between the magnitude and distance factors and the additional examined factor.

Both Eqs. (1) and (2) provide measures of uncertainties. We use the variance about the models in relation to the observed values, and examine the variance reduction associated with different factors. This allows us to sort different factors according to their relevance and improvements of the GMPEs. Having a current best performing GMPE with the lowest variance, we use the residuals defined as $Y_{observed}/Y_{predicted}$ to search for additional event or station factors that may reduce the variance further.

Using the same factor-hierarchy we then determine GMPEs for the set of PGV and PSA values, resulting in our estimated GMPEs for the SJFZ area. In the following sections we provide additional details on the main methods used in the data analysis.

4.1. Regression Method

JOYNER and BOORE (1993) introduced two methods for applying regression analysis for finding GMPE coefficients: the one-stage maximum-likelihood method and the two-stage method. In the one-stage method the different uncertainties of the records and events are calculated in one step with a combined uncertainty. In the two-stage method a random record-uncertainty is first calculated, mainly accounting for the event-station distance, and then the event uncertainty is iterated considering the magnitude of each event. Both methods have shown similar success (JOYNER and BOORE 1993), and in many of the GMPE studies one of these methods was used as the basic approach (e.g., BOORE and ATKINSON 2008). In this work we use the one-stage method to solve Eqs. (1) and (2). The iteration scheme utilizes the golden section search (e.g., PRESS *et al.* 1992), allowing a fast convergence to the set of coefficients defining the GMPEs.

4.2. Directivity Method

Rupture directivities can produce significant variations of ground motion parameters, with higher frequencies and amplitudes in the direction of rupture propagation compared to the orthogonal and opposite directions (e.g. DOUGLAS *et al.* 1988; SPUDICH and CHIOU 2008; CALDERONI *et al.* 2013; BAKER *et al.* 2012, and references therein, as part of the NGA-West 2 project models). Directivity has also been observed and discussed in relatively low magnitudes, down to M 3 events (BOATWRIGHT 2007; SEEKINS and BOATWRIGHT 2010), showing that small to moderate events may have quite strong directivities. In this study we extend the observations down to M 1.5, naturally corresponding with higher frequencies. In order to quantify the effect of rupture directivities on GMPE, we use the observed peak ground motion values corrected for geometrical spreading. This is

Figure 3 ▶

The directivity tool—examples. Here we show two cases: Case 1— M_L 5.9 (**a, c, e** plots) and Case 2—M_L 1.6 (**b, d, f** plots). The **a** and **b** plots are geographical maps, showing the stations recording the event and some of the corresponding waveforms; events are shown as *blue-yellow stars*, and stations are marked as *grey triangles*. The **c** and **d** plots show the normalized A_{corr} in each slice, as a function of $\cos(\phi)$, where ϕ describes the orientation of the 30° slices in reference to the rupture direction. The **e** and **f** plots are a projection of the normalized A_{corr} on a map view, in which θ is the event-station azimuth, and the distances of the stations from the center (location of event) reflect the normalized A_{corr} per slice. These last two plots help to quantify and define in a statistical manner the direction of rupture, showing an unilateral NE solution for Case 1, and a bilateral SE solution for Case 2 (See text for further details)

done by multiplying each peak amplitude value, A, by its corresponding hypocentral distance r:

$$A \propto 1/r \Rightarrow A_{corr} = A \cdot r. \qquad (3)$$

The correction (3) is applied up to a distance of 60–80 km to avoid Pn and Sn phases that can lead to overweighing distant stations. We recognize that the correction for geometric spreading is simplistic, being based on the whole-space spreading rate in the frequency domain, whereas the actual geometric spreading is more complex. This is intended only as a simple first-order correction.

In order to estimate rupture directivity we use 30° wide slices rotated around 360° in 10° increments. For each slice orientation we calculate the average $\tilde{A}_{corr}$ of the A_{corr} values within the slice and search for the slice orientation with the maximum averaged $\tilde{A}_{corr}^{max}$. Since there are other factors that might affect the amplitudes besides distance and directivity (such as radiation pattern or site characteristics), we consider only slices having at least three stations and coefficient of variation below 0.75. The orientation of the slice with $\tilde{A}_{corr}^{max}$ satisfying these requirements is regarded as the rupture direction. Once this direction has been determined, we divide the horizontal space, surrounding the event, into 12 slices of 30° and use φ to denote the angles between the center of each slice and the assumed rupture direction, i.e., $\cos(\varphi)$ ranges from 1 for the slice of $\tilde{A}_{corr}^{max}$ to -1 for the opposite direction. Finally, we calculate for each slice $\tilde{A}_{corr}/\tilde{A}_{corr}^{max} = \tilde{A}_{corr}^{norm}$ where $\tilde{A}_{corr}^{norm}$ ranges from 1 for $\tilde{A}_{corr}^{max}$ to 0 for slices that do not satisfy the forgoing quality criteria.

Figure 3 illustrates the directivity tool using data of the M_L 5.9 July 7, 2010 event (Fig. 3a) and a July

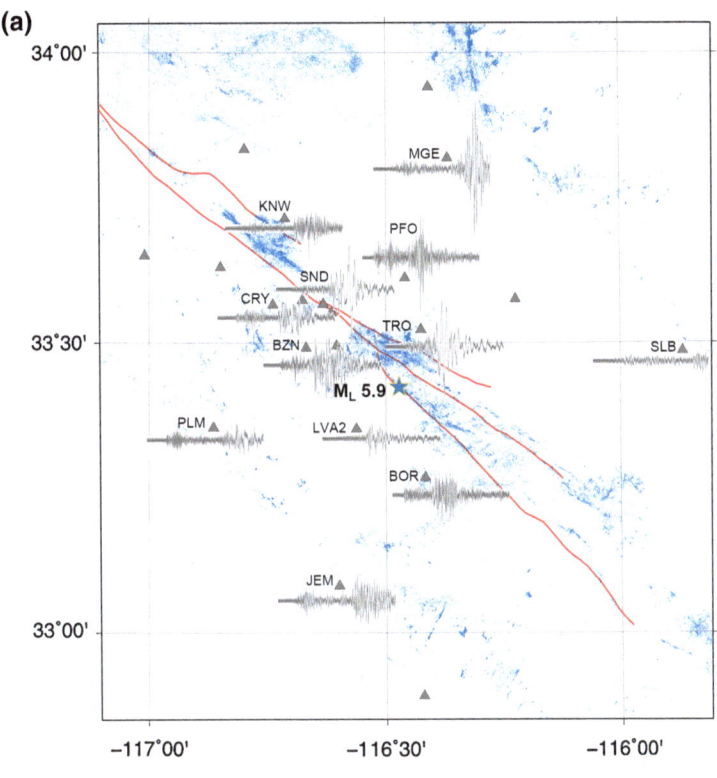

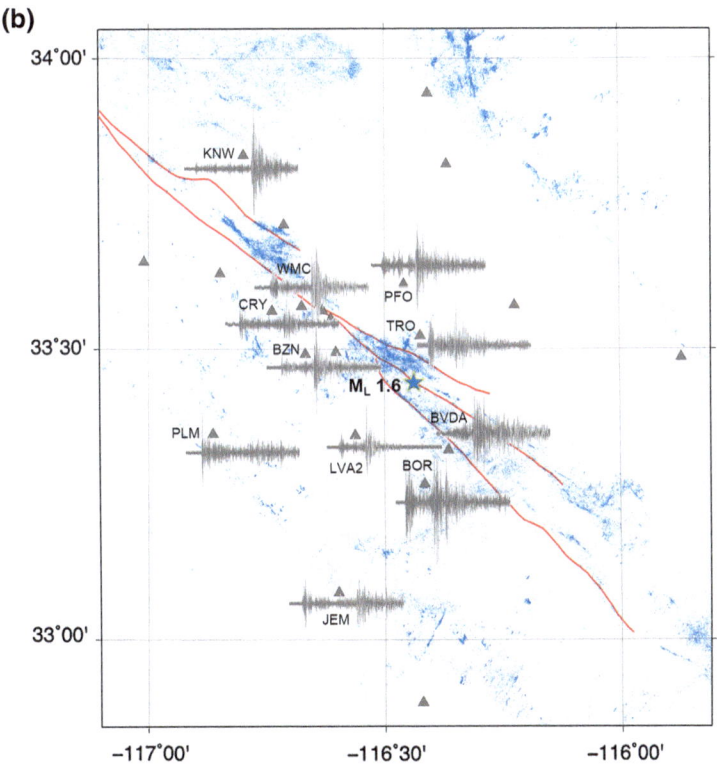

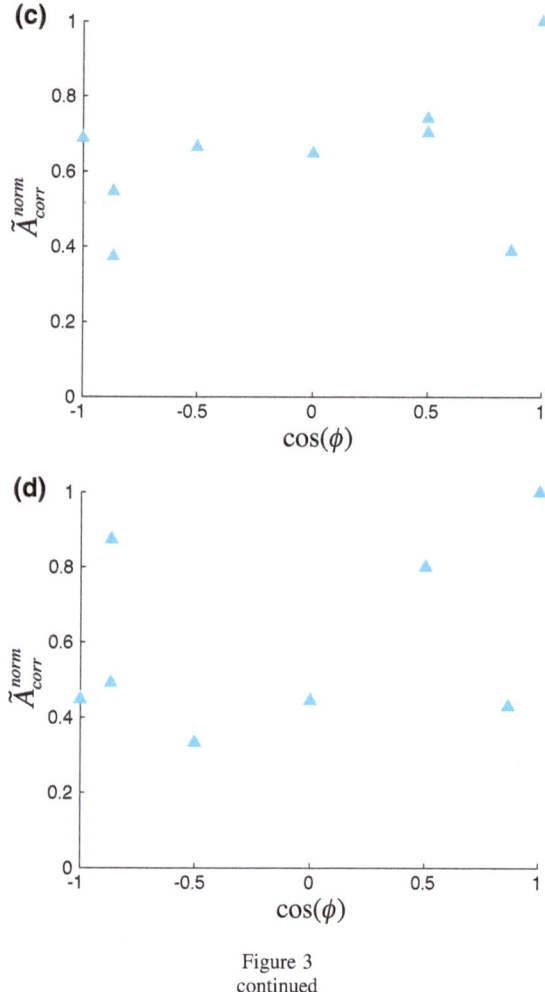

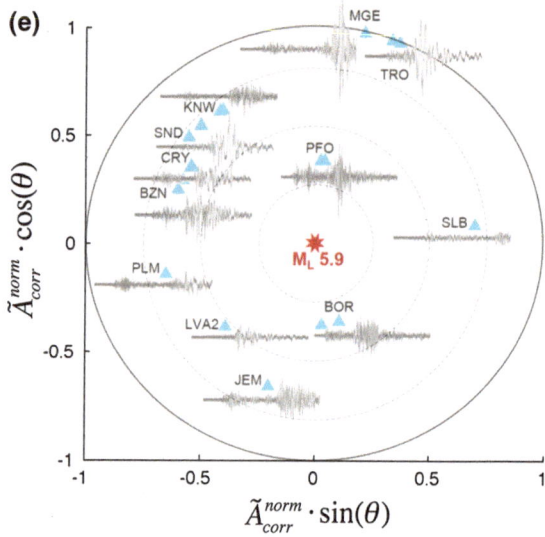

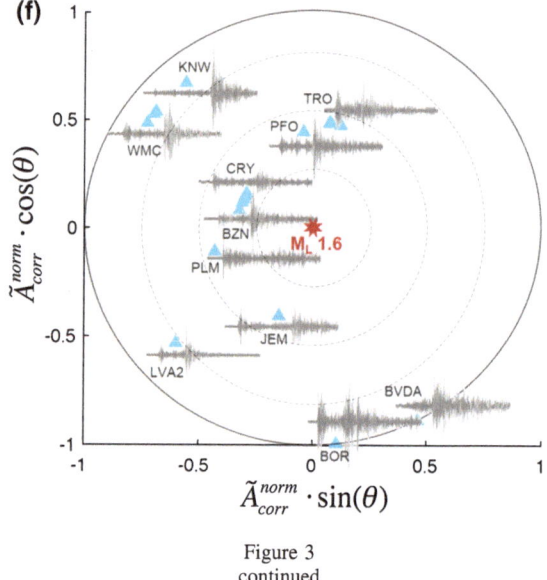

Figure 3
continued

Figure 3
continued

10, 2010 M_L 1.6 aftershock (Fig. 3b). The waveforms are normalized to the maximum recorded amplitude, A, in the stations, without applying corrections for geometrical spreading and slice averaging. A close examination shows that the M_L 5.9 event has slightly larger amplitudes to the NE than to the NW, while the M_L 1.6 event exhibits higher amplitudes to the SE as well as to the NW. The $\tilde{A}_{corr}^{norm}$ vs. $\cos(\varphi)$ plots suggest (see also Fig. 4) an approximately unilateral behavior for the M_L 5.9 event (Fig. 3c) and close to bilateral behavior for the M_L 1.6 event (Fig. 3d); we note that all slices excluding $\cos(\varphi) = 1$ or -1 may have two different $\tilde{A}_{corr}^{norm}$ values for the same $\cos(\varphi)$. In Fig. 3e, f, $\tilde{A}_{corr}^{norm}$ in each slice is assigned back to the stations within that slice and projected onto the map, according to the event-station azimuth θ. The obtained circular map ranges from -1 to 1 on south–

north and west–east axes. The results show the orientation of the stations in relation to the event located in the center of the map, and the event to stations distances reflecting the values of $\tilde{A}_{corr}^{norm}$ in each slice. The orientations of stations with a radius near 1, on the circumference of the circular map, reflect the rupture directivity defined by the azimuth to the middle of the slice containing the stations. This indicates that the M_L 5.9 event ruptured to the NNE, while the M_L 1.6 event ruptured to the southeast, in agreement with the visual characteristics of the waveforms (Fig. 3a, b, e, f).

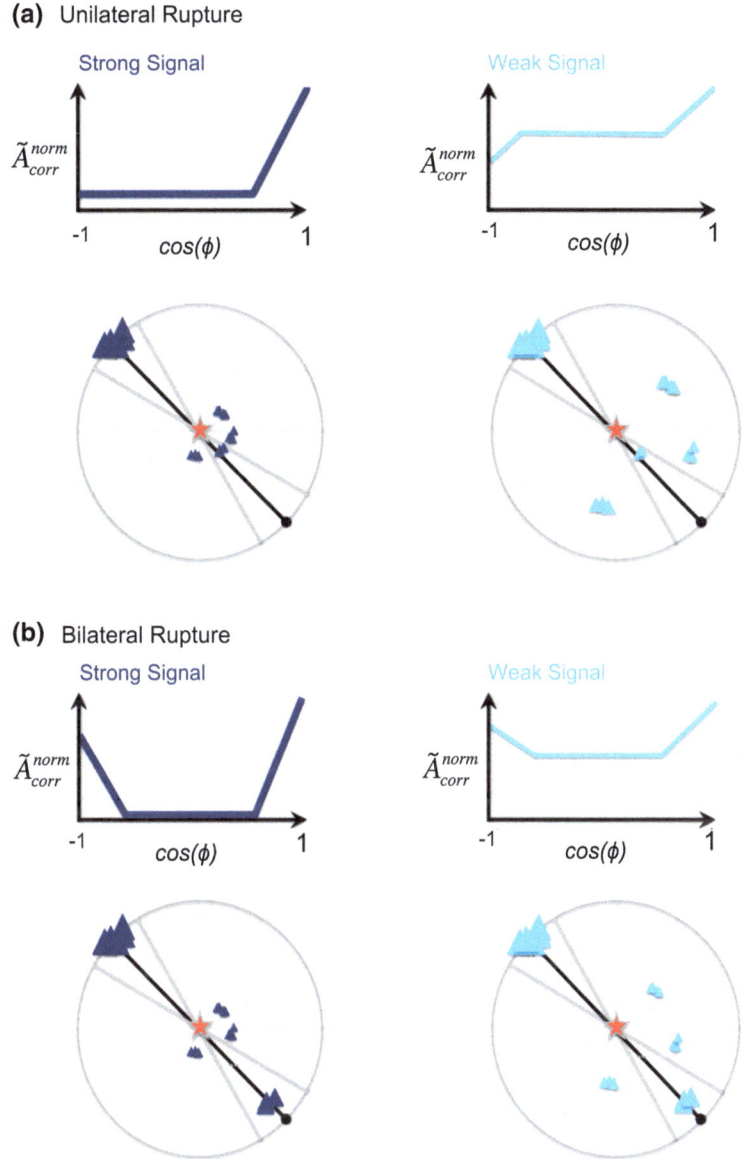

Figure 4
End-member rupture solutions reflected by the directivity tool: **a** unilateral ruptures, and **b** bilateral ruptures, both with strong or alternatively weak directivity options. The strength of the directivity is the reciprocal of the average of the normalized A_{corr} in all slices. Since in the direction of rupture the value is 1, then the lower the values in the other slices, the lower is the average and the signal would be stronger. For a bilateral rupture, the opposite side to the maximum will also show close values to 1. Case 1 in Fig. 3 would be classified as an unilateral rupture with weak directivity signal, and Case 2 as close to a bilateral rupture with intermediate directivity signal

Figure 4 presents schematically four end-member cases that can help in understanding the rupture properties: unilateral (Fig. 4a) and bilateral (Fig. 4b) ruptures, each with weak and strong directivity signals. The strength of the directivity signal is represented by $\langle \tilde{A}_{corr}^{norm} \rangle$ defined by the average of $\tilde{A}_{corr}^{norm}$ on all slices. High values of $\langle \tilde{A}_{corr}^{norm} \rangle$ reflect a weak directivity signal, since many of the slices besides the one with $\tilde{A}_{corr}^{max}$ have relatively high $\tilde{A}_{corr}^{norm}$, possibly reflecting other factors affecting the amplitudes. Low values of $\langle \tilde{A}_{corr}^{norm} \rangle$ reflect a strong directivity signal, since the slice of $\tilde{A}_{corr}^{max}$ dominates over other factors affecting the recorded amplitudes. The value of $\langle \tilde{A}_{corr}^{norm} \rangle$ is also the measure defining the color code in subsequent related figures, with darker blue indicating stronger directivity signal.

We define an index of directivity I_{Dir} based on strength of the directivity signal as:

$$I_{Dir} = 10 \cdot \tilde{A}_{corr}^{norm} / \langle \tilde{A}_{corr}^{norm} \rangle. \quad (4)$$

The highest I_{Dir} values per event will be for stations that are in the direction of rupture, with stronger signals having higher I_{Dir} values. Events with strong directivity signals will have a wide range of I_{Dir} values, while events with weak directivity signals will show a narrower range of I_{Dir} values.

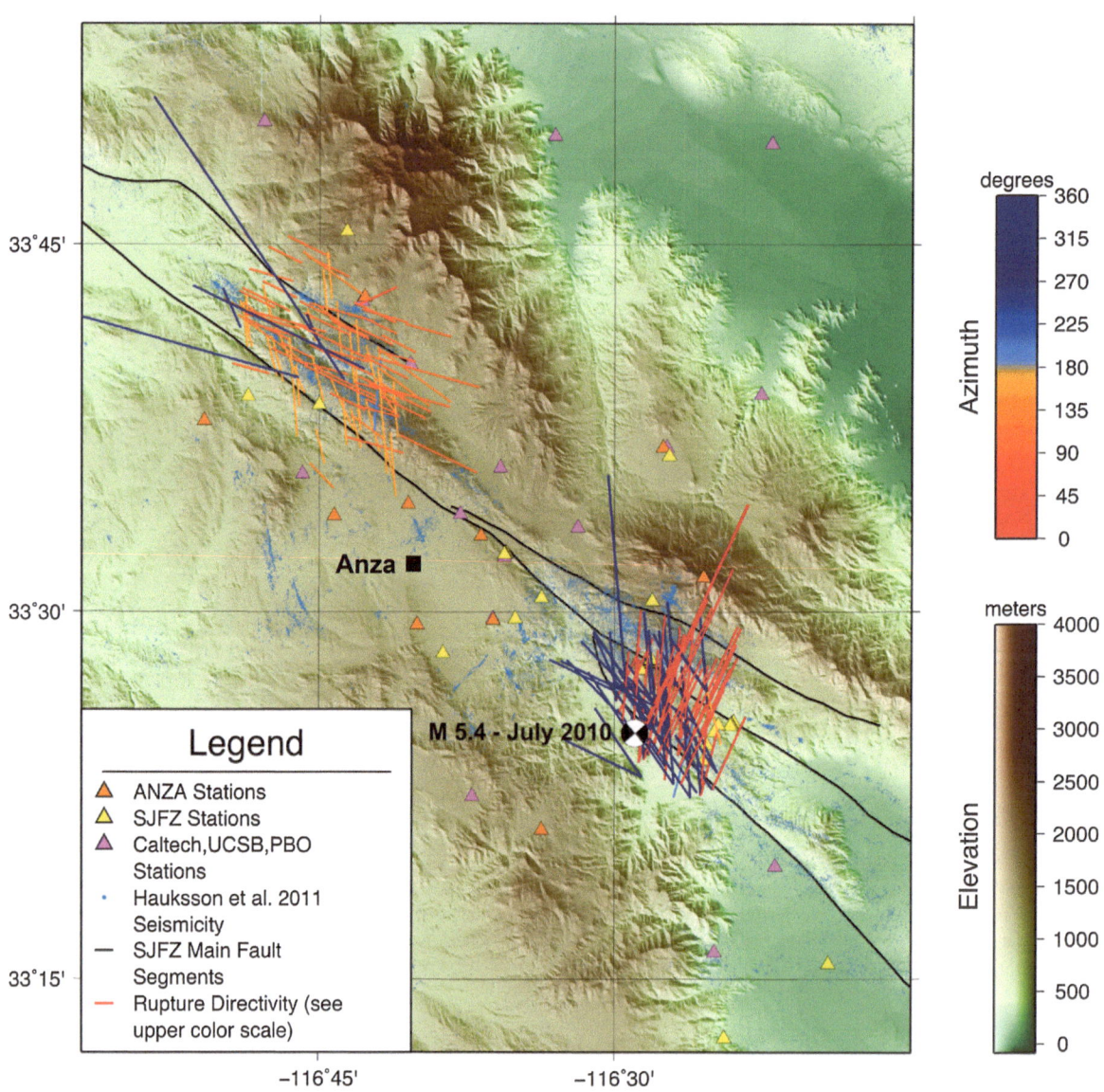

Figure 5

Analysis of rupture directivity. Here we show two clusters of seismic activity: (1) the M_W 5.4 (M_L 5.9) July 2010 and its aftershocks, and (2) December 2011 events in the Hot Springs cluster. Each *line* represents an event, with the length reflecting the strength of the directivity signal, and the color reflecting the directivity orientation. While the July 2010 event shows two main directions, to the NE and to the NW (parallel to the fault), the December 2011 event shows a SE trend, parallel and sub-parallel to the fault. The NE trend of the M_W 5.4 event fits the focal mechanism of the main shock and reflects the observations seen before in Fig. 3

I. Kurzon et al. — Pure Appl. Geophys.

Reprinted from the journal

192

Below we incorporate the variable I_{Dir} in Eq. (2) to generate directivity-dependent GMPEs.

Before applying this tool for the GMPE analysis we demonstrate its utility by translating I_{Dir} to lengths that are plotted with indication of the azimuth of maximum directivity. This is shown in Fig. 5 for the M_L 5.9 July 2010 event and its aftershocks, along with seismicity from December 2011 at the Hot Springs cluster. The results show that each region has distinct rupture directions that are mainly sub-parallel and sub-normal to surface traces within statistical scatter. The waveform observations of the M_L 5.9 July 2010 mainshock (Fig. 3) indicate a similar NE directivity as the results in Fig. 5 in the trifurcation area. We note that a statistical preference of earthquake propagation to the NE in this area is also consistent with observed asymmetry of rock damage (LEWIS *et al.* 2005; WECHSLER *et al.* 2009), combined with theoretical calculations of damage generation during ruptures on a bimaterial fault (BEN-ZION and SHI 2005; XU *et al.* 2012) and recent detailed tomographic images (ALLAM and BEN-ZION 2012; ALLAM *et al.* 2014). The other Hot Springs cluster of seismicity in Fig. 5 is off the main San Jacinto fault and requires additional information to interpret. A more detailed analysis of the directivity tool and its possible implications will be given in a follow up paper (KURZON *et al.* 2014).

4.3. Obtaining PSA Values

The response spectra providing the 5 % damped pseudo-spectral acceleration (PSA) values is calculated using the classical algorithm of NIGAM and JENNINGS (1969) for solving the equation of motion of a simple oscillator:

$$\ddot{\chi} + 2\beta\omega\dot{\chi} + \omega^2\chi = -a(t), \qquad (5)$$

where t is time, ω is the natural frequency of the oscillator, β is the damping coefficient, and $a(t)$ is the base acceleration the oscillator is subjected to. A segmental linear approximation of $a(t)$ is given by:

$$\ddot{\chi} + 2\beta\omega\dot{\chi} + \omega^2\chi = -a_i - (t - t_i)$$
$$\cdot (a_{i+1} - a_i)/(t_{i+1} - t_i) t_i \leq t \leq t_{i+1}. \qquad (6)$$

Since χ and $\dot{\chi}$ are known at $t = 0$, one can solve (6) analytically and obtain displacement, velocity and acceleration for each examined ω (NIGAM and

JENNINGS 1969), which, in accordance with the 1–30 Hz Butterworth filter applied in this study, spans the period range of 0.03–1 s.

Phase 2—merging the M_L 5.1 March 2013 aftershock sequence.

The M_L 5.1 March 2013 aftershock sequence was recorded on the complete network of the SJFZ project, providing many more stations within the fault zone, including two additional linear arrays that did not exist in the Phase 1 dataset; altogether, 30 more stations are within 1 km distance from fault segments. In the Phase 2 dataset there are additional 150 new events and 15,000 new records augmenting the 650 events and 20,000 records of the Phase 1 dataset. Naturally, this has changed some of the statistical balances within the merged dataset, and with a combination of a very different source behavior (discussed later in the text), we had to modify our approach. The method applied in Phase 2 is described in Sect. 6—Results Phase 2.

5. Results Phase 1

We first examine the PGA dataset and by a combination of regression and residual analysis obtain a best-fit basic GMPE for PGA values. We then use the factor hierarchy found during the process of the PGA analysis to develop a complete set of GMPEs for the PGV and PSA values.

5.1. Stage 1—PGA Analysis

5.1.1 Basic Regression for Magnitude and Distance

Figure 6 shows results of the basic regression in four magnitude bins, obtained by solving Eq. (1) with the one stage algorithm of JOYNER and BOORE (1993), along with an uncertainty measurement of $\sigma_t = 0.968$. The uncertainty is quantified by the standard deviation values σ_t, σ_e and σ_r, where σ_t is the total uncertainty of all records given by the sum of variability within the events σ_e and random variability of the records σ_r. The value $\sigma_t = 0.968$, which is our basic uncertainty, serves as a reference for analysis of additional factors made in the following regression steps. The significant weight of the small

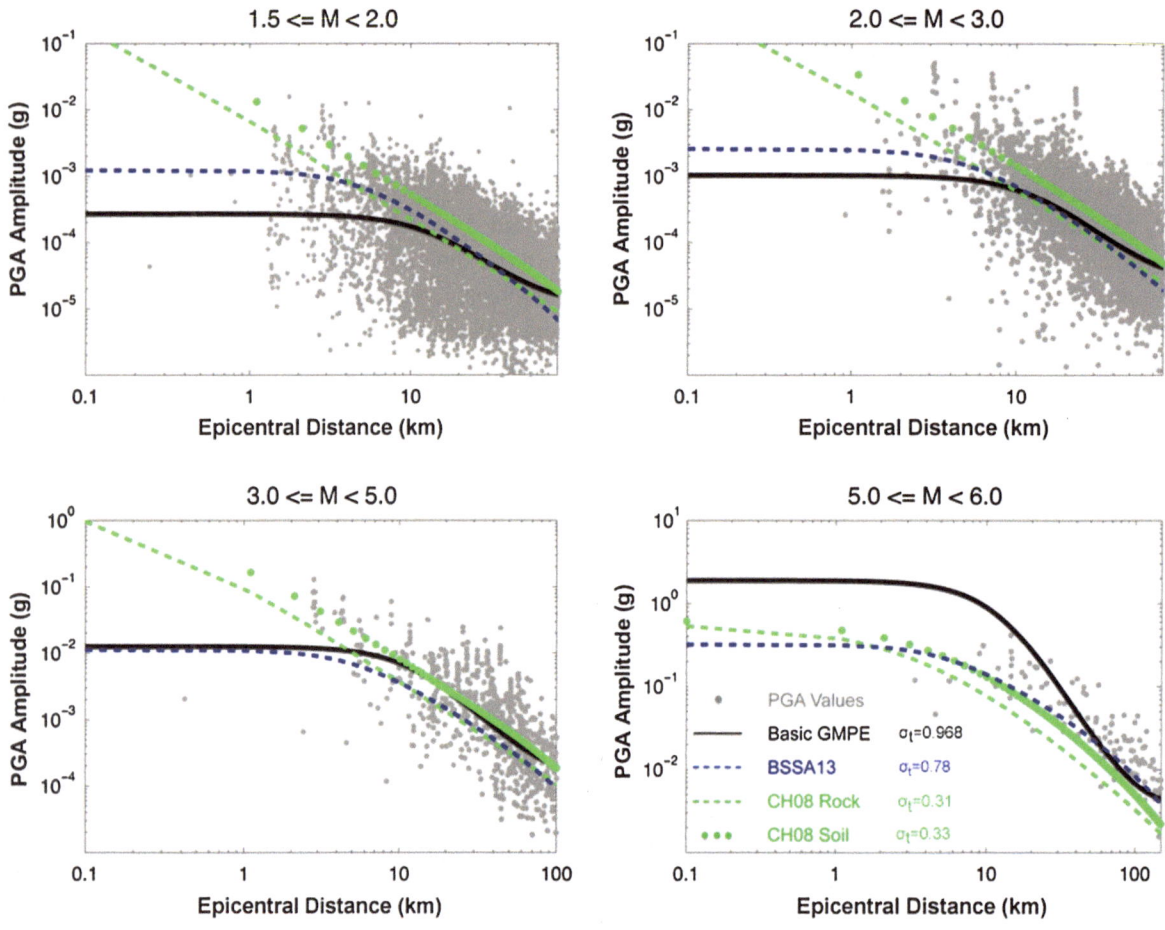

Figure 6
The Phase 1—*basic GMPE* for PGA values, and comparison to other models. The results are divided into four magnitude bins: 1.5–2, 2–3, 3–5, and 5–6. The different curves show the different models: the *black line* is our model, and the other three are models generated from other smaller datasets: BSSA13 is for Boore *et al.* for M 3–8 (2013), and CH08 is for Cua and Heaton (2008), for Rock and Soil sites

magnitude events, reflected in the strongly magnitude-dependent uncertainty, can also be seen in the up-bending of the GMPEs at the far distance (>80 km). This is due to the curvature coefficient, c_3, which is overweighed by the small magnitude events ($M < 3$) at distances beyond 80 km; this artifact is improved in our final Phase 2 GMPEs.

Figure 6 and Table 2 compare our results with GMPEs generated for other datasets in regions around our study area: Boore *et al.* (2013) denoted BSSA13 and Cua and Heaton (2008) denoted CH08. The results of BSSA13 were developed for the western US over a magnitude range of $3 \le M \le 8$. The results of CH08 are for Southern California over

a magnitude range of $2 \le M \le 7.3$, and have two solutions depending on V_{s30} value of site characteristics corresponding to soil and rocks, where V_{s30} denotes the average shear wave velocity in the top 30 m of the crust. The amount of events and records examined in our work, BSSA13 and CH08 are different, with total number of records increasing from CH08 to BSSA13 to our work (Table 2).

5.1.2 Factor Analysis

We consider three effects that can be significant for recorded ground motion in the SJFZ area. The first is site characteristics, mainly measured by V_{s30},

Table 2

Comparison between our basic model and other models with similar magnitude ranges

Models	Magnitudes	N_{events}	$N_{records}$	Frequency band	σ_t	h (km)
CH08	2–7.3	75	958 Rock 2,630 Soil	1 s window envelopes	0.31 Rock 0.33 Soil	3
BA08	5–8	60	1,800	Different per record	0.566	1.35
BSSA13	3–8	350	15,000	Different per record	0.78	4.9
Phase 1—basic GMPE	1.5–6	650	20,000	1–30 Hz	0.968	13.45
Phase 2—GMPE$_{Dir}$	1.5–6	800	30,000	1–30 Hz	0.924	3.54

BA is for BOORE and ATKINSON (2008); BSSA13 is for BOORE *et al.* (2013); and CH08 is for CUA and HEATON (2008). Our GMPEs show higher variability in amplitudes than the other models, probably due to the higher number of events and records, the extension to lower magnitudes (seen also when comparing BA08 to BSSA13) and possibly additional scattering effects once we get below M 2. Our Phase 2—GMPE considers rupture directivity as a first-order factor, as magnitude and distance

reflecting the soil conditions at the recording station. The second is fault zone amplification not captured by V_{s30}, reflecting path effects involving rock damage in a region around the fault. The third is the rupture directivity, exhibiting higher amplitudes for stations in the direction of rupture. The analysis is done by applying Eq. (2) for each of the above factors, as well as color-coding the records for each factor and looking for a clear color-signature.

5.1.2.1 Site Characteristics The classical characterization of site conditions is done via V_{s30}, with higher values corresponding to rock sites and lower ones related to soil (or highly weathered rock) sites. There are many ways of quantifying V_{s30}, either by measurements, or by spatial estimations based on geology or topography. For the stations in our study there are only a handful of actual V_{s30} measurements, so we test the following geological and topographical V_{s30} estimations of the region: (1) Geological estimation according to WILLS *et al.* (2000), (W00), (2) geological estimation according to WILLS and CLAHAN (2006), (WC06), (3) topographical slope estimation according to WALD and ALLEN (2007), (WA07) using (a) SRTM30 (Shuttle Radar Topography Mission) with 30 arc-sec resolution and (b) SRTM3 with 3 arc-sec resolution, and (4) a statistical compilation of measurements and estimations in Southern California (YONG *et al.* 2012).

Our analysis indicates that none of these estimations has a significant effect on the recorded amplitudes in our SJFZ data set. Figure 7 shows

results for estimates based on SRTM30 and YONG *et al.* (2012). The SRTM30 estimation does not produce a clear observable signal (Fig. 7a). The YONG *et al.* (2012) V_{s30} estimation has a weak observable signature, indicating, as expected, that rock sites tend to have higher amplitudes than soil sites (Fig. 7b). This signature is seen for the magnitude range of $2 \leq M \leq 4$ and is probably masked in the lower range by scattering effects. Applying Eq. (2) and regressing for both estimates, we find (Table 3) that the uncertainty reduction in both cases is very small: 0.01 % for the SRTM30 and under 0.2 % for the YONG *et al.* (2012) estimation. This is perhaps not very surprising, given that many of the sites are on rocks in an arid and mountainous terrain, where soil development is relatively poor.

5.1.2.2 Fault Zone Amplification Our use of a very dense network of stations around the SJFZ (Fig. 1) allows us to test for statistical changes in the ground motion amplitudes related to distance from the fault. Starting with the initial deployment of the NSF SJFZ CD project in September 2010, there has been a progressive significant increase in the number and density of near-fault stations. We, therefore, do not expect a clear effect prior to that time; specifically, the two $M_L > 5$ events in our database are not expected to show this effect. Indeed, Fig. 8 shows a clear signature of fault zone amplification in the $2 \leq M \leq 4$ range, but not in higher magnitudes. The lower range $M < 2$ might suffer from scattering effects; namely, the higher involved frequencies may

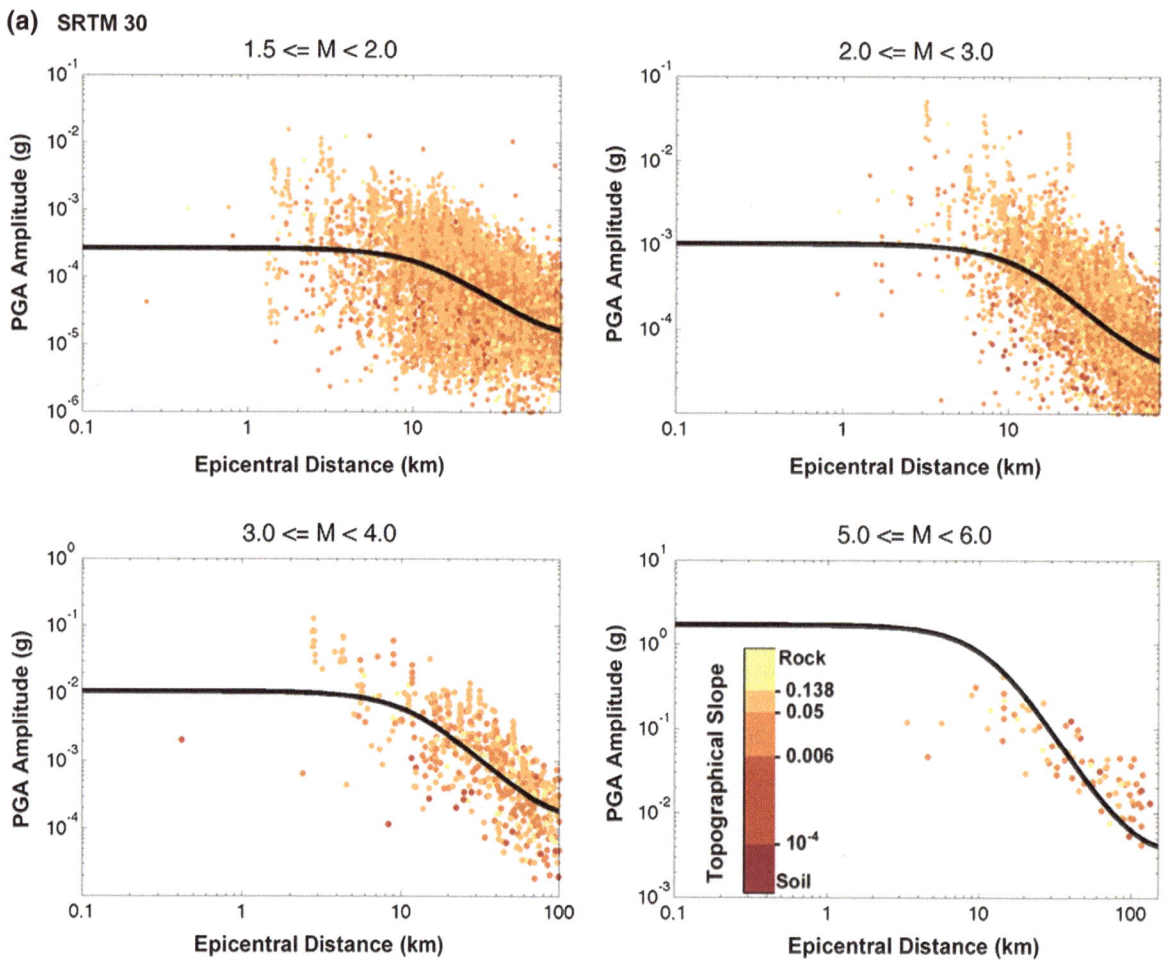

Figure 7
Site Characteristics. Two of the site characteristics models are presented: **a** topographical slope estimation using SRTM 30, and **b** Vs30 estimation according to a combined statistical approach (YONG *et al.* 2012). Color codes are in general: *yellow* for rock-sites and *brown* for soil-sites. The SRTM 30 estimation seems to have a very poor signature, while the YONG *et al.* (2012) solution has clearly an observable signature. Although observable, the actual uncertainty reduction after applying YONG *et al.* (2012) in Eq. (2) is only 0.2 %. The *color* codes for all the following attenuation curves will be located in the the $5 \leq M \leq 6$ magnitude bin

be masked by scattering which is typically in the same high frequency band. Table 3 shows that fault zone amplification is overall more effective than what was found for the V_{s30} estimates with an uncertainty reduction of 1 %.

5.1.2.3 Rupture Directivity We implement the directivity tool of Sect. 4 within the regression algorithm, assigning index of directivity I_{Dir} for each PGA record corresponding to each station orientation in each event. The color-coded values of I_{Dir} show (Fig. 9) a clear signature in the

complete range of magnitudes for many of the records satisfying the statistical requirements stated in Sect. 4. Table 3 indicates that the directivity index has the most significant effect among the three examined factors, producing a 7.4 % uncertainty reduction.

5.1.2.4 Hierarchical Ordering of Factors To separate and quantify different factors, we regress each factor in hierarchical order based on their separate significance. Since the site characteristic signature is very small we neglect it and perform sequential

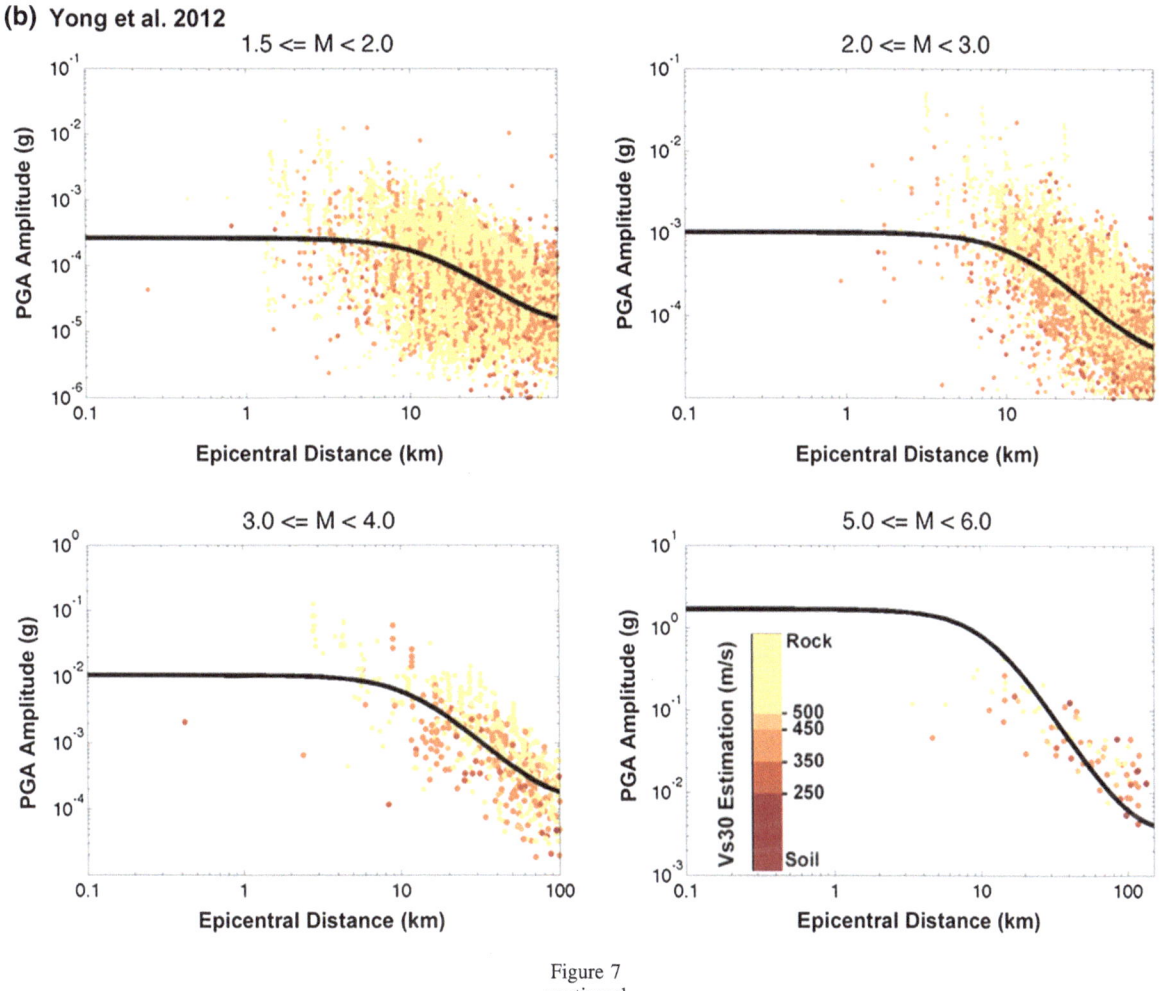

(b) Yong et al. 2012

Figure 7
continued

regressions first for directivity and then for fault zone amplification.

5.1.3 The Addition of Directivity to the GMPE

Figure 10 displays the original records, along with the basic GMPE and the values calculated by GMPE$_{Dir}$, providing the new GMPE including the directivity factor. Since for each record we assign an index of directivity, we obtain the same number of points as the original records with vertical spread reflecting how well the GMPE$_{Dir}$ values cover (or explain) the original data. A wide spread of the GMPE$_{Dir}$ values around the basic GMPE (within the original spread) means that the revised GMPE provides a significantly modified (i.e. better)

Table 3

Comparison between the different factors, controlling peak ground motions, examined in Phase 1

Regression	N_{Coeff}	σ_t	$\delta\sigma_t$ (%)
Basic	7	0.968	–
SRTM 30	10	0.9679	0.01
Yong *et al.* (2012)	10	0.966	0.2
Fault zone	10	0.958	1
Directivity	10	0.896	7.4
GMPE$_{Dir+FZBH}$	13	0.853	11.9

The basic regression accounts for magnitude and distance, the SRTM 30 and Yong *et al.* (2012) account for the addition of site characteristics into the regression, Fault zone considers the normal-to-fault distance of the stations, and directivity accounts for the addition of rupture directivity into the GMPEs. σ_t is the uncertainty quantified by the standard deviation, and $\delta\sigma_t$ is the reduction due to the addition of each of these factors to the basic regression. GMPE$_{Dir+FZBH}$ represents the final solution (see text for details)

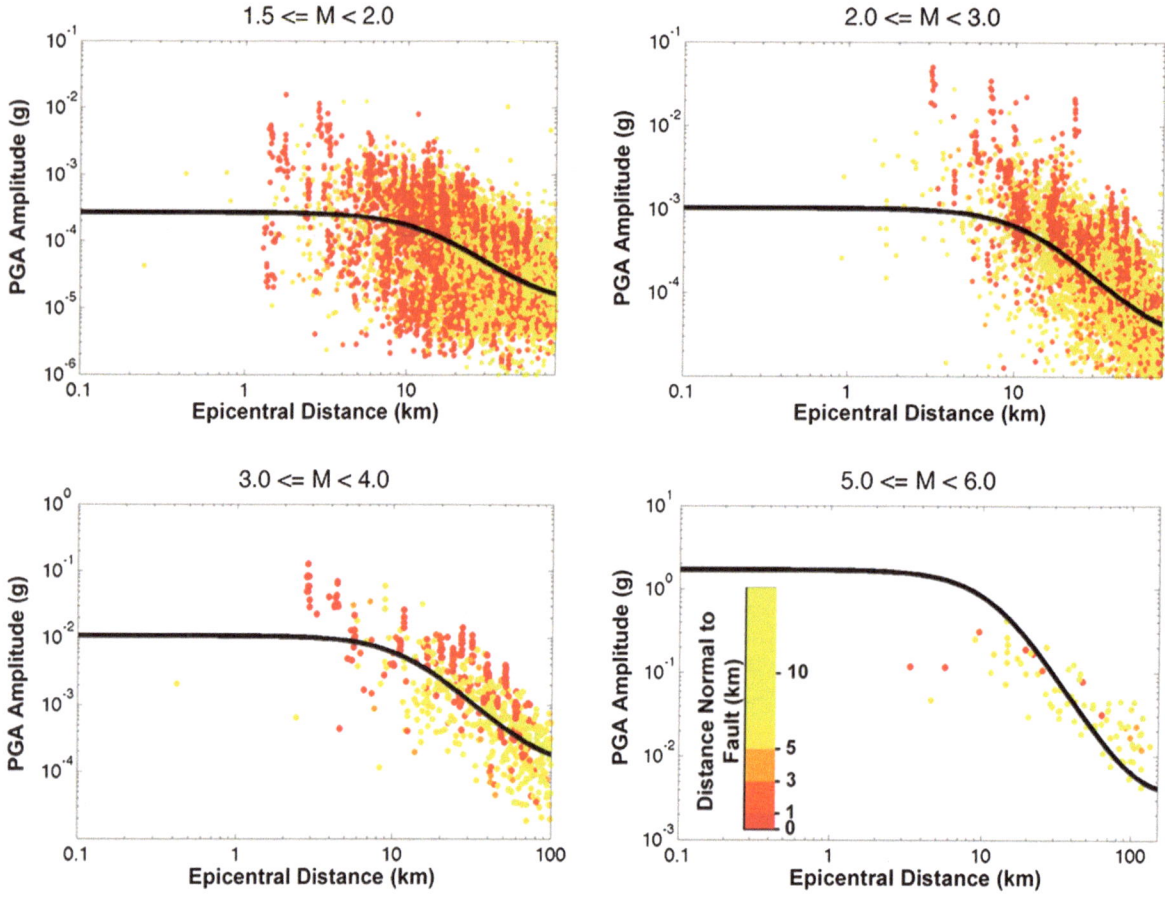

Figure 8

Fault Zone Amplification. The normal-to-fault distances are categorized and used for color-coding the observations with reference to the Phase 1—basic GMPE (*black line*). The end-member *colors* are: *red* for close stations (<1 km from the fault), and *yellow* indicates stations at distances larger than 10 km from the fault. The effect seen as higher amplitudes at closer stations (*red*), is quite significant for the recordings in the mid-magnitude range of *M* 2–4 events

explanation of the dataset, and, therefore, provides lower variance. As seen, the directivity can explain in our case almost an order of magnitude of the two orders of variability in the dataset, leaving the remaining variability to be explained by other factors.

5.1.4 Residual Analysis

To clarify effects of additional factors affecting the PGA values we perform a residual analysis. We define residuals as the ratio between the observed PGA and values provided by the $GMPE_{Dir}$: $res = PGA/GMPE_{Dir}$, and examine the residuals per

station in the same magnitude bins as in the previous sections. In each magnitude bin we identify three classes of stations: (1) those showing the whole range of residuals, (2) stations dominated by low residuals, and (3) stations dominated by large residuals. To simplify the pattern we display in Fig. 11 stations based on the three ranges of minimum and maximum residuals (res_{min} and res_{max}): (1) stations with $res_{min} \geq 0.5$ and $res_{max} \geq 2$ designated as amplifiers and marked in red, (2) stations with $res_{min} \leq 0.5$ and $res_{max} \leq 2$ designated as dampers and marked in blue, and (3) all the rest considered good-fit and marked in green. In the magnitude range $1.5 \leq M \leq 4$ there are stations showing a clear

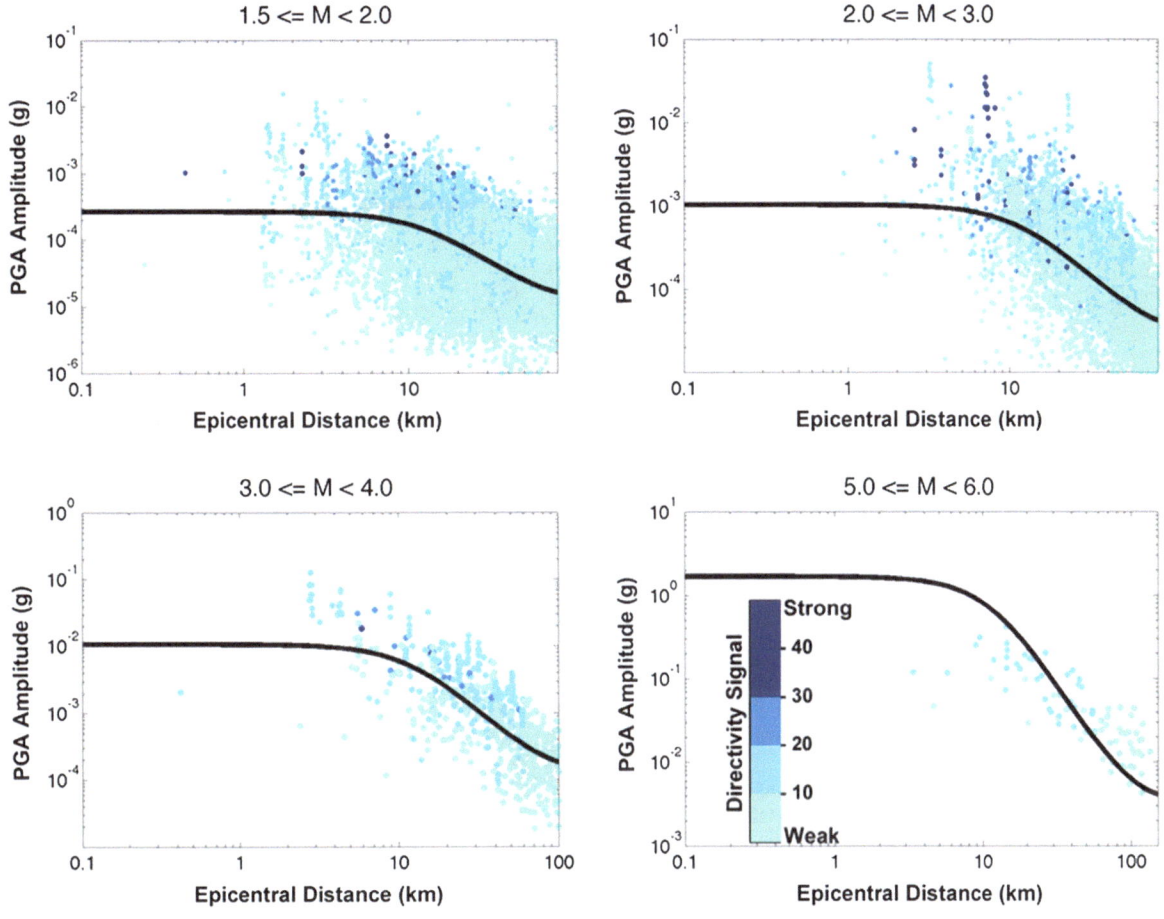

Figure 9

Rupture Directivity. Applying the directivity tool we color–coded the observations according to their index of directivity. The *dark blue* end-member represents recordings near the direction of rupture, that have strong directivity signals. The signature of this factor in the data could be seen in the magnitude range of *M* 1.5–4; higher amplitudes represent in many cases the direction of rupture. It is absent in the *M* 5–6 range, due to the weak signals (as in Case 1 in Fig. 3). We suspect that this is due to the fault size of large events and how it compares to the event-station distances, and/or due to non-linear amplification related to site characteristics, which becomes significant for larger events

tendency for being amplifiers, dampers, or good-fit (Fig. 11). The $M > 5$ bin has the smallest amount of data and probably largest statistical fluctuations reflecting detailed factors of individual events not examined here.

To highlight the general tendencies of stations, we calculated an average that considers all the residuals within the $1.5 \leq M \leq 4$ range where we have ample data (Fig. 12). The rule of thumb is that amplifying stations have at least one magnitude bin, showing amplification, and no magnitude bin showing damping, and the opposite rule defines the dampers. All the other options are considered good-fit, either because they are consistently green (good-fit, Fig. 11), or

because they show no clear tendency. This definition is done automatically in a quantitative manner, considering also the average values of residuals in each station. Examination of the amplifiers and dampers reveals two clear effects (Fig. 12). One involves eight dampers associated with borehole or posthole stations in a depth range of 10–200 m. This may be since these stations are located below the low shear-velocity surface layer ($\sim$5–10 m depth, e.g., LOUIE 2001; WU and HUANG 2013), in which the benefit of removing high-frequency surface noise is bundled with the drawback of losing some of the high amplitudes of high-frequency seismic signals, resulting in some cases with lower amplitudes (AL-SHUKRI

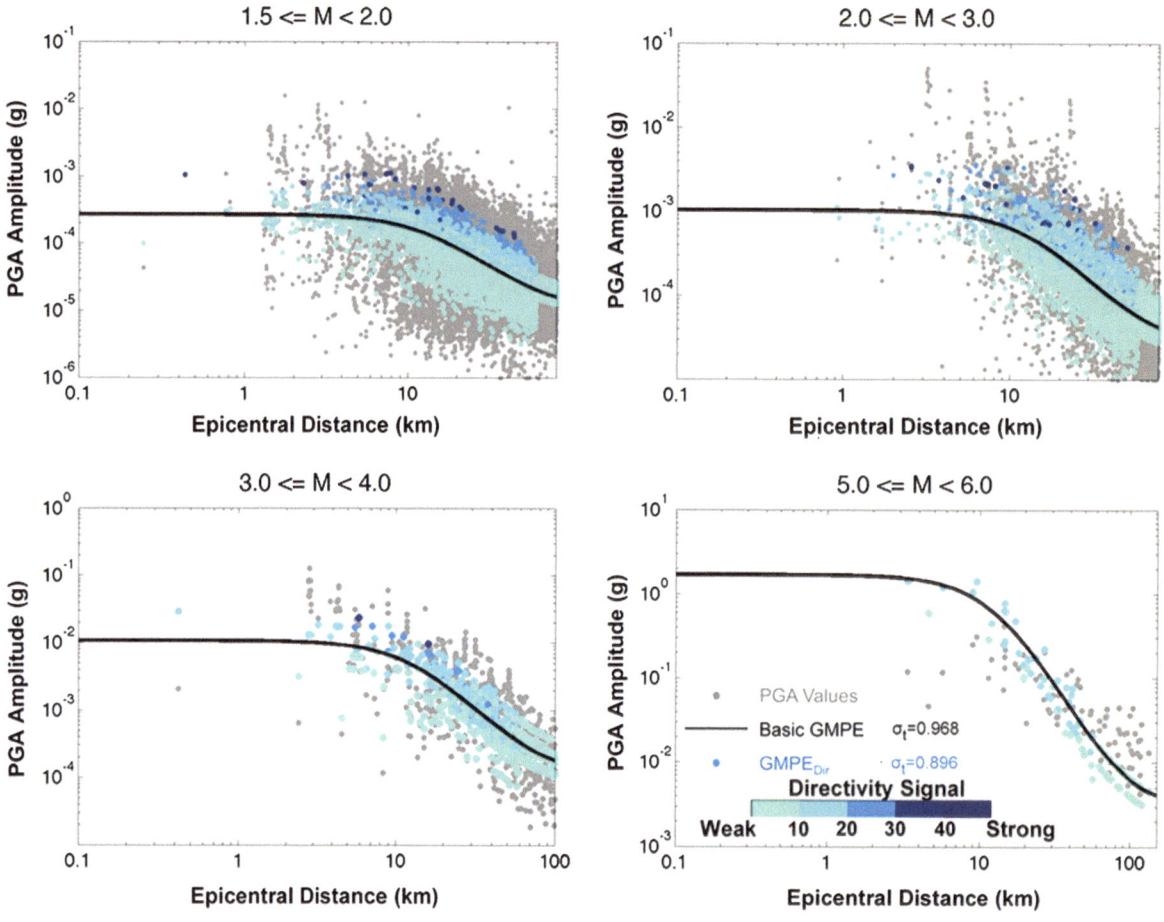

Figure 10

Phase 1—GMPE$_{Dir}$ plot. Here we show the original PGA observations (*grey dots*), the basic GMPE (*black line*), and the GMPE$_{Dir}$ accounting also for rupture directivity (*dark to light blue*). Note that the GMPE$_{Dir}$ is spread around the basic GMPE, since it is calculated for each event-station combination: narrow spreads would reflect small uncertainty reduction and wide spreads would reflect significant uncertainty reduction. A perfect GMPE would have a similar spread as the observations. In our case the uncertainty reduction is 7.4 %

et al. 1995; VERNON *et al.* 1998). The second category involves surface stations that are within 1 km normal-to-fault distances including the three linear arrays crossing the Anza/Clark segment. Almost all stations within that distance are amplifiers and probably reflects reduced elastic moduli associated with damaged fault zone rocks (e.g., BEN-ZION and SAMMIS 2003, and references therein). These two effects can be treated separately in the process of improving the GMPE, by assigning the following station indices: 100 for amplifiers, 0.1 for dampers and 1 for the rest. Using these station values in the regression process according to Eq. (2) leads to:

$$Ln(Y_3) = b_{11} + b_{12}Ln(X) + b_{13}Ln(Y_2), \quad (7)$$

where $Y_2 = $ GMPE$_{Dir}$, X is the vector of the assigned indices and Y_3 is the resulting GMPE. Figure 13 presents the results where the original PGA values are plotted together with the basic and new GMPE, GMPE$_{Dir+FZBH}$, accounting for the directivity effect, the borehole damping (BH) and fault zone amplification (FZ). The color code corresponds to the amplifiers, dampers and good-fit stations. The uncertainty is reduced (Table 3) from $\sigma_t = 0.896$ for the GMPE$_{Dir}$ solution to $\sigma_t = 0.852$ for this new solution GMPE$_{Dir+FZBH}$, reflecting additional uncertainty reduction of 4.9 % and a total reduction of 11.9 % relative to the basic GMPE. This is the final result based on the uncorrected database used in our first stage analysis.

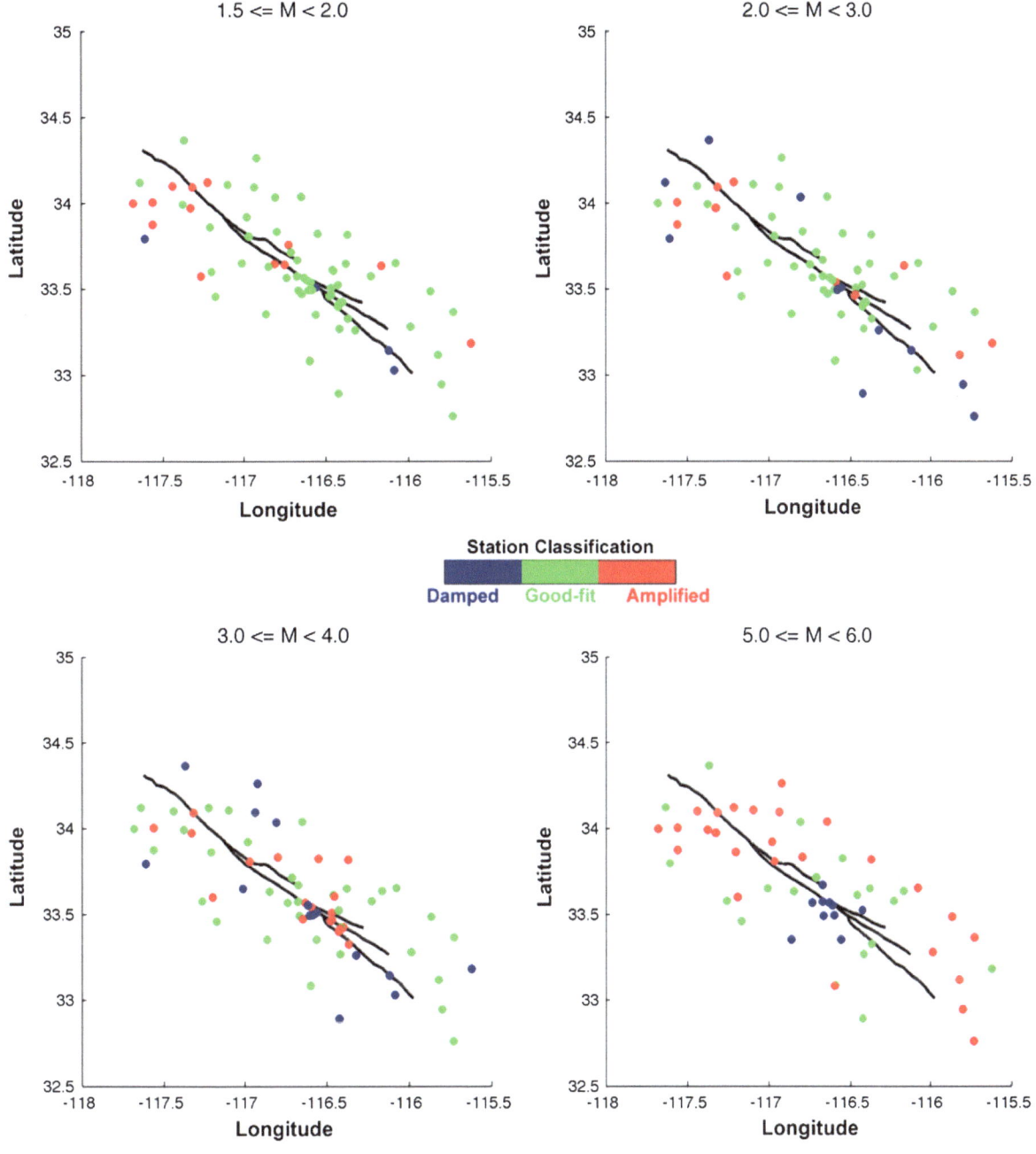

Figure 11

Station classification according to residuals. This step is done still in magnitude range, since we wanted to examine whether we see residuals relatively insensitive to *M. Red* are amplified stations, *blue* are damped and green show good-fit between observations and GMPE$_{Dir}$. It is evident that in the magnitude range of *M* 1.5–4 station classification tend to show consistency. We use the *M* 1.5–4 range to average the residuals and classify the stations for that magnitude range

5.2. Stage 2—PGV and 5 % Damped PSA Analysis

The procedure used for the PGA analysis has the following steps: (1) regression for magnitude and distance, (2) regression for rupture directivity according to the PGA analysis, (3) additional regression for borehole damping and fault zone

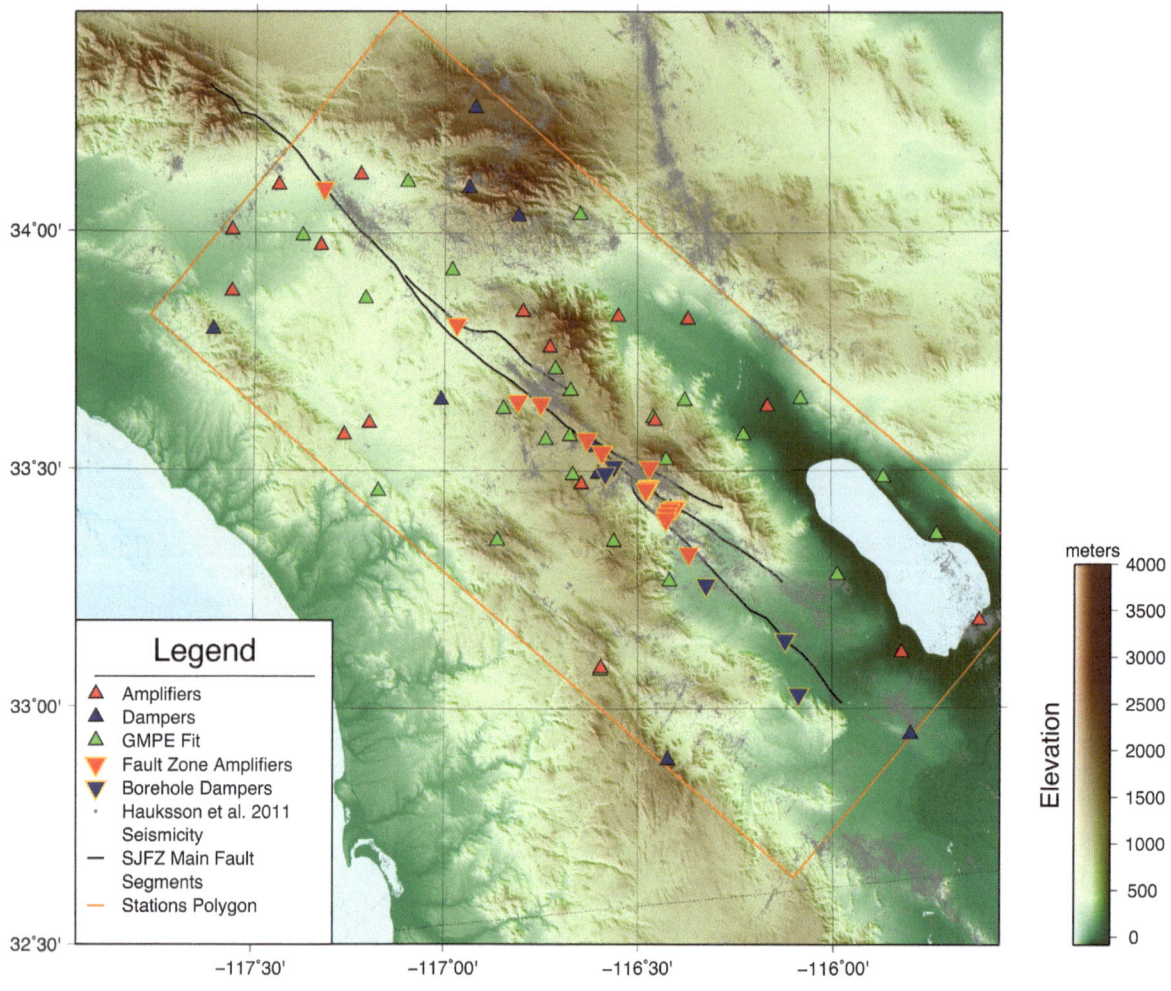

Figure 12
Amplifiers and Dampers. Here we show the station classification following the residual analysis (see text for further details). *Red triangles* are for stations that tend to amplify, *blue triangles* are for stations that damp the signals, and *green triangles* show good-fit between observations and GMPE$_{Dir}$. A closer look would reveal that many of the stations close to the fault segments are amplifiers (*red inverted triangles*). The *blue* dampers close to the fault are borehole or posthole sites buried at least 10 m below the surface (*blue inverted triangles*)

amplification according to the amplifiers and dampers. Applying the same procedure to the PGV and PSA datasets we obtain a full set of GMPEs presented in Tables 4, 5, 6. The different GMPEs presented in these tables have several interesting features outlined below.

There is a general trend of σ_t, the uncertainty of the basic regression, showing higher values for shorter periods, corresponding to noise coming from higher frequencies. In addition, both the apparent depth, h_A, and the uncertainty, σ_t, are showing similarity to the PGA values, between 0.1 and 0.4 s

(~ 2.5–10 Hz), reflecting the frequency-band that dominates the PGA values. The uncertainty reduction due the addition of directivity, borehole damping and fault zone amplification reveals an interesting effect. The PGA and shorter period PSA show high $\delta\sigma_t$ in relation to the directivity (Table 5) and lower $\delta\sigma_t$ in relation to the borehole damping and fault zone amplification (Table 6). For longer periods the tendency flips, with lower values of $\delta\sigma_t$ in relation to directivity than in relation to BHFZ. This may be explained partially by the fact that the higher frequency range works the best for the directivity

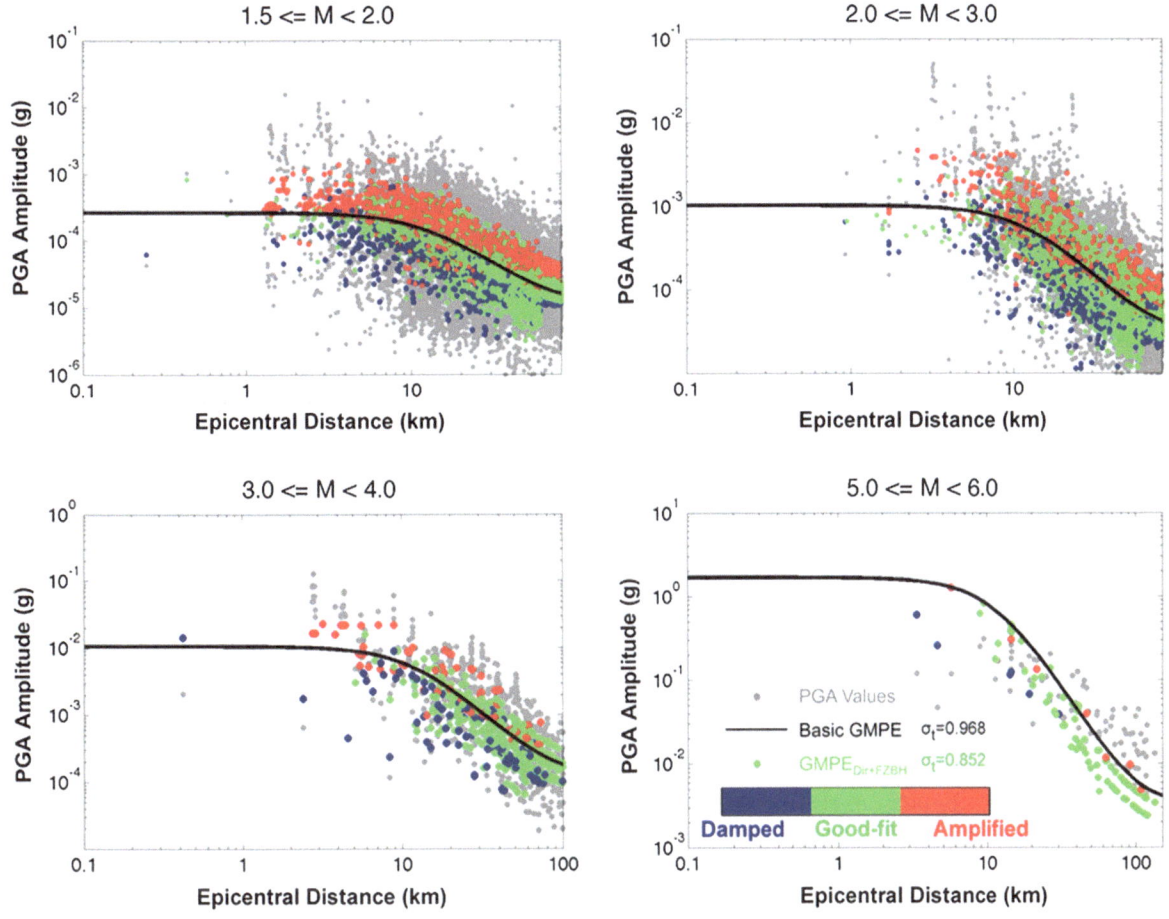

Figure 13
The Phase 1—GMPE$_{Dir+BHFZ}$. Here we show the observations (*grey dots*), the basic GMPE (*black line*) and the GMPE$_{Dir+BHFZ}$ color-coded according to the station classification. The overall improvement of this model is reflected by the 11.9 % uncertainty reduction and the wider spread of the predicted values

tool (which might be also due to the large portion of small events), and the dominance of fault zone amplification for lower frequencies may reflect trapped waves and other phases in fault damage zones (e.g. BEN-ZION and AKI 1990; LI *et al.* 2002; BEN-ZION *et al.* 2003).

6. Results Phase 2

The aftershock sequence of the M_L 5.1 from March 2013 was recorded by a denser network than the previous events, adding ~40 stations to the network, with 20 being part of two linear arrays across

the San Jacinto Fault. The initial merging of the data failed (did not converge) the basic regression for magnitude and distance (Eq. 1) used in the first phase, and required a different approach. Several subsets excluding stations and/or events were attempted and led to large uncertainties ($\sigma_t \sim 1.2$) when regression succeeded. The initial efforts indicated that there is something different about this new dataset, and that the additional fault zone stations might also contribute to the diversity of results. This suggested that there is an additional factor (or factors) that should be treated as a first-order factor like magnitude and distance. Both the directivity and fault zone amplification were examined, using a similar equation as

Table 4

Distance (c₁, c₂, c₃) and magnitude (e₁, e₂, e₃) coefficients of the complete set of Phase 1—GMPEs, for PGA, PGV and 5 % damped PSA

	e_1	e_2	e_3	c_1	c_2	c_3	h_A	σ_t
PGA	0.210	3.220	0.036	−2.840	−0.361	0.021	13.452	0.968
0.03 s	4.452	3.404	0.044	−3.973	−0.446	0.033	18.886	0.982
0.05 s	3.011	3.517	0.008	−3.430	−0.467	0.026	17.264	0.987
0.075 s	3.652	3.804	−0.040	−3.630	−0.503	0.031	18.645	0.969
0.1 s	0.704	3.546	−0.091	−2.746	−0.402	0.021	14.447	0.934
0.15 s	−1.321	3.483	−0.122	−2.246	−0.339	0.017	12.014	0.941
0.2 s	−1.691	3.608	−0.119	−2.293	−0.344	0.019	12.077	0.943
0.25 s	−1.639	3.728	−0.098	−2.467	−0.360	0.023	12.553	0.955
0.3 s	−1.465	3.818	−0.095	−2.641	−0.372	0.026	13.364	0.958
0.4 s	−1.880	3.825	−0.078	−2.690	−0.359	0.027	13.860	0.943
0.5 s	−1.498	3.911	−0.045	−2.980	−0.375	0.032	15.537	0.926
0.75 s	−0.209	4.005	0.009	−3.655	−0.407	0.041	19.401	0.909
1 s	1.247	3.364	−0.015	−3.129	−0.385	0.024	16.376	0.922
PGV	0.997	2.836	−0.049	−2.081	−0.166	0.013	10.405	0.928

h_A is the apparent depth calculated by the regression, and σ_t is the uncertainty quantified by the standard deviation

Table 5

Directivity (b₁, b₂, b₃) coefficients of the complete set of Phase 1— GMPEs, for PGA, PGV, and 5 % damped PSA

	b_1	b_2	b_3	σ_t	$\delta\sigma_t$ (%)
PGA	−2.558	0.912	0.952	0.896	7.4
0.03 s	−2.482	0.918	0.961	0.920	6.26
0.05 s	−2.472	0.913	0.959	0.925	6.24
0.075 s	−2.329	0.850	0.959	0.914	5.69
0.1 s	−2.071	0.748	0.962	0.889	4.79
0.15 s	−1.783	0.633	0.966	0.910	3.32
0.2 s	−1.569	0.549	0.970	0.920	2.49
0.25 s	−1.519	0.522	0.971	0.934	2.19
0.3 s	−1.485	0.507	0.972	0.938	2.07
0.4 s	−1.385	0.466	0.974	0.926	1.79
0.5 s	−1.388	0.463	0.975	0.909	1.83
0.75 s	−1.459	0.483	0.975	0.890	2.03
1 s	−6.694	0.506	0.789	0.938	−1.81
PGV	−2.330	0.918	0.969	0.873	5.9

σ_t is the uncertainty after the addition of the directivity into the GMPEs, and $\delta\sigma_t$ is the corresponding uncertainty change. While for the short period range there is a reduction in the uncertainty, for increasing periods this reduction diminishes, indicating that the effect of directivity is dominant for the higher frequency range

Table 6

Borehole Damping and Fault Zone Amplification (a₁, a₂, a₃) coefficients of the complete set of Phase 1—GMPEs, for PGA, PGV, and 5 % damped PSA

	a_1	a_2	a_3	σ_t	$\delta\sigma_t$ (%) to Basic	$\delta\sigma_t$ (%) to Directivity
PGA	−0.538	0.130	0.957	0.852	11.9	4.9
0.03 s	−0.718	0.346	0.976	0.886	9.80	3.78
0.05 s	−0.738	0.359	0.975	0.887	10.06	4.07
0.075 s	−0.719	0.350	0.976	0.878	9.47	4.01
0.1 s	−0.697	0.336	0.976	0.854	8.54	3.93
0.15 s	−0.700	0.334	0.977	0.876	6.88	3.69
0.2 s	−0.678	0.324	0.979	0.888	5.80	3.39
0.25 s	−0.670	0.319	0.980	0.904	5.29	3.17
0.3 s	−0.647	0.308	0.981	0.911	4.94	2.93
0.4 s	−0.648	0.307	0.982	0.899	4.73	2.99
0.5 s	−0.652	0.308	0.982	0.881	4.89	3.12
0.75 s	−0.653	0.302	0.983	0.863	5.08	3.11
1 s	−0.878	0.298	0.966	0.913	0.96	2.73
PGV	−0.329	0.108	0.966	0.843	9.1	3.4

σ_t is the corresponding uncertainty, and $\delta\sigma_t$ is the uncertainty change in relation to the basic GMPE (Table 4) and to GMPE$_{Dir}$ (Table 5)

Eq. (1), only with an additional term, $d_1 Ln(X)$, for the contribution of the examined factor. While the fault zone amplification term failed in the regression, the directivity proved very useful, providing new GMPEs with reduced variance for the different subsets examined. The new formulation used for the regression is:

$$Ln(Y_1) = e_1 + e_2(M - M^h) + e_3(M - M_h)^2 + [c_1 + c_2(M - M_{ref})] \cdot Ln\left[\sqrt{R_e^2 + h_A^2}\Big/ R_{ref}\right] + c_3\left[\sqrt{R_e^2 + h_A^2}\Big/ R_{ref}\right] + d_1 Ln(I_{Dir}), \tag{8}$$

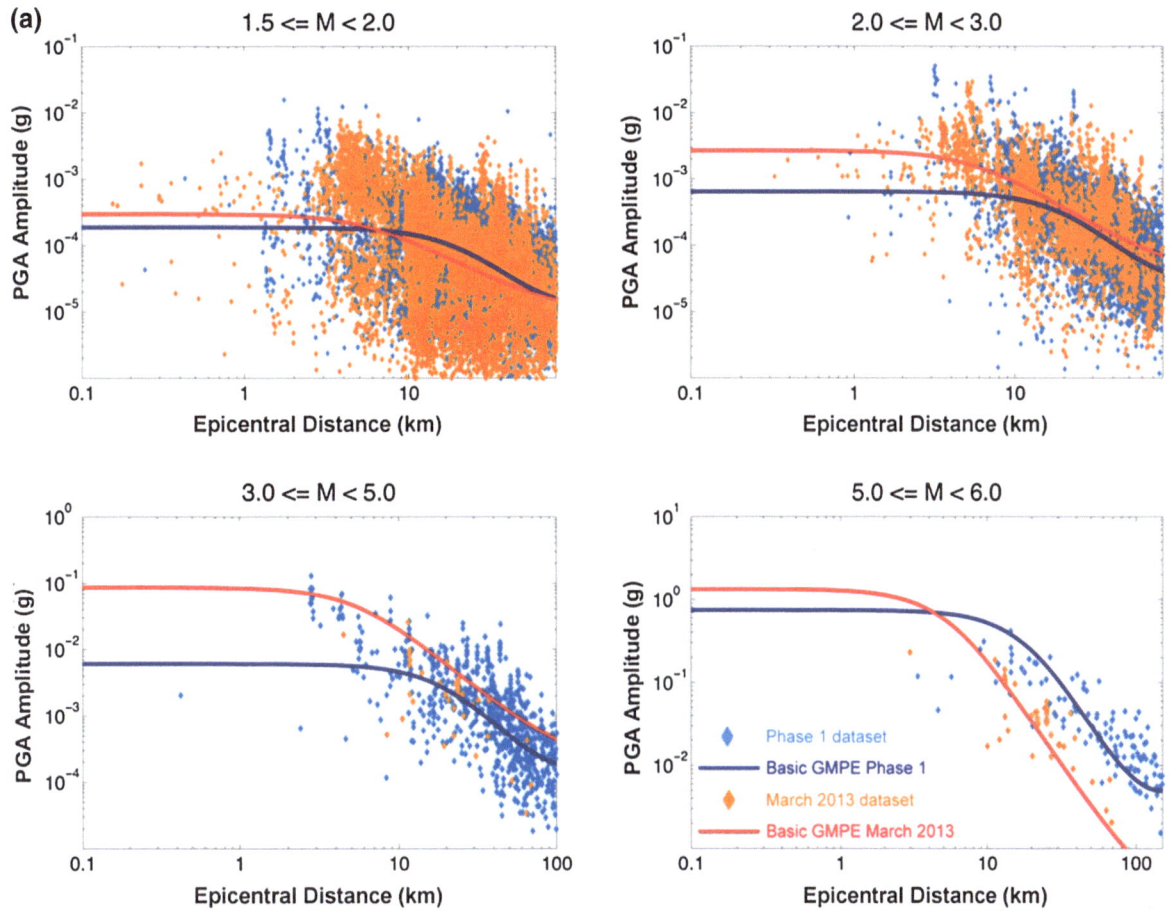

Figure 14
Phase 1 vs. March 2013 datasets: blue for the Phase 1 dataset, and orange-red for the March 2013 dataset. **a** observations (*diamonds*) and the basic GMPE (*curves*) calculated for both datasets; **b** the basic GMPE (*curves*) and the GMPE$_{Dir}$ (*diamonds*) calculated for both datasets. The March 2013 dataset shows a clear tendency for higher amplitudes and amplified GMPE curves in comparison with the Phase 1 dataset (see text for further details). The lower curve seen for the M_L 5.1 of March 2013, is probably due its significant lower magnitude in comparison with the M_L 5.6 of June 2005 and M_L 5.9 of July 2010

where d_1 is the coefficient for the directivity term, and generates GMPE$_{Dir}$ solutions as in Phase 1. We can reverse now and replace the last term in Eq. (8) with $d_1 Ln(\tilde{I}_{Dir})$, in which $\tilde{I}_{Dir}$ is the average of the I_{Dir} values, providing a discreet basic GMPE as was generated in the initial stage of Phase 1.

Figure 14 shows a comparison between two datasets, the one used for Phase 1 (blue), and the additional dataset of March 2013 (orange–red). Figure 14a presents the PGA records together with their corresponding basic GMPEs generated by Eq. (8), replacing the last term by $d_1 Ln(\tilde{I}_{Dir})$, and Fig. 14b shows the basic GMPEs together with their corresponding GMPE$_{Dir}$. Although the spread

of PGA values is overlapping for both datasets, there is a larger variance for the March 2013 dataset, seen in the lower magnitude range, especially $1.5 \leq M \leq 2$. The basic GMPE for March 2013 shows higher-amplitude attenuation curve than the ones produced for the Phase 1 dataset, especially in the closer epicentral distance Re < 10 km. The GMPE$_{Dir}$ seen in Fig. 14b accounts for magnitude, distance and rupture directivity, explaining a significant proportion of the variance of the original PGA records: (a) for the dataset of March 2013— beginning with no conversion when using Eq. (1), through $\sigma_t = 1.2$ for pre-averaging of the arrays, to $\sigma_t = 1.044$ when using Eq. (8), and (b) for the

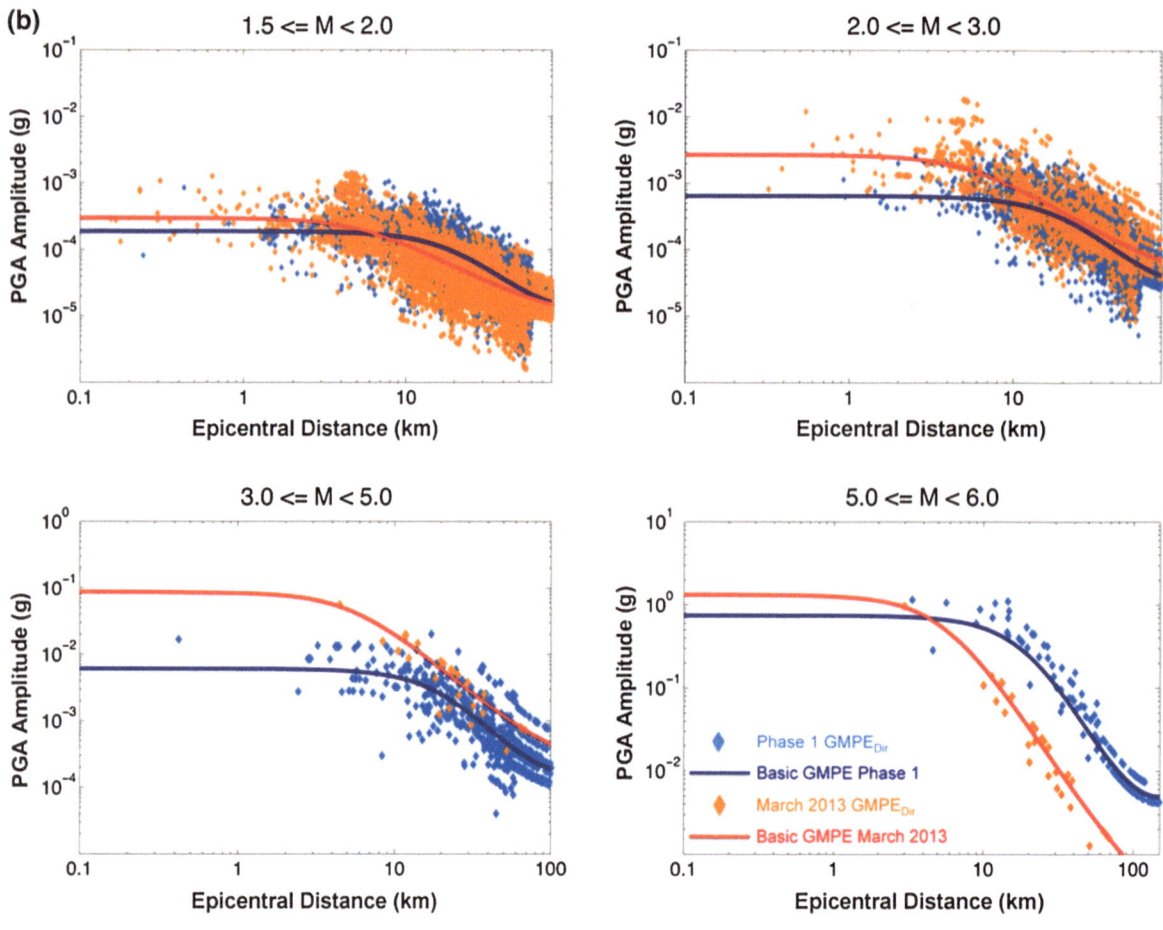

Figure 14
continued

Phase 1 dataset—from $\sigma_t = 0.968$ when using Eq. (1) to $\sigma_t = 0.894$ when using Eq. (8), which is similar to the $\sigma_t = 0.896$ of the Phase 1—GMPE$_{Dir}$ (see Table 7).

Merging the March 2013 with the Phase 1 dataset, we have taken one more measure to improve our GMPEs, also for treating the c_3 problem discussed above, setting epicentral distances cutoff for different ranges of magnitudes. We have set the epicentral distances for the new merged dataset (Phase 2) to be $M < 3$ up to 80 km, $3 \leq M < 5$ up to 100 km, and $5 \leq M < 6$ up to 150 km (Fig. 15a). These settings decrease the weight of small events in large distances, allowing the larger distances to be dominated by the larger events, and reducing some of the noise affecting the generation of the GMPEs. Figure 15b shows a comparison between several different GMPEs for the PGA records of Phase 2, in relation to

the Phase 1—basic GMPE represented by a dashed black line. The up-bending, c_3 related, of the GMPEs at the far distance (>80 km), seen in the Phase 1—basic GMPE, disappears in the Phase 2 solution. The GMPE$_{Dir}$ spread (blue color scale) is similar to what was seen in phase 1 with a $\sigma_t = 0.92$ (slightly higher than the Phase 1—GMPE$_{Dir}$ solution) and the bold black curve is the basic GMPE derived from replacing the last term in Eq. (8) with $d_1 Ln(\tilde{I}_{Dir})$. We added another solution here, basic GMPE$_{HighDir}$ (blue curve), obtained by considering only the GMPE$_{Dir}$ values larger than $\tilde{I}_{Dir}$ and averaging them. This provides an amplified GMPE that should be applied for sites that are in the direction of rupture. For hazard assessment this kind of prediction might be more meaningful than the basic GMPE, or the GMPE$_{Dir}$. In fact, it does something similar to the directivity predictive models implemented in GMPEs

Table 7

Regression results for the Phase 1 and Phase 2 datasets, and some of the subsets examined during the process of developing the Phase 2—GMPEs

Datasets	Equation	N_{Coeff}	$N_{records}$	N_{events}	σ_t
Phase 1—basic GMPE	1	7	19,993	655	0.968
Phase 1—GMPE$_{Dir}$	1 + 2	10	19,993	655	0.896
Phase 1—GMPE$_{Dir}$	8	8	19,993	655	0.894
March 2013, averaged arrays	1	7	7,737	154	1.199
March 2013	8	8	12,874	154	1.044
March 2013, old Stations	8	8	8,886	154	1.035
Phase 2—GMPE$_{Dir}$	**8**	**8**	**29,474**	**809**	**0.924**
Phase 2— GMPE$_{HighDir}$	**8**	**8**	**18,253**	**809**	**0.985**
June 2005 July 2010	8	8	5,525	171	0.752
Main shocks	8	8	14,867	487	0.926

The table shows the equation used for the regression, the number of derived coefficients, N_{Coeff}, the number of records and events used per regression, $N_{records}$ and N_{events}, and the total uncertainty, σ_t. The March 2013 dataset showed a significant improvement only when the directivity factor was treated as a first-order factor, just as magnitude and distance, using Eq. (8). The Phase 2 results (bold) are the final outcome of merging the Phase 1 and March 2013 datasets (see text for further details)

(e.g., SPUDICH *et al.* 2012 and references therein), providing an upper attenuation curve estimation for regions within the direction of rupture.

Calculating the residuals of the Phase 2—GMPE$_{Dir}$ in relation to the observed PGA records, we examined whether we see special tendencies in station or events distribution. An initially surprising result was that we could not see any similar station distribution, indicating Fault Zone amplification, as we have seen in the Phase 1 residual analysis (Fig. 12). This may stem from two possible reasons. First, the new GMPE curves are amplified in relation to the Phase 1—GMPEs, partly due to the increasing weight of stations and records within the Fault Zone, so that the GMPE already includes (at least partially) fault zone amplification. Second, there is something different in a significant portion of the March 2013 ground motion records, providing PGA values with a larger spread per magnitude-distance, so that any Phase 1—tendencies are masked by the new records; this is suggested by the large uncertainty of the March 2013 data, which hardly changed, even for different subsets of stations (Table 7).

Figure 16 presents a residual analysis for the events, where the variance of the residuals per event, res$_{var}$, is plotted; large res$_{var}$ reflects large variance of the residuals and is associated with amplified events, and small res$_{var}$ reflects small variance of the residuals and is associated with damped events. We use a color scale in which $0.5 \leq res_{var} \leq 2$ is a reasonable value per event (green), and beyond these limits res$_{var}$ might indicate a spatial or source signature. Since we are interested in explaining some of the difficulties we had regressing the March 2013 dataset and understanding the large uncertainty associated with it, we examine mainly the large res$_{var}$ (orange to red) values. If a group of events are clustered in space and/or time and show a clear tendency in their res$_{var}$ values, this would indicate that something about the source mechanisms or source region is dominating the amplitudes recorded for the event. Figure 16a shows the complete Phase 2—residual analysis in all magnitude bins. The higher range of magnitudes, $3 \leq M < 6$, seem to have small res$_{var}$ values, especially the M_L 5.1 of March 2013 (blue), so we should examine closer the lower magnitude range of $1.5 \leq M < 3$. Fig. 16b, e compare results for three subsets: (1) the main shocks, which are events with no clear aftershock sequences within the dataset, (2) the June 2005 and July 2010 aftershock sequences, and (3) the March 2013 aftershock sequence. The main shocks in Fig. 16b have low to high res$_{var}$ values, in which possible spatial tendencies could be identified, such as the low res$_{var}$ in the SE of the fault, and in two clusters: to the NW of the Trifurcation, and to the SE of the Hot Springs cluster. The June 2005 and July 2010 (Fig. 16c) are mainly dominated by mid-range res$_{var}$ values, which are also reflected by the corresponding low value of uncertainty (Table 7). The March 2013 dataset (Fig. 16d) displays the entire range of res$_{var}$, but events in the main cluster in the close vicinity of the M_L 5.1 have relatively high values of res$_{var}$ (orange to red). This indicates that the main source region and/or mechanisms are dominating the large res$_{var}$ values and the large uncertainty seen for the March 2013 dataset. Even after considering only the stations that existed also in the Phase 1 dataset (Fig. 16e), there is still a clear signature of large res$_{var}$ values, which is especially pronounced for the $1.5 \leq M < 2$ range. Removing all the new stations out

(a)
Magnitude
Epicentral Distance (km)
Number of PGA values
Depth (km)
(b)
1.5 <= M < 2.0
2.0 <= M < 3.0
3.0 <= M < 5.0
5.0 <= M < 6.0
PGA Amplitude (g)
Epicentral Distance (km)
Phase 1 - Records
Phase 1 - Basic GMPE
Phase 2 - Basic GMPE
Phase 2 - Basic GMPE$_{HighDir}$
Phase 2 - GMPE$_{Dir}$
Directivity Signal
Weak 10 20 30 40 Strong

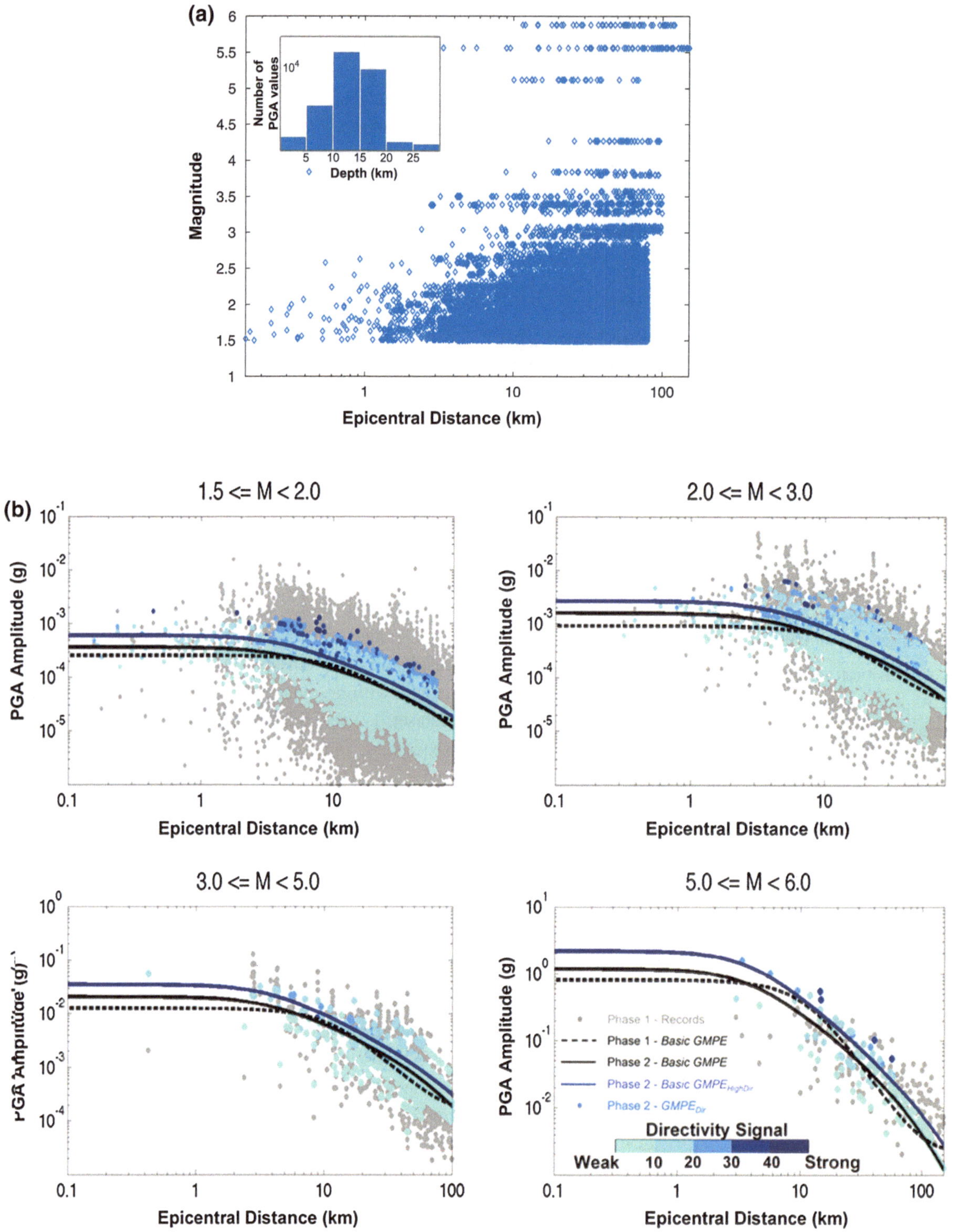

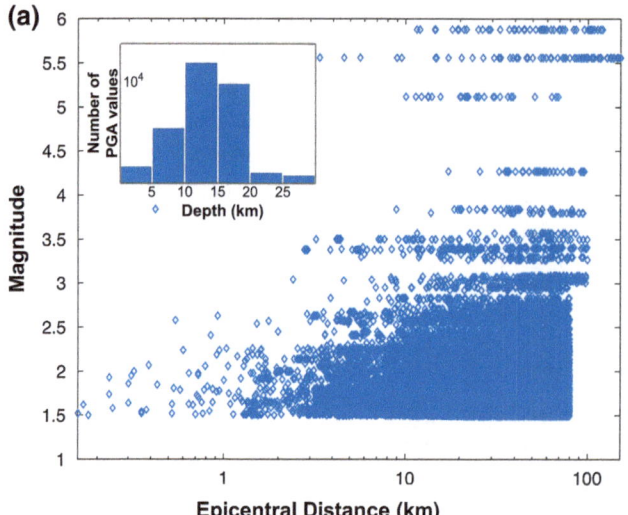

Figure 15
Phase 2—GMPEs. **a** Magnitude-distance distribution after merging the Phase 1 and March 2013 datasets, and selecting records according to epicentral distance-magnitude categories: *M* 1.5–3 up to 80 km, *M* 3–5 up to 100 km, and *M* 5–6 up to 150 km. **b** Phase 2 to Phase 1 GMPEs comparison. Here we show the observations in grey, the Phase 2—GMPE$_{Dir}$ in the blue scale, the Phase 1—basic GMPE (*dashed black curve*), and the Phase 2—GMPE$_{Dir}$ (the *bold black curve*). The Phase 2 shows slightly higher curves, and has better fit to the complete Phase 2 dataset. In addition, the *blue curve* represents the GMPE$_{HighDir}$ appropriate for the direction of rupture, and showing higher estimation of attenuation curves (see text for more details)

of the March 2013 dataset leads to uncertainty reduction by less than 1 % (from $\sigma_t = 1.044$ to $\sigma_t = 1.035$).

7. Discussion

The two phases of this research illuminate different aspects of the development of GMPEs in a local system, and what considerations should be taken into account for the main factors controlling ground motions. The first phase emphasized the roles of rupture directivity, fault zone amplification, posthole damping and site characteristics, in determining the amplitude of ground motions. The second phase showed the effects of (a) including different source mechanisms and/or source regions, and (b) shifting the station balance towards including more data from the fault zone. Both phases were essential for the insights presented in this paper discussed further below.

7.1. Removing the Vs30 for the SJFZ

One of the interesting outcomes of this study is the insignificant role of site characteristics for GMPEs in the SJFZ, unlike many other studies (e.g., CUA and HEATON 2008). Although we have shown that site-characteristics have a signature within the data (Fig. 7b), this signature did not translate into a significant factor affecting a large proportion of the data. Furthermore, it could be seen (Fig. 7b) that the majority of records were classified as rock sites according to YONG *et al.* (2012), which might explain the small effect the minority of soil sites had on the overall data. This is not surprising since most of the sites of the Anza and SJFZ project stations were

installed in granite or weathered granite (SHARP 1967) that, while being softer, seems to have similar site characteristics as un-weathered granite. For these reasons we have chosen to remove this factor from our analysis, although it should be considered in a study including minor effects. Perhaps for $M > 5$ events, some of the soil sites and maybe even the weathered granite would show non-linear amplifications as seen in other works (e.g., BOORE and ATKINSON 2008). However, for that range we do not have enough data for analysis.

7.2. Directivity of Rupture

Besides the ability of the directivity tool to identify the rupture directivity (Fig. 5), we have seen that implementing the tool within the GMPE regression scheme reduces the uncertainty of the peak ground motion values. This notion, derived initially through the Phase 1 analysis, is implemented in Phase 2, allowing the regression to generate successfully the GMPEs (which without this factor was failing to converge). In addition, as seen in the Phase 2 analysis, the directivity factor is significant enough in the SJFZ to be considered a first-order factor, similar to magnitude and distance.

Deriving the basic GMPE$_{HighDir}$ for the direction of rupture (see Fig. 15b) provides a possibly more significant attenuation curve for the use of engineers, estimating the median GMPE that would apply if one considers the worst directivity case. This curve, which is higher by a factor of 2 from the Phase 2—basic GMPE, has a similar effect as the directivity models implemented in the NGA-West 2 GMPEs (e.g., SPUDICH *et al.* 2012, and references therein), estimating the upper shift of the attenuation curves in the direction of rupture. The advantage in our model is that we do not assume anything about the source; the directivity tool reveals the nature of the seismic events directly from the peak ground motion values. This was shown to be effective for $M < 5$; for the small number of high-magnitude events in the dataset, the directivity tool shows weak directivity signals, which might be a result of masking by other factors (such as non-linear site response or a dominant radiation pattern). For $M > 5$ events, the source-model-based approach (e.g., SPUDICH and

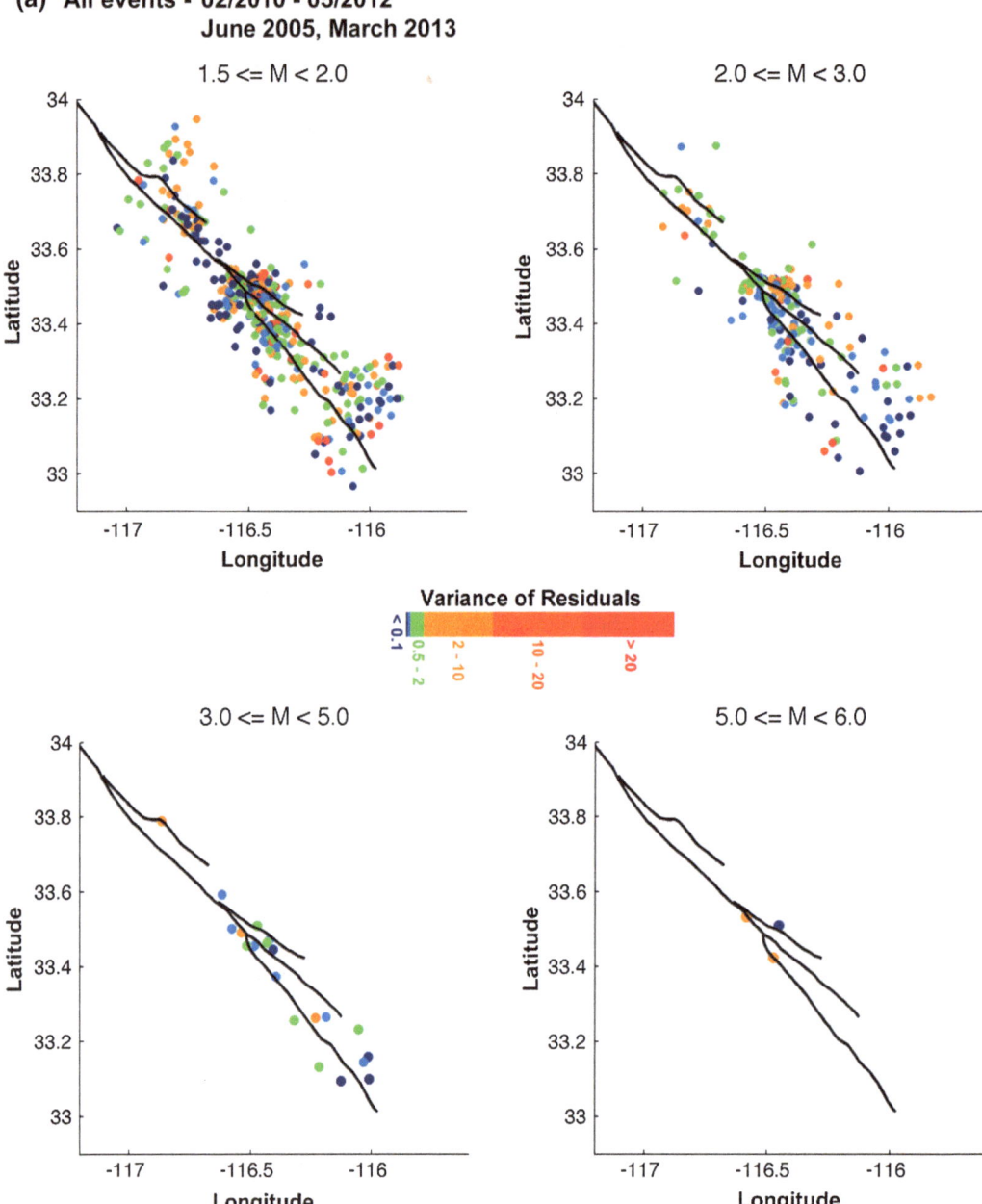

Figure 16

Phase 2—Residual analysis for events. The residuals are analyzed for each event and color-coded according to res$_{var}$, the variance of the residuals. *Green dots* show moderate variance and associate with events showing good fit to predictions, *blue dots* show small variance and associate with damped events (those events with overestimation of the predictions), and orange to *red dots* show large variance, associating with amplified events, in which the GMPEs underestimate the actual observations. **a** All events, **b–e** *M* 1.5–3: **b** main shocks, **c** aftershocks June 2005 and July 2010, **d** aftershocks March 2013 all stations, and **e** aftershocks March 2013 old stations. The main feature of these figures is that the March 2013 dataset shows very high variance in the source region for *M* 1.5–3, which seems to be a combination of source characteristics and fault zone new stations, and explains the high uncertainty seen in the March 2013 dataset

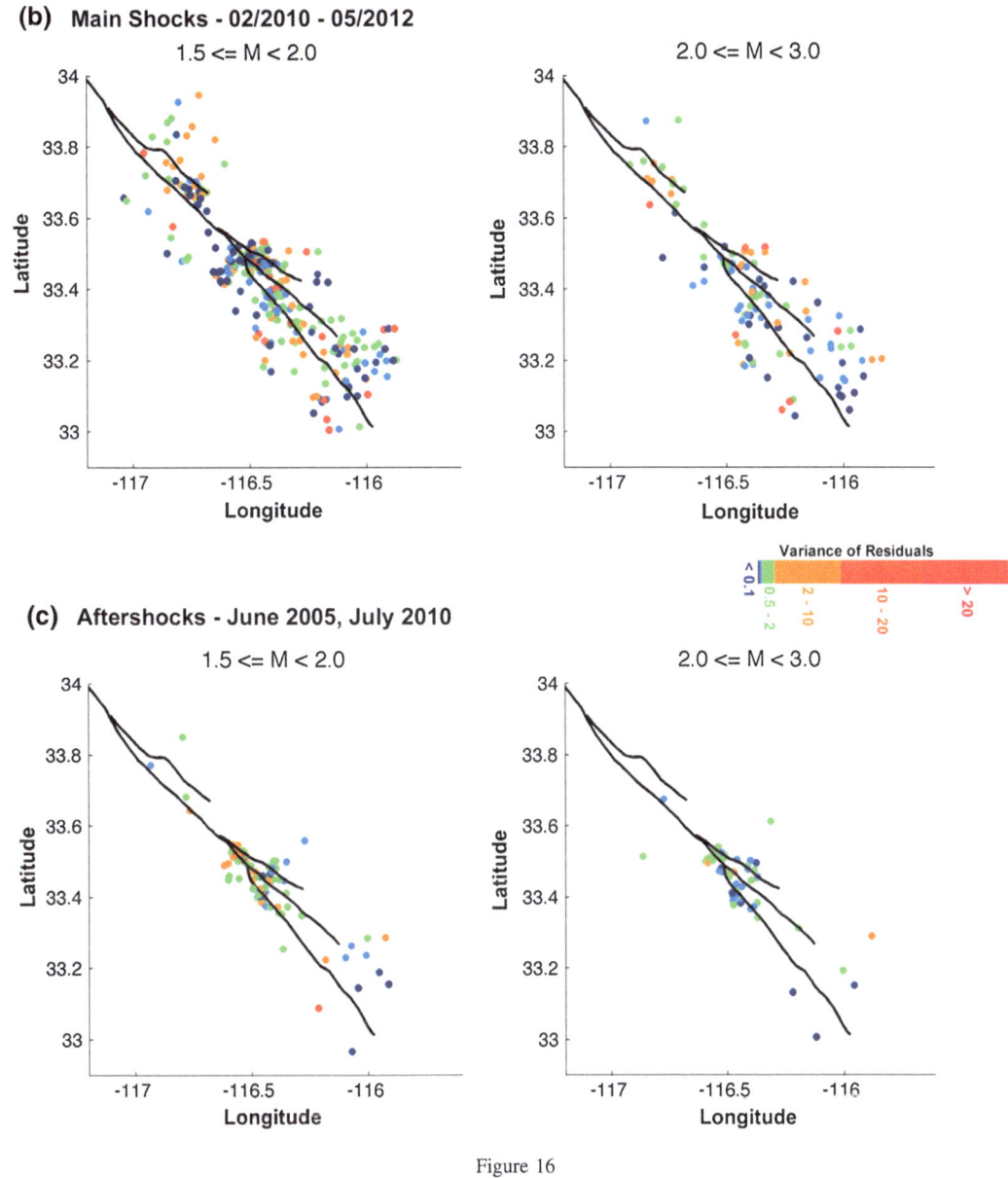

Figure 16
continued

CHIOU 2008) might provide a more reliable estimation for sites in the direction of rupture.

The larger uncertainty seen in the basic GMPE$_{HighDir}$ may originate in the fact that the directivity tool is mainly sensitive to the difference between stations in the direction of rupture, as opposed to stations in other directions. It is not as sensitive to the differences in between the records with high directivity signals. Note that for the PGA and PGV we could set even higher levels of I_{Dir} to generate higher curves of basic GMPE$_{HighDir}$ but for consistency reason we used the minimum I_{Dir} possible so that also the PSA records would converge.

The persistent directivity observed for the trifurcation area (Fig. 5) is consistent with theoretical predictions for bimaterial ruptures with preferred propagation direction (e.g., ANDREWS and BEN-ZION

Reprinted from the journal

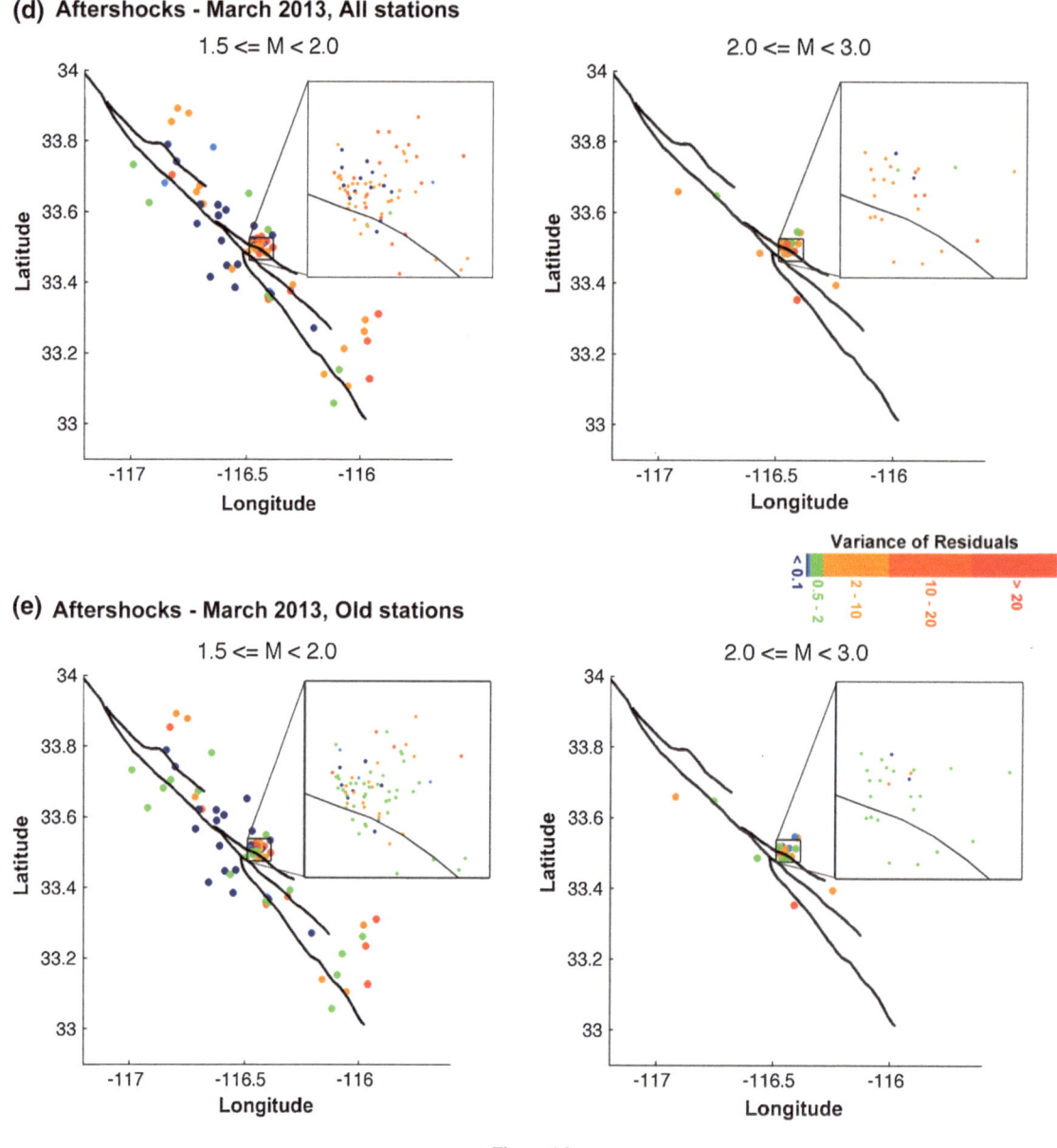

Figure 16
continued

1997; Ben-Zion and Shi 2005; Ampuero and Ben-Zion 2008) associated with the local velocity structure (Allam and Ben-Zion 2012; Allam *et al.* 2014) and observed asymmetry of rock damage across the SJFZ (Lewis *et al.* 2005; Wechsler *et al.* 2009). A more detailed comparison between observed directions of earthquake ruptures in different sections of the SJFZ and theoretical predictions for bimaterial ruptures is left for future work.

7.3. Fault Zone Amplification

Fault Zone Amplification shows a very dominant role in the Phase 1 analysis. In fact, as noted above, since many of the new stations (the additional ones of March 2013) are located in the fault zone, the weight of fault zone stations contributes to the higher curves of Phase 2—GMPEs (Fig. 15b).

Amplification of ground motions in the fault zone could be the outcome of several reasons. (1)

Table 8

Coefficients of the complete set of Phase 2—GMPEs, for PGA, PGV, and 5 % damped PSA. c_1, c_2, c_3 are the distance coefficients, e_1, e_2, e_3 are the magnitude coefficients, h_A is the apparent depth, and σ_t is the uncertainty; d_1 is the directivity coefficient seen in Eq. (8)

	e_1	e_2	e_3	c_1	c_2	c_3	d_1	h_A	σ_t
PGA	−6.822	2.549	−0.121	−0.953	−0.138	−0.013	0.892	3.547	0.924
0.03 s	−5.345	2.453	−0.082	−1.220	−0.165	−0.012	0.842	4.127	0.962
0.05 s	−5.666	2.594	−0.129	−1.046	−0.186	−0.013	0.882	3.253	0.962
0.075 s	−6.001	2.726	−0.187	−0.863	−0.174	−0.015	0.842	3.379	0.950
0.1 s	−6.215	2.877	−0.228	−0.805	−0.171	−0.014	0.772	3.316	0.936
0.15 s	−6.754	2.990	−0.254	−0.731	−0.151	−0.013	0.684	3.298	0.966
0.2 s	−6.971	3.073	−0.247	−0.752	−0.141	−0.011	0.610	3.954	0.961
0.25 s	−7.539	3.093	−0.226	−0.671	−0.129	−0.013	0.591	3.368	0.970
0.3 s	−7.840	3.134	−0.221	−0.675	−0.130	−0.013	0.577	3.703	0.970
0.4 s	−8.312	3.138	−0.200	−0.677	−0.116	−0.013	0.530	4.016	0.942
0.5 s	−8.684	3.152	−0.163	−0.715	−0.116	−0.012	0.525	4.427	0.914
0.75 s	−9.856	3.094	−0.089	−0.587	−0.117	−0.016	0.514	3.606	0.892
1 s	−10.395	3.057	−0.031	−0.609	−0.134	−0.016	0.516	3.937	0.884
PGV	−4.102	2.606	−0.150	−0.879	−0.058	−0.011	0.884	4.224	0.894

The value of d_1 indicates that in the short period range the directivity term is dominant, similar to the PGA equation. The uncertainty and apparent depth do not seem to reflect any clear period-dependent signature

Table 9

Same as Table 8, only that the regression was done for stations in the direction of rupture (those having higher A_{corr} values)

	e_1	e_2	e_3	c_1	c_2	c_3	d_1	h_A	σ_t
PGA	−7.532	2.557	−0.105	−1.121	−0.142	−0.008	1.308	3.749	0.985
0.03 s	−5.934	2.425	−0.048	−1.427	−0.166	−0.005	1.248	4.071	1.016
0.05 s	−6.801	2.524	−0.096	−1.131	−0.169	−0.010	1.395	3.765	1.007
0.075 s	−6.747	2.805	−0.163	−1.015	−0.198	−0.011	1.266	3.383	0.997
0.1 s	−6.771	3.042	−0.205	−0.979	−0.215	−0.011	1.158	3.437	0.979
0.15 s	−7.497	3.073	−0.238	−0.866	−0.171	−0.009	1.131	3.401	1.006
0.2 s	−8.206	3.192	−0.230	−0.895	−0.171	−0.008	1.264	3.842	0.998
0.25 s	−8.839	3.143	−0.219	−0.795	−0.141	−0.010	1.258	3.400	1.008
0.3 s	−9.062	3.367	−0.197	−0.825	−0.191	−0.010	1.224	3.194	1.005
0.4 s	−9.546	3.167	−0.190	−0.764	−0.120	−0.010	1.146	3.684	0.971
0.5 s	−9.888	3.304	−0.142	−0.834	−0.152	−0.010	1.149	3.929	0.937
0.75 s	−11.02	3.202	−0.064	−0.718	−0.142	−0.013	1.136	3.245	0.913
1 s	−11.88	3.135	−0.006	−0.713	−0.152	−0.013	1.230	3.036	0.911
PGV	−4.580	2.772	−0.124	−1.015	−0.101	−0.008	1.185	4.092	0.964

This GMPE$_{HighDir}$ shows higher amplitude curves than the GMPE$_{Dir}$ of Table 8 and larger uncertainties; however, it provides a good estimation of the upper bound of peak ground motions in the direction of rupture (see text for further details)

Body wave amplification resulting from the radiation pattern of the source with different orientations for the P and S phases (e.g., AKI and RICHARDS 2002). (2) Trapped waves and other internal reflections within the damage zone (e.g., BEN-ZION and AKI 1990; LI et al. 2002; BEN-ZION et al. 2003). (3) Reduction in stiffness within the damage zone might cause amplification on the normal component close to the fault (e.g., PISCHIUTTA et al. 2012, 2013). (4) Rupture directivity, discussed above, is in most cases parallel or sub-parallel to the fault zone orientation (when slip surfaces are also parallel to the fault zone orientation), leading to amplification in the direction of rupture for near-fault stations. The specific factors or combination of factors for fault zone

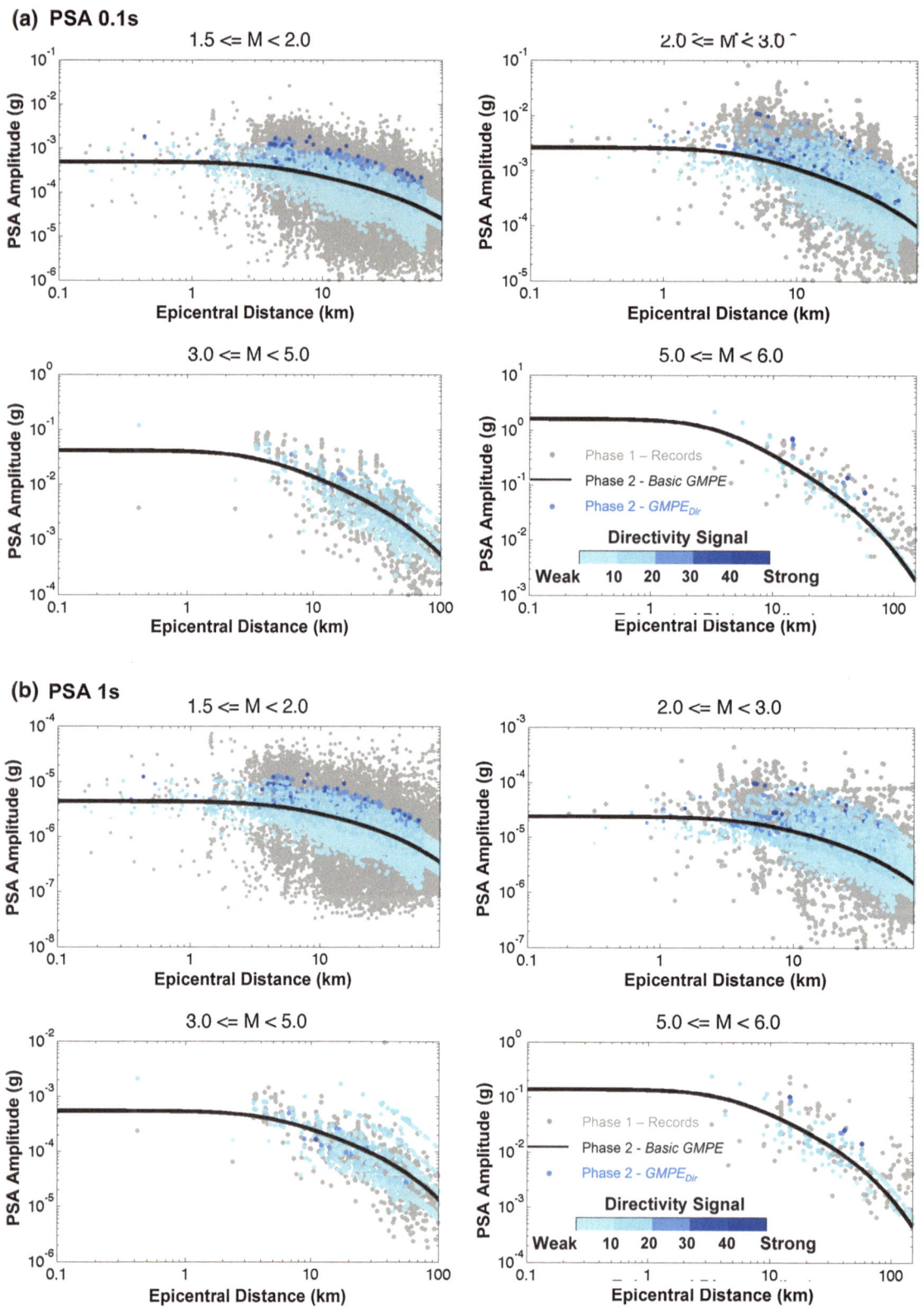

◀

Figure 17

Phase 2—GMPEs for two PSA periods: **a** 0.1 s and **b** 1 s, for **c** PGV, and **d** a PSA to PGA comparison. The records (*grey*) in the first three plots are adjusted according to the magnitude of the reference attenuation curve (*black line*); see text for more details. As in the previous plots, we also show the Phase 2—GMPE$_{Dir}$ spread (*blue color scale*). The last plot (**d**) summarizes Table 8, showing all the PSA curves from 1 to 0.03 s and in relation to the PGA curve (*dashed*). Note that the PGA corresponds with periods of 0.1–0.4 s. See text for details

amplification in the SJFZ could be revealed by a careful examination of the normal—parallel and radial—transversal coordinate systems, under different frequency bands. For example, the directivity tool could be set in different levels of confidence, using the radial—transversal—vertical coordinate system, and analyzing separately the P and S phases. Such detailed analysis may be done in a follow-up work.

From the Phase 1 analysis of the PGA and PSA values we observe (Tables 5, 6) an interesting trend. The PGA and shorter period (≤ 0.1 s) PSA values are more affected by the directivity factor than by the fault zone amplification and borehole-damping factor. For longer periods (>0.1 s), in which the effect of directivity is small, fault zone amplification and borehole damping become more significant. We do not observe any obvious period-dependent signature related to the effect of fault zone amplification and borehole damping; the uncertainly reduction of the PSA GMPEs is of the order of 3 % with no clear trend. The Phase 2—GMPEs for the PSA values support the above insight, showing that the directivity effect, d_1, is more dominant in the shorter periods than in the longer ones (Tables 8, 9).

Figure 17 presents the GMPEs for PGV and for two periods of PSA, 0.1 and 1 s, reflecting the two spectral bands discussed above. In these plots we adjusted the records (grey) according to the reference curve calculated within each magnitude bin. This adjustment adds a slight compaction of the records towards the attenuation curve (black bold line), correcting the magnitude spread of the records within each magnitude range. Also shown is the Phase 2—GMPE$_{Dir}$ spread (blue color scale), expressing the extent of variance that is explained by the GMPEs. Figure 17d represents Table 8, showing the GMPEs for all the calculated PSA datasets (0.03 s—1 s), in

relation to the PGA curve. The PSA curves range from 1 to 1.5 orders of magnitudes in the M 3–6 range, and increase for smaller magnitudes up to 2.5 orders of magnitudes. This is not surprising, since smaller magnitudes reflect higher frequencies, and smaller amplitudes, which might partly originate from scattering and noise effects.

8. Conclusions

We have developed two sets of GMPEs for horizontal peak ground motions, for PGA, PGV, and 5 % damped PSA records, appropriate for the SJFZ area. These equations account for magnitude, distance, rupture directivity, fault zone amplification and borehole damping. We have shown that besides the obvious distance and magnitude factors, rupture directivity and fault zone amplification have a significant role in explaining some of the variations seen in the observations. According to the Phase 1 analysis, the overall uncertainty reduction due to both factors ranges from 1 to 12 %, for the PGA and PSA values. In the Phase 2 analysis, the fault zone amplification seems to shift up the peak ground motion values, showing higher GMPE curves, and rupture directivity explains a significant amount of the data-variation. In Phase 1, there is a shift in the impact of directivity and fault zone amplification, from a more dominant directivity effect for the PGA and high frequencies PSA, to a more dominant fault zone amplification and borehole damping for the lower frequency range of the PSA (0.2–1 s). A similar tendency is seen in the Phase 2 dataset, in which the directivity coefficient (Table 8) is larger for PGA and short-period PSAs (<0.2 s).

The directivity tool has been found to provide useful results, not only for the generation of GMPEs, but also for a closer study of the fault segments and their tendency to rupture in specific directions. Fault zone amplification seems also to play an important role in affecting the peak ground motion, and further work should be done to identify the specific sources of this effect. The roles of both factors are significantly more important in the examined data than the traditional consideration of site characteristics, maybe because many of the sites, even in soft ground, could be considered rock sites.

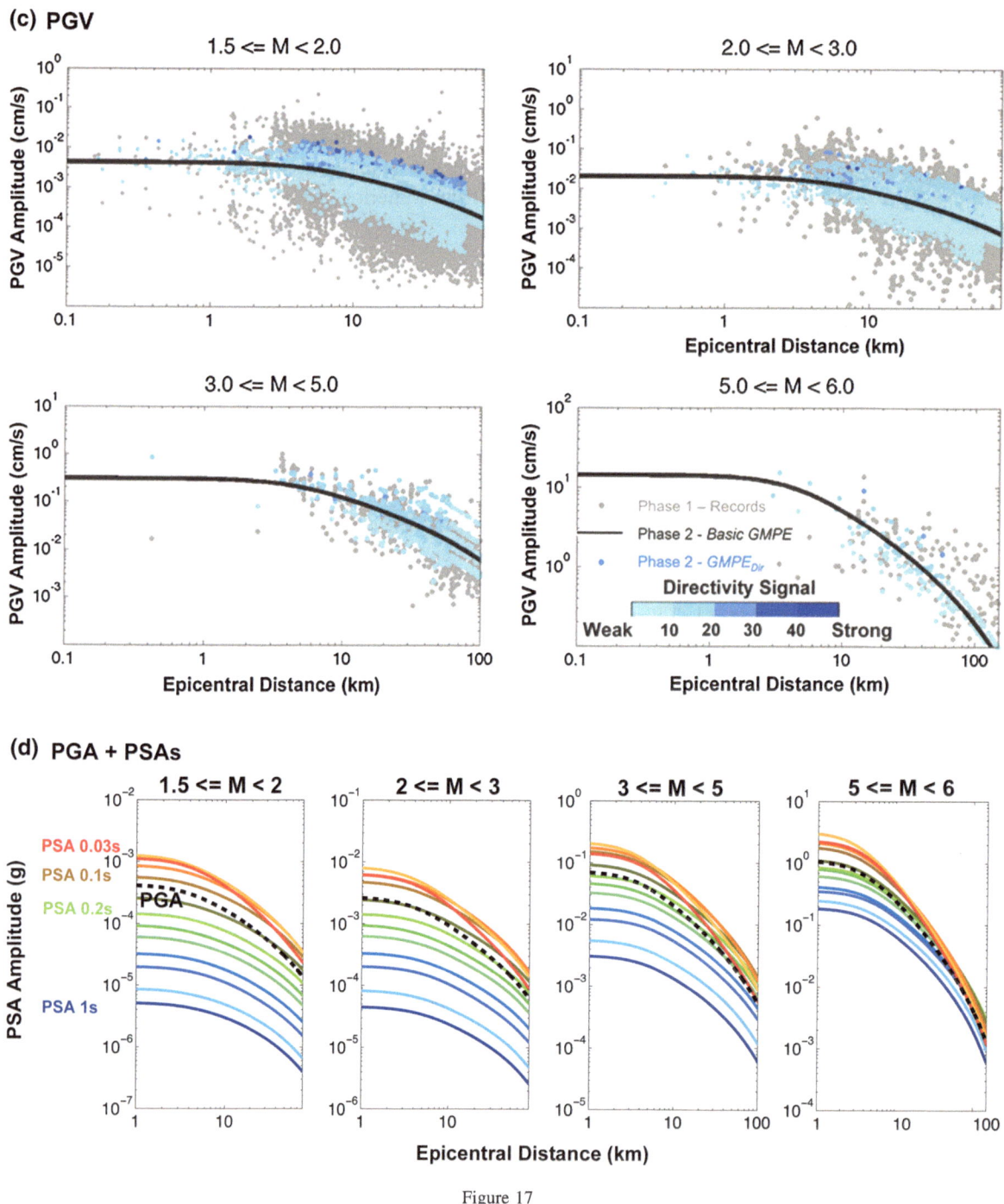

Figure 17
continued

The Phase 1—GMPEs, although illuminating nicely the main factors controlling ground motions, is incomplete. We believe that the more statistically robust Phase 2—GMPE$_{Dir}$ is the most representative of the region, accounting for larger variability in station location, and source mechanisms, and providing a better prediction for the sites near the events. In fact, by masking fault zone amplification due to the large weight of fault zone stations, the Phase 2—GMPE$_{Dir}$ reflects the narrow band around the fault segments (~1 km wide). The Phase 2—basic GMPE_{HighDir} provides even a higher estimation for sites in the direction of rupture, and should also be considered. Currently, there are no data that would allow us to extend the analysis of ground motion parameters in the SJFZ environment for larger earthquakes ($M > 6$), which is a significant practical limitation of the present study. However, such events are expected to occur in the not-too-distant future (ZÖLLER and BEN-ZION 2014; Working Group on California Earthquake Probabilities http://www.wgcep.org/) and could be used to test the generality and extent of our results.

The large difference seen in the GMPEs from Phase 1 to Phase 2, should serve as a cautionary note in using GMPEs to predict future motions, comparing different datasets, and applying models to new datasets. Earthquake ground motions are controlled by a complex interaction among multiple factors, and a model developed for a specific region based on given data may not necessarily perform better than a model developed for another dataset and region.

Data Sources

Seismic data used in this study are gathered and managed by the following networks:

1. The Anza network (AZ) and the SJFZ (YN) network are operated by IGPP, University of California, San Diego http://eqinfo.ucsd.edu/.
2. The SCSN network (CI) is the Southern California Seismic Network operated by Caltech and USGS http://www.scsn.org/.
3. The PBO network (PB) is the Plate Boundary Observatory operated by UNAVCO http://www.earthscope.org/science/observatories//pbo.
4. The SB network is operated by University of California, Santa Barbara http://nees.ucsb.edu/publications.
5. Gaps in data were filled by the IRIS Data Management Center (DMC) http://www.iris.edu/dms/dmc/.
6. Instrumentation for seismic stations was provided by IRIS PASSCAL http://www.passcal.nmt.edu/.

Grids for maps were downloaded from the webtool:

http://www.marine-geo.org/tools/maps_grids.php.

Acknowledgments

Ittai Kurzon thanks Luciana Astiz for her kind guidance and for suggestions on the first draft of this paper, and also Alan Yong from the USGS for the Vs30 estimations per site presented in this paper. The study was supported by the National Science Foundation (Grant EAR-0908903) and Ittai Kurzon was partly supported by a post-doctoral fellowship from the Geological Survey of Israel. The manuscript benefitted from constructive comments by Dave Boore, an anonymous referee, and Editor Antonio Rovelli.

REFERENCES

ABRAHAMSON, N. A., G. M. ATKINSON, D. M. BOORE, Y. BOZORGNIA, K. W. CAMPBELL, B. CHIOU, I. M. IDRISS, W. J. SILVA, and R. R. YOUNGS (2008). *Comparisons of the NGA Ground-Motion Relations*, Earthquake Spectra, 24(1), 45–66.

ABRAHAMSON, N. A. and W. J. SILVA (2008). *Summary of the Abrahamson and Silva NGA Ground-Motion Relations*, Earthquake Spectra, 24(1), 67–97.

AKI, K., & P. G. RICHARDS (2002). *Quantitative Seismology* (second edition), University Science Books.

ALLAM, A. A. and Y. BEN-ZION (2012). *Seismic Velocity Structures in the Southern California Plate-Boundary Environment from Double-Difference Tomography*, Geophys. J. Int., 190, 1181–1196, doi:10.1111/j.1365-246X.2012.05544.x.

ALLAM, A. A., Y. BEN-ZION, I. KURZON and F. L. VERNON (2014). *Seismic velocity structure in the Hot Springs and Trifurcation Areas of the San Jacinto Fault Zone, California, from double-difference tomography*, Geophys. J. Int., in review.

AL-SHUKRI, H. J., G. L. PAVLIS and F. L. VERNON (1995). *Site Effects Observations from Broadband Arrays*, Bull. Seismol. Soc. Am., 85(6), 1758–1769.

ANDREWS, D. J. and Y. BEN-ZION (1997). *Wrinkle-like Slip Pulse on a Fault Between Different Materials*, J. Geophys. Res., 102, 553–571.

AMPUERO, J.-P. and Y. BEN-ZION (2008). *Cracks, pulses and macroscopic asymmetry of dynamic rupture on a bimaterial interface with velocity-weakening friction*, Geophys. J. Int., 173, 674–692, doi:10.1111/j.1365-246X.2008.03736.x.

BAILEY, I. W., Y. BEN-ZION, T. W. BECKER, and M. HOLSCHNEIDER (2010). *Quantifying Focal Mechanism Heterogeneity for Fault Zones in Central and Southern California*, Geophys. J. Int., 183, 433–450.

BAKER, J. W., Y. BOZORGNIA, C. DI ALESSANDRO, B. CHIOU, M. ERDIK, P. SOMERBILLE, and W. SILVA (2012) GEM-PEER Global GMPEs project Guidance for including Near-Fault Effects in Ground Motion Prediction Models, *Proceedings of 15th World Conference on Earthquake Engineering*, Lisbon, Portugal, 10p.

BEN-ZION, Y. and K. AKI (1990). *Seismic radiation from an SH line source in a laterally heterogeneous planar fault zone*, Bull. Seism. Soc. Am., 80, 971–994.

BEN-ZION, Y. and C. G. SAMMIS (2003). *Characterization of Fault Zones*, Pure Appl. Geophys., *160*, 677–715.

BEN-ZION, Y., Z. PENG, D. OKAYA, L. SEEBER, J. G. ARMBRUSTER, N. OZER, A. J. MICHAEL, S. BARIS and M. AKTAR (2003). *Shallow fault zone structure illuminated by trapped waves in the Karadere-Duzce branch of the North Anatolian Fault, western Turkey,* Geophys. J. Int.,*152*, 699–717.

BEN-ZION, Y. and Z. SHI (2005). *Dynamic rupture on a material interface with spontaneous generation of plastic strain in the bulk,* Earth Planet. Sci. Lett., *236*, 486–496.

BLISNIUK, K., T. ROCKWELL, L. A. OWEN, M. OSKIN, C. LIPPINCOTT, M. W. CAFFEE, and J. DORTCH (2010). *Late Quaternary Slip Rate Gradient defined using High-Resolution Topography and 10Be Dating of Offset Landforms on the Southern San Jacinto Fault Zone, California,* J. Geophys. Res., *115*, B08401, doi:10.1029/2009JB006346.

BOATWRIGHT, J. (2007). *The Persistence of Directivity in Small Earthquakes,* Bull. Seismol. Soc. Am., *97* (6), 1850–1861.

BOORE, D. M., and G. M. ATKINSON (2008). *Ground Motion Prediction Equations for the Average Component of PGA, PGV, and 5%-damped PSA at Spectral Periods between 0.01 s and 10.0 s,* Earthquake Spectra, *24*(1), 99–138.

BOORE, D. M., J. P. STEWART, E. SEYHAN, and G. M. ATKINSON (2013). NGA-West 2 Equations for Predicting Response Spectral Accelerations for Shallow Crustal Earthquakes, *PEER Report 2013/xx,* Pacific Earthquake Engineering Research Center, Berkeley, California.

CALDERONI, G., A. ROVELLI, and S. K. SINGH (2013). *Stress drop and source scaling of the 2009 April L'Aquila earthquakes,* Geophys. J. Int. (2013) *192*, 260–274

CAMPBELL, K. W. and Y. BOZORGNIA (2008). *NGA Ground Motion Model for the Geometric Mean Horizontal Component of PGA, PGV, PGD and 5 % Damped Linear Elastic Response Spectra for Periods Ranging from 0.01 to 10 s,* Earthquake Spectra, *24*(1), 139–171

CHIOU, B. S. J. and R. R. YOUNGS (2008). *An NGA Model for the Average Horizontal Component of Peak Ground Motion and Response Spectra,* Earthquake Spectra, *24*(1), 173–215.

CUA, G. and HEATON, T. H. (2008). *Characterizing Average Properties of Southern California Ground Motion Amplitudes and Envelopes,* Bull. Seismol. Soc. Am., submitted.

DOUGLAS, A., J. A. HUDSON and R. G. PEARCE (1988). *Directivity and the Doppler effect,* Bull. Seismol. Soc. Am., *78*(3), 1367–1372.

DOUGLAS, J., (2001). A Critical Reappraisal of Some Problems in Engineering Seismology. PhD thesis, University of London.

DOUGLAS, J., (2003). *Earthquake Ground Motion Estimation Using Strong-Motion Records: a Review of Equations for the Estimation of Peak Ground Acceleration and Response Spectral Ordinates,* Earth-Science Reviews, *61*, 43–104.

DORSEY, R. J. and J. J. ROERING (2006). *Quaternary Landscape Evolution in the San Jacinto Fault Zone, Peninsular Ranges of Southern California: Transient Response to Strike-Slip Fault Initiation,* Geomorphology, *73*, 16–32.

FAY, N., and G. HUMPHREYS (2005). *Fault Slip Rates, Effects of Elastic Heterogeneity on Geodetic Data, and the Strength of the Lower Crust in the Salton Trough region, Southern California,* J. Geophys. Res., *110*, B09401, doi:10.1029/2004JB003548.

FIALKO, Y. (2006). *Interseismic Strain Accumulation and the Earthquake Potential on the Southern San Andreas Fault System,* Nature, *441*, doi:10.1038/nature04797, 968–971.

HARTSE, H. E., M. FEHLER, R. C. ASTER, J. S. SCOTT and F. L. VERNON (1994). *Small-scale heterogeneity in the anza seismic gap, southern California,* J. Geophys. Res., *99*(B4), 6901–6818.

HAUKSSON, E., W. YANG and P. SHEARER (2012). *Waveform Relocated Earthquake Catalog for Southern California (1981 to June 2011),* Bull. Seismol. Soc. Am., *102*(5), 2239–2244.

HUTTON, K., J. WOESSNER and E. HAUKSSON (2008). *Earthquake Monitoring in Southern California for Seventy-Seven Years (1932–2008),* Bull. Seismol. Soc. Am., *100*(2), 423–446, doi:10.1785/0120090130.

IDRISS, I. M. (2008). *An NGA Empirical Model for Estimating the Horizontal Spectral Values Generated by Shallow Crustal Earthquakes,* Earthquake Spectra, *24*(1), 217–242.

JOYNER, W. B. and D. M. BOORE (1993). *Methods for Regression Analysis of Strong-Motion Data.* Bull. Seismol. Soc. Am., *83*, 469–487.

KENDRICK, K. J., D. M. MORTON, S. G. WELLS, and R.W. SIMPSON (2002). *Spatial and Temporal Deformation along the Northern San Jacinto Fault, Southern California: Implications for Slip Rates,* Bull. Seismol. Soc. Am., *92*, 2782–2802.

KLINGER, R. E. and T. K. ROCKWELL (1989). *Flexural-Slip Folding along the Eastern Elmore Ranch Fault in the Superstition Hills Earthquake Sequence of November 1987,* Bull. Seismol. Soc. Am., *79*(2), 297–303.

KURZON, I., F. L. VERNON, Y. BEN-ZION and G. M. ATKINSON (2014). A New Tool for Inferring Rupture Directivity: Implementation for the San Jacinto Fault Zone (in preparation).

LEWIS, M. A., Z. PENG, Y. BEN-ZION and F. L. VERNON (2005). *Shallow Seismic Trapping Structure in the San Jacinto fault zone near Anza, California,* Geophys. J. Int, *162*, 867–881.

LEWIS, M. A, Y. BEN-ZION and J. MCGUIRE (2007). *Imaging the deep structure of the San Andreas Fault south of Hollister with joint analysis of fault-zone head and direct P arrivals,* Geophys. J. Int., *169*, 1028–1042.

LI, Y. G., J. E. VIDALE, S. M. DAY, D. M. OGLESBY and THE SCEC FIELD WORKING TEAM (2002). *Study of the 1999 M 7.1 Hector Mine, California, Earthquake Fault Plane by Trapped Waves,* Bull. Seismol. Soc. Am., *92*, 1318–1332.

LOUIE, J. N. (2001). *Faster, Better: Shear-Wave Velocity to 100 Meters Depth from Refraction Microtremor Arrays,* Bull. Seismol. Soc. Am., *91*(2), 347–364.

NIGAM, N. C. and P. C. JENNINGS (1969) *Calculation of Response Spectra from Strong-Motion Earthquake Records,* Bull. Seismol. Soc. Am., *59*, 909–922.

ONDERDONK, N., T. ROCKWELL, S. MCGILL and G. MARLIYANI (2013). *Evidence for seven surface ruptures in the past 1600 years on the Claremont fault at Mystic Lake, northern San Jacinto fault zone, California.* Bull. Seismol. Soc. Am., *103*, 519–541.

PISCHIUTTA, M., F. SALVINI, J. FLETCHER, A ROVELLI and Y. BEN ZION (2012). *Horizontal polarization of ground motion in the Hayward fault zone at Fremont, California: Dominant fault-high-angle polarization and fault-induced cracks,* Geophys. J. Int., *188*, 1255–1272, doi:10.1111/j.1365-246X.2011.05319.x.

PISCHIUTTA, M., A ROVELLI, F. SALVINI, G. DI GIULIO and Y. BEN ZION (2013). *Directional resonance variations across the Pernicana fault, Mt. Etna, in relation to brittle deformation fields,* Geophys. J. Int., *193*, 986–996, doi: 10.1093/gji/ggt031.

PRESS, W. H., S. A. TEUKOLSKY, W. T. VETTERLING and B. P. FLANNERY (1992). Fortran Numerical Recipes, 2nd Edition, *Cambridge University Press,* Melbourne, Australia. 1447 pages.

ROCKWELL, T.K., R. KLINGER and J. GOODMACHER (1990). Determination of Slip Rates and Dating of Earthquakes for the San Jacinto and Elsinore Fault Zones, in Kooser, M.A., and Reynolds, R.E., eds., Geology around the Margins of the Eastern San Bernardino Mountains, Volume 1: Inland Geological Society, Redlands, p. 51–56.

ROCKWELL, T. K. (2003). *3,000 Years of Ground-Rupturing Earthquakes in the Anza Seismic Gap, San Jacinto Fault, Southern California; Time to Shake it up?* Seismol. Res. Lett., *74*, 236–237.

SALISBURY, J. B., T. K. ROCKWELL, T. J. MIDDLETON and K. W. HUDNUT (2012). *LiDAR and Field Observations of Slip Distribution for the Most Recent Surface Ruptures along the Central San Jacinto Fault*, Bull. Seismol. Soc. Am., *102*, 598–619.

SANDERS, C. O. and H. KANAMORI (1984). *A Seismotectonic Analysis of the Anza Seismic Gap, San Jacinto Fault Zone, Southern California*, J. Geophys. Res., *89*, 5873–5890.

SEEKINS, L. C. and J. BOATWRIGHT (2010). *Rupture Directivity of Moderate Earthquakes in Northern California*, Bull. Seismol. Soc. Am., *100* (3), 1107–1119

SHARP, R. V. (1967). *San Jacinto Fault Zone in Peninsular Ranges of Southern California*, Geol. Soc. Am. Bull., *78*(6), 705–730.

SPUDICH, P. and B. S. J. CHIOU (2008). *Directivity in NGA earthquake ground motions: Analysis using isochrone theory*, Earthquake Spectra, *24*, 279–298.

SPUDICH, P., J. WATSON-LAMPREY, P. G. SOMERVILLE, J. BAYLESS, S. K. SHAHI, J. W. BAKER, B. ROWSHANDEL and B. S. J. CHIOU (2012). Direcitivity Models Produced for the Next Generation Attenuation West 2 (NGA-West 2) project, *Proceedings of 15th World Conference on Earthquake Engineering*, Lisbon, Portugal, 9p.

THATCHER, W., J. A. HILEMAN and T. C. HANKS (1975). *Seismic Slip Distribution along San Jacinto Fault Zone, Southern California, and its Implications*, Geol. Soc. Am. Bull., *86*(8), 1140–1146.

THOMPSON, E. M., L. G. BAISE, R. E. KAYEN, E. C. MORGAN and J. KAKLAMANOS (2011). *Integrated Multiscale Site Response Mapping: A Case Study of Parkfield, California*, Bull. Seismol. Soc. Am., *101*(3), 1081–1100.

VERNON, F. L. (1989). Analysis of Data Recorded on the ANZA Seismic Network. PhD thesis, University of California, San Diego.

VERNON, F. L., G. L. PAVLIS, T. J. OWENS, D. E. McNAMARA and P. N. ANDERSON (1998). *Near-Surface Scattering Effects Observed with a High-Frequency Phased Array at Pinyon Flats, California*, Bull. Seismol. Soc. Am., *88*(6), 1548–1560.

WALD, D. J. and T. I. ALLEN (2007). *Topographic Slope as a Proxy for Seismic Site Conditions and Amplification*, Bull. Seismol. Soc. Am., *97*(5), 1379–1395.

WECHSLER, N., T. K. ROCKWELL and Y. BEN-ZION, (2009). *Application of high resolution DEM data to detect rock damage from geomorphic signals along the central San Jacinto Fault*, Geomorphology, *113*, 82–96, doi:10.1016/j.geomorph.2009.06.007..

WILLS, C. J., M. PETERSEN, W. A. BRYANT, M. REICHLE, G. J. SAUCEDO, S. TAN, G. TAYLOR and J. TREIMAN (2000). *A Site-Conditions Map for California Based on Geology and Shear-Wave Velocity*, Bull. Seismol. Soc. Am., *90*(6B), S187–S208

WILLS, C. J. and K. B. CLAHAN (2006). *Developing a Map of Geologically Defined Site-Condition Categories for California*, Bull. Seismol. Soc. Am., *96*(4A), 1483–1501.

WU, C. F. and H. C. HUANG (2013). *Near-Surface Shear-Wave Velocity Structure of the Chiayi Area, Taiwan*, Bull. Seismol. Soc. Am., *103*(2A), 1154–1164.

XU, S., Y. BEN-ZION and J.-P. AMPUERO, (2012). *Properties of inelastic yielding zones generated by in-plane dynamic ruptures: I. model description and basic results*, Geophys. J. Int., *191*, 1325–1342, doi:10.1111/j.1365-246X.2012.05679.x.

YONG, A., S. E. HOUGH, J. IWAHASHI and A. BRAVERMAN (2012). *A Terrain-Based Site Conditions Map of California with Implications for the Contiguous United States Submitted to* Bull. Seismol. Soc. Am., on 25 September 2010, Accepted on August 1, 2011.

ZÖLLER, G. and Y. BEN-ZION (2014). *Large earthquake hazard of the San Jacinto fault zone, CA, from long record of simulated seismicity assimilating the available instrumental and paleoseismic data*, Pure Appl. Geophys., doi:10.1007/s00024-014-0783-1.

(Received June 19, 2013, revised April 1, 2014, accepted April 23, 2014, Published online May 23, 2014)

Reprinted from the journal

Pure Appl. Geophys. 171 (2014), 3083–3097
© 2014 Springer Basel
DOI 10.1007/s00024-014-0831-x

Wavefield Polarization in Fault Zones of the Western Flank of Mt. Etna: Observations and Fracture Orientation Modelling

F. Panzera,[1,3] M. Pischiutta,[2] G. Lombardo,[1] C. Monaco,[1] and A. Rovelli[2]

Abstract—Ambient noise measurements performed on the western flank of Mt. Etna are analyzed to infer the occurrence of directional amplification effects in fault zones. The data were recorded along short (<500 m) profiles crossing the Ragalna Fault System. Ambient noise records were processed to compute the horizontal-to-vertical noise spectral ratio as a function of frequency and direction of motion. Wavefield polarization was investigated in the time–frequency domain as well. Peaks of the spectral ratios generally fall in the frequency band 1.0–6.0 Hz pointing out directional amplifications that are also confirmed by the results of the time–frequency analysis, the largest amplification occurring with high angle to the fault strike. A variation of the frequency of the spectral peak is observed between the two sides of the fault, possibly related to a damage fault asymmetry. Measurements performed several kilometers away from the fault zone do not show behavior that is as systematic as in the fault zone, and this suggests that the observed directional effects can be ascribed to the fault fabric. We relate the polarization effect to compliance anisotropy in the fault zone, where the presence of predominantly oriented fractures makes the normal component of ground motion larger than the transversal one. In order to test the direction and the type of fractures that are expected in the fault zone, we modeled the brittle deformation pattern of the investigated fault. Theoretical results are in good agreement with field observations of the fracture strike.

Key words: Directional resonance, fault zone, Wavefield polarization, Mt. Etna.

1. Introduction

Fault zone rocks exhibit locally reduced elastic moduli that may lead to amplification of horizontal ground motion during earthquakes (Ben-Zion 1998). These effects were observed, among others, by Spudich and Olsen (2001), within a ∼1–2 km wide low-velocity zone around the rupture of the 1984 Morgan Hill earthquake; by Seeber et al. (2000) and Peng and Ben-Zion (2006) in the rupture zone of the 1999 Izmit earthquake on the Karadere branch of the North Anatolian Fault; and by Calderoni et al. (2010) along the Paganica-San Demetrio fault during the April 2009 L'Aquila earthquake sequences in central Italy.

When the low velocity fault zone layer is coherent over length scales of several km or more, it produces trapped waves that are the effect of constructive interference of critically reflected phases (Ben-Zion and Aki 1990; Li and Leary 1990; Li et al. 1994). Trapped waves have been observed along many active faults (e.g., Li et al. 1990; Mizuno and Nishi-gami 2004; Lewis et al. 2005), as well as near a dormant fault (Rovelli et al. 2002; Cultrera et al. 2003). In trapped waves, the amplified motion is predominantly fault-parallel and vertical (Lewis and Ben-Zion 2010). In several recent studies, amplified motions near faults were found to have a high angle to the fault strike, indicating a mechanism different than for trapped waves. In four faults of the eastern flank of Mt. Etna (the Tremestieri, Pernicana, Moscarello and Acicatena faults), Rigano et al. (2008) observed that seismic signals are strongly polarized, but not fault-parallel as would be expected for trapped waves. Using both volcanic tremor and local earthquakes, Falsaperla et al. (2010) found a strong polarization at seismological stations in the crater area of Mt. Etna, with polarization directions varying site by site, but everywhere transversal to the orientation of the predominant local fracture field.

[1] Dipartimento di Scienze Biologiche, Geologiche e Ambientali, Sezione Scienze della Terra, Università degli studi di Catania, Catania, Italy. E-mail: panzerafrancesco@hotmail.it

[2] Istituto Nazionale di Geofisica e Vulcanologia, Seismology and Tectonophysics, Rome, Italy.

[3] *Present Address*: Physics Department, Icelandic Meteorological Office, Reykjavík, Iceland.

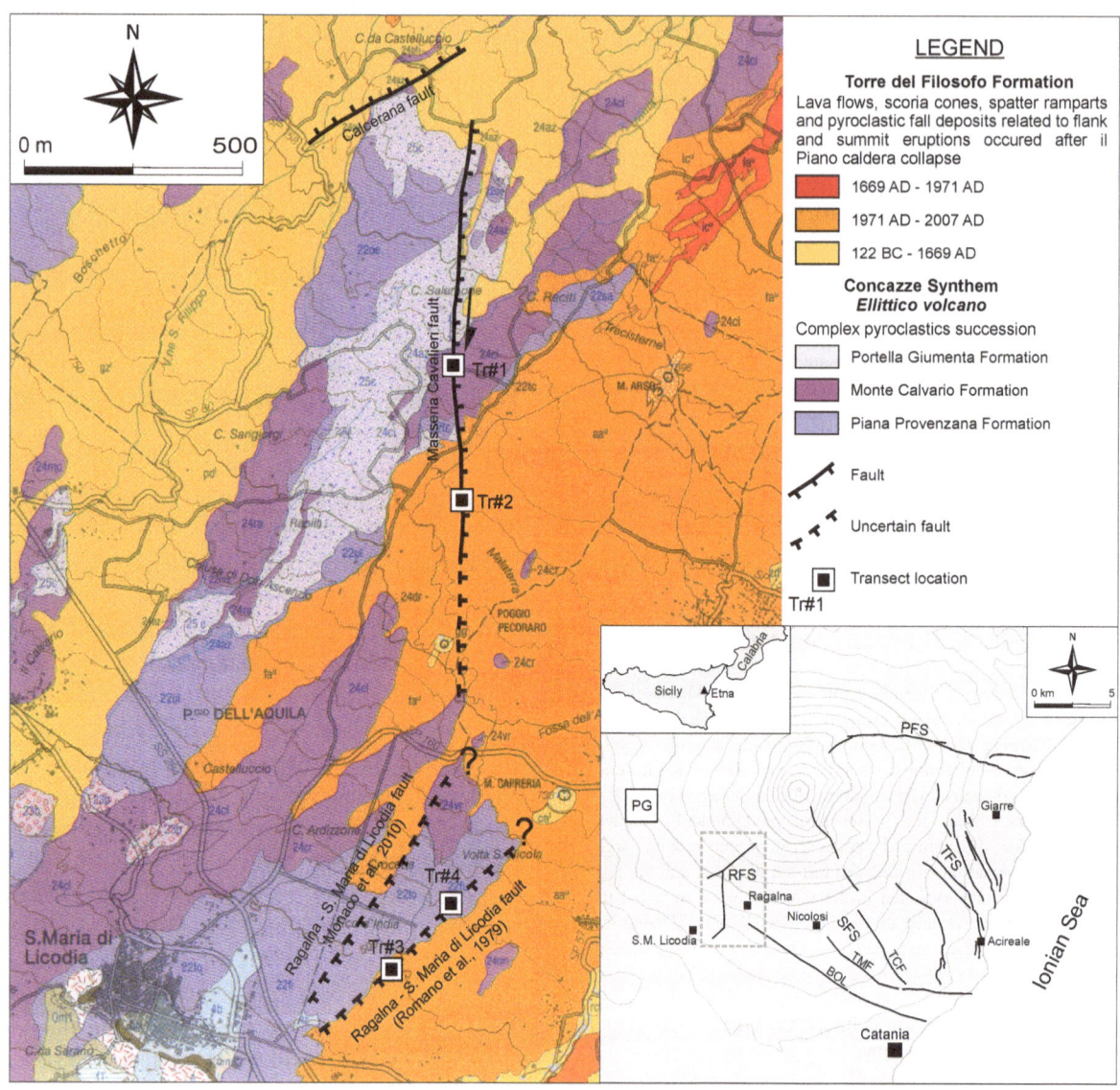

Figure 1
Geological map of the investigated area (modified from BRANCA *et al.* 2011). The *inset* map shows the main structural features of Mt. Etna (modified from NERI *et al.* 2007). The *white square* (PG) indicates the Piano dei Grilli area. From North to South: *PFS* Pernicana Fault System, *TFS* Timpe Fault System, *RFS* Ragalna Fault System, *SFS* Southern Fault System, *BOL* Belpasso-Ognina Lineament, *TMF* Tremestieri-Mascalucia Fault, and *TCF* Trecastagni Fault

Similarly, Dɪ Gɪᴜʟɪᴏ *et al.* (2009) found very stable polarization angles on Mt. Etna, in the NE rift segment and in the Pernicana fault at Piano Pernicana, with horizontal polarization that again was not parallel to the fault strike. Pɪsᴄʜɪᴜᴛᴛᴀ *et al.* (2012, 2013a, b, 2014) interpreted the directional amplification of the horizontal ground motion in many fault zones as due to the crack orientation, which causes a larger rock compliance transversal to fractures. The similar

results obtained in the above mentioned papers by using earthquake records and ambient noise indicate that microtremors are a valid tool to investigate ground motion polarization properties.

In the present study, the results of new measurements of ambient noise performed in the western flank of Mt. Etna are shown and interpreted in the frame of local stress field and fault kinematics. The data were recorded across the tectonic structures of the Ragalna Fault System (RFS). Several measurements were also performed at Piano dei Grilli, up to thousand of meters from the RFS, in order to check the expected change far from the damage fault zone. We followed the approach proposed by PISCHIUTTA et al. (2013a), who interpreted variations of ground motion polarization across the Pernicana fault, on the eastern flank of Mt. Etna, in terms of fracture fields in the fault damage zone, polarization being perpendicular to the predominant fracture direction. Therefore, following PISCHIUTTA et al. (2013a), we have modeled the brittle deformation pattern expected for the investigated fault using the FRAP package (SALVINI et al. 1999).

1.1. Tectonics of the Study Area

Mt. Etna is a 3,300 m high basaltic volcano located along the Ionian coast of Sicily (Fig. 1), at the boundary between the African and European Plates. The volcanic edifice shows a diameter of about 40 km and lies at the front of the Sicilian thrust belt and at the footwall of the northern sector of the Malta escarpment. Its tectonic setting results from the interaction of regional tectonics and local-scale volcano-related processes (McGUIRE and PULLEN 1989).

The eastern and south-eastern sectors of the volcano are the most tectonically active, being affected by normal-oblique faulting, related to WNW-ESE regional extension, and by large sliding processes (NERI et al. 1991; LO GIUDICE and RASÀ 1992; RUST and NERI 1996; MONACO et al. 1997; FROGER et al. 2001; AZZARO et al. 2012). Geodetic surveys indicate a relative movement toward E-ESE (BONFORTE et al. 2008; ALPARONE et al. 2011). The sliding sector is limited to the North by the active Pernicana Fault System (PFS in Fig. 1), a left-lateral strike-slip that represents one of the most pronounced and active tectonic lineaments of the Mt. Etna structure (AZZARO et al. 2001; NERI et al. 2004). Moving to the South, the 30 km long Timpe Fault System (TFS in Fig. 1) is composed by ENE-dipping and NNW-SSE striking segments, characterized by normal-dextral motion and vertical offsets up to 200 m. Most of these faults have a high seismogenic potential and generate shallow earthquakes, as well as

coseismic cracks and creeping (MONACO et al. 1997; AZZARO 1999; Azzaro et al. 2012). The South Fault System (SFS in Fig. 1) is composed of two main, NW–SE trending en-echelon normal-dextral seismogenic segments, the Tremestieri-Mascalucia (TMF) and the Trecastagni (TCF) faults (BARRECA et al. 2013), whose motion is accommodated to the east by an aseismic, W–E striking, fracture alignment. According to geological and geochemical surveys (LO GIUDICE and RASÀ 1992; ACOCELLA et al. 2003; NERI et al. 2004; BONFORTE et al. 2013) and remote sensing data (BORGIA et al. 2000; NERI et al. 2009; SOLARO et al. 2010; BONFORTE et al. 2011), these structures, together with another sub-parallel arc-shaped aseismic alignment, the Belpasso-Ognina Lineament, represent the southern boundaries of the SE-ward sliding sector of Mt. Etna.

Conversely, the western flank of the volcanic edifice is characterized by moderate tectonic activity. The main structure of this sector is represented by the RFS (in Fig. 1). According to NERI et al. (2007), field evidence, together soil radon emission data and geodetic measures, suggest that the RFS is an active right-lateral oblique structure with a minimum long-term dip-slip component. The RFS is formed by two distinct fault segments: the SW–NE striking Calcerana fault and the N–S striking Masseria Cavalieri fault (AZZARO 1999; RUST and NERI 1996). The Calcerana fault shows a smoothed scarp, which is less evident compared to the previous one that develops for about 1.4 km with morphological offsets of about 10 m. The Masseria Cavalieri fault is characterized by a fresh east-facing escarpment up to 20 m high and 5 km long. Along this structure, fractures are well visible on buildings and concrete road walls, and show centimetric-millimetric oblique-dextral offset (see Fig. 2). The tectonic activity reported in the last two centuries seems to be mostly related to fault creep, even though coseismic ground fracturing was observed during the 1982 earthquake (macroseismic magnitude Mm = 3.4; AZZARO 1999). At its southern end, the Masseria Cavalieri fault is characterized by the occurrence of cinder cones aligned in a N–S direction. To the south, the fault loses its morphologic evidence and a 3 km long fracture zone develops with a SSW–NNE direction towards Santa Maria di Licodia. The relationship between the

Figure 2
Evidence of movements linked to the RFS activity, as observed in some man-made structures

distinct fault segments is unclear, and there is lack of apparent continuity in the field. Finally, there is no field evidence of the NE–SW trending fault reported by ROMANO *et al.* (1979) and RASÀ *et al.* (1981) in the area between Ragalna and Santa Maria di Licodia (see also MONACO *et al.* 2010; AZZARO *et al.* 2012).

1.2. Dataset and Polarization Analysis Method

In order to investigate the effect of the damage zone of the RFS on ground motion, ambient noise was recorded along 200–400 m long transects crossing the fault (see Figs. 1, 3). In each of the two fault segments, two transects were carried out, each one consisting of four to six measurement sites (Fig. 3). Measurements were also performed in the Piano dei Grilli (PG) area (Fig. 1) in order to record data in districts significantly far (up to 7 km) from the investigated tectonic structures. This allowed us to observe how directional amplification effects change moving away from the fault zone. Ambient noise was recorded using a 3-component 1-Hz velocimeter Mark L4C 3-D connected to a 12-bit analog-to-digital converter. Time series were 30 min long, with a sampling rate of 100 Hz. The signals were processed computing the horizontal-to-vertical noise spectral ratios (HVNSR). According to the guidelines suggested by SESAME (2004), time windows of 30 s were selected in the stationary part of the records, eliminating transients associated to local disturbances. Fourier spectra were calculated and smoothed using a triangular average on frequency intervals of ± 5 % of the central frequency. HVNSRs were calculated after rotating the NS and EW

Figure 3
Location and name of ambient noise recording sites along the four transects carried out across the RFS

components of motion by bins of 10 from 0° (north) to 180° (south). They are plotted in Fig. 4 using contour plots of amplitude, as a function of frequency (*x*-axis) and direction of motion (*y*-axis). This approach is powerful in enhancing, if any, the occurrence of site-specific directional effects. It was applied to earthquake data (SPUDICH *et al.* 1996; PISCHIUTTA *et al.* 2010) to study directional resonances. A similar procedure was applied by PANZERA *et al.* (2011a, 2012) and BURJÁNEK *et al.* (2012) using ambient noise to identify the site-response directivity on ridges and unstable slopes of rock blocks. Moreover, this technique was adopted by RIGANO *et al.* (2008), DI GIULIO *et al.* (2009) and PISCHIUTTA *et al.* (2012), to study directional resonances in fault zones, using both earthquake and noise recordings.

However, in presence of lateral and vertical heterogeneities or velocity inversion, the HVNSR can be "non-informative" due to the occurrence of amplification on the vertical component of motion (DI GIACOMO *et al.* 2005; PANZERA *et al.* 2011b, 2013).

Thus, in this study, we also applied the time–frequency (TF) polarization analysis proposed by VIDALE (1986) and exploited by BURJÁNEK *et al.* (2012). This technique can provide quite robust results, overcoming the bias that could be introduced by the denominator spectrum in the HVNSR calculation.

Following BURJÁNEK *et al.* (2010, 2012), the continuous wavelet transform (CWT, see KULESH *et al.* 2007) is applied to signals in order to select time windows whose length matches the dominant period: signals are thus decomposed in the time–frequency domain and the polarization analysis is applied. For each time–frequency pair, polarization is characterized by an ellipsoid and is defined by two angles: the strike (azimuth of the major axis projected to the horizontal plane from North) and the dip (angle of the major axis from the vertical axis). Another important parameter is ellipticity that is defined, according to VIDALE (1986), as the ratio between the length of the minor and major axes: this

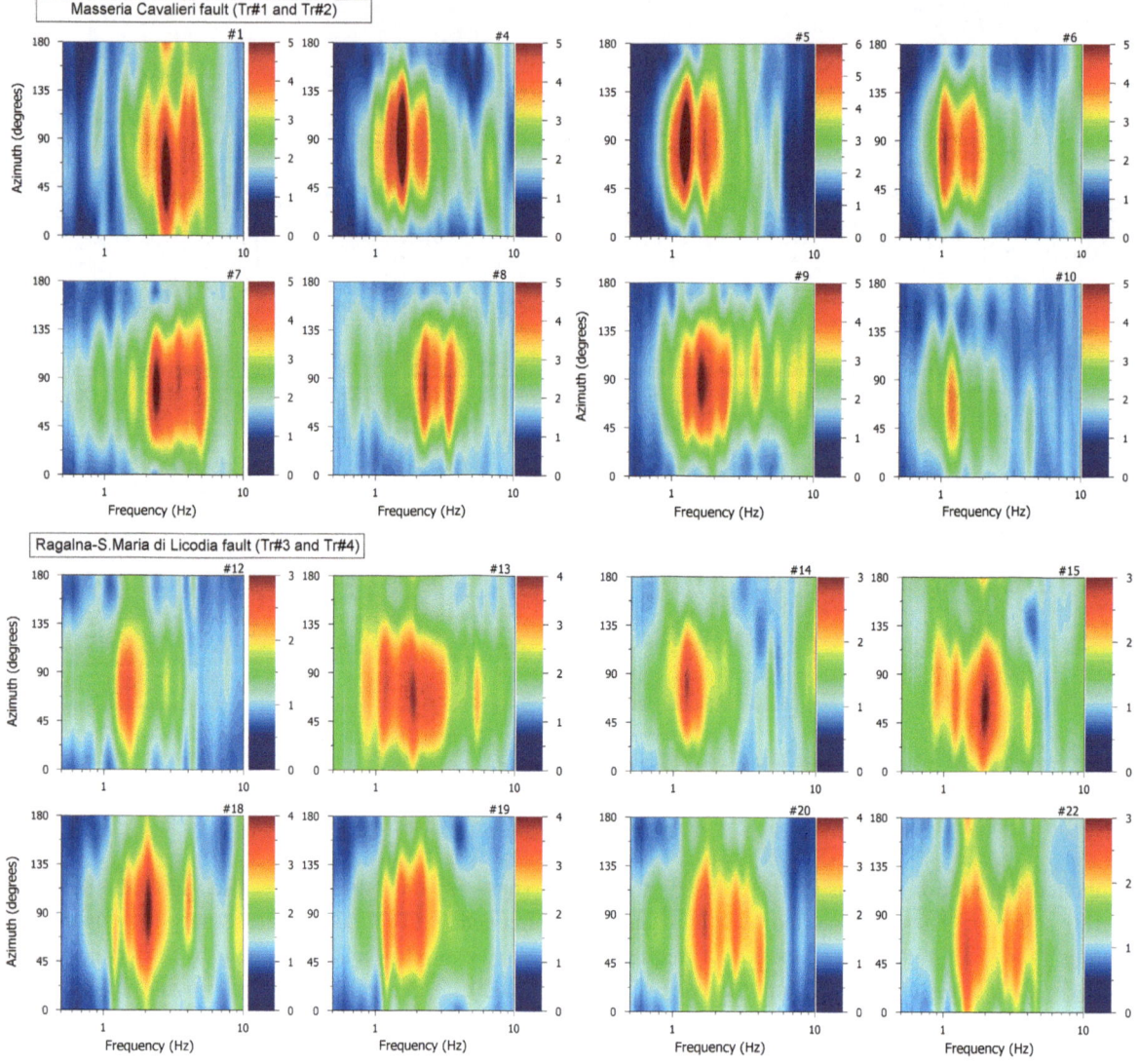

Figure 4
Examples of contours of the spectral ratios geometric mean, as a function of frequency (*x*-axis), direction of motion (*y*-axis) and HVNSR contour plot amplitudes obtained at sites located on the RFS (see locations in Figs. 1 and 3)

parameter approaches 0 when ground motion is linearly polarized. As stressed by BURJÁNEK *et al.* (2010, 2012), the chosen wavelet in the CWF analysis affects all the polarization parameters as well as the analysis resolution. Polarization strike and dip obtained all over the time series analyzed are cumulated and represented using polar plots where the contour scale represents the relative frequency of occurrence of each value, and the distance to the center represents the signal frequency in Hz. In order to assess whether ground motion is

linearly polarized, the ellipticity is also plotted versus frequency.

2. Results

2.1. Observations

In Fig. 4, we show some results of HVNSRs obtained at selected ambient noise recording sites, along the four short profiles crossing the Masseria

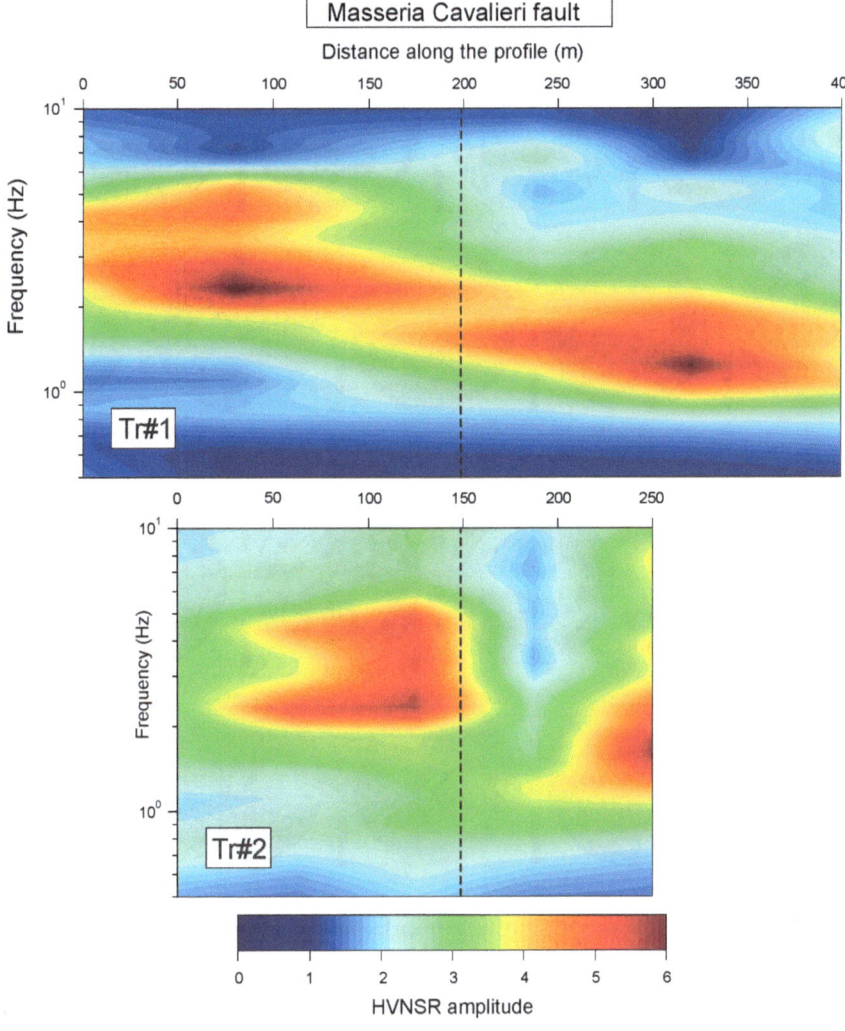

Figure 5

Cross-sections obtained combining the ambient noise measurements along Tr#1 and Tr#2 profiles of the Masseria Cavalieri fault as a function of distance (*x*-axis), frequency (*y*-axis) and HVNSR contour plot amplitudes. The *dotted line* indicates the position of the fault

Cavalieri fault and the inferred Ragalna-S.M. Licodia structure. The HVNSRs depict a significant amplitude increment in the frequency band 1.0–6.0 Hz, at angles of about 80°–90° for the Masseria Cavalieri and 60°–70° for the Ragalna-S.M. Licodia, respectively, showing several adjacent peaks that indicate a preferential and site-dependent direction of horizontal ground motion amplification. It is also interesting to remark that in the Masseria Cavalieri fault, the results related to the stations on the west side of the fault show significant spectral ratio amplitude at frequency values higher than those observed at the stations located on the east side of the fault (Fig. 5). This effect could be an evidence of fault damage asymmetry (Dor *et al.* 2006; Duan 2008 and references therein). Since the peaked frequency is controlled by the ratio of shear velocity by the thickness of the fractured layer, Fig. 5 suggests that, in the Masseria Cavalieri fault, the damage zone has a smaller velocity (higher damage) or the volume of cracks is larger (or both) in the eastern part where lower frequencies are peaked in the spectral ratios.

To get more precise information on the ground motion horizontal polarization, the TF polarization

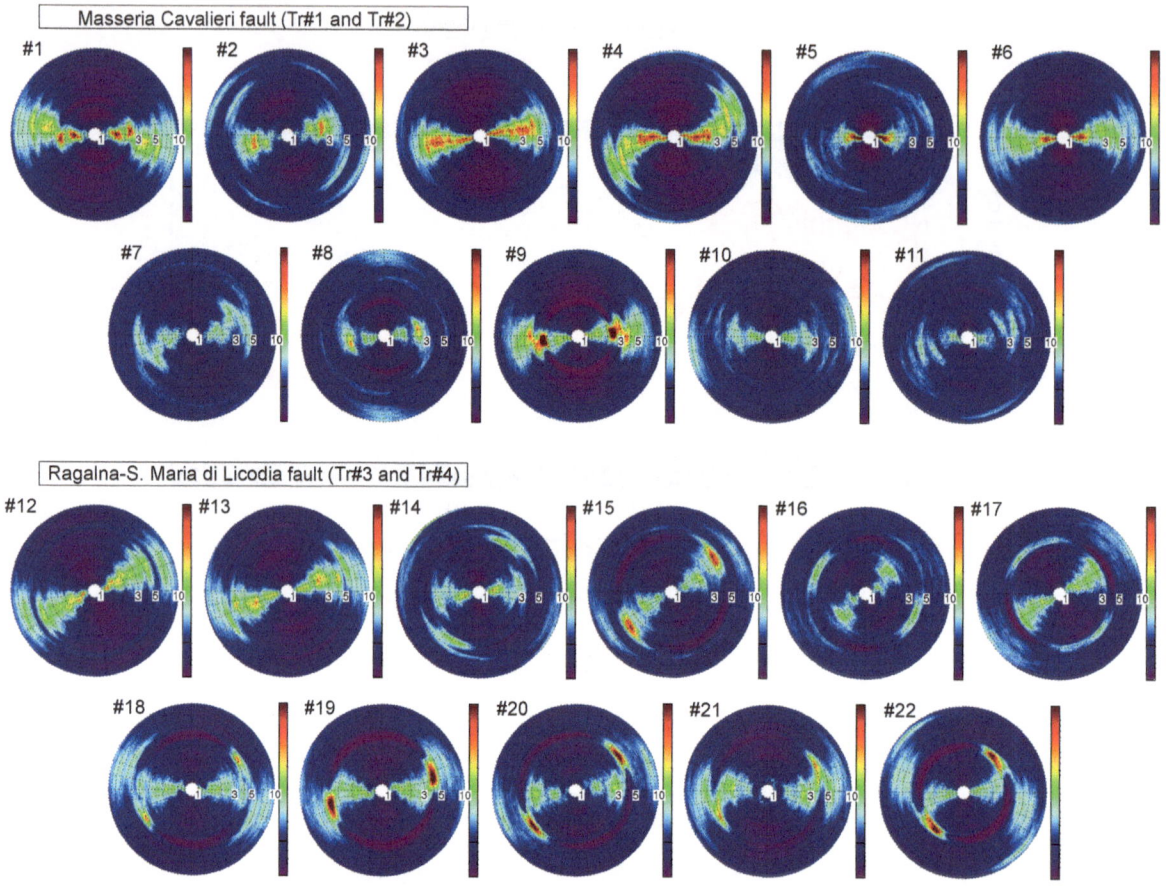

Figure 6
Polar plots, computed through the TF polarization analysis, of the strike angles from ambient noise measurements along the RFS transects (see location in Figs. 1 and 3)

analysis was applied (Fig. 6). It is worth noting that polar plots visibly confirm a pronounced polarization in narrow frequency bands that generally fall in the range 1.0–6.0 Hz, with maxima roughly trending in the W–E and SW–NE direction for the Masseria Cavalieri and the Ragalna-S.M. Licodia segments, respectively. In order to test whether the amplification effects are strictly fault-dependent, we performed some measurements several kilometers away from the RFS, in the Piano dei Grilli area (see location in Figs. 1, 7). Amplification directions obtained from HVNSRs, as well as the polar plots obtained from the time–frequency (TF) polarization analysis, appear to be randomly distributed and/or uniformly scattered (Fig. 7), showing a not-uniform behavior between measuring sites in Piano dei Grilli area. In this way, we can exclude a source-dependence and a time-

dependence of ambient noise in the investigated volcanic area.

The experimental results suggest two different polarization patterns at the two investigated structures. In the Masseria Cavalieri area, we observe quite pronounced amplitudes of the HVNSR peaks (e.g. #1, #4, #5, #8, #10 in Fig. 4). Moreover, the TF polarization analysis revealed values of the ellipticity reaching a minimum in a wide frequency band (1.0–6.0 Hz), as well as dip values showing an horizontal trend in the same frequency band (see examples in Fig. 8a). We stress that in this area, geological investigation found clear evidence of the presence of the fault, as fractures on buildings and concrete road walls.

On the other hand, along the Ragalna-S.M. Licodia alignment we still observe directional effects,

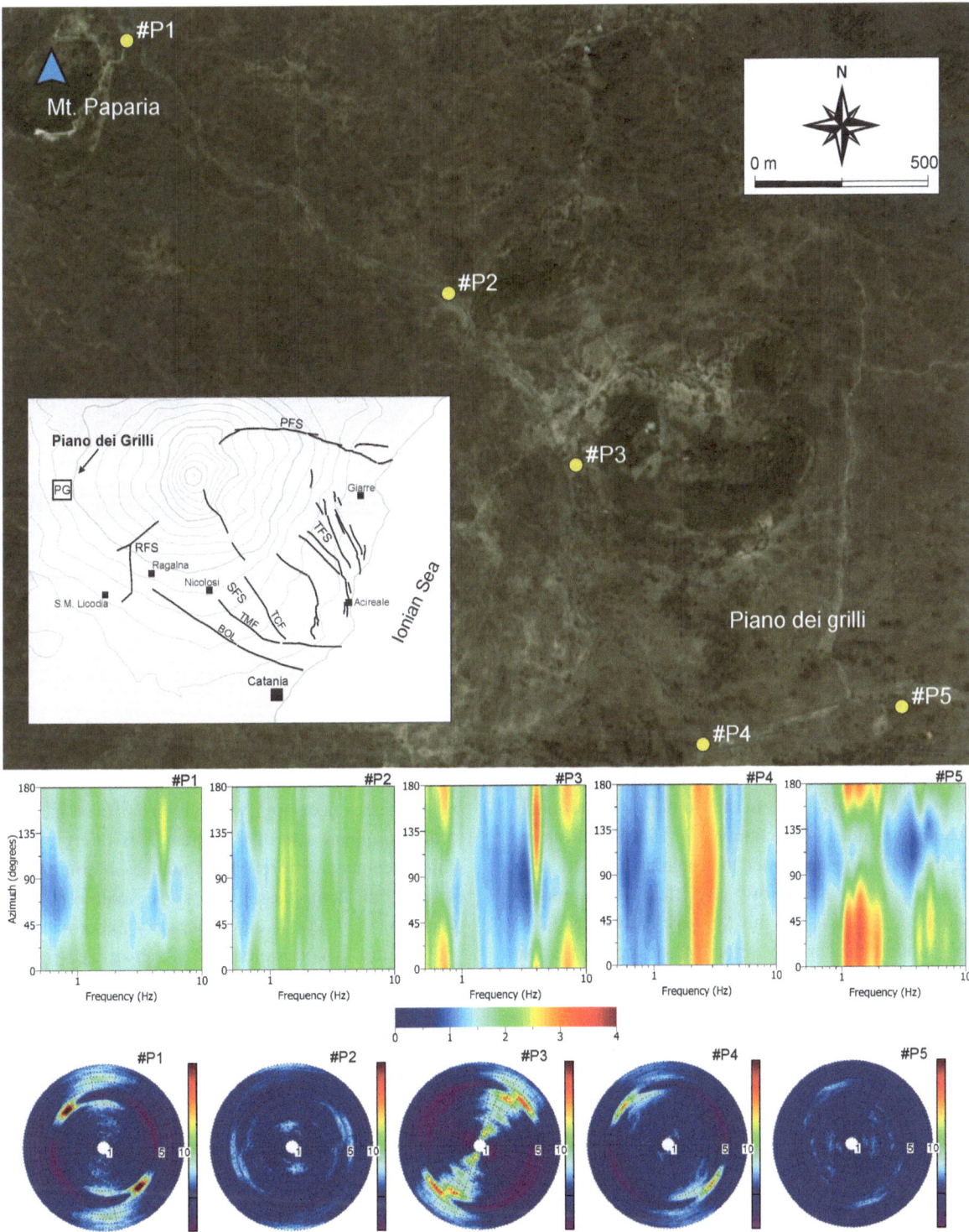

Figure 7
Locations and name of the noise measurement sites at the Piano dei Grilli area and corresponding results obtained through the contours of the spectral ratios geometric mean and the TF polarization analysis (*higher* and *lower* and *panels*, respectively)

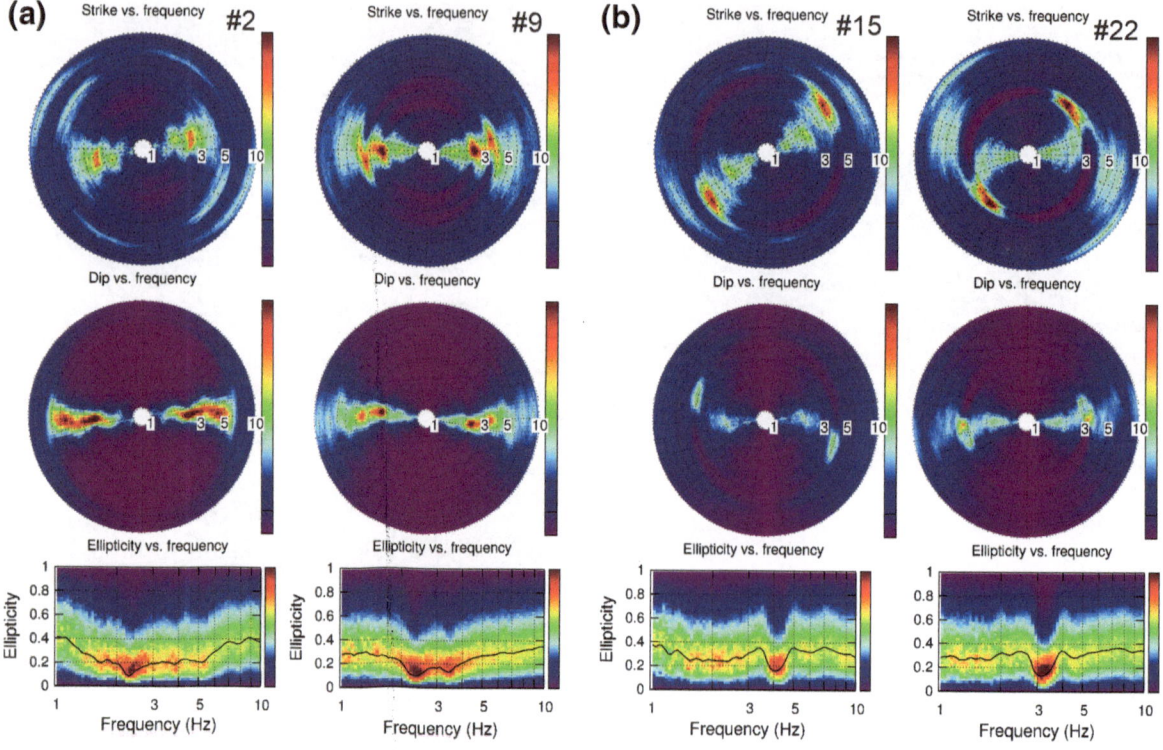

Figure 8
Examples of polar plots and trends of the azimuth, dip and ellipticity in selected sites of the Masseria Cavalieri (**a**) and the Ragalna-S. M. di Licodia structures (**b**)

although they show different features. Indeed, we do not have a decrease of the ellipticity values in a wide frequency range, but a minimum in particular frequency values (see examples in Fig. 8b). Following these findings and since the Ragalna-S.M. Licodia structure does not show any field morphologic evidence, we focus on the Masseria Cavalieri fault. We model the brittle deformation, proposing an interpretation of the observed directional resonance effect in terms of fracture fields.

2.2. Interpretation of Ground Motion Polarization in Terms of Fractures in the Fault-Zone

By analyzing several faults, PISCHIUTTA et al. (2012, 2013a, b) found a near orthogonal relation between the horizontal polarization and the orientation of the main fracture field produced by the fault kinematics. They modeled the expected brittle deformation pattern in the fault damage zone through the Frap Package (SALVINI et al. 1999), that uses a

combination of numeric and analytic tools, and interpreted the directional amplification as produced by stiffness anisotropy, with larger stiffness parallel to the dominant sets of fractures. This interpretation is in agreement with results of numerical simulations of uniaxial compression tests on fractured rocks (GRIFFITH et al. 2009) and laboratory experiments of wet and dry discontinuities (PLACE et al. 2014).

Following the same approach, we investigate the relation between the horizontal polarization and the strike of the fracture field across the Masseria Cavalieri fault. We model the fault as a N–S striking and 80° east-dipping surface, with dimensions 3 km (along strike) × 1 km (along dip). According to the morphotectonic analysis, the fault kinematics is assumed to be characterized by oblique right-lateral strike-slip with a total displacement of 70 m. In order to allow extension to play a concomitant role with strike-slip movement, the kinematic vector is fixed to N145° direction with 45° plunge. The fault surface is discretized into a grid of quadrangular cells, each one

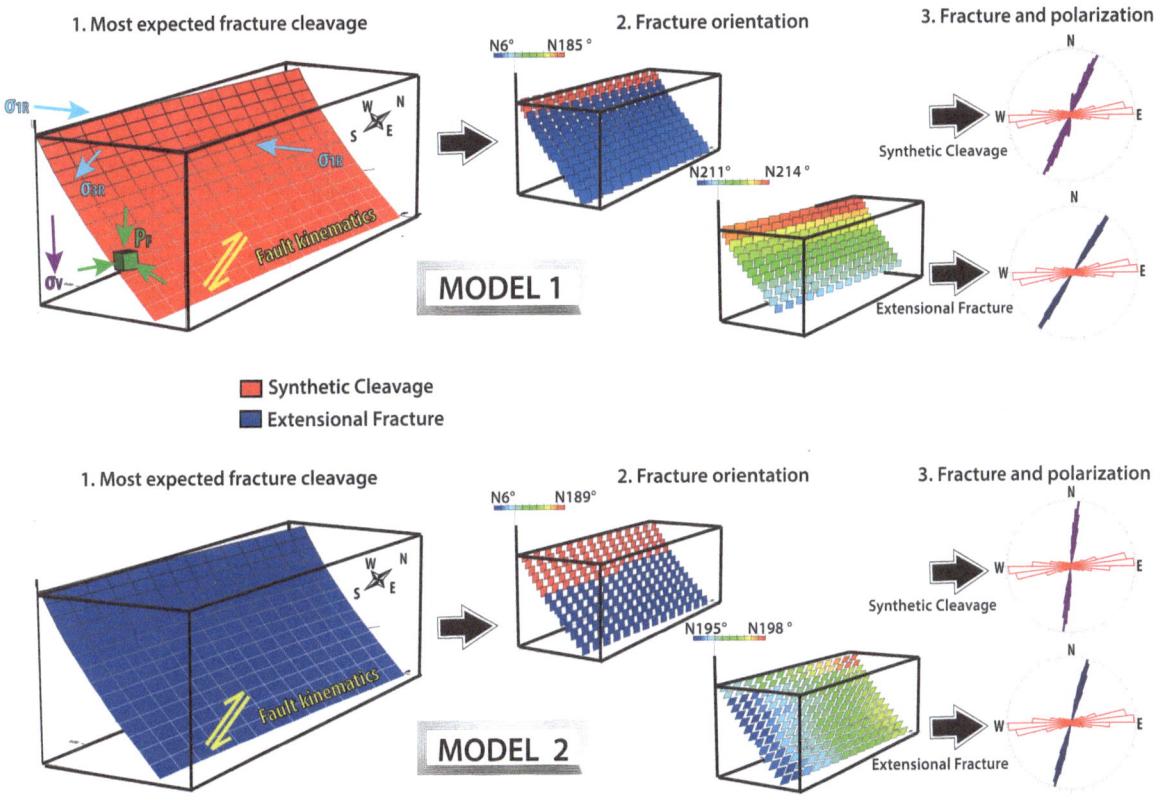

Figure 9

Results of brittle deformation modeling performed for the Masseria Cavalieri fault. The two models differ in the number of stress components (regional stress σ_R, overburden σ_V, fluid pressure P_F and kinematic stress in *Model 1*; only kinematic stress in *Model 2*). The fault surface is discretized through a *grid of quadrangular cells* in a NS–EW reference frame. The cell color depends on the predominant fracture type (*red* for synthetic cleavage and *blue* for extensional fracture). The fault kinematics is assumed to be oblique right-lateral strike-slip and the kinematic vector is fixed to N145° direction with 45° plunge (*yellow arrows*). The *cyan arrows* represent the orientation of the regional stresses $\sigma 1$ and $\sigma 3$, the *violet arrow* the overburden σ_V and the *green arrows* the fluid isotropic pressure P_F. The expected orientation of synthetic cleavage and extensional fractures is drawn using a 3D-plane visualization projected on the fault surface (the *color bars* represent the plane strike). The expected orientation of synthetic cleavage and extensional fractures is visualized through the *wind-rose diagrams* on the *right side* of each panel. The over imposed *red rose diagram* shows the observed polarization across the Masseria Cavalieri fault

being characterized by an attitude and a position in a reference frame. For each cell, the program computes four components of stress: (1) the regional stress tensor σ_R; (2) the overburden σ_V; (3) the fluid isotropic pressure P_F; and (4) the 'kinematic stress' σ_K that results from the brittle strain accumulation due to frictional resistance and failures associated to the fault. The regional stress tensor is set to be consistent with the fault development and is oriented with σ_1 at N35° and σ_3 at N125°, both lying on the horizontal plane (cyan arrows in the top left-hand panels of Fig. 9), and vertical null σ_2 axis.

In our first model, we consider the fracture pattern as produced by all the stress contributors (top panels

of Fig. 9). The resulting stress tensor is calculated at each cell through the sum of the four stress components. Then the resulting stress is compared to the strength in the cell as predicted by the Coulomb–Navier Failure Criterion, in order to evaluate the capability of producing fractures at each cell. This model predicts that synthetic cleavage is the predominating fracture type, with an average strike of N22° (Fig. 9, top left-hand panel). Its distribution is computed using the Daisy Package (SALVINI *et al.* 1999; available at http://host.uniroma3.it/progetti/fralab/), and its strike is visualized through a violet wind-rose diagram (Fig. 9, top central and right-hand panels). Secondary extensional fractures are expected

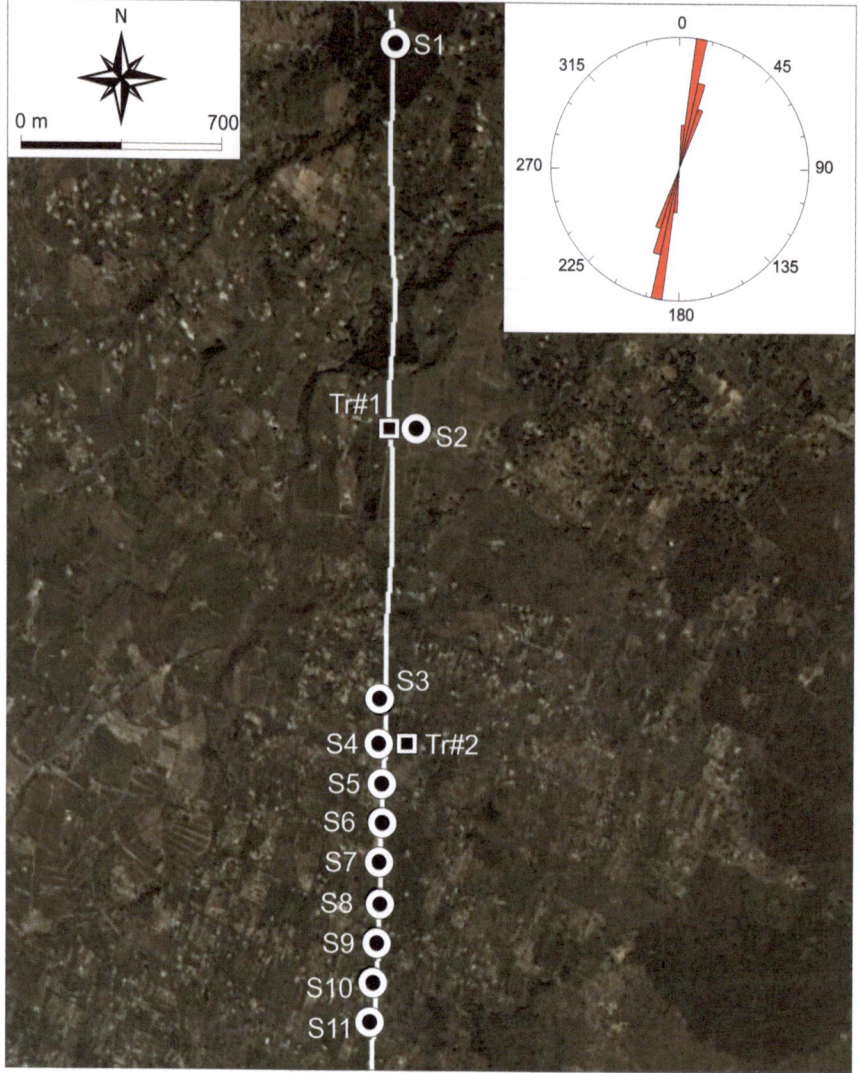

Figure 10
Location of the man-made structures where fracture azimuths were measured and corresponding *rose diagram*

to develop with an average strike of N33° (blue rose diagram).

In our second model, we estimate the expected fracture pattern as produced only by the kinematic stress, which is caused by the frictional motion of the fault (bottom panels of Fig. 9). To evaluate fracture production, the comparison with the strength is here performed only using the kinematic stress values at each cell. This second model yields predominant extensional fractures striking at N15° direction, with synthetic cleavage at N7° direction (Fig. 9, bottom central and right-hand panels).

Considering that in the Masseria Cavalieri fault, observed polarization is oriented roughly E–W (Fig. 9, red rose diagrams), the second model seems to better explain the experimental observations. Moreover, these orientations agree with the fracture strikes as measured in the field along the Masseria Cavalieri fault zone. These extensional fractures were recognized on buildings and concrete road walls in the fault neighborhood, with a centimetric–millimetric oblique-dextral offset. The azimuth distribution through a red rose diagram as well as the location of the 11 measuring sites are shown in Fig. 10.

2.3. Concluding Remarks

We have found new evidence of ground motion polarization in the western flank of Mt. Etna on fault zones associated with the RFS.

The analysis was performed using ambient noise signals through the HVSRs computation and TD polarization analysis. It revealed the occurrence of a directional amplification effect at stations installed in the fault damage zone, and polarization showing a high angle with the fault strike. This is consistent with previous studies on other faults (PISCHIUTTA *et al.* 2012, 2013a, 2014). Noise measurements performed in an area that is thousand meters away from the fault structures did not show any amplification effect, leading to the exclusion of source and time dependence in the investigated volcanic area.

On the two sides of the fault, directional amplification occurs in different frequency bands, peaked frequencies being smaller on the eastern side. This could be related to an asymmetric rock damage.

The observed wavefield polarization is then interpreted in terms of fractures associated with the fault damage zone. Modeling of brittle deformation pattern in the Masseria Cavalieri fault provides a nearly perpendicular relation between the theoretical fracture field and the experimentally observed polarization seismic wavefield. Field observations of fractures on man-made structures match the results of theoretical models well.

Finally, it is worth noting that the present analysis on local amplifications in fault zones gives further support to findings coming out from previous studies (e.g., RIGANO *et al.* 2008; DI GIULIO *et al.* 2009; PISCHIUTTA *et al.* 2012). These results and the evidence of sharp spatial variations of ground motion within small distances could have important implications on shaking hazards at the local scale. Accordingly, they promote the use of ambient noise recordings as a fast technique for preliminary investigations about angular relations between fractures field and directions of amplified ground motion in fault zones.

Acknowledgments

The authors are grateful to Dr. J. Burjánek for having kindly provided access to the time–frequency (TF) polarization analysis software and for useful explanations. The authors also wish to thank anonymous reviewers and the Guest Editor Prof. Yehuda Ben-Zion for constructive comments that contributed to improving the quality of the paper.

REFERENCES

ACOCELLA, V., BEHNCKE, B., NERI, M., and D'AMICO, S. (2003), *Link between major flank slip and eruptions at Mt. Etna (Italy)*, Geophys. Res. Lett., *30*(24), doi:10.1029/2003GL018642.

ALPARONE, S., BARBERI, G., BONFORTE, A., MAIOLINO, V., and URSINO, A. (2011), *Evidence of multiple strain fields beneath the eastern flank of Mt. Etna volcano (Sicily, Italy) deduced from seismic and geodetic data during 2003–2004*, Bull. Volcanol. 7, 869–885, doi:10.1007/s00445-011-0456-1.

AZZARO, R. (1999), *Earthquake surface faulting at Mount Etna volcano (Sicily) and implications for active tectonics*, J. Geodyn., 28, 193– 213, doi:10.1016/S0264-3707(98)00037-4.

AZZARO, R., MATTIA, M., and PUGLISI, G. (2001), *Fault creep and kinematics of the eastern segment of the Pernicana Fault (Mt. Etna, Italy) derived from geodetic observation and their tectonic significance. 333*, 401–415.

AZZARO, R., BRANCA, S., GWINNER, K., and COLTELLI, M. (2012), *The volcano-tectonic map of Etna volcano, 1:100.000 scale: an integrated approach based on a morphotectonic analysis from high-resolution DEM constrained by geologic, active faulting and seismotectonic data*, Ital. J. Geosci. *131*(1), 153–170. doi:10. 3301/IJG.2011.29.

BARRECA, G., BONFORTE, A., and NERI, M. (2013), *A pilot GIS database of active faults of Mt. Etna (Sicily): A tool for integrated hazard evaluation*, J. Volcanol. Geotherm. Res. *251*(1), 170–186, doi:10.1016/j.jvolgeores.2012.08.013.

BEN-ZION, Y. (1998), *Properties of seismic fault zone waves and their utility for imaging low-velocity structures*, J. Geophys. Res., *103*(B6), 12567–12585.

BEN-ZION, Y., and AKI, K. (1990), *Seismic radiation from an SH line source in a laterally heterogeneous planar fault zone*, Bull. seism. Soc. Am. *80*, 971–994.

BONFORTE, A., BONACCORSO, A., GUGLIELMINO, F., PALANO. M., and PUGLISI, G. (2008), *Feeding system and magma storage beneath Mt. Etna as revealed by recent inflation/deflation cycles*, J. Geophys. Res. *113*, B05406, doi:10.1029/2007JB005334.

BONFORTE, A., FEDERICO, C., GIAMMANCO, S., GUGLIELMINO, F., LIUZZO, M., and NERI, M. (2013), *Soil gases and SAR data reveal hidden faults on the sliding flank of Mt. Etna (Italy)*, J. Volcan. Geotherm. Res. *251*, 27–40, doi:10.1016/j.volgeores.2012.08. 010.

BONFORTE, A., GUGLIELMINO, F., COLTELLI, M., FERRETTI, A., and PUGLISI, G. (2011), *Structural assessment of Mt. Etna volcano from Permanent Scatterers analysis*, Geochemistry, Geophysics, Geosystems *12*(2), Q02002, doi:10.1029/2010GC003213.

BORGIA, A., LANARI, R., SANSOSTI, E., TESAURO, M., BERARDINO, P., FORNARO, G., NERI, M., and MURRAY, J.B. (2000), *Actively growing anticlines beneath Catania from the distal motion of Mount Etna's decollement measured by SAR interferometry and GPS*, Geophys. Res. Lett., *27*(20), 3409–3412, doi:10.1029/ 1999GL008475.

Reprinted from the journal

BRANCA, S., COLTELLI, M., GROPELLI, G., and LENTINI, F. (2011), *Geological map of Etna volcano, 1:50000 scale*, Ital. J. Geosci. *130*(3), 265–291.

BURJÁNEK, J., GASSNER-STAMM, G., POGGI, V., MOORE, J. R., and FÄH, D. (2010), *Ambient vibration analysis of an unstable mountain slope*, Geophys. J. Int., *180*, 559–569, doi:10.1111/j.1365-246X.2009.04451.x.

BURJÁNEK, J., MOORE, J. R., MOLINA, F.X.Y., and FÄH, D. (2012), *Instrumental evidence of normal mode rock slope vibration*, Geophys. J. Int., *188*(2), 559–569, doi:10.1111/j.1365-246X.2011.05272.x.

CALDERONI, G., ROVELLI, A., and DI GIOVAMBATTISTA, R. (2010), *Large amplitude variations recorded by an on-fault seismological station during the L'Aquila earthquakes: evidence for a complex fault-induced site effect*, Geophys. Res. Lett. *37*, L24305, doi:10.1029/2010GL045697.

CULTRERA, G., ROVELLI, A., MELE, G., AZZARA, R., CASERTA, A., and MARRA, F. (2003), *Azimuth dependent amplification of weak and strong ground motions within a fault zone, Nocera Umbra, Central Italy*, J. Geophys. Res. *108*(B3), 2156–2170, doi:10.1029/2002JB001929.

DI GIACOMO, D., GALLIPOLI, M.R., MUCCIARELLI, M., PAROLAI, S., and RICHWALSKI, S.M. (2005), *Analysis and modeling of HVSR in the presence of a velocity inversion: the case of Venosa, Italy*, Bull. Seism. Soc. Am. *95*(6), 2364–2372, doi:10.1785/0120040242.

DI GIULIO, G., CARA, F., ROVELLI, A., LOMBARDO, G., and RIGANO, R. (2009), *Evidence for strong directional resonances in intensely deformed zones of the Pernicana fault, Mount Etna, Italy*, J. Geophys. Res., *114*, doi:10.1029/2009JB006393.

DOR, O., ROCKWELL, T. K., BEN-ZION, Y. (2006), *Geological observations of damage asymmetry in the structure of the San Jacinto, San Andreas and Punchbowl faults in southern California: A possible indictor for preferred rupture propagation direction*, Pure Appl. Geophys., *163*, 301–349.

DUAN, B. (2008), *Asymmetric off-fault damage generated by bilateral ruptures along a bimaterial interface*, Geophys. Res. Lett. *35*, L14306, doi:10.1029/2008GL034797.

FALSAPERLA, S., CARA, F., ROVELLI, A., NERI, M., BEHNCKE, B., and ACOCELLA, B. (2010), *Effects of the 1989 fracture system in the dynamics of the upper SE flank of Etna revealed by volcanic tremor data: the missing link?* J. Geophys. Res. *115*, B11306, doi:10.1029/2010JB007529.

FROGER, J.L., MERLE, O., and BRIOLE, P. (2001), *Active spreading and regional extension at Mount Etna imaged by SAR interferometry*, Earth Planet. Sci. Lett. *187*, 245–258, doi:10.1016/S0012-821X(01)00290-4.

GRIFFITH, W.A., SANZ, P.F., and POLLARD, D.D. (2009), *Influence of outcrop scale fractures on the effective stiffness of fault damage zone rocks*, Pure Appl. Geophys., *166*, 1595–1627.

KULESH, M., DIALLO, M.S., HOLSCHNEIDER, M., KURENNAYA, K., KRUGER, F., OHRBERGER, M., and SCHERBAUM, F. (2007), *Polarization analysis in the wavelet domain based on adaptive covariance method*, Geophys. J. Int. *170*(2), 667–678, doi:10.1111/j.1365-246X.2007.03417.x.

LEWIS, M.A., PENG, Z., BEN-ZION, Y. and VERNON, F.L. (2005), *Shallow seismic trapping structure in the San Jacinto fault zone near Anza, California*, Geophys. J. Int., *162*, 867–881, doi:10.1111/j.1365-246X.2005.02684.x.

LEWIS, M., and BEN-ZION, Y. (2010), *Diversity of fault zone damage and trapping structures in the Parkfield section of the San Andreas Fault from comprehensive analysis of near fault seismograms*, Geophys. J. Int., *183*(3), 1579–1595, doi:10.1111/j.1365-246X.2010.04816.x.

LI, Y.G., and LEARY, P. C. (1990), *Fault zone trapped seismic waves*, Bull. Seismol. Soc. Am., *80*, 1245–1271.

LI, Y. G., LEARY, P. C., AKI, K. and MALIN, P. (1990), *Seismic trapped modes in the Oroville and San Andreas fault zones*, Science, *249*, 763– 765, doi:10.1126/science.249.4970.763.

LI, Y.G., AKI, K., ADAMS, D., HASEMI, A., and LEE, W.H.K. (1994), *Seismic guided waves trapped in the fault zone of the Landers, California, earthquake of 1992*, J. Geophys. Res. *99*, 11705–11722.

LO GIUDICE, E., and RASÀ, R. (1992), *Very shallow earthquakes and brittle deformation in active volcanic areas: the Etnean region as example*, Tectonophysics *202*, 257–268.

MCGUIRE, W.J., and PULLEN, A.D. (1989), *Location and orientation of eruptive fissures and feeder-dykes at Mount Etna: influence of gravitational and regional stress regimes*, J. Volcanol. Geotherm. Res. *38*, 325–344.

MIZUNO, T., and NISHIGAMI, K. (2004), *Deep structure of the Mozumi-Sukenobu fault, central Japan, estimated from the subsurface array observation of fault zone trapped waves*, Geophys. J. Int., *159*(2), 622–642, doi:10.1111/j.1365-246X.2004.02458.x.

MONACO, C., TAPPONNIER, P., TORTORICI, L., and GILLOT, P.Y. (1997), *Late Quaternary slip rates on the Acireale–Piedimonte normal faults and tectonic origin of Mt. Etna (Sicily)*, Earth Planet. Sci. Lett. *147*, 125–139.

MONACO, C., DE GUIDI, G., and FERLITO, C. (2010), *The Morphotectonic map of Mt. Etna*, Ital. J. Geosci., *129*, 3, 408–428.

NERI, M., GARDŪNO, V.H., PASQUARÈ, G., and RASÀ, R. (1991), *Studio strutturale e modello cinematico della Valle del Bove e del settore nord-orientale etneo*, Acta Vulcanologica, *1*, 17–24 (in Italian).

NERI, M., ACOCELLA, V., and BEHNCKE, B. (2004), *The role of the Pernicana Fault System in the spreading of Mt. Etna (Italy) during the 2002–2003 eruption*, Bull. Volcanol. *66*, 417–430, doi:10.1007/s00445-003-0322-x.

NERI, M., GUGLIELMINO, F., and RUST, D. (2007), *Flank instability on Mount Etna: Radon, radar interferometry, and geodetic data from the southwestern boundary of the unstable sector*, J. Geophys. Res., *112*, B04410, doi:10.1029/2006JB0047.

NERI, M., CASU, F., ACOCELLA, V., SOLARO, G., PEPE, S., BERARDINO, P., SANSOSTI, E., CALTABIANO, T., LUNDGREN, P., and LANARI, R. (2009), *Deformation and eruptions at Mt. Etna (Italy): a lesson from 15 years of observations*, Geophy. Res. Lett. *36*(2), L02309, doi:10.1029/2008GL036151.

PANZERA, F., D'AMICO, S., LOTTERI, A., GALEA, P., and LOMBARDO, G. (2012), *Seismic site response of unstable steep slope using noise measurements: the case study of Xemxija bay area, Malta*, Nat. Haz. Earth Sci. *12*, 3421–3431, doi:10.5194/nhess-12-3421-2012.

PANZERA, F., LOMBARDO, G., and MUZZETTA, I. (2013), *Evaluation of buildings dynamical properties through in situ experimental techniques and 1D modelling: the example of Catania, Italy*, J. Phys. Chem. Earth, doi:10.1016/j.pce.2013.04.00.

PANZERA, F., LOMBARDO, G., and RIGANO, R. (2011 a), *Evidence of topographic effects analysing ambient noise measurements: the study case of Siracusa, Italy*, Seismol. Res. Lett. *82*(3), 385–391, doi:10.1785/gssrl.82.3.385.

PANZERA, F., RIGANO, R., LOMBARDO, G., CARA, F., DI GIULIO, G., ROVELLI, A. (2011b), *The role of alternating outcrops of sediments and basaltic lavas on seismic urban scenario: the study case of Catania, Italy*, Bulletin of Earthquake Engineering 9(2), 411–439. doi:10.1007/s10518-010-9202-x.

PENG, Z., and BEN-ZION, Y. (2006), *Temporal changes of shallow seismic velocity around the Karadere-Duzce Branch of the North Anatolian Fault and strong ground motion*, Pure Appl. Geophys. *163*, 567–600.

PISCHIUTTA, M., CULTRERA, G., CASERTA, A., LUZI, L., and ROVELLI, A. (2010), *Topographic effects on the hill of Nocera Umbra, central Italy*, Geoph. J. Inter., *182*(2), 977–987, doi:10.1111/j.1365-246X.2010.04654.x.

PISCHIUTTA, M., SALVINI, F., FLETCHER, J., ROVELLI, A., and BEN-ZION, Y. (2012), *Horizontal polarization of ground motion in the Hayward fault zone at Fremont, California: dominant fault-high-angle polarization and fault-induced cracks*, Geophys. J. Int., *188*(3), 1255–1272, doi:10.1111/j.1365-246X.2011.05319.x.

PISCHIUTTA, M., ROVELLI, A., SALVINI, F., DI GIULIO, G., and BEN-ZION, Y. (2013a), *Directional resonance variations across the Pernicana Fault, Mt. Etna, in relation to brittle deformation fields*, Geophys. J. Int., *193*(2), 986–996, doi:10.1093/gji/ggt031.

PISCHIUTTA, M., ANSELMI, M., CIANFARRA, P. ROVELLI, A., and SALVINI, F. (2013b), *Directional site effects in a non-volcanic gas emission area (Mefite d'Ansanto, southern Italy): Evidence of a local transfer fault transversal to large NW–SE extensional faults?* J. Phys. Chem. Earth, 116–123, doi:10.1016/j.pce.2013.03.008.

PISCHIUTTA, M., PASTORI, M., IMPROTA, L., SALVINI, F., and ROVELLI, A. (2014), *Orthogonal relation between wavefield polarization and fast S-wave direction in the Val d'Agri region: an integrating method to investigate rock anisotropy*, J. Geophys. Res, *119*, 1–13, doi:10.1002/2013JB010077.

PLACE, J., BLAKE, O., RIETBROCK. A., and FAULKNER, D. (2014), *Wet fault or dry fault? A laboratory approach to monitor at distance the hydromechanical state of a discontinuity using controlled source seismics*, Pure Appl. Geophys, this volume.

RASÀ, R., ROMANO, R., and LO GIUDICE, E. (1981), *Morphotectonic map of Mt. Etna, 1:100.000 scale*, Progetto Finalizzato Geodinamica, Istituto Internazionale di Vulcanologia, *CNR*, Catania, Italy.

RIGANO, R., CARA, F., LOMBARDO, G., and ROVELLI, A. (2008), *Evidence of ground motion polarization on fault zones of Mount Etna volcano*, J. Geophys. Res. *113*, B10306, doi:10.1029/2007-JB005574.

ROMANO, R., STURIALE, C., and LENTINI, F. (1979), *Geological map of Mt. Etna, 1:50.000 scale*, Progetto Finalizzato Geodinamica, Istituto Internazionale di Vulcanologia, *CNR*, Catania, Italy.

ROVELLI, A., CASERTA, A., MARRA, F., and RUGGIERO, V. (2002), *Can seismic waves be trapped inside an inactive fault zone? The case study of Nocera Umbra, central Italy*, Bull. Seism. Soc. Am., *92*(6), 2217–2232, doi:10.1785/0120010288.

RUST, D., and NERI, M. (1996), *The boundaries of large-scale collapse on the flanks of Mount Etna, Sicily. In: Volcano Instability on the Earth and Other Planets*, edited by W.M. McGuire, A.P. Jones and J. Neuberg, Spec. Pub. Geol. Soc. London, *110*, 193–208.

SALVINI, F., BILLI, A., and WISE, D.U. (1999), *Strike-slip fault-propagation cleavage in carbonate rocks: the Mattinata Fault Zone, Southern Apennines, Italy*, J. Struct. Geol., *21*, 1731–1749.

SEEBER, L., ARMBRUSTER, J. G., OZER, N., AKTAR, M., BARIS, S., OKAYA, D., BEN-ZION, Y., and FIELD, E. (2000), *The 1999 Earthquake Sequence along the North Anatolia Transform at the Juncture between the Two Main Ruptures, in The 1999 Izmit and Duzce Earthquakes: preliminary results*, Edited by Barka A., Kazaci, O., Akyuz, S., and Altunel, E., Istanbul Technical University, 209–223.

SESAME (2004), *Guidelines for the implementation of the H/V spectral ratio technique on ambient vibrations: Measurements, processing and interpretation*, SESAME European Research Project WP12, deliverable D23.12, at http://sesame-fp5.obs.ujf-grenoble.fr/Deliverables, 2004.

SOLARO, G., ACOCELLA, V., PEPE, S., RUCH, J., NERI, M., and SANSOSTI, E. (2010), *Anatomy of an unstable volcano from InSAR: multiple processes affecting flank instability at Mt. Etna, 1994–2008*, J. Geophys. Res., *115*, B10405, doi:10.1029/2009JB000820.

SPUDICH, P., HELLWEG, M., and LEE, W.H.K. (1996), *Directional topographic site response at Tarzana observed in aftershocks of the 1994 Northridge, California, earthquake: implications for mainshock motions*, Bull. Seism. Soc. Am., *86*(1B), S193–S208.

SPUDICH, P., and OLSEN, K.B. (2001), *Fault zone amplified waves as a possible seismic hazard along the Calaveras Fault in central California*, Geophys. Res. Lett. *28*(13), 2533–2536, doi:10.1029/2000GL011902.

VIDALE, J.E. (1986), *Complex polarization analysis of particle motion*. Bull. Seism. Soc. Am. *76*, 1393–1405.

(Received September 17, 2013, revised March 10, 2014, accepted March 12, 2014, Published online March 27, 2014)

Reprinted from the journal

Pure Appl. Geophys. 171 (2014), 3099–3123
© 2014 Springer Basel
DOI 10.1007/s00024-014-0845-4

A Continuum Damage–Breakage Faulting Model and Solid-Granular Transitions

VLADIMIR LYAKHOVSKY[1] and YEHUDA BEN-ZION[2]

Abstract—We present a thermodynamically-based formulation for mechanical modeling of faulting processes in the seismogenic brittle crust using a continuum damage–breakage rheology. The model combines previous results of a continuum damage framework for brittle solids with continuum breakage mechanics for granular flow. The formulation accounts for the density of distributed cracking and other internal flaws in damaged rocks with a scalar damage parameter, and addresses the grain size distribution of a granular phase in a failure slip zone with a breakage parameter. The stress–strain relation and kinetics of the damage and breakage processes are governed by the total energy function of the system, which combines the energy of the damaged solid with the energy of the granular material. A dynamic brittle instability is associated with a critical level of damage in the solid, leading to loss of convexity of the solid energy function and transition to a granular phase associated with lower energy level. A non-local formulation provides an intrinsic length scale associated with the internal damage structure, which leads to a finite length scale for damage localization that eliminates the unrealistic singular localization of local models. Shear heating during deformation can lead to a secondary finite-width internal localization. The formulation provides a framework for studying multiple aspects of brittle deformation, including potential feedback between evolving elastic moduli and properties of the slip localization zone and subsequent rupture behavior. The model has a more general transition from slow deformation to dynamic rupture than that associated with frictional sliding on a single pre-existing failure zone, and gives time and length scales for the onset of the dynamic fracturing process. Several features including the existence of finite localization width and transition from slow to rapid dynamic slip are illustrated using numerical simulations. A configuration having an existing narrow slip zone with localized damage produces for appropriate loading conditions an overall cyclic stick–slip motion. The simulated frictional response includes transitions from friction coefficient of ∼0.7 at low slip velocity to dynamic friction below 0.4 at slip rates above ∼0.1 m/s, followed by rapidly increasing friction for slip rates above ∼1 m/s, consistent with laboratory observations.

Key words: Mechanics of faulting, fault zone rheology, friction, fracture, brittle damage, granular flow, phase transitions.

List of symbols

Thermodynamics state variables

T	Temperature
α	Damage parameter
B	Breakage parameter
ε_{ij}	Elastic strain tensor

Strain components

$\varepsilon_{ij}^{(t)}$	Total strain tensor
$\varepsilon_{ij}^{(p)}$	Irreversible (plastic) strain tensor
e_{ij}	Strain-rate tensor

Stress

σ_{ij}	Stress tensor

Material properties

ρ	Density
λ, μ, γ	Damage-dependent elastic moduli of solid phase
a_0, a_1, a_2, a_3	Elastic moduli of granular phase
C_p	Specific heat capacity
κ	Thermal conductivity

Kinetics of damage and breakage

ξ_o	Strain invariants ratio at the onset of damage accumulation/healing
ξ_1	Strain invariants ratio at the transition between mode-I (pseudo-gas) and mode-II (pseudo-liquid)
ξ_d	Strain invariants ratio at the onset of breakage healing
C_d	Rate constant for the damage accumulation
C_1, C_2	Two constants controlling the rate of damage healing
C_B	Rate constant for the breakage accumulation
C_{BH}	Rate constant for the breakage healing

[1] Geological Survey of Israel, 95501 Jerusalem, Israel.
E-mail: vladi@geos.gsi.gov.il
[2] Department of Earth Sciences, University of Southern California, Los Angeles, CA 90089-0740, USA. E-mail: benzion@usc.edu

Fluidity of the granular phase

C_g, m_1, m_2 Rate constant and two power-law indexes for breakage-dependent flow

A, n, Q Rate constant, power-law indexes and activation energy for the temperature-dependent ductile flow component

1. Introduction

Rocks in the brittle crust are damaged in the sense that they have joints, cracks and other flaws over wide ranges of scales. This is especially the situation in active tectonic regions with high deformation rates. Laboratory experiments with increasing differential strain of rock samples not dominated by a single frictional interface show, universally, generation of acoustic emission beyond the elastic limit, and related evolution of elastic moduli, wave speeds and other properties of damaged solids (e.g., JAEGER and COOK, 1979; LOCKNER et al., 1992; STANCHITS et al., 2006). The evolution of various fields and properties in laboratory experiments of brittle deformation have been modeled by various damage rheology frameworks (e.g. ASHBY and SAMMIS, 1990; HAMIEL et al., 2004, 2009; BHAT et al., 2011).

Fault models based on a fixed pre-existing frictional interface may be used to study the seismic response of individual fault zones over time scales of several large earthquake cycles, but they become increasingly inappropriate as the spatio-temporal scales of interest increase (e.g. BEN-ZION, 2008). Crustal regions have fault systems with complex geometries including intersections, stepovers, and many other deviations from planarity. The material around such geometrical irregularities is subjected during ongoing deformation to large stress concentrations, which lead to evolution of the elastic properties and geometry of the deforming regions. A quantitative understanding of seismicity over broad regions of space and time requires a framework that accounts for fault networks with evolving geometrical and material properties, as well as time-dependent interactions between the seismogenic zone and underlying viscoelastic substrate.

To provide such a framework, LYAKHOVSKY et al. (1997a, 2001, 2011), BEN-ZION et al. (1999), HAMIEL et al. (2004), LYAKHOVSKY and BEN-ZION (2008, 2009) and others developed a thermodynamically-based viscoelastic continuum damage model that is discussed and extended in the present paper. Various other damage models were used in numerous engineering and geophysical applications (e.g., KACHANOV, 1986; KRAJCINOVIC, 1996; LEMAITRE, 1996; BERCOVICI and RICARD, 2003; TURCOTTE et al., 2003; ALAVA et al., 2006; SUZUKI, 2012). An overview of applications illustrating the utility of a "local" version of the employed damage model to coupled evolution of earthquakes and faults is given in BEN-ZION (2008, section 6). In the local damage model, the free energy of the brittle material, $F(T, \varepsilon_{ij}, \alpha)$, is a function of the temperature T, the elastic strain tensor $\varepsilon_{ij} = \varepsilon_{ij}^{(t)} - \varepsilon_{ij}^{(p)}$ denoting the difference between the total and irreversible strain components, and a damage state variable $0 \leq \alpha \leq 1$ representing the local crack density (LYAKHOVSKY et al., 1997a). Since both the elastic strain ε and damage state variable α are non-dimensional functions of time and space, the local damage model has no intrinsic length scale.

The local dependency of variables on the damage leads for ranges of rheological parameters and boundary conditions to unrealistic singular strain localization. In various two-dimensional or three-dimensional cases the strain becomes completely localized, i.e., $\varepsilon(x - x_0) \rightarrow \delta(x - x_0)$ around x_0, where $\alpha(x_0) \rightarrow 1$ (LYAKHOVSKY et al., 1997a). Under constant stress boundary conditions, the damage accumulation follows a power law relation (BEN-ZION and LYAKHOVSKY, 2002) with monotonically increasing rate of damage growth that becomes unbounded $(d\alpha/dt \rightarrow \infty)$ toward the macroscopic failure $(\alpha \rightarrow 1)$. The unrealistic singularities of the local damage formulation are eliminated in "non-local" formulations of integral or gradient type that account for properties in "neighborhoods" (e.g., ERICKSEN, 1998; BAZANT and JIRASEK, 2002). Similar extreme strain localization during flow of granular material is also avoided by non-local formulation (e.g., KAMRIN and KOVAL, 2012). In the context of brittle rock deformation, LYAKHOVSKY et al. (2011) formulated a gradient-type damage model that takes non-locality into account by generalizing the free energy, $F(T, \varepsilon_{ij}, \alpha, \nabla_i\alpha)$, to include a dependency on the gradient of the damage state variable $\nabla_i\alpha$. The gradient term $\nabla_i\alpha$ has length units, providing intrinsic length scale associated with internal material structure.

When the level of damage reaches a critical value, the material sustains macroscopic brittle instability involving dynamic rupture and localization of deformation into a narrow slipping zone consisting of granulated material. The generation of granulated material in the slip zone behind propagating ruptures, and various other theoretical and observational results, suggest that transitions between solid and granular states of rocks play important roles in the physics of earthquakes and faults (BEN-ZION, 2008; section 7). Multiple episodes of brittle deformation associated with earthquakes lead to material evolution from a competent solid through damaged rock to granular material, while material healing in the interseismic periods produces evolution in the opposite direction.

Some theoretical aspects of the solid-granular transition and related dynamic phenomena were discussed by LYAKHOVSKY *et al.* (2011) using non-local continuum damage and BEN-ZION *et al.* (2011) using statistical physics results. LYAKHOVSKY and BEN-ZION (2014) proposed recently a continuum damage–breakage rheology model to describe the transition from damaged solid to granular phase near the brittle instability. Their formulation augments the non-local continuum damage model for solid of LYAKHOVSKY *et al.* (2011) with aspects of the breakage mechanics for granular material of EINAV (2007a, b). Both the damage and breakage models are thermodynamically-based continuum frameworks that can be naturally combined into a continuum *damage–breakage* model (CDBM). In the breakage mechanics, the free energy of the granular material, $F(T, \varepsilon_{ij}, B)$, is a function of the temperature, the elastic strain tensor, and a breakage state variable, $0 \leq B \leq 1$, that measures the relative distance of the current grain size distribution of the granular material between the initial and ultimate states (EINAV, 2007a, b).

LYAKHOVSKY and BEN-ZION (2014) focused primarily on the transition from a solid phase into pseudo-liquid (granular) or pseudo-gas (fragmented) phases. In the present paper we analyze in detail the reversed transition from a granular phase to damaged solid. We also extend the CDBM model of LYAKHOVSKY and BEN-ZION (2014) with explicit considerations of thermal effects, detailed descriptions of the dynamic stress and energy release in a granulated slip zone, and the arrest of dynamic motion due to material healing and back-transition from the granular to solid phase. The extended model provides a more complete framework for multiple failure episodes with two-ways transitions between slowly deforming damaged solid and dynamic shear flow of granular material leading to stress relaxation.

In the next two sections (and Appendix) we provide a general background material. We then use theoretical considerations and numerical simulations to illustrate several features of the developed CDBM. These include spontaneous formation of a narrow slipping zone; effective steady-state friction at low shear flow velocities with transition to low effective dynamic friction at high rates; scaling relation controlling the width of the slipping zone, and nucleation size of a dynamically-failed patch of the damage zone. Shear heating and temperature-dependent effective viscosity of the granular phase during deformation can lead to secondary finite-width strain localization within a slipping zone.

2. Thermodynamics of Complex Non-Local Material

The first law of classical thermodynamics postulates that any change of the total energy of the system, E, is equal to the mechanical power, P, plus heat input rate, Q, (e.g., MALVERN, 1969):

$$\frac{dE}{dt} = P + Q. \qquad (1)$$

A complete energy balance equation should also include a kinetic energy term (see Eq. 46 in the Appendix), which is omitted here for mathematical simplicity. Stokes noted in 1851 that the mechanical power is equal to the rate of the kinetic energy change plus the total stress power expressed by the volume integral of the stress tensor, σ_{ij}, multiplied by the strain rate tensor, e_{ij} $\left(\int_V \sigma_{ij} e_{ij} dV\right)$. The heat input rate consists of conduction flux and distributed internal heat sources, r, per unit mass (e.g., MALVERN, 1969). According to classical thermodynamics (e.g., ONSAGER, 1931; PRIGOGINE, 1955; TRUESDELL, 1966), restrictions on the constitutive equations can be obtained from the principles of thermodynamics. In the context of single phase continuum thermodynamics for local or simple materials, the forms of the energy and entropy balance equations are universal. The first law of thermodynamics for such materials is given by

$$\rho\frac{\mathrm{d}U}{\mathrm{d}t} = \sigma_{ij}e_{ij} - \nabla_i J_i^{(q)} + \rho r, \qquad (2)$$

where ρ is mass density, U is specific internal energy and $J_i^{(q)}$ is the heat flux. DUNN and SERRIN (1985) noted that for complex non-local materials, the balance equation for the internal energy (2) is, in general, not compatible with the second law of thermodynamics. As a possible remedy to such incompatibility, they proposed an extension of the classical balance of energy through an additional rate of mechanical energy supply, the interstitial working, $J_i^{(i)}$, involving long-range interactions among the molecules (see also Eq. 47 in the Appendix):

$$\rho\frac{\mathrm{d}U}{\mathrm{d}t} = \sigma_{ij}e_{ij} - \nabla_i J_i^{(q)} + \rho r - \nabla_i J_i^{(i)}. \qquad (3)$$

Such modification allows achieving compatibility with the second law of thermodynamics (e.g., TRIANI and CIMMELLI, 2012). However, the presence of the extra flux $J_i^{(i)}$ in (3) requires a posteriori definition of the energy partitioning between local entropy production and interstitial working. This implies that thermodynamic formulations for non-local materials are not unique, since additional assumptions are needed (e.g., FABRIZIO et al., 2011). For granular materials the interstitial work flux is associated with the surface energy of the grain boundaries and its change during collisions and crushing of grains (e.g., GIOVINE, 1999). It is reasonable to assume that in damaged solids this flux is associated with the surface energy change of nucleated and growing microcracks. Following these ideas, we suggest that the interstitial working in the continuum damage–breakage model involves the time-variations of both the damage and breakage state variables of the system.

Part of the dissipated energy associated with conductive heat transport and accumulated plastic or irreversible deformation (shear heating) is converted into heat, while the other part goes to entropy. The heat balance equation controlling the rate of temperature change includes two terms involving conductive heat transport and shear heating (Eq. 58 of the Appendix):

$$\rho C_p\frac{\partial T}{\partial t} = k\nabla^2 T + \rho\frac{\partial F}{\partial \varepsilon_{ij}}\frac{\mathrm{d}\varepsilon_{ij}^{(p)}}{\mathrm{d}t}, \qquad (4)$$

where C_p is specific heat capacity, k is thermal conductivity and ρ is again material density.

According to the second law of thermodynamics, the energy dissipation associated with damage and breakage evolution must be non-negative. It might be represented as a product of a damage-related ($\mathrm{d}\alpha/\mathrm{d}t$) and breakage-related ($\mathrm{d}B/\mathrm{d}t$) thermodynamic fluxes multiplied by conjugate thermodynamic forces (corresponding derivatives of the free energy). Adopting ONSAGER (1931) reciprocal relations between thermodynamic forces and fluxes, we write phenomenological kinetic equations for the evolution of the damage and breakage state variables as a set of two coupled differential equations (Eq. 60 of the Appendix):

$$\begin{aligned}\frac{\mathrm{d}B}{\mathrm{d}t} &= C_{BB}\frac{\partial F}{\partial B} + C_{B\alpha}\left(\frac{\partial F}{\partial \alpha} - \nabla_i\frac{\partial F}{\partial(\nabla_i\alpha)}\right)\\[4pt]\frac{\mathrm{d}\alpha}{\mathrm{d}t} &= C_{\alpha B}\frac{\partial F}{\partial B} + C_{\alpha\alpha}\left(\frac{\partial F}{\partial \alpha} - \nabla_i\frac{\partial F}{\partial(\nabla_i\alpha)}\right).\end{aligned} \qquad (5)$$

This is the most important equation of the damage–breakage model controlling the coupled evolution of the state variables. The kinetic coefficients (or functions) C_{BB}, $C_{B\alpha}$, $C_{\alpha B}$, and $C_{\alpha\alpha}$ connect the thermodynamic forces associated with certain state variables and thermodynamic fluxes. The diagonal terms with C_{BB} and $C_{\alpha\alpha}$ are associated with a given state variable, while the off-diagonal terms with $C_{\alpha B}$ and $C_{B\alpha}$ provide coupling between evolving damage with breakage and vice versa. To assure non-negativity of the entropy production, the matrix of the kinetic coefficients has to satisfy special conditions discussed by DEGROOT and MAZUR (1962), MALVERN (1969) and others. These coefficients will be specified in the next section after the energy form of the damage–breakage material is described.

The general form of the stress tensor, σ_{ij}, given by the derivative of the free energy, provides the relevant constitutive relation for the discussed system (Eq. 53 of the Appendix):

$$\sigma_{ij} = \rho \left[\frac{\partial F}{\partial \varepsilon_{ij}} - \frac{1}{2} \left(\frac{\partial F}{\partial (\nabla_i \alpha)} \nabla_j \alpha + \frac{\partial F}{\partial (\nabla_j \alpha)} \nabla_i \alpha \right) \right].$$

(6)

In addition to the usual term $\partial F / \partial \varepsilon_{ij}$ (e.g., MALVERN, 1969), there are terms associated with $\nabla \alpha$ corresponding to "structural stresses". In a deformable system, the material time derivative is given by $d(\cdot)/dt = \partial(\cdot)/\partial t + v_i \nabla_i(\cdot)$ leading to additional stress components associated with heterogeneous damage distribution (see Appendix). These stresses are important only in narrow zones with high damage gradients, or damage fronts separating between areas with intact and highly damaged material.

Equations (4)–(6) are the major general relations of the continuum damage–breakage mechanics model. Further model development can be done only after the free energy of the system is specified. An appropriate energy form, as well as the kinetic parameters (C_{BB}, $C_{B\alpha}$, $C_{\alpha B}$, $C_{\alpha\alpha}$) and relations controlling the rate of irreversible strain accumulation ($d\varepsilon_{ij}^{(p)}/dt$), are discussed in the next section.

3. Continuum Damage–Breakage Model

The macroscopic stability condition for deforming materials is convexity of the elastic strain energy, which is necessary for the existence of a unique solution of the static problem (e.g., EKELAND and TEMAM, 1976). This condition implies positivity of the second derivative of the energy function in the one dimensional case, or positivity of all the eigenvalues of the Hessian matrix ($\partial^2 F_S / \partial \varepsilon_{ij} \partial \varepsilon_{kl}$) in the complete three dimensional formulation. Starting with the classical *van der Waals theory*, non-convex energy functions are commonly used in theoretical models of multi-phase systems known as *Cahn–Hillard theory* and *Landau–Ginzburg theory* (e.g., ERICKSEN, 1998; FREMOND, 2012). If the energy is non-

convex, as in the case of phase transitions, one must introduce additional physical properties of the system and provide regularization of the energy function to obtain a unique solution at the continuum level. Such regularization may be achieved by adding to the local constitutive relations the first or higher gradients of state variables (e.g., NGAN and TRUSKINOVSKY, 2002) and introducing additional collective variables (e.g., TYUPKIN, 2007).

LYAKHOVSKY et al. (2011) connected the loss of convexity of the energy function at a critical level of damage signifying macroscopic failure with transition from a highly damaged solid to a granular material. To describe a two-phase system including solid-granular transition, LYAKHOVSKY and BEN-ZION (2014) proposed a continuum damage–breakage mechanics that includes aspects of two formulations: continuum damage mechanics for solid material and continuum breakage mechanics for granular material. This is discussed in more detail below.

3.1. Free Energy Function of Damaged Material

The continuum damage–breakage model combines the damage-dependent energy of the solid phase, F_S, and the breakage-dependent energy of the granular phase, F_B, into complete free energy of the system. EINAV (2007a, b) suggested that the free energy of the Continuum Breakage Mechanics is a linear function of a breakage parameter, B. LYAKHOVSKY and BEN-ZION (2014) wrote the total free energy of the CDBM as a linear superposition of F_S characterizing the solid phase and F_B of the granular phase:

$$F(T, \varepsilon_{ij}, \alpha, \nabla_i \alpha, B) = (1 - B) \cdot F_S(T, \varepsilon_{ij}, \alpha, \nabla_i \alpha) + B \cdot F_B(T, \varepsilon_{ij}).$$

(7)

To clarify basic properties of the energy function, we first discuss the energy form, $F(\varepsilon_{ij})$, for a local isothermal case, and then discuss the more general expression that includes non-local gradient term for the free energy of the solid phase and the energy of the granular phase.

The free energy of a simple isotropic elastic solid is a function of three invariants of the elastic strain tensor ε_{ij} (e.g., NOLL, 1958; TRUESDELL and NOLL, 2004), which for small deformation are $I_1 = \varepsilon_{ij} \delta_{ij}$, $I_2 = \varepsilon_{ij} \varepsilon_{ij}$, and $I_3 = \det(\varepsilon_{ij})$. The strain energy function of a Hookean elastic solid has two second order (quadratic) terms I_1^2 and I_2. MURNAGHAN (1951) extended the Hookean expression for the strain energy by including higher order terms of the three strain invariants and their combinations. The Murnaghan model can produce gradual reduction of the slope of the stress–strain curve at relatively low stress levels, as observed in deformation of soft sediments (e.g., EXADAKTYLOS, 2006) or materials such as graphite and soil (e.g. KAJI et al., 2001; LI and DING, 2002). Models with high-order Murnaghan terms can be successful in large strain analysis of the Earth's interior (e.g., BIRCH, 1952). However, HAMIEL et al. (2011) demonstrated that the Murnaghan model can not fit with the same set of parameters the extended linear regime of stiff crystalline rocks under conditions of brittle deformation. Moreover, the Murnaghan model does not explain the observed abrupt changes of stress–strain curves upon reversal of loadings from tension and compression conditions, which may be attributed to opening and closure of micro-cracks, as well as the vanishing of the elastic stiffness of the highly damaged and granular materials under tension.

HAMIEL et al. (2011) modeled experimentally observed nonlinear rock deformation using a more general function than classical models requiring analyticity (polynomial functional form of the strain invariants). Similar to the assumption of a Hookean solid, they dropped the dependency of the elastic potential on the third invariant, I_3, and wrote the second-order energy function in the form:

$$F(\varepsilon_{ij}) = \frac{1}{\rho} I_2 \cdot f(\xi), \qquad (8)$$

where $f(\cdot)$ is a general function of the strain invariants ratio $\xi = I_1/\sqrt{I_2}$. The complete form of the strain potential can be obtained by using a polynomial Taylor expansion of the function $f(\xi)$:

$$f(\xi) = a_0 + a_1\xi + a_2\xi^2 + a_3\xi^3 + \cdots. \qquad (9)$$

LYAKHOVSKY et al. (1997b) and HAMIEL et al. (2011) showed that a basic local energy function for a damaged crystalline solid phase is given by incorporating the first three terms from (9), up two the second order of ξ, and using the notation $a_0 = \mu$, $a_1 = -\gamma$, and $a_2 = \lambda/2$:

$$F_S(\varepsilon_{ij}, \alpha) = \frac{1}{\rho}\left(\frac{\lambda}{2}I_1^2 + \mu I_2 - \gamma I_1 \sqrt{I_2}\right), \qquad (10)$$

where the first two quadratic terms are usual for a perfectly Hookean elastic solid with Lamé constants λ and μ, and the additional third term couples volumetric and shear strain via a third modulus γ of a damaged solid. Another derivation of (10) can also be obtained by assuming a general second-order function of I_1 and I_2 for the energy function, and eliminating terms with singularities or behavior not consistent with dilatational strain during brittle deformation (BEN-ZION and LYAKHOVSKY, 2006).

Taking the derivative of the energy form (10) with respect to the strain leads to the following stress–strain relation:

$$\sigma_{ij} = (\lambda - \gamma/\xi)I_1\delta_{ij} + (2\mu - \gamma\xi)\varepsilon_{ij}. \qquad (11)$$

A specific relation between the energy (10) and the damage state variable is established by making the elastic moduli functions of α. Following AGNON and LYAKHOVSKY (1995), we assume for mathematical simplicity that the moduli μ and γ are linear functions of α and that λ is constant:

$$
\begin{aligned}
\lambda &= \lambda_0, \\
\mu &= \mu_0 + \alpha\xi_0\gamma_r \\
\gamma &= \alpha\gamma_r,
\end{aligned}
\qquad (12)
$$

where λ_0, μ_0 and $\gamma = 0$ are the elastic moduli of a damage-free material ($\alpha = 0$), and $\lambda = \lambda_0$, $\mu = \mu_0 + \xi_0\gamma_r$ and $\gamma = \gamma_r$ give the moduli values at maximum damage level ($\alpha = 1$). The parameter ξ_0 is a critical value of the strain invariants ratio ξ at the onset of damage accumulation and is connected to the internal friction angle of the BYERLEE (1978) friction law for rocks (AGNON and LYAKHOVSKY, 1995). During damage accumulation the modulus γ increases and the shear modulus μ decreases (ξ_0 is negative). This leads to material evolution from linear elastic solid ($\alpha = 0$) to strongly non-linear behavior and macroscopic brittle instability at a critical damage level (α_{cr}).

The simplest extension of the previous local model (10) to a gradient-type non-local isotropic model involves incorporating quadratic term of the damage gradient:

$$F_S^*(\varepsilon_{ij}, \alpha, \nabla\alpha)$$

$$= \frac{1}{\rho}\left(\frac{\lambda}{2}I_1^2 + \mu I_2 - \gamma I_1\sqrt{I_2} + \frac{\vartheta}{2}\nabla_i\alpha \cdot \nabla_i\alpha\right). \quad (13)$$

The additional dimensional coefficient ϑ with Pa m^2 units characterizes the connection between stress and length scales of the non-local formulation. Additional gradient term appears in the stress–strain relation given from the derivative of the energy function:

$$\sigma_{ij} = (\lambda - \gamma/\xi)I_1\delta_{ij} + (2\mu - \gamma\xi)\varepsilon_{ij} - \vartheta\nabla_i\alpha \cdot \nabla_j\alpha. \quad (14)$$

The last term in (14) defines the "structural stresses" associated with the heterogeneous damage distribution.

3.2. Brittle Failure of Solid Material

The macroscopic stability condition or convexity of the elastic strain energy (10) is expressed by the following inequalities derived from the eigenvalues calculation for the Hessian matrix (LYAKHOVSKY et al., 2011):

$$(2\mu - \gamma\xi)^2 + (2\mu - \gamma\xi)(3\lambda - \gamma\xi)$$
$$+ (\lambda\gamma\xi - \gamma^2)(3 - \xi^2) > 0, \quad (15a)$$

$$2\mu - \gamma\xi > 0. \quad (15b)$$

These inequalities define the critical damage level ($\alpha = \alpha_{cr}$) shown in Fig. 1. The γ_r value is defined using the scaling condition $\alpha = \alpha_{cr} = 1$ for $\xi = \xi_0$ (LYAKHOVSKY et al., 1997a). Within the interval of the strain invariants ratio $-\sqrt{3} \leq \xi \leq \xi_0$, corresponding to loading from 3-D compaction with gradually increased deviatoric stress until the onset of fracturing, both conditions (15) are satisfied for every $\alpha \leq 1$. With further increase of the deviatoric stress, the damage increases toward its critical value ($\alpha = \alpha_{cr}$) and condition (15a) is violated first. This occurs within the interval of the strain invariants ratio up to a certain value ξ_1 ($\xi_0 \leq \xi \leq \xi_1$). This type of loss of convexity is associated with mode-II failure and transition to a granular flow under compression (Fig. 1). The granular material under these conditions may be referred to as pseudo-liquid (LYAKHOVSKY and BEN-ZION, 2014). At higher values of the strain

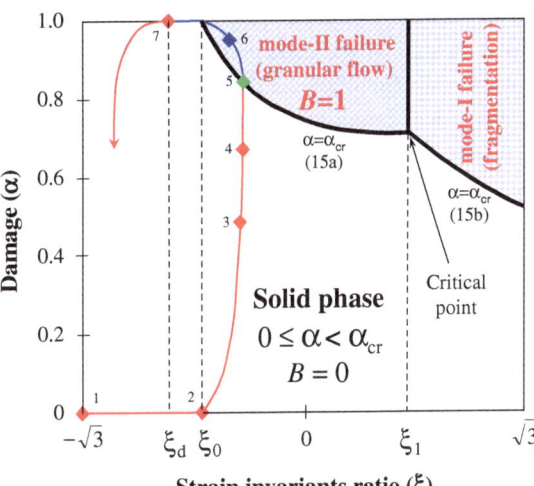

Figure 1
Schematic diagram illustrating the critical level of damage versus the strain invariants ratio ($\xi = I_1/\sqrt{I_2}$) for $\lambda_0 = \mu_0$. Solid black lines separate regions where different phases exist and correspond to convexity loss of the solid energy function (Eqs. 15a and 15b). The parameter ξ_0 prescribes the onset of damage accumulation. The material is stable for $\xi < \xi_0$. The parameter ξ_1 defines the location of the critical point where all three phases, solid, pseudo-liquid and pseudo-gas exist for $\alpha = \alpha_{cr}$. Red line with markers represents a loading path leading to a solid-granular transition; blue line (with marker 6) is its extension in the granular state; final red line segment represents transition back to solid phase and healing

invariants ratio ($\xi > \xi_1$), condition (15b) is violated first. In this case at least one of the principal stress values becomes tensile, and there is mode-I failure that leads to fragmentation (Fig. 1); the granular material may be referred to in this case as pseudo-gas.

At the transition between mode-I and mode-II failures associated with the point $\alpha = \alpha_{cr}$ and $\xi = \xi_1$, all eigenvalues of the Hessian matrix vanish simultaneously and conditions (15) are reduced to:

$$\lambda - \gamma/\xi = 0, \quad (16a)$$

$$2\mu - \gamma\xi = 0. \quad (16b)$$

Combining (16) with (12) leads to a solution for values of α_{cr} and ξ_1:

$$\xi_1 = \xi_0 + \sqrt{\xi_0^2 + 2\frac{\mu_0}{\lambda_0}}, \quad (17)$$

$$\alpha_1 = \alpha_{cr}(\xi_1) = \frac{\lambda_0\xi_1}{\gamma_r}. \quad (18)$$

A comparison of (16) with the stress–strain relation (11) indicates that all eigenvalues of the Hessian matrix vanish simultaneously when both effective elastic moduli of the damaged solid are zero (LYAK-HOVSKY and BEN-ZION, 2014). This resembles the mathematical definition of a critical (triple) point of a classical thermodynamic system. For a pure substances (say liquid or gas), there is an inflection point in the critical isotherm (constant temperature line) on a P–V diagram, where both the first and second partial derivatives of the pressure with respect to the volume at constant temperature are zero.

3.3. Free Energy of the Granular Phase

The total free material energy in Eq. (7) is a sum of two contributions characterizing the solid and the granular phases. In Sect. 3.1 we specified the general form of the energy function (8, 9) and defined the energy (10) of a damaged solid phase. Here we follow the same ideas and specify the energy function for the granular phase.

The constitutive relations for cohesionless granular material should account for vanishing tensile modulus while the compressive modulus remains finite (e.g., NGUYEN et al., 2003; KARRECH et al., 2007). MYASNIKOV and OLEINIKOV (1991) discussed a mathematical model of granular material as a limit case of media that respond differently to tension and compression using the energy form (8). Since the granular phase energy appears in (7) as $B \cdot F_B$, the function $F_B(T, \varepsilon_{ij})$ depends only on the temperature and elastic strain tensor, and it does not depend on the damage value which is relevant for the solid phase only. LYAKHOVSKY and BEN-ZION (2014) showed that a thermodynamically consistent model formulation accounting for two-way solid-granular transitions requires energy function obtained using the third order polynomial expansion (9) with constant coefficients, a_0, a_1, a_2, a_3:

$$F_B(\varepsilon_{ij}) = \frac{1}{\rho}\left(a_0 I_2 + a_1 I_1 \sqrt{I_2} + a_2 I_1^2 + a_3 \frac{I_1^3}{\sqrt{I_2}}\right).$$

(19)

Taking the derivative of the energy form (19) with respect to strain, leads to the following stress–strain relation:

$$\sigma_{ij} = (2a_2 + a_1/\xi + 3a_3\xi)I_1\delta_{ij} + (2a_0 + a_1\xi - a_3\xi^3)\varepsilon_{ij}.$$

(20)

The transition between the pseudo-liquid (granular flow) and pseudo-gas (fragmentation) phases is associated with loss of convexity of F_B represented by a vertical line ($\xi = \xi_1$) in the (ξ, α) space (Fig. 1). Since all three phases, solid, pseudo-liquid and pseudo-gas, co-exist in the triple (critical) point (ξ_1, α_1), all eigenvalues of the Hessian matrix associated with F_B should be equal zero for $\xi = \xi_1$. This condition is satisfied only if both effective elastic moduli are equal to zero for $\xi = \xi_1$:

$$2a_2 + a_1/\xi_1 + 3a_3\xi_1 = 0$$
$$2a_0 + a_1\xi_1 - a_3\xi_1^3 = 0.$$

(21)

Lengthy calculations (not shown here) of the eigenvalues of the Hessian matrix ($\partial^2 F_B/\partial\varepsilon_{ij}\partial\varepsilon_{kl}$) demonstrate that the energy potential (19) is convex for the pseudo-liquid phase ($\xi < \xi_1$). The energy of the granular phase should be below that of the solid phase for $\alpha = \alpha_{cr}(\xi)$, providing thermodynamically consistent conditions for transition from a higher energy solid state to a lower energy pseudo-liquid granular phase (LYAKHOVSKY and BEN-ZION, 2014). This transition from solid to granular state is expected within the interval $\xi_0 < \xi < \xi_1$ of the strain invariant ratio values. The energy difference between these states is equivalent to the latent heat of phase transition in classical thermodynamics. The scaling condition $F_B/F_S = \chi < 1$ may be defined for any prescribed ξ-value within the interval involving the transition from solid to pseudo-liquid granular phase. The simplest mathematical form of this scaling relation is obtained for $\xi = 0$ (pure shear):

$$a_0 = \chi \cdot \mu(\alpha_{cr}|_{\xi=0}).$$

(22)

The three linear equations (21, 22) for the elastic coefficients of the granular media could be solved for the second order polynomial function (8) assuming $\alpha_3 = 0$. In such case the energy of the granular phase is always below that of the solid phase; this condition prohibits the reversed transition from granular to solid, even under 3-D compaction. This model feature is corrected here by incorporating a third order term with $\alpha_3 \neq 0$. We add another condition that requires the energy of the granular phase to be above that of the solid phase within the interval $\xi < \xi_d < \xi_0$ (Fig. 1). This condition is equivalent to $F_B = F_S$ for $\xi = \xi_d$:

$$a_0 + a_1\xi_d + a_2\xi_d^2 + a_3\xi_d^3 = \mu - \gamma\xi_d + \frac{\lambda}{2}\xi_d^2, \quad (23)$$

where elastic moduli of the solid phase λ, μ, γ are calculated using (12) with $\alpha = 1$. The four linear equations (21, 22, 23) provide a unique solution for the elastic coefficients of the granular media.

3.4. Kinetics of the Coupled Damage–Breakage Evolution

The kinetics of the coupled evolution of the damage and breakage state variables is defined by two differential equations (5). To specify these relations, we calculate the partial derivatives of the energy with respect to α, $\nabla_i\alpha$, and B and then constrain the kinetic functions C_{BB}, $C_{B\alpha}$, $C_{\alpha B}$, and $C_{\alpha\alpha}$. Combining (12) with (13), using (19) and then substituting into (7), the partial derivatives of the energy $F(T, \varepsilon_{ij}, \alpha, \nabla_i\alpha, B)$ with respect to α, $\nabla_i\alpha$, and B are:

$$\frac{\partial F}{\partial B} = F_B - F_S$$
$$= \frac{I_2}{\rho}\left[(a_0 - \mu) + (a_1 + \gamma)\xi + \left(a_2 - \frac{\lambda}{2}\right)\xi^2 + a_3\xi^3\right]$$
$$\frac{\partial F}{\partial \alpha} - \nabla_i\frac{\partial F}{\partial(\nabla_i\alpha)} = -\frac{1-B}{\rho}\left[\gamma_r I_2(\xi - \xi_0) + \vartheta\nabla^2\alpha\right].$$
$$(24)$$

The diagonal kinetic coefficients (or functions) $C_{\alpha\alpha}$ and C_{BB}, in (5) should always be negative to ensure unconditional non-negative entropy production (e.g., DEGROOT and MAZUR, 1962; MALVERN, 1969). The off-diagonal kinetic coefficients are often assumed to be symmetric ($C_{B\alpha} = C_{\alpha B}$), anti-symmetric ($C_{B\alpha} = -C_{\alpha B}$), or zero ($C_{B\alpha} = C_{\alpha B} = 0$) for many practical applications (e.g., DEGROOT and MAZUR, 1962; MALVERN, 1969). The simplest mathematical assumption of zero off-diagonal terms

means that the change of a given state variable mostly depends on the energy gradient associated with that variable. We adopt this approximation and write the equation for damage evolution in a form similar to that discussed by LYAKHOVSKY et al. (2011):

$$\frac{d\alpha}{dt} = \begin{cases} (1-B)[C_d I_2(\xi - \xi_0) + D\nabla^2\alpha] & \text{for } \xi \geq \xi_0, \\ (1-B)\left[C_1\exp\left(\frac{\alpha}{C_2}\right)I_2(\xi - \xi_0) + D\nabla^2\alpha\right] & \text{for } \xi < \xi_0. \end{cases}$$
$$(25)$$

The Laplacian term with damage diffusion coefficient D is associated with non-local gradient energy term. The coefficient C_d gives the rate of positive damage evolution (material degradation) for $\xi > \xi_0$ and is constrained by laboratory fracturing experiments (LYAKHOVSKY et al., 1997a; HAMIEL et al., 2004, 2009). The rate of damage recovery (material healing) for states of strain $\xi < \xi_0$ is assumed to depend exponentially on α. This produces logarithmic healing with time in agreement with laboratory experiments (e.g., DIETERICH and KILGORE, 1996; SCHOLZ, 2002; JOHNSON and JIA, 2005) with rocks and other materials. LYAKHOVSKY et al. (2005) showed that the above damage model reproduces the main phenomenological features of rate- and state-dependent friction, and constrained the healing parameters C_1, C_2 by comparing model calculations with laboratory frictional results. The value $\xi = \xi_0$ controlling the transition from healing to damage accumulation is directly related to the internal friction of intact rock (LYAKHOVSKY et al., 1997a). An additional multiplier, $(1 - B)$, is associated with energy partitioning between the solid and granular phases.

Substituting expression (24) for the partial derivative of the energy with respect to breakage into the kinetic relations (5), and using again zero values of the off-diagonal kinetic coefficients ($C_{B\alpha} = C_{\alpha B} = 0$), lead to the following equation for the breakage evolution:

$$\frac{dB}{dt} = \begin{cases} C_B P(\alpha)(1-B)I_2 \times \left[(\mu - a_0) - (a_1 + \gamma)\xi + \left(\frac{\lambda}{2} - a_2\right)\xi^2 - a_3\xi^3\right] & \text{for } \xi \geq \xi_d, \\ C_{BH}(\alpha, B)I_2 \quad\times \left[(\mu - a_0) - (a_1 + \gamma)\xi + \left(\frac{\lambda}{2} - a_2\right)\xi^2 - a_3\xi^3\right] & \text{for } \xi < \xi_d. \end{cases} \quad (26)$$

245

The transition from breakage accumulation ($dB/dt > 0$) to recovery ($dB/dt < 0$) according to (26) occurs at $\xi = \xi_d$, when the difference between solid and granular energy, $F_B - F_S$, changes its sign. Similarly to the kinetics for damage evolution (25), we assume that breakage accumulation for $\xi > \xi_d$ and healing for $\xi < \xi_d$ are not symmetric, and the values or functional forms of the rate coefficients may be different.

LYAKHOVSKY et al. (2011) suggested that the rate of evolving material properties is proportional to the probability, $P(\alpha)$, for a material element to be in a granular phase. Since the phase transition occurs at $\alpha = \alpha_{cr}(\xi)$, the probability that a material element with low values of damage is in granular phase should be very low ($P(\alpha \ll \alpha_{cr}) \rightarrow 0$). On the other hand, a highly damaged material element should be in a granular phase ($P(\alpha > \alpha_{cr}) \rightarrow 1$). Similar to classical statistical mechanics the probability, $P(\alpha)$, represents distribution of the states of a system. In analogy to the Fermi–Dirac distribution, which is widely applied for complex systems with phase transitions (e.g., LANDAU and LIFSHITZ, 1980), we assume that the probability, $P(\alpha)$, has the functional form (LYAKHOVSKY et al., 2011):

$$P(\alpha) = \frac{1}{\exp\left(\frac{\alpha_{cr}(\xi) - \alpha}{\beta}\right) + 1}, \qquad (27)$$

where the β-value defines the width of the transitional region. For $\beta \rightarrow 0$, $P(\alpha)$ approaches the Heaviside (step) function that abruptly changes its value from zero to one. Together with the fast kinetics of the breakage growth ($C_B \rightarrow \infty$), the transition from the solid to granular phase is instantaneous without any mixture zone. With finite values of β and C_B, accumulating damage toward the critical value leads to a gradual increase in the breakage value. These results provide general guidelines for describing the deformation processes near the unstable regime $\alpha \rightarrow \alpha_{cr}(\xi)$.

The transition back from the granular to solid phase occurs for $\xi \leq \xi_d \leq \xi_0$. For this range of the ξ-values, the solid phase is always stable and the material element is expected to be in a solid state ($P(\alpha) \rightarrow 0$). Therefore, the probability term may be dropped in the kinetic function controlling breakage healing. At present the C_{BH} function is not well constrained by laboratory data. Some experimental observations (e.g., LUBE et al., 2004) demonstrate that during a period of lowering velocity, instead of gradual deceleration, the granular flow comes to an abrupt halt. The slide-hold-slide experiments of RECHES and LOCKNER (2010) show fast and efficient strengthening of a lubricated powder during a hold stage, no slip period. We interpret this behavior as fast kinetics of the transition from granular to solid phase, which requires relatively high C_{BH} values, but its functional form is still not clear. It might be exponential function similar to the damage healing, producing logarithmic healing in time, as suggested by frictional experiments with granular media (e.g. MARONE, 1998).

3.5. Flow of the Pseudo-Liquid Granular Phase

Starting with the pioneering works by von Mises and von Karman, several models involving yield conditions and flow rules attempted to account for the observed solid-like and fluid-like phenomenology of granular material (e.g., JAEGER et al., 1996; SAVAGE, 1998; LIU and NAGEL, 2001; FORTERRE and POULIQUEN, 2008). Several authors noted that the bulk flow parameters of granular media are extremely sensitive to the average grain size and the size distribution (e.g., RAO and NOTT, 2008). Fluidity of a fine powder with sub-micron particle size may be several orders of magnitude above the fluidity values for a granular material with particle sizes of millimeters to centimeters (e.g., WORNYOH et al., 2007). HESHMAT (1995), and later on HESHMAT and HESHMAT (1999), developed an equivalent viscosity model for lubricated fine powder hydrodynamics and demonstrated that non-Newtonian power-law constitutive relation fits well the experimental observations.

In the context of our model, the fluidity of the granular phase should strongly depend on the value of the breakage parameter which measures the grain size distribution between the initial coarse granular material and ultimate state corresponding to a fine powder. Adopting a power-law constitutive relation, we approximate the rate of granular flow by the breakage-dependent relation given in scalar form as:

$$\frac{d\varepsilon^{(p)}}{dt} = C_g \cdot B^{m_1} \cdot \tau^{m_2}, \qquad (28a)$$

or in tensor form as:

$$\frac{d\varepsilon_{ij}^{(p)}}{dt} = C_g \cdot B^{m_1} \cdot \tau^{m_2-1} \cdot \tau_{ij}. \qquad (28b)$$

The dimensional coefficient C_g is a material parameter; the two power-law indexes m_1 and m_2 controls respectively, the fluidity of the granular material and its sensitivity to the breakage value and applied load.

Plastic strain during flow of granular material is mostly accommodated by slip along grain boundaries. In coarse media, granular particles undergo nearly elastic collisions and fracturing, while in fine powder, they undergo completely inelastic collisions and ductile deformation (e.g., WORNYOH et al., 2007). Ductile flow may become the dominant deformational mechanism under elevated temperatures associated with local heat production at the sliding surfaces. According to general idea of the Maxwell visco-elastic rheology, the total strain is a sum of components representing different deformation mechanisms (e.g., REGENAUER-LIEB and YUEN, 2003). A power-law rheology is widely accepted for expressing the strain rate of the temperature-activated ductile flow (e.g. WEERT-MAN, 1978). To account for temperature-dependent effects, the overall strain rate of the granular material is represented in our model by a sum of two components, the breakage dependent term (28) and an additional classical temperature-dependent component:

$$\frac{d\varepsilon^{(p)}}{dt} = C_g \cdot B^{m_1} \cdot \tau^{m_2} + A\tau^n \exp\left(-\frac{Q}{RT}\right). \qquad (29)$$

The material parameters A and n are empirical constants, Q is activation energy, T is temperature, and R is the gas constant. For relatively low pressures corresponding to depths less than 100 km, the effects of pressure and activation volume (PV term in Arrhenius exponent) is negligible (e.g., REGENAUER-LIEB and YUEN, 2003).

4. Model Results

4.1. Instantaneous Transition from Solid to Granular Phase

In this section we discuss some general features of the solid-granular transition, assuming a very fast

kinetic of the breakage evolution that leads to rapid transition when the solid becomes unstable. The schematic red line with markers in Fig. 1 represents a realistic loading path with evolving damage. The external work increases the energy of the system, which is schematically shown by the red dash line with markers in Fig. 2. The loading starts with a 3-D compaction (confining pressure) associated with $\xi = -\sqrt{3}$ and $\alpha = 0$ (first marker in Fig. 1). The strain invariants ratio, ξ, increases with a differential load toward its critical value, $\xi = \xi_0$, corresponding to the onset of damage accumulation (second marker in Fig. 1). The elastic moduli of the material remain constant up to that point, and the elastic energy change over small strain perturbations (around points #1 and #2 in Fig. 2) retains parabolic shape shifted to a higher level. With further load, the progressive damage accumulation decreases the effective elastic moduli (markers 3 and 4 in Fig. 1), reducing the convexity of the energy function. The energy change over a small strain perturbation (around points #3 and #4 in Fig. 2) is still parabolic, but its shape is significantly flattened.

At the critical damage value (marker 5 in Fig. 1), the energy of the damaged solid losses convexity and becomes flat (blue line around marker 5 in Fig. 2). However, at the same strain conditions the energy of the granular phase ($B = 1$) is stable and below that of the solid phase (red parabola near marker 6 in Fig. 2). The damage–breakage system jumps from the unstable solid state to the stable granular state with lower energy (marker 6 in Fig. 1). Since the state of the system around point #4 is meta-stable, the system may jump over a small energy barrier due to small stress perturbations and transition may occur from #4 to #6 (insert in Fig. 2). Further system evolution is controlled by the flow rules (29) for the granular media (blue path with marker 6 in Fig. 1). The granular flow is halted with the onset of breakage and damage healing (marker 7 in Fig. 1).

Details of the energy change during the system evolution in the granular phase are shown in Fig. 3. The black line represents the energy of the solid phase, F_S, and the red line represents the energy of the granular phase, F_B. For $\xi > \xi_d$, the energy of the solid phase is above that of the granular phase ($F_S > F_B$) leading to a transition from solid to

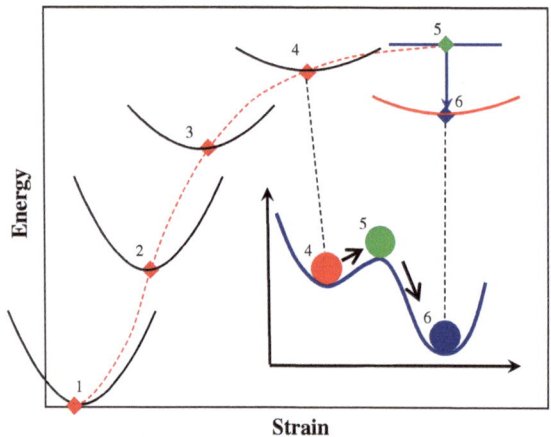

Figure 2

Schematic energy change along the loading path shown in Fig. 1 (*dashed line*) and its change under small strain perturbation around points with markers (not in scale). The flattening of the parabolas around *markers 3* and *4* represents material degradation due to damage increase compared with the damage free material (*markers 1* and *2*). At the critical damage level ($\alpha = \alpha_{cr}$) the energy curve loses convexity (*marker 5*) leading to a transition to a granular phase with lower and convex energy (*marker 6*). The state of the system around point #4 is metastable (*insert*). See text for additional explanation

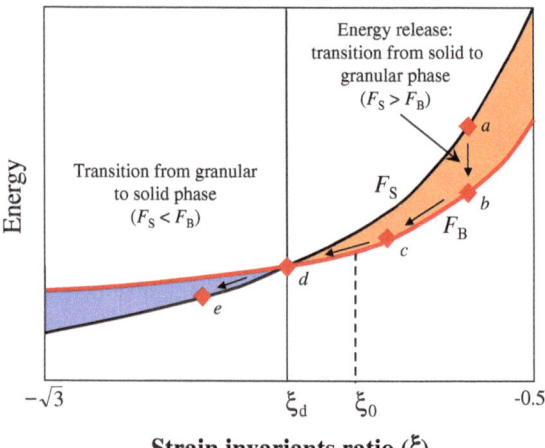

Figure 3

Schematic diagram demonstrating energy change during phase transition from solid to granular phase, relaxation and transition back to the solid phase. *Black line* represents the energy of the solid phase, *red line* energy of the granular phase. *Markers with letters* represent the evolving path during the phase transition from solid to granular and back to solid state

granular state. For $\xi < \xi_d$, the energy of the solid phase becomes less than the energy of the granular phase ($F_S < F_B$) leading to a reversed transition to the solid phase. Marker "a" represents the state of the system just at the onset of the transition from solid to granular phase (corresponding to point #5 in Figs. 1, 2). The assumption of instantaneous transition facilitates a distinction between different types of energy release. The energy release due to breakage change from $B = 0$ (marker "a" in Fig. 3) to $B = 1$ (maker "b") involves crossing the red polygon in Fig. 3. This energy form or the interstitial working discussed earlier is equivalent to latent heat of phase transition in classical thermodynamics and does not lead to a temperature change.

The system is heated when it evolves along the red line through markers "b", "c" to "d", during which all the energy release is converted into heat. In more realistic cases with finite rate of the breakage kinetics, the energy partitioning is a complex process and interstitial working occurs simultaneously with heating. The transition from the granular phase back to the solid state is smoothed, without additional

stress drop or energy release, since $F_B = F_S$ for $\xi = \xi_d$. The system evolution in the solid phase (between markers "d" and "e") is controlled by the rate of the damage and breakage healing. The duration of the healing stage depends on the ratio between the rate of healing and rate of on-going loading that increases ξ toward its critical value leading again to damage accumulation. As shown by BEN-ZION *et al.* (1999, 2011) and LYAKHOVSKY *et al.* (2001), the same non-dimensional ratio also controls on larger scales different dynamic regimes of earthquake and fault patterns.

4.2. Rate- and State-Dependent versus Dynamic Friction in Cyclic Evolution

The system evolution discussed in the previous section (Figs. 1, 2, 3) has different stages including onset of loading, damage accumulation, two-way solid-granular transitions, and healing (breakage and damage decrease). This is schematically shown in a 3-D diagram using the (ξ, B, α) coordinates (Fig. 4). Markers 1–7 along the colored line correspond to stages equivalent to those shown in Figs. 1 and 2. The material recovery enhanced by confining pressure and low differential stress lasts until the ongoing load

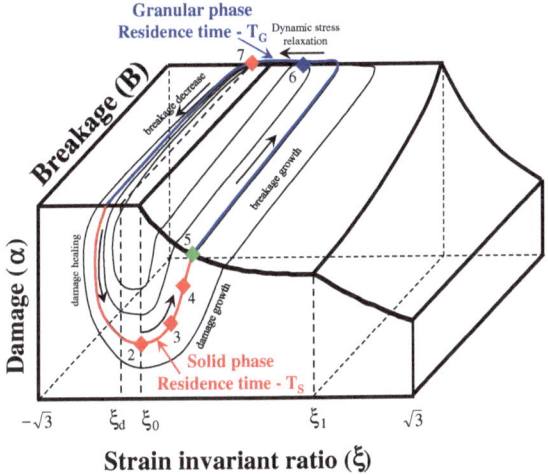

Figure 4
Schematic 3-D diagram in (ξ, α, B) coordinates demonstrating cyclic evolution of the system under slow shear load. *Markers 2–7 along the cycle correspond to the stages equivalent to those shown in Figs. 1 and 2*

They analyzed the evolution of a narrow damage zone located between two moving elastic blocks transmitting shear stress to the strip. Each element of the damage zone exhibits cyclic behavior, with the most prolonged part being the healing stage (Fig. 5). The duration of this stage is controlled by the velocity of the prescribed block motion and the rate of material strengthening. In the case of homogeneous damage zone, the evolution of all elements is synchronized and stress drop occurs simultaneously along the whole strip. The stress in the system is oscillating with a period equal to the duration of the cycle of the evolving element. In more realistic cases, stress drop occurs only in a small part of the strip (e.g., asperity) and the average stress in the block remains about constant. LYAKHOVSKY *et al.* (2005) showed that the steady-state friction or the ratio between the shear stress and normal load averaged

overcomes a threshold value corresponding to the onset of damage accumulation at $\xi = \xi_0$ (marker 2). The damage increases (markers 3 and 4) and reaches its critical value ($\alpha = \alpha_{cr}$), where the transition into granular phase occurs (marker 5). Fluidity of the granular phase with high breakage values allows fast stress relaxation (stress drop) keeping the confining pressure (marker 6). At ξ values below ξ_d the transition from the granular back to solid phase occurs (marker 7), the damage level is reduced due to healing, and shear stress accumulates again according to the rate of on-going loading.

The system behavior may be repeated in a cyclic manner (Fig. 4). The shape of each cycle depends on the kinetic parameters of the model, confining pressure, and rate of differential load. Adopting for simplicity fast breakage evolution (instantaneous transition as above), we ignore the duration of the solid-granular transition and represent the duration of the whole cycle as a sum of two residence time values, T_S for the solid state and T_G for the granular state. The behavior of the system is defined by the ratio between these values. For $T_S \gg T_G$, the system is in a solid state most of the time and the process of stress drop and irreversible strain accumulation in the granular phase is very rapid. This approximation was discussed previously by LYAKHOVSKY *et al.* (2005).

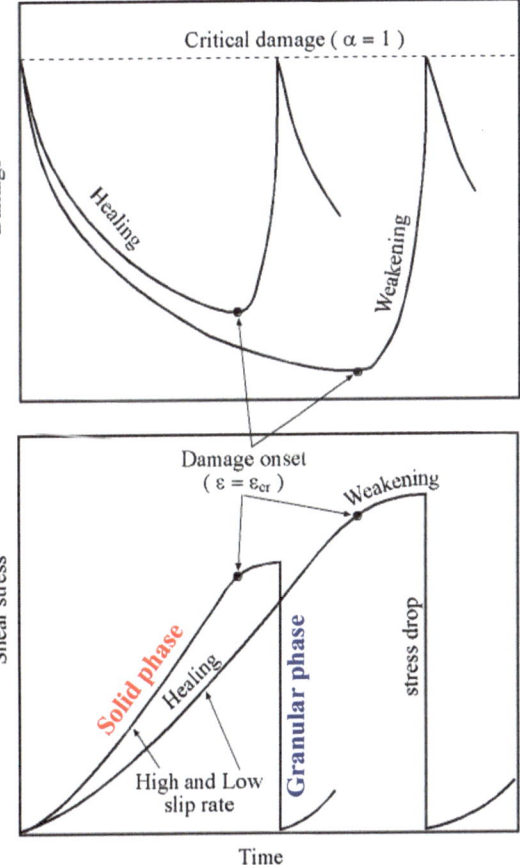

Figure 5
Calculated changes of damage and stress during deformation cycle with two different slip rates

along the strip, decreases proportionally to the logarithm of the slip rate.

We now extend the results of LYAKHOVSKY et al. (2005) to transient ($T_S \sim T_G$) and granular flow ($T_S \ll T_G$) regimes. The three lines in Fig. 6 show simulated steady-state friction versus slip rate for a model with damage zone that is 0.1 mm wide, 10 MPa normal stress, and 10 GPa elastic moduli of the bounding solid blocks. The rate of breakage accumulation is an order of magnitude above the rate of damage accumulation ($C_B = 10 \cdot C_d$), and the rate of breakage healing is higher by another order of magnitude ($C_{BH} = 10 \cdot C_B$). The transition between solid to granular phases in the system with these extremely high rates of the breakage kinetics is practically instantaneous. The onset of damage accumulation is controlled by the material strength corresponding to static friction coefficient, or Byerlee law with $\mu_S = 0.8$. The flow rate of the granular phase is calculated using (29) with $C_g = 10^3$ s^{-1} MPa^{-3} and power-law index $m = 3$. Thermal effects during the flow of the granular phase are not incorporated at this stage and will be discussed below. We use the stress balance equations controlling the deformation of the damage zone and the numerical method developed by LYAKHOVSKY et al. (2005).

In agreement with previous analytical and numerical results, the steady-state friction for slow slip rates (below $\sim 10^{-2}$ m/s) decreases with almost constant slope (Fig. 6), $\partial \mu_{SS} / \partial \ln(V)$, equal to the healing rate

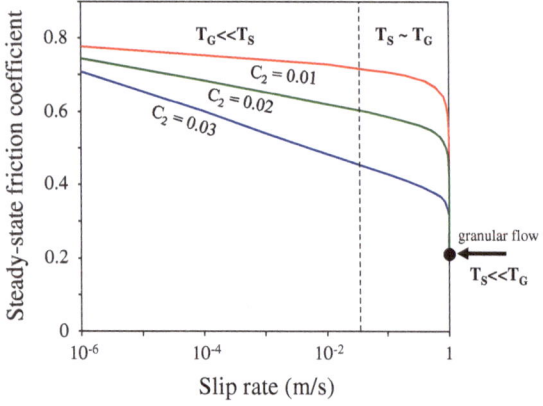

Figure 6
Results of numerical simulations of steady-state friction coefficient versus slip rate

coefficient C_2 of (25). This solid-like regime similar to the rate- and state-frictional behavior is valid when the residence time in the solid phase is well above the residence time in the granular phase ($T_S \gg T_G$). The residence time in the solid phase is controlled by the ratio between the rate of healing and rate of loading, and therefore it decreases with increased slip rate. The residence time in the granular phase practically remains constant. Therefore, the ratio (T_S/T_G) between these time scales decreases and the steady-state friction coefficient drops to significantly lower values controlled by the fluidity of the granular phase. With the chosen flow parameters of the granular phase, the material permanently stays in the granular phase at slip rates of 1 m/s and higher. In this regime the steady-state friction or "dynamic" friction is defined by Couette flow of the granular material in a narrow channel between two solid blocks.

4.3. Length Scale in the Non-Local Model

Basic scaling aspects of the non-local damage rheology model can be illustrated by examining the damage evolution in a long strip subjected to constant load. In a 1-D case of uniform deformation, the kinetic equation for the damage evolution (25) is reduced to:

$$\frac{\partial \alpha}{\partial t} = C_d \varepsilon^2 + D \frac{\partial^2 \alpha}{\partial x^2}. \tag{30}$$

We consider the solid phase only ($B = 0$) and assume a constant loading condition ($\sigma = $ Const.), linear damage-moduli relation (12), $\mu(\alpha) = \mu_0(1 - \alpha)$, and corresponding stress strain relation $\sigma = 2\mu_0(1 - \alpha)\varepsilon$. First we ignore the damage diffusion term ($D = 0$). This leads to a power law damage accumulation previously discussed by BEN-ZION and LYAKHOVSKY (2002):

$$\alpha(t) = 1 - \left(1 - \frac{3C_d \sigma^2}{4\mu_0^2} t\right)^{1/3}. \tag{31}$$

This solution allows introducing a time scale called time-to-failure, t_f, given by

$$t_f = \frac{4\mu_0^2}{3C_d \sigma^2}. \tag{32}$$

The parameter t_f controls the time scale of all the processes associated with accumulated material damage, including the growth rate of narrow fracture zones and accelerated seismic release (see also TURCOTTE et al., 2003).

The "diffusion" term ($D > 0$ in Eq. 30) prevents strong damage localization and produces damage delocalization into a finite width of the generated damage zone. Even if the initial damage has a localized δ-function perturbation, the damage grows in a zone with final width, w, which scales with the square root of the time-to-failure (LYAKHOVSKY et al., 2011):

$$w = \sqrt{Dt_f}. \tag{33}$$

To refine this scaling relation we apply Fourier stability analysis to define the characteristic width of the damage zone. We search for a solution of (30) as a sum of homogeneous solution (31) and small periodic perturbation with a length scale w:

$$\alpha(x,t) = \alpha(t) + f(t) \cdot \exp\left(\frac{i\pi x}{w}\right). \tag{34}$$

Substituting (34) into (30) and using (31) leads to an expression for the smallest growing mode ($df/dt > 0$) controlling the effective width of the damage zone:

$$w^2 = \frac{2\pi^2 D\mu_0^2}{9C_d\sigma^2}. \tag{35}$$

This, and the general relation (33) for the diffusive length scale, demonstrate that accounting for damage diffusion yields a reciprocal scaling relation between the stress and width of the damage zone, $\sigma \propto w^{-1}$. Such scaling means that relatively wide damage zones are expected to be formed under low stresses at shallow depths, while higher stress at larger depths within the seismogenic zone should lead to narrower zones. LYAKHOVSKY et al. (2011) obtained a similar scaling relation for a critical fracture size, consistent with a size effect of crack initiation or scaling between stress and micro-crack length (e.g., BAZANT, 2005). The overall damage zone shape with depth is predicted from (35) to be flower-like, in agreement with geological and seismological observations (e.g., SYLVESTER, 1988; ALLAM and BEN-ZION, 2012).

4.4. Transition Between Quasi-Static and Dynamic Regimes

The scaling analysis of the previous section is extended here to approach the conditions controlling transition from quasi-static to dynamic fracturing, and critical size of dynamically failed patch of the damage zone (nucleation size). From (35), the width of the narrow damage zone between two moving blocks is controlled by the diffusive length scale w. The strain in the material outside the narrow zone may be represented as a superposition of two components (e.g., LYAKHOVSKY et al., 2005): static strain ($\varepsilon_s = \sigma/2\mu_0$) and dynamic strain ($\varepsilon_d = w/V_S \cdot d\varepsilon/dt$). The latter is a combination of radiation damping term (e.g., RICE and BEN-ZION, 1996) with shear wave velocity, V_S, and strain rate, $d\varepsilon/dt$, associated with slip within the damage zone.

The transition from quasi-static to dynamic regime occurs when the ratio between the dynamic and static strain components in the surrounding material ($R_S = \varepsilon_d/\varepsilon_s$) achieves a threshold value. The R_S parameter provides a measure of the ratio between inertial forces and static forces acting in the system, and quantifies the relative importance of waves radiated from the propagating fracture. In this sense it can be viewed as a solid-mechanics analog to the Reynolds number of fluid mechanics.

Taking the time derivative of the stress–strain relation in the damage zone, $\sigma = 2\mu_0 \cdot (1 - \alpha) \cdot \varepsilon$, the strain rate is:

$$\frac{d\varepsilon}{dt} = \frac{1}{(1-\alpha)^2}\frac{d\alpha}{dt}\frac{\sigma}{2\mu_0} + \frac{1}{2\mu_0 \cdot (1-\alpha)}\frac{d\sigma}{dt}. \tag{36}$$

Under quasi-static loading conditions with slow motion of the external blocks, the first term in (36) is dominant. Close to failure ($\alpha \to 1$) it has a stronger singularity and dominates the rate of strain change. Ignoring the second term, the dynamic strain component is:

$$\varepsilon_d = \frac{w}{V_S}\frac{d\varepsilon}{dt} = \frac{w}{V_S}\frac{1}{(1-\alpha)^2}\frac{d\alpha}{dt}\frac{\sigma}{2\mu_0}. \tag{37}$$

For these conditions, the ratio between the dynamic and static strain components is:

$$R_S = \frac{w}{V_S} \frac{1}{(1-\alpha)^2} \frac{d\alpha}{dt}. \qquad (38)$$

Using a threshold R_S value, Eq. (38) leads to a definition of critical dynamic damage value, α_d, at the onset of dynamic failure as a function of the damage growth:

$$\alpha_d = \alpha_{cr} - \sqrt{\frac{w}{R_S V_S} \frac{d\alpha}{dt}}. \qquad (39)$$

For the more generic form, the damage value at the macroscopic failure ($\alpha = 1$ in the examined 1-D case) used in (36–38), is substituted by a critical damage, α_{cr}, depending on the loading conditions.

Previous derivations based on general dynamic stability analysis of the local damage rheology formulation (BEN-ZION and LYAKHOVSKY, 2006) provided the square root correction to the critical damage at the macroscopic failure to a dynamic value in the form of (39), $\alpha_d = \alpha_{cr} - \sqrt{\tau_d \cdot \dot{\alpha}}$. This dynamic correction allowed simulating the transition from quasi-static to dynamic fracturing, and reproducing various types of seismicity patterns with quasi-static simulations. The dynamic weakening, τ_d, controls the efficiency of a run-away mode of quasi-static propagating damage front during simulations, and its value was calibrated using the Bath law for aftershock sequences (e.g., BEN-ZION and LYAKHOVSKY, 2006). However, the connection between the τ_d value and physical properties of the damage material is not defined in the local damage model formulation. The extended non-local formulation includes intrinsic length scale and leads to basic definition of the dynamic weakening parameter as:

$$\tau_d = \frac{w}{R_S V_S}. \qquad (40)$$

The dynamic correction to the critical damage value (39) means that the dynamic regime associated with seismic wave generation may start well before the actual transition to the granular phase. In the case $\tau_d d\alpha/dt \to 1$ the dynamic fracturing starts just with onset of damage accumulation and the size of the process zone in front of the propagating dynamic crack is negligibly small (e.g., LYAKHOVSKY, 2001). In the case of propagating quasi-static crack, the dynamic correction controls its critical length, L_C,

at the transition to the dynamic regime. The stress and strain distributions around a quasi-static crack are linearly scaled with its length. This means that the rate of damage accumulation increases proportionally to the squared crack length:

$$\frac{d\alpha}{dt} = C_d \cdot \varepsilon^2 = C_d \cdot \varepsilon_0^2 \frac{L^2}{w^2}, \qquad (41)$$

where ε_0 is the strain applied to the initial crack, whose length is scaled to the diffusive length scale. Using the condition $\tau_d d\alpha/dt \to 1$, for the transition to the dynamic fracturing, and combining (40, 41) together with the time-to-failure (32), leads to an expression for a critical crack length, L_C,

$$L_C^2 = \frac{3}{4} \cdot R_S \cdot w \cdot V_S \cdot t_f = \frac{\sqrt{2\pi}}{3} R_S \cdot \frac{\mu_0^3}{\sigma^3} \cdot \sqrt{\frac{D}{C_d}} \cdot \frac{V_s}{C_d}. \qquad (42)$$

This result is a product of two length scales, i.e., diffusive length scale, w, multiplied by the distance $V_S t_f$ that a shear wave travels during time equal to time-to-failure. Another form of (42) is obtained using (35) for the diffusive length scale and (32) for time-to-failure.

4.5. Coupling with Thermal Effects During Granular Flow

The granular material produced from highly damaged solid sheared between two blocks accommodates irreversible strain according to its fluidity. In the idealized case of a straight channel with constant width, the velocity profile is linear following the classical Couette flow solution. Several mechanisms may lead to deviation from a linear velocity profile and flow localization within the channel. For example, variation of the channel width and its curvature can lead to a rotating Couette flow. In this case, the flow of power law fluids is unstable and is localized in a narrow part of the channel (PASCAL and PASCAL, 1994). Another possible mechanism involves variations of material properties across the channel, as expected for (bi-material) faults separating different rock bodies. Even in the idealized homogeneous straight channel, the flow may become localized as a result of coupling between thermal effects and the granular flow.

We consider here a model with 1 mm width zone, assuming that the transition from solid to granular phase already occurred. A high dynamic slip rate (1 m/s) prevents the transition from the granular phase back to solid. Since the channel is straight, the Couette flow solution dictates that the shear stress across the channel is constant. The strain rate, $e(z, t)$, depends not only on the stress, but also on the temperature profile, $T(z, t)$. Assuming for mathematical simplicity $m_2 = n$ in (29), the strain rate–stress relation is reduced to:

$$e(z, t) = \left[C_g + A \cdot \exp\left(-\frac{Q}{RT(z, t)}\right) \right] \tau^n. \quad (43)$$

Integrating the strain rate (43) across the channel width and using constant slip rate, v, lead to a solution for the stress level

$$v = \left[w \cdot C_g + A \cdot \int_0^w \exp\left(-\frac{Q}{RT(z, t)}\right) dz \right] \tau^n. \quad (44)$$

Equations (43, 44) provide a complete solution for the flow for a given temperature profile. The heat equation (4) inside the slipping zone is reduced to:

$$\rho C_p \frac{\partial T}{\partial t} = k \frac{\partial^2 T}{\partial z^2} + \tau \cdot e. \quad (45)$$

The same heat equation with zero source term holds for the outer solid blocks surrounding the channel. The heat problem is solved numerically using an explicit finite difference scheme with constant initial temperature, which remains the same far from the channel. We use a one dimensional numerical domain with 10,000 cells of same size. The channel is represented by 1,000 cells and the other cells correspond to the outer blocks, so the numerical domain is ten times larger than the channel width. The integral over the width of the slipping zone on the right side of (44) is calculated numerically, and combined with (43) defines the source term for the heat production in (45). This setting reproduces heating of the material within the channel due to its flow (shear heating), along with cooling due to heat transfer to the outer solid blocks.

Typical material properties representing an upper crust at about 10 km depth are used for the numerical modeling. The initial temperature $T(z, 0) = 500$ K;

density, $\rho = 2.6 \cdot 10^3$ kg/m^3; heat capacity, $C_p = 10^3$ J/kg/K; heat conductivity, $k = 1$ W/m/K; activation energy, $Q = 200$ kJ/mol. This activation energy corresponds to the onset of melting at about 1,100 K (~ 800 °C), typical for granite composition. The power law index, $n = 3$, is typical for upper crust rocks. A shear stress in the slipping zone during seismic event about 50 MPa corresponds to a dynamic friction value between 0.2 and 0.3. The average strain rate in the simulated zone is about $e = 10^3$ 1/s for slip velocity of about 1 m/s along 1 mm thick zone. These strain rates are achieved under 50 MPa shear stress if the fluidity of the granular material is $C_g = 5 \cdot 10^{-21}$ Pa^{-3} s^{-1}. The thermally activated flow component with pre-exponent value $A = 10^{-5}$ Pa^{-3} s^{-1} is negligibly small at relatively low temperatures and becomes dominant approaching the melting temperatures. Heat production in the slipping zone ($\tau e = 50 \cdot 10^3$ MPa/s) is very efficient leading to temperature increase >100 K during the first 0.01 s. Cooling due to conductive heat transfer into outer solid blocks is controlled by the thermal diffusivity, $k/\rho C_p \sim 10^{-6}$ m^2/s, which is relatively slow. The sequential thermal profiles for every 0.01 s are shown in Fig. 7a. The initial strain rate is constant across the slipping zone (Fig. 7b). After 0.02 s the strain rate starts to self-localize in the center, where the temperature is maximal. When the temperature in the center of the zone is about 800 K (time ~ 0.03 s), the strain rate is strongly localized. The difference between the strain rate in the center and along the walls of the zone is above three orders of magnitude. During the late stages of the motion (time >0.04 s) the heating is only slightly above the cooling, so both temperature and strain profiles (Fig. 7) show only minor changes.

The symmetry of the presented results (Fig. 7) is associated with homogeneous initial conditions and material parameters. In more realistic cases, we expect a certain gradient of the breakage distribution across the zone. This gradient may be associated with different properties of the solid blocks on different sides of the slip zone (bi-material faults). In the case of linear initial gradient of the fluidity, C_g, the heating is more efficient on the side of the zone where the C_g value is higher. This asymmetry is amplified with time and can lead to very asymmetric strain rate

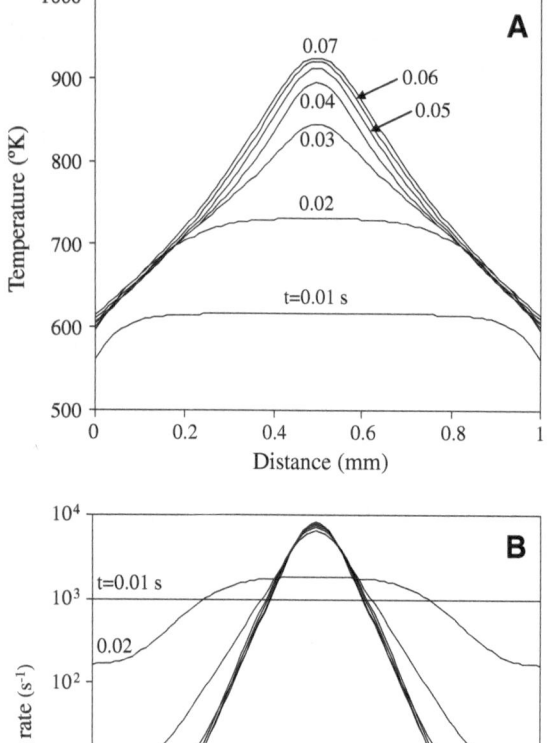

Figure 7
Calculated temperature (**a**) and strain rate (**b**) profiles across a narrow slipping zone for every 0.01 s

stable distributed brittle cracking in a solid with evolving elastic moduli, to strain localization and dynamic rupture in a narrow zone with granular material. The non-local damage formulation leads to an intrinsic length scale associated with the damage structure, while the incorporation in the CDBM of solid-granular phase transition facilitates the analysis of dynamic system evolution. The variables and parameters used to model different mechanical and thermal processes are summarized in the Notation list.

Most material properties (e.g., density, elastic moduli, heat capacity, thermal conductivity) are well constrained by laboratory experiments. Three threshold values of the strain invariants ratio ξ control transitions between different modes of material evolution. The onset of damage accumulation, or transition from weakening to healing, occurs at ξ_0. This value is constrained experimentally by the onset of acoustic emission and can also be calculated using the internal friction angle of intact rocks (e.g. LYAKHOVSKY *et al.*, 1997a; HAMIEL *et al.*, 2004). The model predicts that $\xi = \xi_1$ at the transition between mode-I (pseudo-gas) and mode-II (pseudo-liquid) types of failure (LYAKHOVSKY and BEN-ZION, 2014). The third ratio value, ξ_d, controls the onset of damage healing

profile after 0.05 s (Fig. 8). The strain rate close to the right side of the zone is above 10^4 1/s, which is about four orders of magnitude above the strain rate values in the middle and left side of the zone.

5. Discussion

The developed continuum damage–breakage mechanical model (CDBM) addresses basic aspects of phase transitions between solid and granular states of material near brittle instability. The model provides a detailed description of rock deformation from

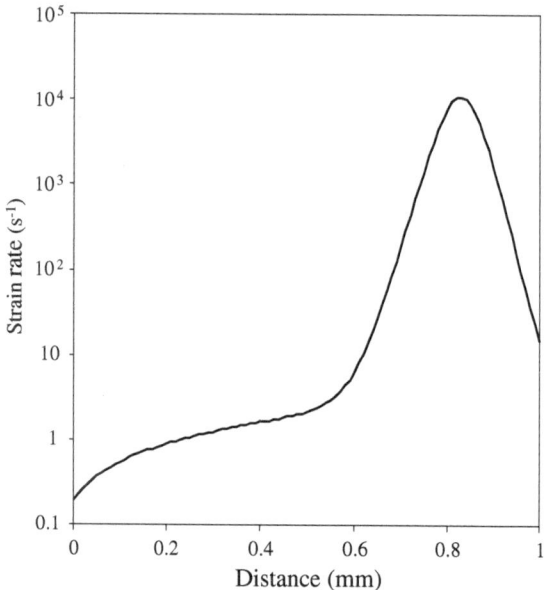

Figure 8
Asymmetric strain rate distribution in heterogeneous slip zone

and abrupt locking of the granular flow (Fig. 1). A group of parameters controls the kinetics of the damage and breakage processes. Only the rate of damage accumulation, C_d, is relatively well constrained by the available laboratory data (e.g., HAMIEL et al. 2004, 2009). Finally, several parameters control the fluidity of the granular material. The uncertainty on these is currently very large and different values may significantly influence the signals generated during dynamic rupture.

The CDBM model results capture the main observed features of frictional sliding along a narrow slipping zone, including rapid transition to dynamic frictional regime. The experimentally measured coefficient of friction drops from values between 0.6–0.8 to dynamic values below 0.4 at slip rates between 10^{-2} to 1 m/s (or slightly above). This feature was observed by many laboratory experiments with various types of rocks (see Fig. 3 of DI TORO et al., 2011 and references therein). As illustrative example, we compare in Fig. 9 model calculations (red line) with experimental results (blue line) of the friction coefficient during a rotary shear experiment with a tonalite (Sierra White granite) sample from experiment #616 of RECHES and LOCKNER (2010). The measured friction coefficient drops from ~ 0.7 to ~ 0.4 at slip velocity slightly above 0.01 m/s, remains about constant up to velocities about 0.1 m/s, and then abruptly increases. RECHES and LOCKNER (2010) and SAMMIS et al. (2011) associated the frictional increase at high velocities with a temperature rise (green line).

Simulations of slip along narrow zone similar to those shown in Fig. 3 with almost the same parameters fit well the experimental observations. A model run with $C_2 = 0.01$ reproduces closely (red line in Fig. 9) the small rate-dependent friction at slow slip rates below 0.01 m/s. During the stick–slip motion accommodated by the gouge zone, and cyclic transition between solid and granular states, the residence time of a material element in the granular phase is negligibly small. In this regime the model behavior is equivalent to previous simulations of evolution and organization of damage zones obtained in a framework of local damage rheology and instantaneous stress drop (LYAKHOVSKY et al., 2005), which

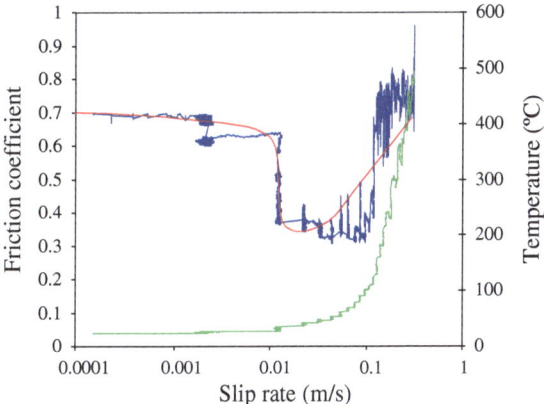

Figure 9
Friction coefficient (*blue line*) and temperature (*green line*) during a rotary shear experiment (#616 of RECHES and LOCKNER, 2010), plotted with the predicted friction coefficient of the CDBM model (*red line*). The experiment used a tonalite sample at normal stress of 3.05 MPa with total slip of 65 m at 30 slip velocity steps. See text for the used model parameters

reproduced the key features of rate- and state-dependent friction in slow-rate experiments (e.g., DIETERICH and KILGORE, 1996; MARONE, 1998; SCHOLZ 2002). At higher slip rates, the duration of the stress drop or residence time of a material element in the granular phase becomes comparable with the residence time in the solid phase. Under lower normal stress (3 MPa instead 10 MPa used for Fig. 3), and with different model parameters controlling the breakage evolution and fluidity of the granular phase, the granular residence time can be significantly extended.

The simulated transition to the dynamic regime at about 0.01 m/s (Fig. 9) is achieved by decreasing the fluidity value (C_g) and decreasing the breakage kinetic coefficients. These parameters increase the duration of the transition to the granular phase, as well as the stress release in the granular state. Following the complete transition to the granular phase at high slip rates, which prevents a transition back to the solid phase, the friction coefficient is controlled by the fluidity. The power law constitutive relation with index $m = 3$ (Eq. 28) and a constant channel width predicts that the shear stress increases as slip rate to the power $1/m$. This means that the friction coefficient is about doubled (factor $10^{1/3}$) when the slip rate increases by one order of magnitude. This model prediction provides an alternative explanation to the pure temperature effect discussed by RECHES and LOCKNER (2010) and SAMMIS et al. (2011). The modified constitutive relation (29) accounting for the temperature-activated ductile flow leads to the secondary slip localization in a small internal portion of the formed damage zone (Figs. 7, 8).

Another important factor affecting the simulated friction coefficient is the width of the channel assumed constant during the model run. This model simplification ignores a wear process which may be crucial under high slip rates. LYAKHOVSKY et al. (2014) discusses the widening of the gouge zone by wear during steady-state slip motion, using the same model formulation and including the damage diffusion of the non-local formulation. They assume that the wear process reflects micro-fracturing or damage front that propagates from the gouge zone into the solid rock, and consider the front velocity as the steady-state wear-rate. Their model connects the measured steady-state friction coefficient with the wear rate and successfully fits the laboratory experiments.

The non-local gradient type damage model formulation with intrinsic length scale not only eliminates unphysical singular damage and strain localization, but gives rise to several scaling relations. One basic relation predicts (Eq. 38) that the width of the damage zone satisfies $w \propto \sigma^{-1}$. This scaling means that relatively wide damage zones are expected under low stresses at shallow depths, while higher stress within the seismogenic zone should lead to the formation of narrow zones. The competition between loading-controlled damage growth and diffusion-controlled damage dissociation leads to a scaling relation for a critical fracture size for failure, consistent with a size effect at crack initiation or scaling between stress and micro-crack length (LYAKHOVSKY et al., 2011).

The scaling analyses presented in this study clarify the conditions controlling the transition from quasi-static to dynamic fracturing, and provide a fracture-based estimate of the critical size of failing damage zone (nucleation size) leading to dynamic instability. The threshold value of the ratio between dynamic strain components associated with radiated seismic waves and static strain ($R_S = \varepsilon_d/\varepsilon_s$) can be viewed as an analog to the Reynolds number of fluid mechanics.

This "damage-mechanics Reynolds number" quantifies the relative importance of waves radiated from the propagating fracture which, together with values of material properties, leads to an expression (Eq. 42) for critical crack length, L_C, at the onset of dynamic fracture. The R_S value at the onset of dynamic rupture is scaled as $R_S \propto L^2$ and $R_S \propto \sigma^3$. This implies that relatively small increase of crack length and loading can significantly increase the R_S value, leading to rapid transition from quasi-static to dynamic regime. The results explain why the transitional regime to dynamic rupture can hardly be observed in laboratory experiments. There are sets of material parameters and loading conditions for which the nucleation size of a quasi-static rupture approaches the critical length of dynamic rupture. In such cases every nucleated fracture grows dynamically without a significant process zone in front of the propagating crack tip, similar to classical dynamic fracture mechanics.

The scaling in the CDBM between the critical nucleation length for dynamic fracture and stress (Eq. 42) predicts a power-law relation $L_C \propto \sigma^{-3/2}$. This relation differs from the expression of friction-based nucleation $h^* = CD_C \cdot \mu/[(b-a) \cdot \sigma]$, where C is a dimensionless constant of order 1, a quantifies the initial change of the friction coefficient following a velocity jump, b gives the amplitude of a gradual strength evolution over a characteristic slip distance D_c, μ is the rigidity and σ is the normal stress (e.g. RICE and BEN-ZION, 1996; BEN-ZION, 2008; Sections 3 and 5). However, the predicted nucleation sizes in the CDBM and friction models have overlapping ranges of values. This is seen by estimating L_C values using the following ranges of properties: $V_S \sim 10^3$ m/s, time-to-failure under loading in the seismogenic zone, $t_f \sim 10^2-10^3$ s, diffusive length scale $L_D \sim 10^{-5}-10^{-3}$ m, and damage Reynolds number, $R_S \sim 10^{-4}-10^{-3}$. For these ranges of parameters, the critical length values are $L_C \sim 10^{-2}-1$ m, similar to the typical range of estimated frictional nucleation size $h^* \sim 10^{-2}-10$ m.

Very narrow zones with sub-micron or even nanometer-scale particles are observed in the cores of exhumed fault zones (e.g., CHESTER et al., 2005; SAMMIS and BEN-ZION, 2008) suggesting that very fine powder is formed in the dynamically-slipping core region of the fault zone. The core, which accommodates most of the slip, is surrounded by wider zones of fault gouge and breccia with particle sizes ranging from microns to centimeters (e.g., ROCKWELL et al., 2009; WECHSLER et al., 2011). Direct connections between values of the breakage parameter, grain-size distribution and fluidity of the granular material in the dynamically slipping and surrounding gouge zone are not well constrained. Such connections should be addressed in future studies using additional laboratory experiments, field observations and related theoretical results.

Acknowledgments

We thank A. Ilchev, Z. Reches and A. Sagy for discussions. The manuscript benefitted from useful comments by Editor Antonio Rovelli and two anonymous referees. We acknowledge support by the US–Israel Bi-national Science Foundation (Grant 2008248) and the National Science Foundation (Grant EAR-0908903).

Appendix: Thermodynamic Formulation

The total energy of a solid with a unit mass includes internal and kinetic components:

$$E = E_k + U. \tag{46}$$

The specific kinetic energy of the solid is $E_k = v_i v_i / 2$ with v_i being velocity and the specific internal energy, U, is expressed through the specific free energy F, temperature T and entropy S as $U = F + TS$. The energy balance equation dictates that the change in the energy of a system consists of a sum of different terms representing different energy forms. These terms are the total stress power, $\sigma_{ij} e_{ij}$, divergence of the heat flux $J_i^{(q)}$, internal heat source per unit mass, r, and interstitial work flux $J_i^{(i)}$:

$$\rho \frac{dU}{dt} = \rho \frac{d}{dt}(F + TS)$$
$$= \sigma_{ij} e_{ij} - \nabla_i J_i^{(q)} + \rho r - \nabla_i J_i^{(i)}. \tag{47}$$

The entropy balance equation includes entropy flux $J_i^{(s)}$, internal heat source, and non-negative local entropy production Γ:

$$\rho \frac{dS}{dt} = -\nabla_i J_i^{(s)} + \frac{\rho r}{T} + \Gamma, \quad \Gamma \geq 0. \tag{48}$$

The non-negative local entropy production results from all the dissipative irreversible processes in the medium including internal friction and damage evolution. From Eq. (7) of the main text, the change in the free energy, Gibbs equation, can be expressed as:

$$dF = -S dT + \frac{\partial F}{\partial \varepsilon_{ij}} d\varepsilon_{ij} + \frac{\partial F}{\partial \alpha} d\alpha + \frac{\partial F}{\partial (\nabla_i \alpha)} d(\nabla_i \alpha)$$
$$+ \frac{\partial F}{\partial B} dB. \tag{49}$$

In a system with internal motions, the material time derivative is given by $d(\cdot)/dt = \partial(\cdot)/\partial t + v_i \nabla_i(\cdot)$. Similarly,

$$\nabla_i \left(\frac{d\alpha}{dt}\right) = \nabla_i \left(\frac{\partial \alpha}{\partial t} + v_k \nabla_k \alpha\right)$$
$$= \frac{d(\nabla_i \alpha)}{dt} + (\nabla_i v_k) \nabla_k \alpha. \tag{50}$$

From (50), the time derivative of the last term in (49) can be written as:

$$\frac{\partial F}{\partial (\nabla_i \alpha)} \cdot \frac{d(\nabla_i \alpha)}{dt} = -\left[\nabla_i \left(\frac{\partial F}{\partial (\nabla_i \alpha)}\right) \frac{d\alpha}{dt} + e_{ik} \frac{\partial F}{\partial (\nabla_i \alpha)} \nabla_k \alpha\right]$$
$$+ \nabla_i \left(\frac{\partial F}{\partial (\nabla_i \alpha)} \frac{d\alpha}{dt}\right). \tag{51}$$

Combining the above equations, the non-negative energy dissipation ($D_E \geq 0$) is represented as the sum of the local entropy production and interstitial work flux:

$$D_E = T\Gamma + \nabla_i J_i^{(i)} = \left(\sigma_{ij} - \rho \frac{\partial F}{\partial \varepsilon_{ij}} + \rho \frac{\partial F}{\partial (\nabla_i \alpha)} \nabla_j \alpha\right) e_{ij}$$
$$+ T\nabla_i J_i^{(s)} - \nabla_i J_i^{(q)}$$
$$- \rho \nabla_i \left(\frac{\partial F}{\partial (\nabla_i \alpha)} \frac{d\alpha}{dt}\right) + \rho \frac{\partial F}{\partial \varepsilon_{ij}} \frac{\partial \varepsilon_{ij}^{(p)}}{\partial t}$$
$$- \rho \left(\frac{\partial F}{\partial \alpha} - \nabla_i \frac{\partial F}{\partial (\nabla_i \alpha)}\right) \frac{d\alpha}{dt} - \rho \frac{\partial F}{\partial B} \frac{dB}{dt}. \tag{52}$$

If all the dissipative processes in the system are frozen, the local entropy production and interstitial working are zero. This condition leads to the definition of the stress tensor:

$$\sigma_{ij} = \rho\left[\frac{\partial F}{\partial \varepsilon_{ij}} - \frac{1}{2}\left(\frac{\partial F}{\partial(\nabla_i\alpha)}\nabla_j\alpha + \frac{\partial F}{\partial(\nabla_j\alpha)}\nabla_i\alpha\right)\right]. \tag{53}$$

In addition to the usual term $\partial F/\partial\varepsilon_{ij}$ (e.g., MALVERN, 1969), there are terms associated with $\nabla\alpha$, or "structural stresses". These stress components are associated with heterogeneous damage distribution and can be important in zones with high $\nabla\alpha$ values.

The presence of the extra flux $J_i^{(i)}$ in (47) requires a posteriori definition of the energy partitioning between local entropy production converted into heat and heat-free interstitial working. We assume that the two components of the energy dissipation involving conductive heat transport and irreversible deformation are associated with heat, while the interstitial working associated with the damage–breakage evolution does not affect the heat balance of the system. This means that the local entropy production is:

$$\Gamma = \Gamma_H + \Gamma_V. \tag{54}$$

Using the definition $J_i^{(s)} = J_i^{(q)}/T$, the term related to the conductive heat transport is:

$$\Gamma_H = -\frac{J_i^{(q)}\cdot\nabla_i T}{T^2}, \tag{55}$$

and the term related to the irreversible deformation is:

$$\Gamma_V = \frac{\rho}{T}\frac{\partial F}{\partial\varepsilon_{ij}}\frac{d\varepsilon_{ij}^{(p)}}{dt}. \tag{56}$$

Non-negativity of the local entropy production gives rise to the Fourier law for the thermal conductivity

$$J_i^{(q)} = -k\cdot\nabla_i T, \tag{57}$$

where k is thermal conductivity. The non-negativity of the second term related to the irreversible deformation also known as shear heating, requires positively defined scalar or tensor of the effective viscosity in the constitutive relations connecting stress with the rate of irreversible strain accumulation. Finally, the heat balance equation controlling the rate of the temperature change includes two usual terms, i.e., conductive heat transport and shear heating:

$$\rho C_p\frac{\partial T}{\partial t} = k\nabla^2 T + \rho\frac{\partial F}{\partial\varepsilon_{ij}}\frac{d\varepsilon_{ij}^{(p)}}{dt}. \tag{58}$$

The interstitial working flux associated with damage–breakage evolution is defined as:

$$\nabla_i J_i^{(i)} = -\rho\left[\left(\frac{\partial F}{\partial\alpha} - \nabla_i\frac{\partial F}{\partial(\nabla_i\alpha)}\right)\frac{d\alpha}{dt} + \frac{\partial F}{\partial B}\frac{dB}{dt}\right] - \rho\nabla_i\left(\frac{\partial F}{\partial(\nabla_i\alpha)}\frac{d\alpha}{dt}\right). \tag{59}$$

Since the heat conduction, shear heating and interstitial working are independent physical processes, each should provide non-negative energy dissipation. This implies that the interstitial working is also non-negative ($\nabla_i J_i^{(i)} \geq 0$). The divergence term on the right side of (59) is associated with transport of the heterogeneous damage production. Being integrated over the whole system volume, according to Gauss theorem, its impact is equal to the flux through the external boundary of the system. This quantity should be defined as an additional boundary condition in the non-local system. Therefore, the first term in (59) standing for the energy dissipation associated with the damage–breakage evolution should be non-negative. Adopting ONSAGER (1931) reciprocal relations between thermodynamic forces and fluxes, we write phenomenological equations for the kinetics of the state variables α and B as a set of two coupled differential equations:

$$\begin{aligned}
\frac{dB}{dt} &= C_{BB}\frac{\partial F}{\partial B} + C_{B\alpha}\left(\frac{\partial F}{\partial\alpha} - \nabla_i\frac{\partial F}{\partial(\nabla_i\alpha)}\right)\\
\frac{d\alpha}{dt} &= C_{\alpha B}\frac{\partial F}{\partial B} + C_{\alpha\alpha}\left(\frac{\partial F}{\partial\alpha} - \nabla_i\frac{\partial F}{\partial(\nabla_i\alpha)}\right).
\end{aligned} \tag{60}$$

The kinetic coefficients (or functions), C_{BB}, $C_{B\alpha}$, $C_{\alpha B}$, and $C_{\alpha\alpha}$, connect the thermodynamic forces associated with certain state variable with thermodynamic fluxes. The diagonal terms with C_{BB} and $C_{\alpha\alpha}$ are associated with a given state variable, while the off-diagonal terms with $C_{\alpha B}$ and $C_{B\alpha}$ provide coupling between evolving damage with breakage and vice versa.

REFERENCES

AGNON, A. and LYAKHOVSKY, V., (1995), Damage distribution and localization during dyke intrusion. In: The Physics and Chemistry of Dykes, edited by G. Baer and A. Heimann, Balkema, Rotterdam, 65–78.

ALAVA, M. J., NUKALA, P., and ZAPPERI, S., (2006), *Statistical models of fracture*, Adv. Phys. *55*, 349–476.

ALLAM, A.A. and BEN-ZION, Y., (2012), *Seismic velocity structures in the Southern California plate-boundary environment from double-difference tomography*, Geophys. J. Int., *190*, 1181–1196, doi:10.1111/j.1365-246X.2012.05544.x.

ASHBY, M.F. and SAMMIS, C.G. (1990), *The damage mechanics of brittle solids in compression*, Pure Appl. Geophys., *133*, 3, 489–521.

BAZANT, Z.P., (2005), *Scaling of structural strength*. Elsevier, 327 pp.

BAZANT, Z.P. and JIRASEK, M., (2002), *Nonlocal integral formulations of plasticity and damage: Survey of progress*. J. Engineering Mechanics, *128*, 1,119–1,149.

BEN-ZION, Y., (2008), *Collective Behavior of Earthquakes and Faults: Continuum-Discrete Transitions, Evolutionary Changes and Corresponding Dynamic Regimes*, Rev. Geophysics, *46*, RG4006, doi:10.1029/2008RG000260.

BEN-ZION, Y., LYAKHOVSKY, V., (2002), *Accelerating seismic release and related aspects of seismicity patterns on earthquake faults*. Pure Appl. Geophys. *159*, 2385–2412.

BEN-ZION, Y., and LYAKHOVSKY, V., (2006). *Analysis of aftershocks in a lithospheric model with seismogenic zone governed by damage rheology*. Geophys. J. Int. *165*, 197–210, doi:10.1111/j.1365-246X.2006.02878.x.

BEN-ZION, Y., DAHMEN, K., LYAKHOVSKY, V., ERTAS, D. and AGNON, A., (1999), *Self-driven mode switching of earthquake activity on a fault system*, Earth Planet. Sci. Lett., *172*, 11–21.

BEN-ZION, Y., DAHMEN, K. A. and UHL, J.T., (2011), *A unifying phase diagram for the dynamics of sheared solids and granular materials*, Pure Appl. Geophys., doi:10.1007/s00024-011-0273-7.

BERCOVICI, D., and RICARD, Y., (2003), *Energetics of a two-phase model of lithospheric damage, shear localization and plate-boundary formation*, Geophys. J. Int., *152*, 581–596, doi:10.1046/j.1365-246X.2003.01854.x.

BHAT, H. S., SAMMIS, C.G. and ROSAKIS, A.J., (2011), *The micromechanics of Westerley Granite at large compressive loads*, Pure Appl. Geophys., *168*. 12, 1–18, doi:10.1007/s00024-011-0271-9.

BIRCH, F., (1952), *Elasticity and constitution of the Earth's interior*, J. Geophys. Res, *57*, 221–286.

BYERLEE, J.D., (1978), *Friction of rocks*. Pure Appl. Geophys., *116*, 615–626.

CHESTER, J.S., CHESTER, F.M. and KRONENBERG, A.K., (2005), *Fracture surface energy of the Punchbowl fault, San Andreas system*, Nature, *437*, 133–136.

DEGROOT, S.R., and MAZUR, P., (1962), *Nonequilibrium thermodynamics*, North-Holland Publishing Co., Amsterdam.

DI TORO, G., HAN, R., HIROSE, T., DE PAOLA, N., NIELSEN, S., MIZOGUCHI, K., FERRI, F., COCCO, M., and SHIMAMOTO, T., (2011), *Fault lubrication during earthquakes*. Nature, *471*, 494–498, doi:10.1038/nature09838.

DIETERICH, J.H. and KILGORE, B.D., (1996), *Imaging surface contacts; power law contact distributions and contact stresses in quarts, calcite, glass, and acrylic plastic*, Tectonophysics, *256*, 219–239.

DUNN, J.E. and SERRIN, J. (1985) *On the thermodynamics on interstitial working*, Arch. Rational Mech. Anal., *88*, 95–133.

EINAV I., 2007a. *Breakage mechanics - Part I: Theory*, J. Mech. Phys. Solids, *55*, 1274–1297.

EINAV I., 2007b. *Breakage mechanics - Part II: Modeling granular materials*, J. Mech. Phys. Solids, *55*, 1298–1320.

EKELAND, I., and TEMAM, R., (1976), *Convex analysis and variational problems*. Elsevier.

ERICKSEN, J.L., (1998), *Introduction to the thermodynamics of solids*. Springer-Verlag, NY.

EXADAKTYLOS, G.E. (2006), Nonlinear rock mechanics. In: (P.P. Delsanto, ed.) *Universality of nonclassical nonlinearity*, Springer, New York, p. 71–90.

FABRIZIO, M., LAZZARI, B. and NIBBI, R., (2011), *Thermodynamics of non-local materials: extra fluxes and internal powers*. Continuum Mech. Thermodyn. *23*: 509–525. doi:10.1007/s00161-011-0193-x.

FORTERRE Y., and POULIQUEN, O., (2008), *Flows of dens granular media*. Ann. Rev. Fluid. Mech. *40*, 1–24.

FREMOND, M., (2012), *Phase change in mechanics*. Springer-Verlag, Berlin.

GIOVINE, P., (1999), *Nonclassical thermomechanics of granular materials*. Math. Phys., Anal. Geom., *2*: 179–196.

HAMIEL, Y., LIU, Y., LYAKHOVSKY, V., BEN-ZION. Y. and LOCKNER, D., (2004), *A visco-elastic damage model with applications to stable and unstable fracturing*. J. Geophys. Int. *159*, 1155–1165.

HAMIEL, Y, LYAKHOVSKY, V., STANCHITS, S., DRESEN, G. and BEN-ZION, Y., (2009), *Brittle deformation and damage-induced seismic wave anisotropy in rocks*, Geophys. J. Int., *178*, 901–909, doi:10.1111/j.1365-246X.2009.04200.x.

HAMIEL, Y., LYAKHOVSKY, V. and BEN-ZION, Y., (2011), *The elastic strain energy of damaged solids with applications to nonlinear deformation of crystalline rocks*. Pure Appl. Geophys. *168*, 2199–2210. doi:10.1007/s00024-011-0265-7.

HESHMAT, H., (1995), *The quasi-hydrodynamic mechanism of powder lubrication. 3: On theory and rheology of triboparticulates*. Tribol Trans. *38*, 209–276.

HESHMAT, H. and HESHMAT, C.A., (1999), On the rheodynamics of powder lubricated journal bearing: theory and experiment. In: Lubrication at the Frontier: The Role of the Interface and Surface Layers in the Thin Film and Boundary Regime. *Tribology series* 36. D. Dowson (Ed.). Elsevier, p 537–550.

JAEGER, J. C. and COOK, N.G.W., (1979), *Fundamentals of rock mechanics*, Chapman and Hall.

JAEGER, H.M., NAGEL, S.R. and BEHRINGER, R.P., (1996), *Granular solids, liquids, and gases*. Rev. Mod. Phys. *68*, 1259–1273.

JOHNSON, P.A. and JIA, X., (2005), *Non-linear dynamics, granular media and dynamic earthquake triggering*, Nature, *437*, 871–874.

KACHANOV, L.M., (1986), *Introduction to continuum damage mechanics*, Martinus Nijhoff Publishers, pp. 135.

KAJI Y., GU W., ISHIHARA M., ARAI T., and NAKAMURA H. (2001), *Development of structural analysis program for non-linear elasticity by continuum damage mechanics*, Nuclear Engineering and Design, *206*, 1–12.

KAMRIN, K. and KOVAL, G. (2012), *Nonlocal constitutive relation for steady granular flow*, Phys. Rev. Lett. *108*, 178301.

KARRECH, A., DUHAMEL, D., BONNET, G., ROUX, J.N., CHEVOIR, F., CANOU, J., DUPLA, J.C. and SAB, K., (2007), *A computational procedure for the prediction of settlement in granular materials under cyclic loading*. Comput. Methods Appl. Mech. Engrg. *197*, 80–94.

KRAJCINOVIC, D., (1996), *Damage Mechanics*, Elsevier, Amsterdam.

LANDAU, L.D. and LIFSHITZ, E.M., (1980), *Statistical Physics, Course of Theoretical Physics*, 3rd Edition, Vol. 5. Pergamon Press, Oxford, 387 p.

LEMAITRE, J., (1996), *A course on damage mechanics*. Springer Verlag, Berlin, 228 pp.

LI, J. and DING, D.W., (2002), *Nonlinear elastic behavior of fiber-reinforced soil under cyclic loading*, Soil Dynamics and Earthquake Engineering, *22*, 977–983.

LIU, A.J., and NAGEL, S.R., (eds), (2001), *Jamming and rheology: constrained dynamics on microscopic and macroscopic scales*. Taylor and Francis, London.

LOCKNER, D. A., BYERLEE, J.D., KUKSENKO, V., PONOMAREV, A. and SIDORIN, A., (1992), Observations of quasi-static fault growth from acoustic emissions. In: Fault mechanics and transport properties of rocks, International Geophysics Series, Vol. 51, 3–31, eds Evans, B. & Wong, T.-f., Academic Press, San Diego, CA.

LUBE, G., HUPPERT, H.E., SPARKS, R.S.J. and HALLWORTH, M.A., (2004) *Axisymmetric collapses of granular columns*. J. Fluid Mech., *508*, 175–199.

LYAKHOVSKY, V., (2001), *Scaling of fracture length and distributed damage*. Geophys. J. Int., *144*, 114–122.

LYAKHOVSKY, V., and Y. BEN-ZION., (2009), *Evolving geometrical and material properties of fault zones in a damage rheology model*, Geochem. Geophys. Geosyst., *10*, Q11011, doi:10.1029/2009GC002543.

LYAKHOVSKY, V., and BEN-ZION Y., (2014), *Damage–Breakage rheology model and solid-granular transition near brittle instability*, J. Mech. Phys. Solids, *64*, 184–197, doi:10.1016/j.jmps.2013.11.007.

LYAKHOVSKY, V. and BEN-ZION., Y., (2008) *Scaling relations of earthquakes and aseismic deformation in a damage rheology model*. Geophys. J. Int. *172*, 651–662.

LYAKHOVSKY, V., BEN-ZION, Y. and AGNON, A., (1997a), *Distributed damage, faulting, and friction*, J. Geophys. Res., *102*, 27635–27649.

LYAKHOVSKY, V., RECHES Z., WEINBERGER, R., SCOTT, T.E., (1997b), *Non-linear elastic behavior of damaged rocks*. Geophys. J. Int. *130*: 157–166.

LYAKHOVSKY, V., BEN-ZION, Y., and AGNON, A., (2001), *Earthquake cycle, fault zones and seismicity pattern s in a rheologically layered lithosphere*. J. Geophys. Res. *106*, 4103–4120.

LYAKHOVSKY, V., BEN-ZION, Y. and AGNON, A., (2005), *A visco-elastic damage rheology and rate- and state-dependent friction*. Geophys. J. Int. *161*, 179–190.

LYAKHOVSKY, V., HAMIEL, Y. and BEN-ZION, Y., (2011), *A non-local visco-elastic damage model and dynamic fracturing*, J. Mech. Phys. Solids, *59*, 1752–1776, doi:10.1016/j.jmps.2011.05.016.

LYAKHOVSKY, V., SAGY, A., BONEH, Y. and RECHES, Z., (2014), *Fault wear by damage evolution during steady-state slip*, Pure Appl. Geophys. doi:10.1007/s00024-014-0787-x.

MALVERN, L.E., (1969), *Introduction to the mechanics of a continuum medium*, Prentice-Hall, Inc., New Jersey.

MARONE, C., (1998), *Laboratory-derived friction laws and their application to seismic faulting*, Annu. Rev. Earth Planet. Sci., *26*, 643–649.

MURNAGHAN, F.D. (1951), *Finite deformation of an elastic solid*, John Wiley, Chapman, New York, 140 pp.

MYASNIKOV, V.P. and OLEINIKOV, A.I., (1991), *Deformation model of a perfectly free-flowing granular medium*, Soviet Physics Doklady, *36*, 51–53.

NGAN, S.C. and TRUSKINOVSKY, L., (2002), *Thermo-elastic aspects of dynamic nucleation*. J. Mech. Phys. Solids. *50*, 1193–1229.

NGUYEN, V-H., DUHAMEL, D., and NEDJAR, B., (2003), *A continuum model for granular materials taking into account the no-tension effect*. Mechanics Materials, *35*, 955–967.

NOLL, W., (1958), *A mathematical theory of the mechanical behavior of continuous media*. Arch. Ration. Mech. Anal. 2, 198–226.

ONSAGER, L., (1931), *Reciprocal relations in irreversible processes*. Phys. Rev., *37*, 405–416.

PASCAL, H. and PASCAL, J-P., (1994), *Similarity solutions to rotating Couette flow with power law fluids*. Acta Mechanica, *107*, 93–100.

PRIGOGINE, I., (1955), *Introduction to thermodynamics of irreversible processes*, Springfield, Illinois.

RAO, K.K. and NOTT, P.R., (2008), *An introduction to granular flow*. Cambridge Univ. Press, Cambridge, 490 p.

RECHES, Z., and LOCKNER, D.A., (2010), *Fault weakening and earthquake instability by powder lubrication*. Nature, *467*, 452–456, doi:10.1038/nature09348.

REGENAUER-LIEB, K., and YUEN, D.A., (2003), *Modeling shear zones in geological and planetary sciences: solid-and fluid-thermal–mechanical approaches*, Earth-Sci. Rev., *63*, 295–349.

RICE, J.R. and BEN-ZION, Y., (1996), *Slip complexity in earthquake fault models*, Proc. Natl. Acad. Sci. U.S.A., *93*, 3811–3818.

ROCKWELL, T., SISK, M., GIRTY, G., DOR, O., WECHSLER, N., and BEN-ZION, Y., (2009), *Chemical and Physical Characteristics of Pulverized Tejon Lookout Granite Adjacent to the San Andreas and Garlock Faults: Implications for Earthquake Physics*, Pure Appl. Geophys., *166*, 1725–1746, doi:10.1007/s00024-009-0514-1.

SAMMIS, C. and BEN-ZION, Y., (2008), *Mechanics of grain-size reduction in fault zones*. J. Geophys. Res., *113*, B02306. doi:10.1029/2006JB004892.

SAMMIS, C., LOCKNER, D., and RECHES, Z., (2011), *The role of adsorbed water on the friction of a layer of submicron particles*. Pure Appl. Geophys. *168*, 2325–2334.

SAVAGE, S.B., (1998), *Analyses of slow high-concentration flows of granular materials*. J. Fluid Mech. *377*, 1–26.

SCHOLZ, C.H., (2002), *The mechanics of earthquakes and faulting*, 2nd ed. Cambridge University Press, 471 p.

STANCHITS, S., VINCIGUERRA, S., and DRESEN, G., (2006), *Ultrasonic Velocities, Acoustic Emission Characteristics and Crack Damage of Basalt and Granite*, Pure Appl. Geophys., *163*, 974–993, doi:10.1007/s00024-006-0059-5.

SUZUKI, T. (2012), *Understanding of dynamic earthquake slip behavior using damage as a tensor variable: Microcrack distribution, orientation, and mode and secondary faulting*, J. Geophys. Res., *117*, B05309, doi:10.1029/2011JB008908.

SYLVESTER, A.G., (1988), *Strike-slip faults*. Geol. Soc. Am. Bull., *100*, 1666–1703.

TRIANI, V. and CIMMELLI, V.A., (2012), *Interpretation of second law of thermodynamics in the presence of interfaces*, Continuum Mech. Thermodyn., *24*, 165–174, doi:10.1007/s00161-011-0231-8.

TRUESDELL, C.A., (1966), *The elements of continuum mechanics*. Springer, New York.

TRUESDELL, C., and NOLL, W., (2004), *The non-linear field theories of mechanics*. 3ed., Edited by S.S. Antman, Springer-Verlag, Berlin, 627 pp.

TURCOTTE, D.L., NEWMAN, W.I. and SHCHERBAKOV, R., (2003), *Micro and macroscopic models of rock fracture*. Geophys. J. Int., *152*, 718–728.

TYUPKIN, Yu.S., (2007), *Earthquake source nucleation as self organizing process.* Tectonophysics, *431*, 73–81.

WECHSLER, N., ALLEN, E.E., ROCKWELL, T., GIRTY, G., CHESTER, J. S., and BEN-ZION, Y., (2011), *Characterization of Pulverized Granitoids in a Shallow Core along the San Andreas Fault, Little Rock, CA*, Geophys. J. Int., *186*, 401–417, doi:10.1111/j.1365-246X.2011.05059.x.

WEERTMAN, J., (1978), *Creep laws for the mantle of the Earth.* Philos. Trans. R. Soc. London A, *288*, 9–26.

WORNYOH, E.Y.A., JASTI, V.K., and HIGGS III, C.F., (2007), *A review of dry particle lubrication: powder and granular materials.* J. Tribology, *129*, 438–449.

(Received September 6, 2013, revised March 30, 2014, accepted April 2, 2014, Published online May 4, 2014)

Pure Appl. Geophys. 171 (2014), 3125–3141
© 2014 Springer Basel
DOI 10.1007/s00024-014-0801-3

Evolution of Wear and Friction Along Experimental Faults

Y. Boneh,[1,3] J. C. Chang,[1] D. A. Lockner,[2] and Z. Reches[1]

Abstract—We investigate the evolution of wear and friction along experimental faults composed of solid rock blocks. This evolution is analyzed through shear experiments along five rock types, and the experiments were conducted in a rotary apparatus at slip velocities of 0.002–0.97 m/s, slip distances from a few millimeters to tens of meters, and normal stress of 0.25–6.9 MPa. The wear and friction measurements and fault surface observations revealed three evolution phases: A) An initial stage (slip distances <50 mm) of wear by failure of isolated asperities associated with roughening of the fault surface; B) a running-in stage of slip distances of 1–3 m with intense wear-rate, failure of many asperities, and simultaneous reduction of the friction coefficient and wear-rate; and C) a steady-state stage that initiates when the fault surface is covered by a gouge layer, and during which both wear-rate and friction coefficient maintain quasi-constant, low levels. While these evolution stages are clearly recognizable for experimental faults made from bare rock blocks, our analysis suggests that natural faults "bypass" the first two stages and slip at gouge-controlled steady-state conditions.

1. Introduction

Rock faulting is a complex process that occurs by brittle fracturing, coalescence and sliding of multiple fractures, and crushing of fracture-bounded wedges (e.g., Peng and Johnson 1972; Hallbauer et al. 1973; Hadley 1975; Tapponnier and Brace 1976; Aydin 1978; Sammis et al. 1987; Reches and Lockner 1994; Katz and Reches 2004). Naturally, fault surfaces that form by these processes are irregular, fragmented and rough (Lockner et al. 1992; Heesakkers et al. 2011a). Field, experimental, and theoretical studies suggest that such irregular and rough surfaces of immature faults evolve into smoother, mature surfaces by a combination of rock wear, shear localization, and flow (Ben-Zion and Sammis 2003; Sagy et al. 2007). On the other hand, fault roughness may increase by dynamic branching (Sagy et al. 2001) or by linking of fault segments (Candela et al. 2011). In the present work, we focus on the transient wear during initial slip along experimental faults.

Wear is the process of material removal from sliding surfaces by mechanical actions (Rabinowicz 1965). Early investigators of wear processes focused primarily on wear of machinery components made of metals (Bowden and Tabor 1939). Nevertheless, it was found later that the basic concepts of metals wear could be applied to brittle rocks wear because asperities interaction was considered the controlling process for both metals and rocks (Wang and Scholz 1994). Archard (1953) suggested that wear between sliding blocks occurs at touching asperities that comprise the real contact area, which is a small fraction of the nominal sliding area (Bowden and Tabor 1939; Dieterich and Kilgore 1994). The asperity-controlled model is a two-body configuration in which the wear is dominated by adhesion, abrasion, and plowing at the touching asperities (Wang and Scholz 1995). Two stages of wear intensity were experimentally recognized: an early stage with transient, high wear-rate that is followed by a steady-state stage with quasi-stable, lower wear-rate (Queener et al. 1965; Levy and Jee 1988; Power et al. 1988; Wang and Scholz 1994). The early slip, called "running-in", was well-described by Queener et al. 1965: "If two new machine parts are subjected to sliding...the wear process...is usually characterized by a high initial wear-rate which gradually diminishes to some steady-state value. ...the wear occurring before steady-state...from this 'running-in' process frequently

---
[1] School of Geology and Geophysics, University of Oklahoma, 100 E Boyd St., Norman, OK 73019, USA. E-mail: reches@ou.edu
[2] US Geological Survey, 345 Middlefield Rd., Menlo Park, CA 94025, USA.
[3] Earth and Planetary Sciences, Washington University, One Brooking Drive, St. Louis, MO 63130, USA.

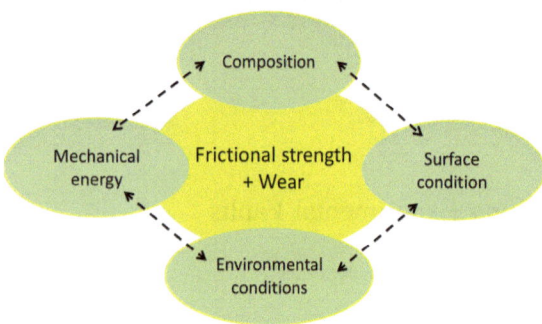

Figure 1
Schematic illustration of conditions and processes that control the frictional strength and fault wear, which are interdependent

represents a significant fraction of the total wear." Queener et al. assumed that during running-in, the wear volume is proportional to the volume available for wear, V, which is the volume bound by the sliding blocks roughness. This assumption led to an expression of the wear-rate, dV/dL,

$$dV/dL = -n \times V$$

where L is the slip distance, and n is a constant parameter that incorporates the loading intensity, slip conditions, and material properties. This equation implies that during running-in, the wear volume is

$$V = V_0 \exp(-nL)$$

where V_0 is the initial volume available for wear. QUEENER et al. (1965) and others (WANG and SCHOLZ 1994) showed that the last equation fits well with experimental results of running-in in both the exponential decrease of wear-rate with slip distance and with proportionality to the initial roughness (equivalent of V_0).

Slip along faults is always associated with frictional resistance that is the integrated effect of adhesion between solid blocks, plastic deformation, and fault wear (Fig. 1). Although wear and friction are two inherently connected processes, and may be seen as two properties of the same process (RABINOWICZ 1965; LYAKHOVSKY et al. 2014), their relations are non-unique (DOBSON and WILMAN 1963; RABINOWICZ 1965). Friction strength and wear are related through asperities failure (WANG and SCHOLZ 1994), pulverization at fault tip (RECHES and DEWERS 2005), evolution of fault surface roughness (OHNAKA, 1973;

SANTNER et al. 2006; BRODSKY et al. 2011), thermal fracturing (HIROSE et al. 2012), gouge welding (NAKATANI 1998), and rock comminution (WILSON et al. 2005; HENDERSON et al. 2010).

We present here an experimental analysis of the transient evolution of wear-rate and correlate it to the frictional strength. Our experimental system (RECHES and LOCKNER 2010) allows for continuous quantification of fault friction and wear-rate. Two types of experiments were run: (1) short slip experiments (3–50 mm) designed to characterize early wear processes, and (2) extended slip experiments (up to tens of meters) designed to characterize the evolution of wear and friction to a steady-state condition.

2. Experimental Setting

2.1. Apparatus and Experimental Procedures

We tested experimental faults in a high-velocity rotary apparatus (RECHES and LOCKNER 2010). The samples comprised two solid cylindrical rock blocks; the rotating, lower block had a planar surface and the stationary, upper block had a raised-ring contact (Fig. 2). We used two ring configurations, one with 2.2 and 5.1 cm as inner and outer diameters, respectively, and the other with 5.4 and 7.6 cm as inner and outer diameters, respectively. The apparatus is capable of unlimited slip distance, normal stress up to 35 MPa, and slip velocity from 0.001 to 2 m/s. We continuously monitored the normal stress, shear stress, slip velocity, distance, and temperature at rates of up to 2,000 samples/s. A critical parameter for the analysis is the fault-normal displacement that was measured by two non-contact, eddy-current sensors (~ 1 μm resolution) mounted 180° from each other on the sample grips. Samples were ground flat, roughened with #600 SiC grit, and dried for 24 h at 100 °C. The experimental fault blocks were made of five rock types: Tonalite (commercial name Sierra White granite), Kasota dolomite, Karoo gabbro, Blue quartzite and Tennessee sandstone. In each experiment, the fault was loaded to a predetermined normal stress that was maintained at a constant level by a gas/oil accumulator piston system. The slip velocity and duration were prescribed in a command script.

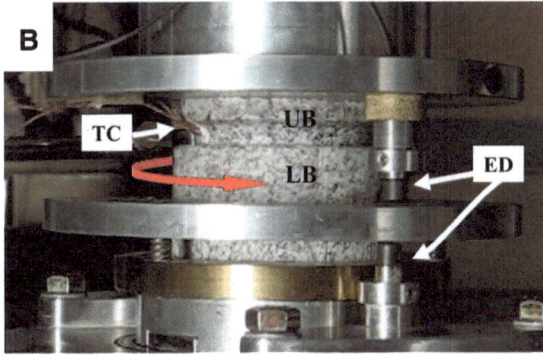

Figure 2

a The rotary shear apparatus, ROGA, of the present study (for details, see text and RECHES and LOCKNER 2010). **b** Sierra White granite sample; UB: upper, stationary block with raised ring shaped bottom that is in contact with LB, the lower, rotating block with planar upper surface; TC: Thermocouples cemented into the upper blocks; ED: Eddy-current sensors that measure the displacements normal to the experimental fault; *red arrow*: rotation sense of the lower block

We present the results of 74 runs at velocities of $V = 0.002$–0.97 m/s, slip distances up to tens of meters, and normal stress of $\sigma_n = 0.25$–6.9 MPa (Table 1). All tests were conducted at room temperature and ambient humidity.

2.2. Wear Measurements

Three methods were used to calculate wear: (1) Weight of wear products collected from the sliding surfaces (WANG and SCHOLZ 1994); (2) wear volume determined from fault-normal displacement (HIRATSUKA and MURAMOTO 2005; HIRD and FIELD 2005; RECHES and LOCKNER 2010); and (3) optical measurements of the worn surfaces. We used the second method since it allows continuous measurement of wear without disruption or modification of the fault surfaces.

We continuously measured fault-normal displacement (FND) in an open fault where excess gouge can be ejected out of the slipping surface. The convergence across the fault (FND) is the sum of fault closure/dilation and thermal expansion of the sample blocks due to frictional heating. We removed the thermal effect by the following procedure. For a period of 60–120 s after slip

terminated, we monitored sample contraction as FND, and temperature at 3 mm from the sliding surface with embedded one or two thermocouples. Then, for each experiment, we calculated an empirical thermal expansion coefficient (μm/°C) from this post-slip contraction that occurred without sample slip or wear. Finally, this coefficient was used in combination with the measured temperature history during each run to remove the thermal expansion of the sample and yield the net FND due to wear (RECHES and LOCKNER 2010). The accuracy of this linearized, simplified procedure, tested by heating a granite sample with an electric heater (neither slip nor wear) while measuring the temperature and FND in the standard way. The RMS of the difference between calculated and measured thermal FND is 3–10 % of its total.

Wear results are presented in geometric units: Wear is specified by microns of thermally corrected FND, and wear-rate is presented in [μm/m], which is the wear per unit slip. This procedure for wear-rate calculations enables effective analysis of wear evolution for high-velocity and long distances that cannot be done in rock wear studies with short slip distances or very slow slip velocities.

Table 1

General experimental conditions of the present experiments

Experiment no.	Velocity (m/s)	Normal stress (Mpa)	Static friction	Friction (steady-state)	Initial wear-rate	Steady-state wear-rate (µm/m)	Slip distance (m)	Comments
D_1040_1*	0.144, 0.12	1.55	0.87	0.63 ± 0.06	180	53 ± 4	1.9	
D_1040_2*	0.144, 0.12	1.55	0.88	0.65 ± 0.06	75	46 ± 5	1.9	
D_1041	0.144	1.6	0.72	0.53 ± 0.05	55	16 ± 1	14.6	
D_1050_1	0.14	1.85	0.70	0.53 ± 0.04	65	11 ± 3	14.6	
D_1050_2	0.14	1.85	0.72	0.56 ± 0.03	100	28 ± 9	14.6	
D_1012	0.06	1.9	0.82	0.81 ± 0.03	800	201 ± 134	0.6	
D_1370	0.01	0.6	0.87	0.92 ± 0.03	800	205 ± 21	1.1	
D_1450_1	0.015	1.5	0.80	0.83 ± 0.02	3500	961 ± 637	0.1	
D_1504	0.05	0.6	0.93	1.02 ± 0.03	700	12 ± 2	9.0	
D_1510	0.05	0.6	0.98	1.02 ± 0.03	250	12 ± 2	4.5	
D_1280	0.05	3.6	0.75	0.76 ± 0.04	700	104 ± 23	2.3	
G_236	0.07	1.05	0.83	0.41 ± 0.04	105	1.5 ± 0.5	13.0	
G_602	0.046	3.1	0.69	0.42 ± 0.05	95	6.7 ± 1.7	28.1	
G_660	0.045	0.48	0.87	0.43 ± 0.07	1000	2.0 ± 1.0	13.4	
G_661	0.045	0.5	0.98	0.41 ± 0.03	50	1.2 ± 0.8	13.3	
G_670	0.045	2.4	0.81	0.64 ± 0.04	20	1.7 ± 0.8	26.9	
G_720	0.05	2.32	0.67	0.31 ± 0.08	33	1.1 ± 0.9	15.1	
G_760_A	0.03	3.9	0.69	0.33 ± 0.07	16	2.6 ± 8.1	25.0	
G_1551	0.048	1.15	0.70	0.40 ± 0.08	48	4.2 ± 3.2	14.5	
G_1558*	0.004, 0.024	0.94	0.66	0.33 ± 0.08	650	1.0 ± 21	7.7	
G_1561*	0.004, 0.024	2.85	0.69	0.42 ± 0.02	125	12.0 ± 30	8.0	
G_1586*	0.002, 0.03	0.37	0.89	0.46 ± 0.04	17	1.2 ± 2.0	5.5	
G_1587*	0.002, 0.04	0.37	0.93	0.30 ± 0.03	22	6.7 ± 2.4	5.5	
G_1588	0.048	0.43	0.76	0.27 ± 0.01	23	1.0 ± 0.0	28.7	
G_1614*	0.003, 0.042	3.15	0.79	0.40 ± 0.01	160	7.6 ± 4.6	5.5	
G_700	0.05	2	0.71	0.71 ± 0.01	110	4.7 ± 0.6	15.1	
G_740	0.06	3.7	0.70	0.69 ± 0.04	600	42 ± 27	9.6	
G_760	0.05	3.9	0.73	0.70 ± 0.02	650	15.9 ± 0.5	16.4	
D_1010	0.011	2.02	0.75	0.85 ± 0.03	90	92 ± 8	1.3	
D_1013	0.171	1.89	0.8	0.60 ± 0.03	150	114 ± 41	1.1	
D_1030	0.010	1.91	0.86	0.83 ± 0.03	250	299 ± 48	1.1	
D_1250.1	0.010	1.10	0.76	0.88 ± 0.03	−600	52 ± 25	1.1	
D_1250.2	0.010	1.08	0.84	0.84 ± 0.03	250	256 ± 28	1.1	
D_1261	0.032	1.08	0.84	0.90 ± 0.03	5	17 ± 5	3.1	
D_1262	0.063	1.07	0.89	0.90 ± 0.03	20	19 ± 3	3.1	
D_1263.1	0.094	1.06	0.90	0.85 ± 0.03	20	21 ± 3	2.4	
D_1263.2	0.094	1.06	0.91	0.85 ± 0.03	15	26 ± 8	2.4	
D_1265	0.048	1.09	0.91	0.91 ± 0.03	20	17 ± 3	28.0	
D_1270	0.047	3.72	0.72	0.67 ± 0.03	−10	68 ± 16	2.3	
D_1290**	0.047	6.95	0.82	0.62 ± 0.03	200	169 ± 50	1.4	No steady
D_1440	0.048	2.37	0.76	0.67 ± 0.03	250	273 ± 29	1.5	
G_651.5	0.022	2.90	0.68	0.40 ± 0.03		7.4 ± 3.4	22.0	No running
G_662	0.045	0.50	0.75	0.42 ± 0.03		0.6 ± 0.2	13.0	No running
G_663	0.045	0.50	0.94	0.56 ± 0.03		0.5 ± 0.1	12.8	No running
G_664	0.045	0.50	0.96	0.50 ± 0.02		1.1 ± 0.7	13.3	No running
G_700.5	0.05	2.10	0.68	0.34 ± 0.02		0.4 ± 0.3	14.9	No running
G_710	0.05	1.96	0.57	0.39 ± 0.02		2.5 ± 1.7	22.0	No running
G_730	0.05	6.80	0.66	0.72 ± 0.02		2.9 ± 0.7	7.6	No running
G_770	0.04	1.86	0.75	0.35 ± 0.02		8.6 ± 4.7	8.0	No running
G_780.91	0.04	1.50	0.70	0.39 ± 0.02		2.3 ± 1.0	8.2	No running
G_780.92	0.04	1.50	0.38	0.22 ± 0.02		2.3 ± 0.5	8.1	No running
G_1552	0.048	1.15	0.58	0.44 ± 0.02		4.3 ± 1.4	14.4	No running
Q_1430	0.08	2.5	0.67	0.69 ± 0.03			2.7	

Table 1 *continued*

Experiment no.	Velocity (m/s)	Normal stress (Mpa)	Static friction	Friction (steady-state)	Initial wear-rate	Steady-state wear-rate (μm/m)	Slip distance (m)	Comments
Short slip distance experiments								
Ga_1620	0.004	4.9	0.45				0.002	
Ga_1650	0.005	2.2	0.46				0.002	
Q_1630	0.004	3.6	0.63				0.004	
Q_1800	0.004	1.0	0.4				0.003	
Q_1801	0.004	1.0	0.5				0.003	
Q_1802	0.006	1.0	0.53				0.01	
Q_1803	0.003	1.0	0.56				0.011	
Q_1804	0.003	3.1	0.6				0.01	
Q_1805	0.003	3.1	0.64				0.01	
TS_1700	0.003	1.4	0.5				0.0019	
TS_1701	0.003	1.4	0.55				0.007	
Ga_1570	0.002	0.42	0.85				0.057	
Ga_1621	0.004	5.0	0.58				0.0023	
Ga_1622	0.004	5.0	0.57				0.0022	
Ga_1623	0.004	5.0	0.56				0.0022	
Ga_1624	0.004	5.0	0.59				0.008	
Ga_1625	0.004	5.1	0.6				0.112	
Q_1400_1	0.01	0.6	0.73				0.31	
Q_1400_2	0.01	0.6	0.74				0.31	
Q_1630	0.006	3.6	0.63				0.0035	
Q_1631	0.007	3.6	0.64				0.0041	
Q_1632	0.007	3.7	0.62				0.0038	
Q_1720	0.003	2.5	0.59				0.0037	

'No running' comment implies an experiment without running-in stage

G Sierra White Granite, *D* Kasota Dolomite, *Ga* Karoo Gabrro, *Q* Blue Quartzite, *TS* Tennessee sandstone

* indicates an experiment with stepping slip velocity

3. Wear Evolution

The present experiments revealed a contemporaneous evolution of wear and frictional strength. We recognized three stages: An initial stage that was observed only in experiments with small slip distances of $D < 50$ mm, and displayed wear-rates as high as 10^4 μm/m; a running-in stage (QUEENER et al. 1965) that is a long transient stage over slip distances of 0.5–3 m during which both the wear-rate and friction coefficient significantly drop; and a steady-state stage that is characterized by lowest, quasi-constant wear-rate and friction coefficient (ARCHARD 1953).

3.1. Initial Stage: Wear of Individual Asperities

3.1.1 Transition from Original to Effective Roughness

The early wear was studied in 12 runs with small slip distances of $D = 2.2$–47.9 mm, at $\sigma_n = 1$–5 MPa,

and slip velocity $V = 0.007$–0.013 m/s (Table 2). Four of these experiments started with bare rock surfaces (fresh after SiC grit roughening), and were opened after the slip for surface inspection and measurement. In eight cases, the initial gouge was removed, and another short distance run was conducted on the same sample without grit roughening. We measured the sample surface roughness, before and after the small slip, with a stylus profilometer Surtronic 3+ (Taylor-Hobson); each measurement included five scan profiles, 12.5 mm long, with typical mean roughness of $R_a = 2.5 \pm 1.2$ μm after the SiC grit roughening (Table 2).

The dominant features in experiments on fresh surfaces are bright, elongated striations (Fig. 3), long, depressed scratch striations, and local, deep pits. The bright striations are composed of powder that is smeared parallel to the slip direction (Fig. 3), and are located next to scratches and pits. For example, the gabbro sample after slip distance of 2.8 mm (Fig. 3a) shows abrasive scratch marks oriented parallel to slip

Table 2

Roughness data of experiments with short slip distance
(D < 50 mm)

Run no.	Rock type	Normal stress (MPa)	Mean roughness before slip (μm)	Mean roughness after slip (μm)	Before– after mean roughness
1620	Gabbro	5.0	1.76 ± 0.20	3.68 ± 0.99	1.92
1650	Gabbro	2.2	0.68 ± 0.22	1.52 ± 0.87	0.84
1660	Gabbro	1.8	2.08 ± 0.92	1.76 ± 1.08	−0.32
1630	Quartzite	3.6	1.84 ± 0.55	2.72 ± 1.17	0.88
1800	Quartzite	1.0	2.43 ± 0.25	2.11 ± 0.29	−0.32
1801	Quartzite	1.0	2.11 ± 0.29	2.16 ± 0.27	0.05
1802	Quartzite	1.0	2.16 ± 0.27	2.56 ± 0.94	0.41
1803	Quartzite	1.0	2.56 ± 0.94	2.22 ± 0.39	−0.34
1804	Quartzite	3.1	2.22 ± 0.39	2.49 ± 0.25	0.27
1805	Quartzite	3.1	2.49 ± 0.25	2.93 ± 0.55	0.44
1700	Sandstone	1.4	3.92 ± 0.61	3.20 ± 1.06	−0.72
1701	Sandstone	1.4	3.20 ± 1.06	2.69 ± 0.74	−0.51

direction with approximately the same length as the slip distance (2 mm). The measured length of the striations is about equal to the slip distance for small slip experiments, $D = 2$–5 mm, and less than total slip for higher slip experiments, $D > 6$ mm (Fig. 4a). The local, deep damage of these scratches and the spatial association between scratch striations and smeared powder striations (Fig. 3), suggest that these features are the product of plowing by hard grains locked on one side of the fault and act as effective asperities. This highly intensive local wear ends when the asperity grain fails, and the elevated stresses migrate to another large grain or asperity. The deep pits that are associated with the bright powder striation suggest that the powder striations formed by plucking a few grains from the fault surface while leaving the deep pit behind.

The roughness of the original bare surface (maximum summits height of ∼0.05 mm, Table 2) is significantly smaller with respect to the localized deep pits and scratch striations (ENGELDER and SCHOLZ 1976). Thus, the initial stage modifies the surface roughness by particle plucking and smearing (JACKSON and DUNN 1974; MOODY and HUNDLEY-GOFF 1980; ROBERTSON 1982). The comparison between the initial and final fault roughness, R_a, (Table 2) during this stage suggests no roughness change for runs at $\sigma_n < 2$ MPa, and a slight roughening for higher σ_n (Fig. 4b).

3.1.2 Wear Mechanisms

The scratching and roughening during the initial stage (Fig. 4a, b) are associated with distinct stress-dilation events (Fig. 5). The records of fault-normal displacements (FND) of several runs revealed short-lived dilation events during the initial stage (black curves in Fig. 5, and horizontal arrow marked E for one event). These events display temporal dilation magnitudes of 3–15 μm that lasted for slip distances of a few mm to a few cm (Fig. 5 shows only the deviation of FND from its absolute value). The dilation amplitude of the events falls between the mean roughness of $R_a \approx 2.5$ μm, and the height difference between lowest trough and highest peak, 13–54 μm (Table 2). A striking feature of these events is the mimicking relations between the dilation variations and small, transient changes in the normal stress and shear stress; the deviation of the stresses from the global stresses are plotted by blue and red curves, respectively, as function of slip (Fig. 5). The stress deviations are smaller than 0.1 MPa (blue and red vertical scales).

We propose that the simultaneous rise and fall of dilation, shear stress and normal stress reflect slip along a rough fault surface. For simplicity, we consider a local mating contact between two surfaces with similar, sinusoidal rough surfaces (Fig. 6a). As the fault blocks are forced to slip by the applied stresses, the upper block climbs the gentle slope of an asperity (Fig. 6b). This climbing leads to dilation between the blocks (open, black arrows in Fig. 6b), and to a temporal increase of the normal stress.[1] The shear stress also temporarily increases due to the normal stress increase. During the rising phase, the asperity is intact and behaves as a small barrier. Next, the asperity fails and disintegrates (Fig. 6c) leading to temporal drop of the stresses and closure associated with brittle failure. We envision that each dilation event reflects stopping by a set of asperities followed by their failure. To test this mechanism, we plotted

---

[1] In the present experimental system, a gas-oil actuator that can maintain constant normal stress with variation about 5 % controls the normal stress. Small stress variations, such as in the present events, are not corrected due to seal friction and oil response time.

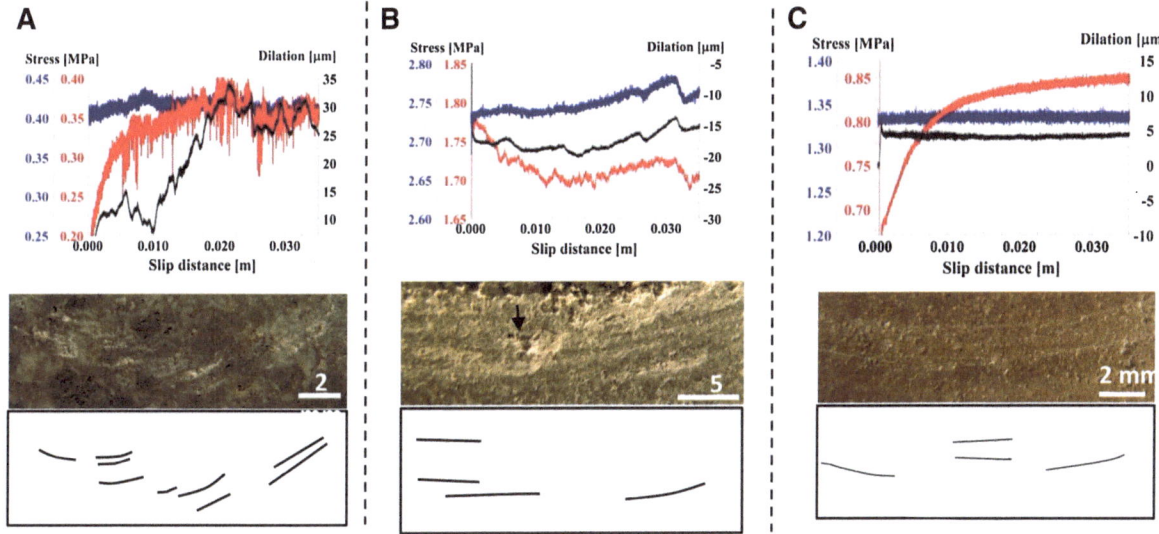

Figure 3
Initial stage observations for the first 35 mm of slip of three sample lithologies: **a** Karoo gabbro, **b** Blue quartzite, **c** and Tennessee sandstone. *Upper panel* displays details of the dilation variations (*black curve*), normal stress (*blue curve*), and shear stress (*red curve*). Note the curves scale as they show the deviations from the global values of the dilation and stresses. Middle panel displays close-up photos of the surfaces of these experimental faults, and the lower panel shows the mapped striations of these fault surfaces. Note that the dilation and stress curves in **a** and **b** are "noisy", whereas the curves are smooth in **c**, and correspondingly, the amount and depth of surface scratches and striations are more pronounced in **a** and **b**; see related discussion in the text on the dilation events during the initial stage

the global shear stress, τ, and normal stress, σ_n, before the events (brown dots, Fig. 4c), and the stress deviations, $\Delta\tau$ and $\Delta\sigma_n$, during the event (local maximum minus local minimum, red dots, Fig. 4c). The slopes on this Mohr diagram, which are the frictional strengths of the rock, indicate $\tau = 0.11 + 0.78 \quad \sigma_n$ before the peak, and $\Delta\tau = 0.03 + 0.67 \; \Delta\sigma$ during the event. Based on the similarity of these strength values, we deduce that similar brittle failure processes control the macroscopic sliding, and the temporary, local failure of asperities.

The above observations suggest that a finite number of touching asperities control the initial wear of a fresh sample. This interpretation is based on the isolated occurrence of the striations and scratches (Fig. 3), and the length similarity of striation and slip distance (Fig. 4a). These asperities were highly loaded (BYERLEE 1967a, b; SCHOLZ and ENGELDER 1976), they failed in a brittle fashion, and their debris was smeared for the total length of the slip distance. This documentation of asperity failure is in agreement with previous studies of wear production

through rupture of asperities by plowing, shearing, fracturing and plucking (BOWDEN and TABOR 1942; BYERLEE 1967a, b; ENGELDER and SCHOLZ 1976; HUNDLEY-GOFF and MOODY 1980; MOODY and HUNDLEY-GOFF 1980; HAGGERT *et al.* 1992; WANG and SCHOLZ 1994; McLASKEY and GLASER 2011).

In summary, the distinct features of the initial stage are:

a. During small slip distances (< 50 mm), clear, separate striations develop on a fresh, bare rock surface (Fig. 3).
b. The length of the striations is similar to the total slip, indicating that they form by failure of the larger contacting asperities (Fig. 4a). The striations disappear later, during the running-in stage when many new asperities come into contact and fail.
c. Dilation events with corresponding stress changes (Fig. 5), were recognized only during the initial stage, and they are likely to be associated by deformation and failure of large asperities (Fig. 6).
d. The wear-rates in the initial stage are typically higher by an order of magnitude than wear-rates of the running-in.

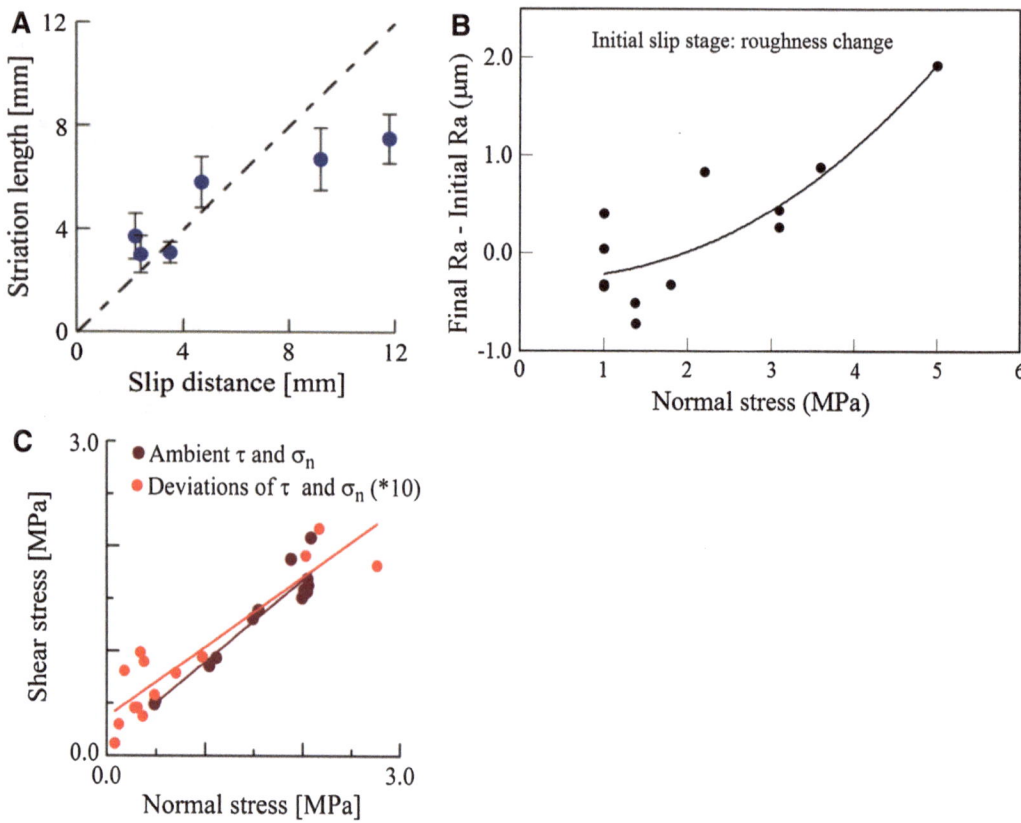

Figure 4

Features of the initial wear stage. **a** Striation lengths as a function of the total slip displacement; each point is the mean value of 10 striations in six separate experiments with standard deviation shown as *error bars*; the *dashed line* represents equality of striation length and slip distance. **b** The change of mean roughness, R_a, during the initial stage as a function of normal stress (Table 2); positive change implies increase of surface roughness. **c** Mohr diagram of the stresses during 13 dilation events (text) in five Kasota dolomite experiments. Global stresses (i.e. the macroscopic stresses during event initiation) are large, *brown dots*, and event stress deviations are small, *red dots*; the corresponding failure envelopes are in *brown* and *red lines*, respectively. The shear and normal stress deviations $\Delta\tau$ and $\Delta\sigma$ are [peak stress—event initial stress]; the smaller stress deviations are multiplied by 10 to allow the same plot for both stress sets. The failure envelopes are similar: $\tau = 0.11 + 0.78\ \sigma_n$, ($R^2 = 0.98$) for the global stresses and $\Delta\tau = 0.03 + 0.67\ \Delta\sigma$ ($R^2 = 0.83$) for the event stress deviations

3.2. Running-in Stage: Wear and Friction Evolution

As mentioned in the Introduction, the running-in stage, as defined by QUEENER *et al.* (1965), is the stage of intense wear during early sliding along new parts. These authors suggested that the total wear, W, is the sum of the wear contributions of the running-in and the steady-state,

$$
\begin{aligned}
W &= W_{\text{running-in}} + W_{\text{steady-state}} \\
&= A\left[1 - \exp(-nL)\right] + KL
\end{aligned}
\tag{1}
$$

where A and n and are parameters characteristic for the running-in stage (materials, hardness, normal stress, temperature, and velocity), K is the corresponding parameter for steady-state wear according

to ARCHARD (1953), and L is the slip distance. Equation (1) indicates that after a slip distance, L_0, the contribution of the running-in approaches a constant value of $W_0 = A\left[1 - \exp(-nL_0)\right]$, which according to QUEENER *et al.* (1965) may be a significant fraction of the total wear. This running-in concept fits well with the present results. We analyzed 52 runs of Kasota dolomite and Sierra White granite samples with slip distances >1 m. In 28 runs, the FND indicated closure (= negative dilation) during running-in; eight runs showed dilation during running-in, and 16 runs showed no running-in stage. Out of these experiments, 17 experiments had no pre-existing gouge at the slip surface (either fresh

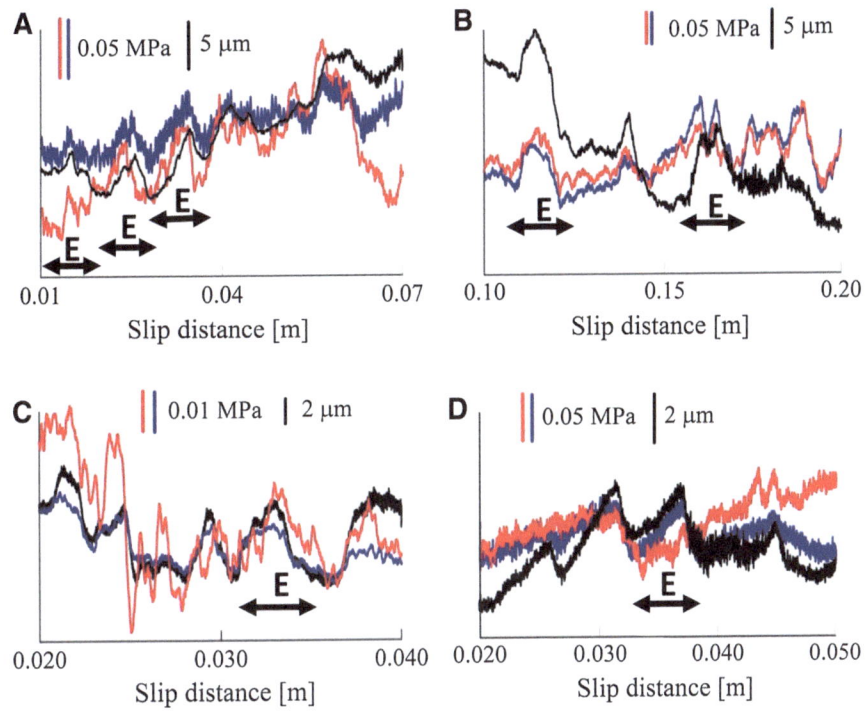

Figure 5

Dilation events during the initial wear stage of four experiments [Kasota dolomite, slip = 10 mm, σ_n = 2.0 MPa (**a**); Kasota dolomite slip = 100 mm, σ_n = 1.1 MPa (**b**); Karoo Gabbro, slip = 20 mm, σ_n = 0.4 MPa (**c**); Blue quartzite, slip = 20 mm, σ_n = 2.7 MPa (**d**)]. Dilation—*black curves*, shear stress—*red curves*, and normal stress—*blue curves*; *horizontal double-head arrows*, **e** indicates one dilation event. The *curves* are arbitrarily shifted, and show only the deviations from the global values; note the *vertical scale bars* in the corresponding colors

samples or samples in which gouge was removed by air pressure); 13 of the 17 experiments displayed closure running-in and four displayed dilational running-in. In these experiments the total wear fits well with Eq. (1) (Fig. 7) showing a clear transition from running-in to steady-state at L_0 = 1–3 m. We consider L_0 the "running-in distance", after which the initial, high wear-rate drops significantly, in agreement with Queener's model. The periodic signal in the wear curves of Fig. 7 (light gray) has a dominant wavelength of ~22 cm, which equals the sample circumference. This signal reflects the sample tilt/ wobble, and it is eliminated from the wear calculation by taking a polynomial fit to the dilation curves (BONEH *et al.* 2013).

A striking feature in the present experiments is the parallel evolution of wear and friction during the running-in stage. The curves of total wear, wear-rate and friction coefficient display high values during running-in that systematically and simultaneously

decrease to lower, steady-state values. For example, Fig. 8a displays a drop from initial wear-rate of 90 to 5 μm/m over L_0 ≈ 1.8 m, and the initial friction coefficient, μ_i = 1.0, drops to μ_{steady} = 0.42 ± 0.05, over a slip-weakening distance of d_W ≈ 2.75 m. The range of 1–3 m of the slip-weakening distance in our experiments was commonly observed in rotary shear experiments (RECHES and LOCKNER 2010; DI TORO *et al.* 2011; BROWN and FIALKO 2012). We determined the L_0 and d_W in all experiments of granite and dolomite with a closure running-in and weakening (20 runs sheared at σ_n = 0.4–4 MPa and V = 0.003–0.14 m/s), and found linear relations between the two parameter (Fig. 8b),

$$L_0 = 0.76 \times d_W, r^2 = 0.70.$$

These observations of wear and friction similar evolution suggest that they are either "cause and effect" or two aspects that depend on the same system conditions (e.g., roughness, mechanical properties, slip

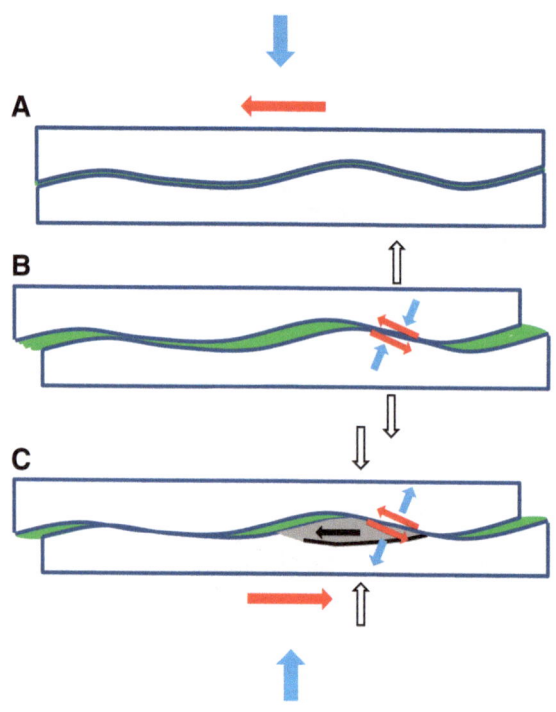

Figure 6
Proposed asperity interaction mechanism for the observed dilation events (Figs. 4c, 5); *green* opening between blocks, *blue arrow* normal stress, *red arrows* shear stress, *open arrows* fault normal displacement. **a** Starting state with locked, mating surfaces. **b** Slip initiates, leading to asperity climbing that causes a temporal dilation and associated stresses increase. **c** Asperity failure (*gray area* with *black arrow*) by shearing off its base leading to temporal closure and stress drops of the event

granite frictional strength under negligible wear conditions, and we attribute the much higher friction coefficients ($\mu = 0.8$, 1.3) of the interlocking surfaces to energy dissipation by asperity breakage and wear. An upper bound on brittle frictional strength is the internal friction coefficient of 1.4–1.8 (LOCKNER and BYERLEE 1993) determined for an intact rock (= total interlocking) in which the shear is governed by intense micro-fracturing (RECHES and LOCKNER 1994). Similarly, CHEN *et al.* (2013) documented friction reduction due to smoothing at the sub-micron scale. This interpretation implies that the weakening during the running-in stage (red curve, Figs. 8a) is primarily due to reduction of wear intensity, in good agreement with the observation of contemporaneous wear and friction evolution. Quantitative relations between frictional work and wear is presented in the "Discussion" below.

3.3. Steady State: The Three-Body Mode

By the end of the running-in stage after slip of L_0, the experimental faults are covered by a continuous gouge layer that fully separates the two rock blocks (Fig. 9a, b). This layer signifies a transition from a two-body frictional mode, which is controlled by asperity wear, to a three-body mode, which is controlled by the gouge frictional strength (RECHES and LOCKNER 2010). This three-body mode is geometrically similar to the well-known y-shear surface in fault zones (GU and WONG 1994). In the present experiments, the steady-state stage is characterized by quasi-constant frictional coefficient with deviations from the mean not exceeding 7 %, and wear-rates with similar variations. The experimental faults have open ring-on-flat configuration, and thus newly worn particles from the sliding surface are free to be ejected. We envision that the gouge layer establishes a quasi-constant thickness during the steady-state stage. In a confined setting, e.g., natural faults, the wear products are trapped and thicken the gouge layer, and consequently may shorten the running-in stage by faster gouge accumulation. Even during steady-state slip, the experimental faults continue to wear by microcracking at the gouge-rock contact (Fig. 9b) in wear-rates that depend on slip velocity, normal stress and lithology (BONEH *et al.* 2013;

velocity, and normal stress) (Fig. 1). The analysis of BYERLEE (1967a) on granite friction casts light on this wear-friction relation. He showed that the friction of polished surfaces of Westerly granite strongly depends on the roughness: The friction coefficient approached $\mu = 0.2$ for granite surfaces of $R_a = 0.6$ μm, and $\mu = 0.1$, when the granite was sheared against smooth sapphire with $R_a = 0.013$ μm (Fig. 5 in BYERLEE 1967a). Byerlee also found that "In contrast to ground surfaces, μ for totally interlocking surfaces of granite...is...$\mu = 1.3$ for $\sigma_n < 6$ MPa, and $\mu = 0.8 + 0.03\sigma_n$ in the range 6 MPa $< \sigma_n < 15$ MPa" (BYERLEE 1967a, Fig. 4). He noted, similarly to our observations, that "In all the experiments the surfaces contained fine white debris after sliding; the amount of debris and the size of the particles increased with the roughness of the surfaces in contact". We propose that the very low friction coefficient of $\mu = 0.1$ of BYERLEE (1967a) is the

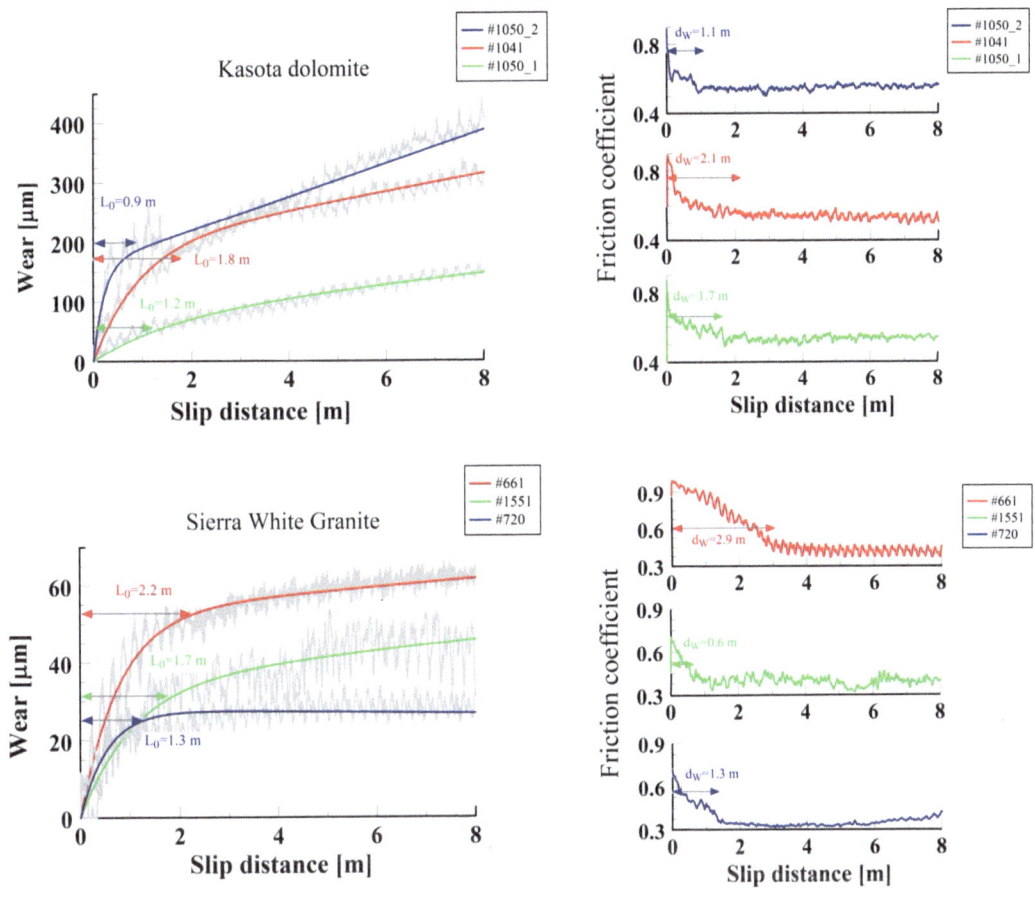

Figure 7

Wear and friction evolution during experiments with Kasota dolomite (*upper plots*), and Sierra White granite (*lower plots*). *Left side* wear-distance evolution (*light gray curves*) with least-square fit of Queener relations (Eq. 1 in text) (*colored, smooth curves*), L_0 the slip distance of the transient running-in stage (text). *Right side* the associated friction-distance evolution of the same experiments (corresponding *curves color*), d_W the weakening slip distance (text). The periodic signal (*light gray*) with dominant wavelength of $\sim$ 22 cm reflects the sample tilt/wobble (text). The KD experiments are #1050_2, σ_n = 1.7 MPa, and V = 0.14 m/s; #1041, σ_n = 1.87 MPa, and V = 0.14 m/s; and #1050_1, σ_n = 1.87 MPa, and V = 0.14 m/s. The SWG experiments are #661, σ_n = 0.5 MPa, and V = 0.045 m/s; #1551, σ_n = 1.1 MPa, and V = 0.048 m/s; and #720, σ_n = 2.3 MPa, and V = 0.05 m/s

LYAKHOVSKY *et al.* 2014). Here, however, we focus on the transient earlier stages of wear-rate and frictional strength.

4. Discussion

4.1. Friction-Wear Relations

The present observations indicate that the reduction of frictional strength strongly correlates with gouge generation and rock comminution. This correlation is manifested by both qualitative similarities (Fig. 7) and quantitative similarities (Fig. 8b), and is supported by the experimental results of BYERLEE (1967a). We envision that the work dissipated by asperity failure and rock comminution significantly contributes to the macroscopic frictional strength. Thus, it is our interpretation that the decrease of wear-rate during running-in controls the observed simultaneous fault weakening by reducing energy dissipation. C. Scholz (written communication) suggested that the experimental documentation of the relations between the evolution of frictional

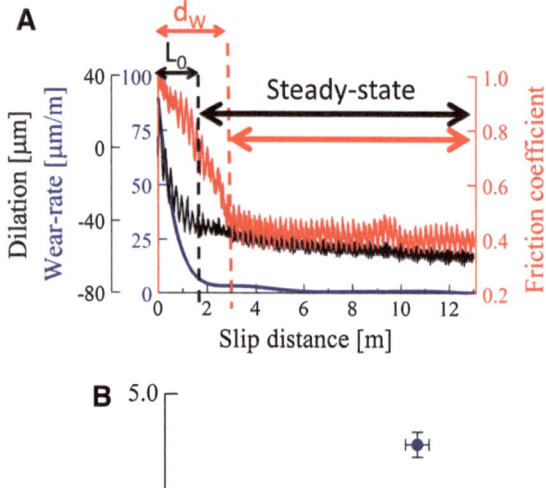

A

Figure 8

a Friction coefficient (*red*), total wear (*black*) and wear-rate (*blue*) in typical Kasota dolomite experiment, and scales in corresponding colors (text). L_0 the slip distance of the transient running-in stage, d_W the weakening slip distance. **b** L_0 and d_W relations for KD (*red*) and SWG (*blue*) experiments displaying $L_0 = 0.76 \times d_W$ (text)

strength and wear intensity (Figs. 7, 8) allows quantification of the wear contribution to the friction work (also: FULTON and RATHBUN 2011). If the experimental shear work, W_f, is the sum of frictional heat, Q, and gouge surface energy, U_S, then,

$$W_f = \tau u = Q + U_S \qquad (2)$$

where τ and u are the shear stress and slip distance, respectively. We assume that at a constant velocity and normal stress, the rate of frictional heating is constant, and it equals the work during steady-state slip, $\tau_1 u$, when the wear-rate is low and may be ignored ($dU_S/du \sim 0$). The surface energy dissipation during the running-in stage can now be calculated from the total wear during this stage. In a typical run of Kasota dolomite in Fig. 7, where $\tau_0 = 1.5$ MPa (initial shear stress), $\tau_1 = 1.0$ MPa (steady-state shear stress), $L_0 = 2$ m (slip distance to

reach steady-state of vanishing wear-rate), and $A = 0.002$ m^2 (experimental fault area). We apply the above assumptions to this typical experiment to calculate, the energy dissipation by wear during the experimental running-in:

$$U_s = 0.5\,(\tau_0 - \tau_1)\,AL_0 \times 10^6 = 10^3 J.$$

This dissipation can be compared to the weight of the wear product (gouge). The compaction normal to the fault surface, FND, is a conservative estimate of the total wear, and it is ~ 50 μm during the running-in of the typical Kasota dolomite experiment (Fig. 7); this compaction corresponds to wear weight of $V_W \sim 0.25$ g. We take a common value for specific surface area of rock minerals, $\gamma = 1$ J/m^2 (KANAMORI and RIVERA 2006), and assume that all the wear energy, U_S, was dissipated by increasing the surface area, S, of the experimental gouge, then,

$$S = (U_S/\gamma)/V_w \sim 4,000 \text{ m}^2/\text{g}$$

This value is orders of magnitude larger than surface area measurements of 10–80 m^2/g (WILSON *et al.* 2005). Thus, our conservative assumption of frictional heat and surface area increase (Eq. 2) cannot explain the energetics of the present observations. This result indicates the activity of additional dissipating processes, e.g., disintegration at the crystal structure and amorphization (YUND *et al.* 1990). Studying these processes is beyond the scope of the present analysis.

4.2. *Wear Evolution Along Natural Faults*

We recognized three evolution stages of the experimental faults. First, an initial stage of small displacements (<50 mm) that is characterized by wear and failure of a few isolated asperities (Figs. 3, 5, 6), and roughening of the fault surfaces (Fig. 4b). Second, a running-in stage of 0.5–3 m slip distance with intense wear (Fig. 7) due to failure at many touching asperities, and simultaneous reduction of the friction coefficient (Figs. 7, 8). Third, a steady-state stage that initiates when the fault surface is covered by a gouge layer (Fig. 9), and the wear-rate and friction coefficient maintain quasi-constant, low levels (Figs. 7, 8). This wear evolution transfers the experimental faults from a two-body shear system to

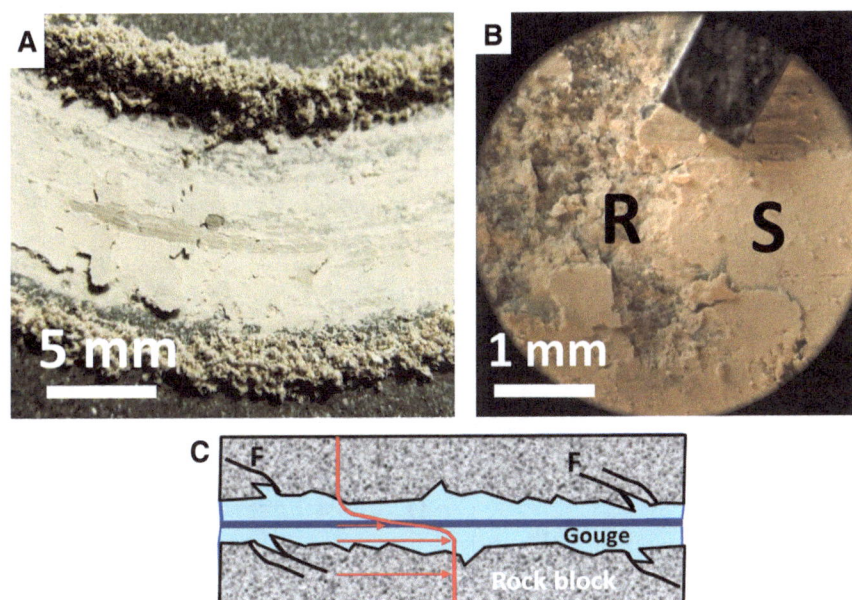

Figure 9

Close-up view of three-body configuration of experimental faults during steady-state slip. **a** Blue quartzite fault, run #1806, after slip of $D = 1.1$ m. Slip surface is covered with smeared, striated gouge with ejected gouge of both side of the slip zone. **b** Close-up view on the sliding surface; S a smooth surface of localized slip within the gouge layer that corresponds to the dark blue line in **c**, R rough rock surface at a site of gouge removal. **c** Conceptual cross-section of a three-body configuration of a fault. *Gray zones* fractured (F) host rock with rough surfaces (R in **b**), *light blue* gouge powder separating the rock blocks, *dark blue* zone of localized slip within the gouge that accommodates most of the slip (*red arrows*) and develops a smooth gouge surface (S in **b**)

a three-body system in which the gouge powder separates the two blocks (Fig. 9c). The relevancy of this evolution to faulting of intact rocks and to natural faults is discussed below.

Failure of intact rocks occurs by coalescence of multiple, interacting fractures during fault propagation and the associated crushing of the blocks that bound the fracture zone (Fig. 10a) (RECHES and LOCKNER 1994; LOCKNER and BYERLEE 1993). This process generates a rough fault with a continuous gouge zone made of the crushed blocks and gouge powder (Fig. 10b) (HEESAKKERS *et al.* 2011b). In this respect, the fault acquires the steady-state geometry (gouge layer in three-body mode) from its onset, in contrast to the 0.5–3 m of slip needed for steady-state along bare, ground rock surfaces. Thus, we anticipate that a new fault in intact rock will display negligible running-in stage, and will slip at quasi-constant friction of the steady-state stage. Servo-controlled triaxial experiments allow exploring the post-failure stage support this prediction. LOCKNER *et al.* (1992) used the rate of acoustic emission events to prevent

catastrophic failure during intact granite faulting. The differential stress in one typical experiment (Fig. 10c) shows about a 30 % drop after peak stress (b–f curve) with an extension of the sample (to prevent catastrophic failure). This drop was associated with sample failure by a through-going fault-zone (LOCKNER *et al.* 1992). The slip along the new fault occurred at fairly constant differential stress of ~330 MPa. WAWERSIK and BRACE (1971) observed similar behavior (Fig. 10d) when they used a manually operated servo-control to stabilize the post-failure slip. We interpret this behavior as indicating slip under steady-state stage without passing through the earlier running-in stage of bare fault surfaces (Fig. 7).

Natural faults are not composed of bare, planar surfaces, and their wear is not likely to be dominated by ploughing and crushing of asperities. We envision that new natural faults in pristine, intact rock nucleate and grow similarly to experimental faults in intact rock samples (Fig. 10). The gouge zone of such faults develops similarly to intact rock experiments by

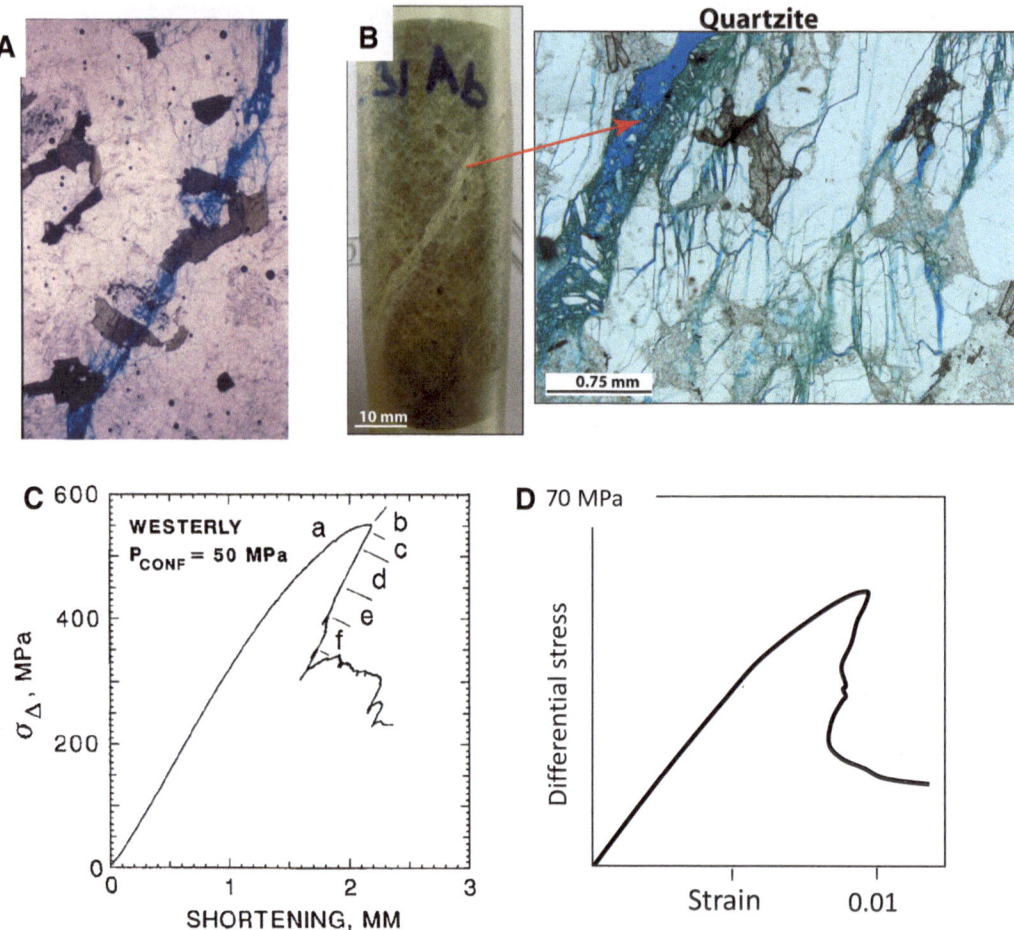

Figure 10

Faulting processes of an intact rock sample. **a** An array of microfractures (*blue* epoxy) at the tip zone of a propagating fault in Westerly granite; thin section view, 2.5 mm wide after RECHES and LOCKNER (1994); $\sigma_n = 50$ MPa, axial shortening in the vertical direction. **b** Experimental fault-zones in quartzite from Pretorius fault, South Africa. Runs under 20 MPa confining pressure. Dilated gouge zone is filled with blue epoxy, and extensive off-fault damage in the quartzite dominated by micro-fractures that branch from the main fault and die away from it; thin-section view after HEESAKKERS *et al.* (2011b); $\sigma_n = 20$ MPa, axial shortening in the vertical direction. **c** Differential stress during a servo-controlled failure experiment of an intact sample of Westerly granite (LOCKNER *et al.* 1992). The post-failure sample extension (**b–f**) was generated by the servo system to prevent total failure. The post failure slip occurs after point f at fairly constant differential stress. **c** Differential stress during a servo-controlled failure experiment of an intact sample of Westerly granite at confining pressure of 17 MPa [redrawn from Fig. 1 WAWERSIK and BRACE (1971)]

fragmentation and coalescence of multiple micro-cracks in the process zone (LOCKNER *et al.* 1992; VERMILYE and SCHOLZ 1998). For this reason, small faults have relatively thick gouge. For example, KATZ *et al.* (2003) mapped small faults in a syenite intrusion, and found that faults with displacements of centimeter scale display gouge zones of millimeter scale (their Fig. 16).

Larger faults wear by more complex processes. The fault-zone undergoes significant healing and cementation between slipping phases, and each slip phase requires the failure of the cemented fault-zone (TENTHOREY *et al.* 2003; MUHURI *et al.* 2003; HEESAK-KERS *et al.* 2011a). A new slip phase not only regenerates the three-body structure of the healed gouge-zone, but also wear parts of the adjacent host rocks. The later wear is controlled by two main mechanisms. First, many fault surfaces are fractal (or self-affine) with roughness at all scales (POWER *et al.* 1988; SAGY *et al.* 2007). The slip along such irregular surfaces is expected to continuously wear the protruding asperities (CHESTER and CHESTER 2000), leading to a general gouge thickening with increasing fault displacement (SCHOLZ 1987). Second, most faults in the upper crust slip unstably, and unstable rupture propagation leads to intense pulverization and damage in the process zone (RECHES and DEWERS 2005) as well as in the surrounding crustal rocks (ANDREWS 2002). This dynamic pulverization may lead to significant widening of the gouge zone (WILSON et al. 2003; 2005). For example, a fresh gouge zone of 1–5 mm thickness was formed during earthquake slip of ~25 mm along the Pretorius fault, TauTona Mine, South Africa (HEESAKKERS et al. 2011a), indicating wear-rate of 4–20×10^7 μm/m. This discussion underscores the complexity of wear along natural faults along which multiple wear mechanisms could operate in during many slip phases.

5. Conclusions

1. Our analysis revealed three evolution stages of the experimental faults:

   a. An initial stage (slip distances <50 mm) of wear by failure of isolated asperities associated with roughening of the fault surface.

   b. A "running-in" stage of slip distances of 1–3 m with intense wear-rate, failure of many asperities, and simultaneous reduction of the friction coefficient and wear-rate.

   c. The steady-state stage initiates when a gouge layer covers the fault surface forming a three-body shear system, and during which both wear-rate and friction coefficient maintain quasi-constant, low levels.

2. The frictional strength and the wear-rate evolves contemporaneously from high initial, high values to lower steady-state levels; this parallel evolution occurs during slip-distances of 1–3 m. We interpret the fault weakening as indicating a reduction of energy dissipation rate by the dropping wear-rate during the running-in stage.

3. The above stages were observed along experimental faults that before shearing were bare rock surfaces, which are nominally planar and

relatively smooth (mean roughness of ~2.5 μm). However, spontaneous faults, both in failure experiments of intact rocks and in the field, are much rougher and contain gouge layers from their incipience. We thus envision that the initial and running-in stages may not be realized along natural faults that always slip with as existing gouge layer.

Acknowledgments

We benefitted from help and advice of Andrew Madden and Xiaofeng Chen, University of Oklahoma. We had fruitful discussions with Emily Brodsky, UC Santa Cruz; Chris Scholz, Lamont-Doherty Institute; Amir Sagy, Israel Geological Survey; Einat Aharonov and Shalev Siman-Tov, the Hebrew University. Eric Ferre of Southern Illinois University kindly provided the Karoo gabbro sample. The manuscript was greatly improved by the constructive comments of Chris Scholz and an anonymous reviewer. The study was supported by the NSF, Geosciences, Equipment and Facilities, Grant No. 0732715, and partial support of NSF, Geosciences, Geophysics, Grant No. 1045414, and ConocoPhillips Foundation grant.

REFERENCES

ANDREWS, D. J. (2002), *A fault constitutive relation accounting for thermal pressurization of pore fluid*, J. Geoph. Res., *107*, 2363.

ARCHARD, J. F. (1953), *Contact and rubbing of flat surfaces*, J. Appl. Phys., *24*, 981–988.

AYDIN, A. (1978), *Small faults formed as deformation bands in sandstone*, Pure Appl. Geophys., *116*, 913–930.

BEN-ZION, Y., and C. SAMMIS (2003), *Characterization of fault zones*, Pure Appl. Geophys., *160*, 677–715.

BONEH, Y., A. SAGY, and Z. RECHES, (2013), *Frictional strength and wear-rate of carbonate faults during high-velocity, steady-state sliding*, Earth Planet. Sci. Lett., V. *381*, P. 127–137.

BOWDEN, F. P., and D. TABOR (1939), *The area of contact between stationary and between moving surfaces*, Proc. R. Soc. London, *169* (938), 391–413.

BOWDEN, F. P., and D. TABOR, (1942), *Mechanism of metallic friction*, Nature, *150*, 197–199.

BROWN, K. M., and Y, FIALKO, (2012), *"Melt welt" mechanism of extreme weakening of gabbro at seismic slip rates,* Nature *488*, 7413, 638–641.

BRODSKY, E. E., J. J. GILCHRIST, A. SAGY, and C. COLLETTINI, (2011), *Faults smooth gradually as a function of slip*, Earth Planet. Sci. Lett., *302*, 185–193.

BYERLEE, J. D. (1967a), *Theory of friction based on brittle fracture*, J. Appl. Phys., *38*, 2928.

BYERLEE, J. D. (1967b), *Frictional characteristics of granite under high confining pressure*, J. Geophys. Res., *72*, 3639–3648.

CANDELA, T., F. RENARD, M. BOUCHON, J. SCHMITTBUHL, and E. E. BRODSKY (2011), *Stress drop during earthquakes: effect of fault roughness scaling*, Bull. Seismol. Soc. Am., *101/5*, 2369–2387.

CHEN, X., MADDEN, A. S., BICKMORE, B. R., and RECHES, Z. (2013) *Dynamic weakening by nanoscale smoothing during high.* Geology. V. *41*, p. 739–742.

CHESTER, F. M., and J. S. CHESTER (2000), *Stress and deformation along wavy frictional faults*, J. Geophys. Res., *105*, 23,421–423,430.

DI TORO, G., R. HAN, T. HIROSE, N. DE PAOLA, S. NIELSEN, K. MIZOGUCHI, F. FERRI, M. COCCO, and T. SHIMAMOTO (2011), *Fault lubrication during earthquakes*, Nature, *471/7339*, 494–498.

DIETERICH, J. H., and B. D. KILGORE (1994), *Direct observation of frictional contacts: New insights for state-dependent properties*, Pure Appl. Geophys., *143*, 283–302.

DOBSON, P. S., and H. WILMAN, (1963), *The friction and wear, and their inter-relationship, in abrasion of a single crystal of brittle nature*, Br. J. Appl. Phys., *14*, 132–136.

ENGELDER, T., and C. H. SCHOLZ (1976), *The role of asperity indentation and ploughing in rock friction- II. Influence of relative hardness and normal load*, Int. J. Rock Mech. Min. Sci. & Geomech., *13*, 155–163.

FULTON, P. M., and A. P. RATHBUN (2011), *Experimental constraints on energy partitioning during stick–slip and stable sliding within analog fault gouge*, Earth Planet. Sci. Lett., *308/1*, 185–192.

GU, Y., and T. F. WONG (1994), *Development of shear localization in simulated quartz gouge: Effect of cumulative slip and gouge particle size*, Pure Appl. Geophys, *143/1–3*, 387–423.

HADLEY, K. (1975), *Dilatancy: Further studies in crystalline rocks*, Ph.D. thesis, Mass. Inst. Technol., Cambridge.

HAGGERT, K., S. J. COX, and M. W. JESSELL (1992), *Observation of fault gouge development in laboratory see-through experiments*, Tectonophysics, *204/1*, 123–136.

HALLBAUER, K., H. WAGNER, and N. G. W. COOK (1973), *Some observations concerning the microscopic and mechanical behaviour of quartzite specimens in stiff, triaxial compression tests*, Int. J. Rock Mech. Min. Sci., *10/6*, 713–726.

HEESAKKERS, V., S. MURPHY, and Z. RECHES (2011a), *Earthquake rupture at focal depth, part I: Structure and rupture of the Pretorius fault TauTona mine South Africa*, Pure Appl. Geophys., *168/12*, 2395–2425.

HEESAKKERS, V., S. MURPHY, D. A. LOCKNER, and Z. RECHES (2011b), *Earthquake rupture at focal depth, Part II: Mechanics of the 2004 M2. 2 Earthquake along the Pretorius Fault, Tautona mine, South Africa*. Pure Appl. Geophys., *168/12*, 2427–2449.

HENDERSON, I. H., G. V. GANERØD, and A. BRAATHEN (2010), *The relationship between particle characteristics and frictional strength in basal fault breccias: Implications for fault-rock evolution and rockslide susceptibility*. Tectonophysics, *486/1*, 132–149.

HIRATSUKA, K. I., and K. I. MURAMOTO (2005), *Role of wear particles in severe–mild wear transition*, Wear, *259/1*, 467–476.

HIRD, J. R., and J. E. FIELD (2005), *A wear mechanism map for the diamond polishing process*, Wear, *258/1*, 18–25.

HIROSE, T., K. MIZOGUCHI, and T. SHIMAMOTO (2012), *Wear processes in rocks at slow to high slip rates*, J. Struct. Geol., *38*, 102–116.

HUNDLEY-GOFF, E. M., and J. B. MOODY (1980), *Microscopic characteristics of orthoquartzite from sliding friction experiments. I. Sliding surface*, Tectonophysics, *62/3*, 279–299.

JACKSON, R. E., and D. E. DUNN (1974), *Experimental sliding friction and cataclasis of foliated rocks*, Int. J. Rock Mech. Sci. & Geomech. Abstr., *11/6*, 235–249.

KANAMORI, H., and L. RIVERA, (2006), *Energy partitioning during an earthquake*. Geophys. Mono. Ser., *170*, 3–13.

KATZ, O., and Z. RECHES (2004), *Microfracturing, damage and failure of brittle granites*. J. Geophy. Res. *109*, doi:10.1029/2002JB001961.

KATZ O., Z. RECHES, and G. BAER (2003), *Faults and their associated host rock deformation: Structure of small faults in a quartz-syenite body, southern Israel*. J. Structural Geology, *25*, 1675–1689.

LEVY, A. V., and N. JEE, (1988), *Unlubricated sliding wear of ceramic materials*, Wear, *121/3*, 363–380.

LOCKNER D. A., J. D. BYERLEE, V. KUKSENKO, A. PONOMAREV, and A. SIDRIN (1992), *Observations of quasi-static fault growthfrom acoustic emissions*, in Fault Mechanics and TransportProperties of Rocks, edited by B. Evans and T.-f. Wong, 3–31.

LOCKNER, D. A., and J. D. BYERLEE (1993), *How geometrical constraints contribute to the weakness of mature faults*, Nature, *363*, 250–252, doi:10.1038/363250a0.

LYAKHOVSKY, V., A. SAGY, Y. BONEH, and Z. RECHES (2014), *Fault wear by damage evolution in a three-body slip mode*, Pure Appl. Geophys, this volume.

MCLASKEY, G. C., and S. D. GLASER (2011), *Micromechanics of asperity rupture during laboratory stick slip experiments*, Geophy. Res. Lett., *38/12*, doi:10.1029/2011GL047507.

MOODY, J. B., and E. M. HUNDLEY-GOFF, (1980), *Microscopic characteristics of orthoquartzite from sliding friction experiments. II. Gouge*, Tectonophysics, *62*, 301–319.

MUHURI, S.K., T. A. DEWERS, T. E. SCOTT, and Z. RECHES (2003), *Interseismic fault strengthening and earthquake-slip instability: Friction or cohesion?*, Geology, *31/10*, 881–884.

NAKATANI, M. (1998), *A new mechanism of slip weakening and strength recovery of friction associated with the mechanical consolidation of gouge*, J. Geophys. Res., *103/B11*, 27239–27256.

OHNAKA, M. (1973), *Experimental studies of stick-slip and their application to the earthquake source mechanism*, J. Phys. Earth, *21*(3), 285–303.

PENG, S., and A. M. JOHNSON (1972), *Crack growth and faulting in cylindrical specimens of chelmsford granite*, Int. J. Rock Mech. Min. Sci., *9/1*, 37–42.

POWER, W. L., T. E. TULLIS, and J. D. WEEKS (1988), *Roughness and wear during brittle faulting*, J. Geophys. Res., *93*, 15268–15278.

QUEENER, C. A., T. C. SMITH, and W. L. MITCHELL (1965), *Transient wear of machine parts*, Wear, *8*, 391–400.

RABINOWICZ, E. (1965), *Friction and wear of materials*, John Wiley, New York.

RECHES, Z., and D. A. LOCKNER (1994), *Nucleation and growth of faults in brittle rocks*, J. Geophys. Res. Solid Earth (1978–2012), *99/B9*, 18159–18173.

RECHES, Z., and T. A. DEWERS, (2005), *Gouge formation by dynamic pulverization during earthquake rupture*, Earth Planet. Sci. Lett., *235*, 361–374.

RECHES, Z., and D. A. LOCKNER (2010), *Fault weakening and earthquake instability by powder lubrication*, Nature, *467*, 452–456.

ROBERTSON, E. C. (1982), *Continuous formation of gouge and breccia during fault displacement*, In The 23rd US Symposium on Rock Mechanics (USRMS).

SAGY, A., Z. RECHES, and I. ROMAN (2001), *Dynamic fracturing: field and experimental observations*, J. Struct. Geol., *23/8*, 1223–1239.

SAGY, A., E. BRODSKY, and G. J. AXEN (2007), *Evolution of fault-surface roughness with slip*, Geology, *35*, 283–286. doi:10.1130/G23235A.1.

SAMMIS, C, G. KING, and R. BIEGEL (1987), *The kinematics of gouge deformation*, Pure Appl. Geophys., *125*, 777–812.

SANTNER, E., D. KLAFFKE, K. MEINE, CH. POLACZYK, and D. SPALTMANN (2006), *Effects of friction on topography and vice versa*, Wear, *261/1*, 101–106.

SCHOLZ, C. H. (1987), *Wear and gouge formation in brittle faulting*, Geology, *15*:, 493–495.

SCHOLZ, C. H., and J. T. ENGELDER, (1976), *The role of asperity indentation and ploughing in rock friction—I: Asperity creep and stick-slip*, Int. J. Rock Mech. Sci. & Geomech. Abstr., *13/5*, 149-154.

TAPPONNIER, P., and B. F. BRACE (1976), *Development of stress induced microcracks in Westerly granite*, Int. J. Rock Mech. Sci. & Geomech. Abstr., *13*, 103–112.

TENTHOREY, E., S. F. COX, and H. F. TODD, (2003), *Evolution of strength recovery and permeability during fluid–rock reaction in experimental fault zones*, Earth Planet. Sci. Lett., *206/1*, 161–172.

VERMILYE, J. M., and C. H. SCHOLZ (1998), *The process zone: A microstructural view of fault growth,* J. Geoph. Res., *103*(B6), 12223–12237.

WANG, W. B., and C. H. SCHOLZ (1994), *Wear Processes During Frictional Sliding of Rock - a Theoretical and Experimental-Study*, J. Geoph. Res., *99*(B4), 6789–6799.

WANG, W.B., and C. H. SCHOLZ (1995), *Micromechanics of rock friction .3. Quantitative modeling*, J. Geoph. Res., *100*, 4243-4247.

WAWERSIK, W. R., and W. F. BRACE (1971), *Post-failure behavior of a granite and diabase*, Rock Mechanics, *3/2*, 61–85.

WILSON, J. E., J. S. CHESTER, and F. M. CHESTER (2003), Microfracture analysis of fault growth and wear processes, Punchbowl fault, San Andreas system, California, J. Struct. Geol., *25*, 1855–1873.

WILSON, B., T. DEWERS, Z. RECHES, and J. BRUNE (2005), *Particle size and energetics of gouge from earthquake rupture zones*, Nature, *434*/7034, 749–752.

YUND, R. A., BLANPIED, M. L., TULLIS, T. E. & WEEKS, J. D. *Amorphous material in high strain experimental fault gouges.* J. Geophys. Res. *95*, 15589–15602 (1990).

(Received August 16, 2013, revised January 31, 2014, accepted February 13, 2014, Published online April 19, 2014)

Pure Appl. Geophys. 171 (2014), 3143–3157
© 2014 Springer Basel
DOI 10.1007/s00024-014-0787-x

Fault Wear by Damage Evolution During Steady-State Slip

Vladimir Lyakhovsky,[1] Amir Sagy,[1] Yuval Boneh,[2] and Ze'ev Reches[3]

Abstract—Slip along faults generates wear products such as gouge layers and cataclasite zones that range in thickness from sub-millimeter to tens of meters. The properties of these zones apparently control fault strength and slip stability. Here we present a new model of wear in a three-body configuration that utilizes the damage rheology approach and considers the process as a micro-fracturing or damage front propagating from the gouge zone into the solid rock. The derivations for steady-state conditions lead to a scaling relation for the damage front velocity considered as the wear-rate. The model predicts that the wear-rate is a function of the shear-stress and may vanish when the shear-stress drops below the microfracturing strength of the fault host rock. The simulated results successfully fit the measured friction and wear during shear experiments along faults made of carbonate and tonalite. The model is also valid for relatively large confining pressures, small damage-induced change of the bulk modulus and significant degradation of the shear modulus, which are assumed for seismogenic zones of earthquake faults. The presented formulation indicates that wear dynamics in brittle materials in general and in natural faults in particular can be understood by the concept of a "propagating damage front" and the evolution of a third-body layer.

Key words: Fault wear, Wear-rate, Friction, Damage rheology.

1. Introduction

Wear is a fundamental process in shearing surfaces that was studied experimentally and theoretically, especially for engineering materials (Archard 1953; Archard and Hirch 1956; Queener et al. 1965; Levy and Jee 1988; Kato and Adachi 2000). However, the mechanics of the process in a three-body configuration is still poorly understood.

The Archard model for wear between sliding solid blocks states that the wear volume linearly increases with applied normal stress (Archard 1953). This prediction is not consistent with recent experimental results of shearing rock faults demonstrating that wear-rates at steady-state conditions strongly depend on slip velocity.

Slip along faults in the upper crust is always associated with wear of the shearing rock blocks (Power et al. 1988; Scholz 2002), as evident by slickenside striations (Petit 1987) and gouge zones (Engelder 1974; Sibson 1977; Chester and Logan 1987). The wear processes strongly affect the fault structure. First, the plucking and crushing of fault surface asperities during shear (Fig. 1a) (Wang and Scholz 1994) modifies the geometry of the fault slip surfaces (Sagy and Brodsky 2009). Second, the wear leads to the establishment of gouge and cataclasite zones (Fig. 1b) that range in thickness from sub-millimeter to tens of meters (Katz et al. 2003; Wilson et al. 2005; Shipton et al. 2006; Wibberley et al. 2008). Experimental works showed that a gouge zone may be established after short slip distances of just a few centimeters (Boneh 2012). Thus, the existence of a gouge-zone transforms the fault slip from a "two-body" mode, in which the shearing occurs at direct contacts between asperities (Fig. 1a), into a "three-body" mode in which shearing occurs with a granular material, powder or fluid layer that separates the blocks (Fig. 1b) (Rabinowicz et al. 1961; Godet 1984; Fillot et al. 2007).

The wear of fault blocks into gouge powder is an energy dissipative process, and thus its intensity contributes to the frictional resistance of the fault (Byerelee 1967; Power et al. 1988; Wilson et al. 2005). Moreover, the gouge properties and its evolution apparently control the strength and stability of the fault (Han et al. 2010; Reches and Lockner

[1] Geological Survey of Israel, 30 Malkhei Israel St., Jerusalem, Israel. E-mail: vladi@geos.gsi.gov.il
[2] Earth and Planetary Sciences, Washington University, One Brooking Drive, St. Louis, MO 63130, USA.
[3] School of Geology and Geophysics, University of Oklahoma, Norman, OK 73019, USA.

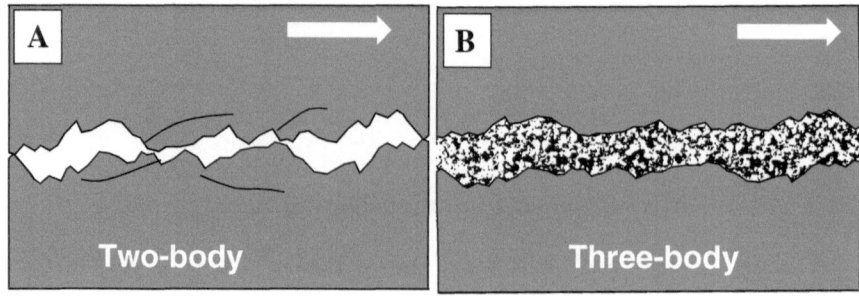

Figure 1
Schematic illustration of a fault zone in two-body (**a**) and three-body modes (**b**). In two-body mode, the sliding blocks interact at touching asperities that reflect the surface roughness. In three-body mode, the sliding blocks are separated by a gouge layer, and wear accumulation occurs at the rock-gouge contact

2010). Whereas wear of bare rock blocks was studied experimentally and theoretically (WANG and SCHOLZ 1994; POWER *et al.* 1988; BRODSKY *et al.* 2011), fault wear in a third-body setting is yet poorly understood; we propose here a wear model for this configuration.

The present model utilizes the damage rheology approach (LYAKHOVSKY *et al.* 1997) for the analysis of rock wear. First, we outline the model's approach and describe the friction and wear-rate observations during slip along experimental faults (BONEH *et al.* 2013; RECHES and LOCKNER 2010). Then, we use the key results of the damage rheology model, which are presented in the "Appendix", to evaluate the wear-rate of the present experiments. Finally, we discuss the implications of the present model for natural faults.

2. Wear Rate Model

2.1. Concepts and Assumptions

The common model for wear between sliding solid blocks (ARCHARD 1953; ARCHARD and HIRCH 1956) states that the wear volume, WV, follows the relation

$$\text{WV} = K \times D \times \frac{P}{H} \qquad (1)$$

where K is a dimensionless probability constant, D—slip distance, P—applied normal load, and H—hardness of the softer component of the slipping system or flow pressure. The Archard model is based

on the assumption that wear occurs at the real contact area between the sliding surfaces, e.g., the touching asperities (Fig. 1a), which is significantly smaller than the nominal area. QUEENER *et al.* (1965) experimentally recognized an initial transient stage with high wear production, which is termed "running-in". The combined wear model, high running-in wear-rate (QUEENER *et al.* 1965) followed by lower steady-state wear-rate (Eq. 1) was applied to many experimental analyses of slip between two solid blocks (e.g., QUEENER *et al.* 1965; LEVY and JEE 1988; KATO and ADACHI 2000, WANG and SCHOLZ 1994; BONEH *et al.* 2014). However, recent experimental analyses of shearing rock faults demonstrated that Archard's model cannot explain some central observations of fault wear. HIROSE *et al.* (2012) found that wear-rates at steady-state conditions strongly depend on slip velocity, which is not considered in Archard's model. They found that granite samples displayed power relations of wear-rate, whereas a negligible wear-rate value was measured for sandstone samples during high velocity runs. BONEH *et al.* (2013) further found that the values of wear-rate and frictional strength in carbonate rocks are interconnected and depend on both slip-velocity and normal stress.

Following these experiments, we present here a new, general model for wear at steady-state slip. The model incorporates the effects of both normal stress and slip velocity, and it shows that Archard's model is a special case of a more general behavior. Our model is based on the following plausible assumptions:

1. Steady-state fault slip occurs in three-body configuration in which a gouge layer separates the rock blocks (Fig. 1b) (Boneh *et al.* 2014).
2. The wear of the rock blocks occurs at the rock-gouge contact (Sagy and Brodsky 2009; Heesakkers *et al.* 2011a, b).
3. The solid rock wears by microfracturing-induced damage, which is controlled by the shear stresses at the rock-gouge contact.
4. Once the microfractures reach a critical density, the solid rock disintegrates and its fragments merge with the gouge zone.

We envision the above process as a microfracturing/damage front that propagates into the solid rock and consider the front velocity as the steady-state wear-rate. Modeling fault wear according to the above assumptions is accomplished here by application of damage mechanics that was developed to quantify brittle rock evolution during continuous deformation (e.g., Krajnovic 1996; Allix and Hild 2002). We show below that (a) wear evolution can be modeled in the framework of phase transition, while the shear kinematics is controlled by the resistance to shear (Lu *et al.* 2007; Lyakhovsky and Ben-Zion 2014a, b); and (b) wear-rate is a function of the system shear and the normal stresses. Finally, we test the validity of the present model by comparing the theoretical predictions with wear-rate observations of experimental faults made of carbonate rock (Boneh *et al.* 2013) and tonalite (Reches and Lockner 2010).

2.2. Damage Rheology

Models of damage rheology account for evolving elastic/plastic properties of a deforming body in terms of a "damage-state" variable that most likely represents the local density of micro-cracks. This concept is supported by the observations of gradual accumulation of distributed micro-cracks, their coalescence and localization in a narrow, highly damaged zone with strong micro-crack interaction prior to failure (e.g., Lockner *et al.* 1992; Zang *et al.* 2000; Reches and Lockner 1994). The continuum damage mechanics, which provides a general framework for the rheological behavior during failure processes, has been based on pioneering works by

Robinson (1952), Hoff (1953), Kachanov (1958, 1986) and Rabotnov (1959, 1988) and been further developed in engineering (e.g., Hansen and Schreyer 1994; Kachanov 1994; Krajcinovic 1996; Lemaitre 1996; Allix and Hild 2002) and the earth sciences (e.g., Newman and Phoenix 2001; Bercovici *et al.* 2001; Bercovici and Ricard 2003; Shcherbakov and Turcotte 2003; Turcotte *et al.* 2003; Ricard and Bercovici 2009; Karrech *et al.* 2011). The most important advantage of these and other continuum damage mechanics model formulations is that they account for the time-dependent gradual micro-crack accumulation. This is the main difference between damage mechanics and classical fracture mechanics or elasto-plastic models which postulate that failure occurs at given yielding stress conditions ignoring time-dependency of the fracture process. The damage rheology model is capable in reproducing the main stages of the faulting process starting from subcritical crack growth at very early stages of loading, material degradation due to increasing crack concentration, macroscopic brittle failure, post failure deformation, and healing. This physical framework allows modeling the evolution of various fields and properties in laboratory experiments of brittle deformation as well as simultaneous evolution of damage and its localization into narrow highly damaged zones (faults) at crustal scale, earthquakes and associated deformation fields.

Strong micro-crack interaction in a small, intensely damaged volume prior to failure (e.g., Reches and Lockner 1994) raised the need for a non-local model, in which the constitutive relation at a given position is substituted by a law accounting for the spatial distribution of the state variable over a selected neighborhood. The concept of non-local continuum was first introduced to model small-scale effects and heterogeneities in elastic solids (e.g., Eringen 1966; Kroner 1968; Bazant 1991). Non-local models, either integral or gradient type (e.g., Bazant and Jirasek 2002), account for strong micro-crack interaction in a highly damaged area prior to total failure and are capable of reproducing size effect (e.g., Bazant 2005). In this study we use the simplified 1-D version of the model formulated by Lyakhovsky *et al.* (2011) and further developed in Lyakhovsky and Ben-Zion (2014a, b), where complete thermodynamic derivations are presented. LYAKHOVSKY and BEN-ZION (2014b) utilized a three-body configuration and simulated frictional response including transitions from high quasi-static friction values at low slip velocity to dynamic friction at high slip rates, but ignored the wear process addressed in the present study.

2.3. Damage Front Propagation

LYAKHOVSKY et al. (2011) recently derived a gradient-type damage formulation that incorporates non-local behavior by enriching the local constitutive relations with a gradient of the damage-state variable. Damage accumulation, $\alpha(x, t)$, in their simplified, 1-D case is

$$\frac{\partial \alpha}{\partial t} = \frac{\partial^2 \alpha}{\partial x^2} + f(\alpha), \qquad (2)$$

This equation has the form of the well-known Fisher-KPP reaction–diffusion type equation, or diffusion equation with non-linear source function $f(\alpha)$ (FISHER 1937; KOLMOGOROV et al. 1937) that was extensively applied in a wide range of models in biology, social science, phase transition and critical phenomena (e.g., GRINDROD 1996; MURRAY 2002; LIFSHITZ and PITAEVSKII 1981; MA 2000). The solution of the Fisher-KPP Eq. (2) exhibits traveling waves or fronts switching between equilibrium states. Travelling fronts propagating with the speed c exist in homogeneous media when $c \geq c_* \equiv 2\sqrt{f'(0)}$, where c_* is a critical speed. The front profile satisfies exponential decay (with algebraic correction) when $c = c_*$ and may fail propagating in heterogeneous media (e.g., NOLEN et al. 2012). The general form of the source term $f(\alpha)$ in the equation for the damage evolution (2) is discussed by LYAKHOVSKY and BEN-ZION (2014b) and is given here in the "Appendix" (Eq. 4). The calculated rate of damage accumulation under the conditions that mimic the experimental set-up ("Appendix") leads to the analytical estimation for the speed of the propagating damage front.

2.4. Wear Model Predictions

In the present model, we assume that a narrow, highly damaged band (wear zone) exists between the shearing rock blocks, and thus the shear occurs in a "three-body" mode (Fig. 1b). We model the fault wear as a front that separates between the intensely damaged zone (gouge) and the intact rock. In this 1-D configuration, the wear-rate is the travelling speed, c (measured by volume per unit contact area per time), of the propagating front into the intact rock under steady-state conditions. The present derivations ("Appendix") expresses the wear-rate, WR as a function of the normal stress, σ_n, and friction coefficient, μ_S, during steady-state slip (3a); this relation can also be written as a function of the steady-state shear stress, τ_S (3b):

$$\text{WR} = F \times \sigma_n \sqrt{\mu_S^2 - \mu_{cr}^2} \qquad (3a)$$

$$\text{WR} = F \times \sqrt{\tau_S^2 - \tau_{str}^2} \qquad (3b)$$

This expression includes two adjustable material parameters, F, and μ_{cr} (or τ_{str}), which is a critical strength parameter of the rock blocks. For constant slip velocity the wear rate represents widening of the damaged band (wear layer) per unit slip distance instead of time. For contact area this is equivalent to wear volume rate per slip distance (ARCHARD 1953). The scaling dimensional parameter F is in some sense equivalent to K in Eq. 1. The rock strength parameter, τ_{str}, is the minimum shear stress necessary for micro-fracturing of the host rock. The values of F and μ_{cr} (or $\tau_{str} = \mu_{cr} \times \sigma_n$) are expected to depend only on the rock type, and to be independent of normal stress and slip velocity, while the confining pressure effect is very weak.

Equations (3a, b) reveal two wear regimes: (1) fault wear when μ_S exceeds a critical, threshold value, $\mu_S \geq \mu_{cr}$ (or applied stress exceeds the material strength, $\tau_S \geq \tau_{str}$), which provides a positive expression under the radical, and (2) no fault wear at lower friction ($\mu_S < \mu_{cr}$) (or lower stress $\tau_S < \tau_{str}$), which implies that the damage front does not propagate at the specified stress conditions. We test the quality of Eqs. (3a, b) by its application to relevant experimental results.

3. Model Application

3.1. Experimental Case Study

We apply our model (Eqs. 3a, b, "Appendix") to the results of BONEH et al. (2013) who measured the wear-rates and friction coefficients of experimental fault made of carbonate rocks. They used a high-velocity, rotary shear apparatus (RECHES and LOCKNER 2010) to shear solid rock samples with ring-shaped contact (5.4 and 7.6 cm inner and outer diameters). Three rock types were used: Kasota dolomite, Dover limestone and a sample containing an upper Kasota dolomite and a lower Blue quartzite. BONEH *et al.* (2013) performed 87 experiments with total slip of 2–28 m at normal stresses between 0.25 and 6.9 MPa. The experiments included 72 constant-velocity experiments and 15 stepping-velocity experiments at a slip velocity range of 0.002–0.96 m/s.

BONEH *et al.* (2013) found that the steady-state friction coefficient μ_S of the tested carbonate faults correlates best with the experimental power-density (shear stress times slip velocity). Figure 2a presents the steady-state friction coefficients, μ_S, measured as a function of normal stress and slip velocity in all tests of the Kasota dolomite and Dover limestone, whereby both upper and bottom sheared surfaces are from the same lithology. Under slow slip velocity (below ~ 0.01 m/s) the μ_S is high in the range of 0.6–0.8 with a minor decrease under elevated normal stress. The most significant feature of Fig. 2a is the steep decrease of μ_S as the slip velocity increases from ~ 0.01 to ~ 0.3 m/s.

BONEH *et al.* (2013) followed RECHES and LOCKNER (2010) and calculated the continuous wear-rate along the experimental faults in terms of wear-rate = [fault-normal shortening]/[slip distance], with units of [micron/m]. BONEH *et al.* (2013) found that the wear-rates during steady-state slip also depend on both slip velocity and normal stress. At low velocity, $V < 0.1$ m/s, the wear-rate linearly depends on the normal stress, whereas at high velocities, $V > 0.3$ m/s, the wear-rate decreases with increasing normal stresses. In these experiments, the wear-rates approach zero at the highest velocities, $V \sim 1$ m/s, and normal stresses, $\sigma_n \sim 7$ MPa.

3.2. Model Adjustments

To apply the present model to the above experimental results, we first generalized the trend of the experimental frictional results (Fig. 2a) to a simpler trend (Fig. 2b), in which high, constant friction values are set at low velocity and a steep friction drop is set for 0.01–0.1 m/s velocity increase. The

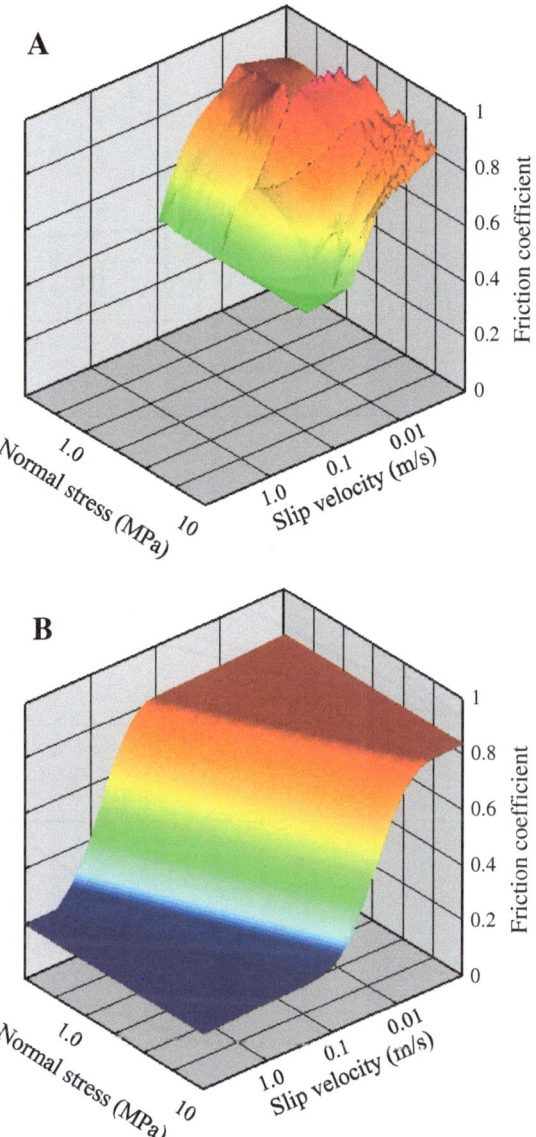

Figure 2

Steady-state friction coefficient as a function of slip velocity and normal stress. **a** Interpolation of 87 results of two sets of dynamic shear experiments (dolomite on dolomite and limestone on limestone), modified after BONEH *et al.* (2013). **b** Generalized function pointing to significant friction decrease in the transition zone and about constant value at higher slip-rate values

friction values are about constant at high slip velocities. A similar steep drop of friction values was observed in laboratory experiments with different rock types (e.g., DI TORO *et al.* 2011). A cross-section of this function (Fig. 3a, black line) displays the relations between slip velocity and steady-state

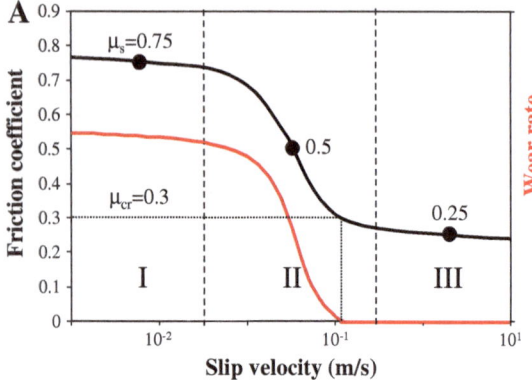

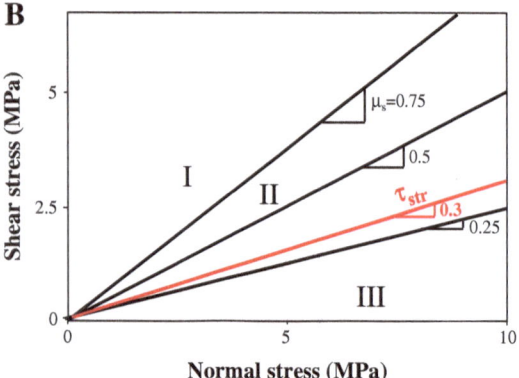

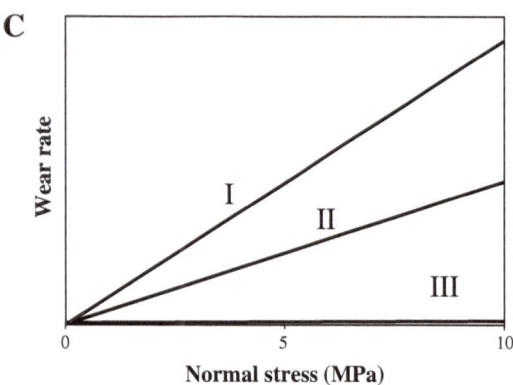

Figure 3

Present model predictions. **a** Schematic relationship between steady-state friction and slip velocity at constant normal stress (*black line*). *Red line* shows calculated wear-rate. Three regimes of steady-state frictional strength are marked: (*I*)—high friction at low slip velocity; (*II*)—steep drop with increasing slip rate; (*III*)—low friction at high slip-rates. Wear-rate drops to zero at critical friction coefficient μ_{cr}. **b** Shear versus normal stress for frictional values representing three different regimes. *Red line* shows shear stress corresponding to material strength for $\mu_{cr} = 0.3$. **c** Wear-rate versus normal stress for the same regimes

friction at constant normal stress. Three model regimes of steady-state frictional strength are built into Figs. 2b and 3a:

Regime I—high, quasi-constant μ_S of 0.7–0.8 at low slip velocity;
Regime II—steep drop rate of μ_S with increasing slip-rate;
Regime III—low, quasi-constant μ_S at high slip-rates.

Our model predicts that the wear-rate is proportional to the shear stress (Eq. 3b); this dependence is a natural outcome of model assumption #3 above. Thus, under constant normal stress, the wear-rate (red curve in Fig. 3a) dependence on slip velocity is predicted to mimic the dependence of friction coefficient μ_S (black curve in Fig. 3a). In this respect, similarly to the FILLOT *et al.* (2007) model, our model links the fault wear to the shear resistance of the fault during steady state. The model predicts significantly different wear-rates associated with three frictional regimes: high wear-rates in regime I, transitional wear-rates in regime II and vanishing wear-rates in regime III. Furthermore, the wear-rate drops when $\mu_S \rightarrow \mu_{cr}$ (Eqs. 3a, b) and may even vanish when the frictional strength decreases below the defined critical value ($\mu_S < \mu_{cr}$) (Fig. 3a). These wear-rate variations may be expressed in terms of the applied stress (Fig. 3b). The red line in Fig. 3b shows the material strength versus normal stress for the selected $\mu_{cr} = 0.3$. Under loading conditions corresponding to $\mu_S = 0.75$ or 0.5 the shear stress is above the material strength (lines I and II in Fig. 3b). These stress conditions favor micro-fracturing in the form of a propagating damage front into the solid rock. In the case of low shear stresses in regime III with $\mu_S = 0.25$, the shear stress is below the material strength, which is the threshold for micro-crack nucleation and growth; with no rock damage, fault slip might continue without wear.

Figure 3c displays the expected relations between normal stress and wear-rate for the three regimes of Fig. 3a. The wear-rate is expected to linearly increase with respect to the normal stress, as in Archard's

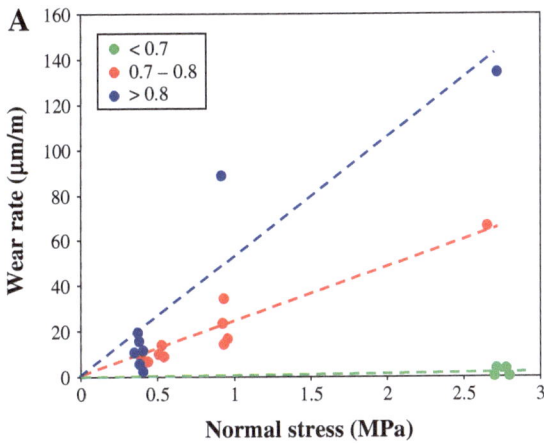

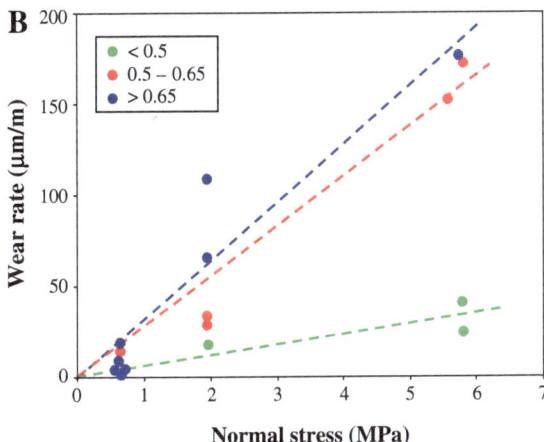

Figure 4

Experimentally measured wear-rates under steady-state conditions (after BONEH *et al.* 2013) for Dover limestone faults (**a**), and for samples composed of an upper Kasota dolomite and a lower Blue quartzite (**b**). *Colored points* are grouped according to the range of the measured friction coefficient (*inset*). *Dashed lines* with the same color are the model-predicted wear-rates according to the frictional value for each group

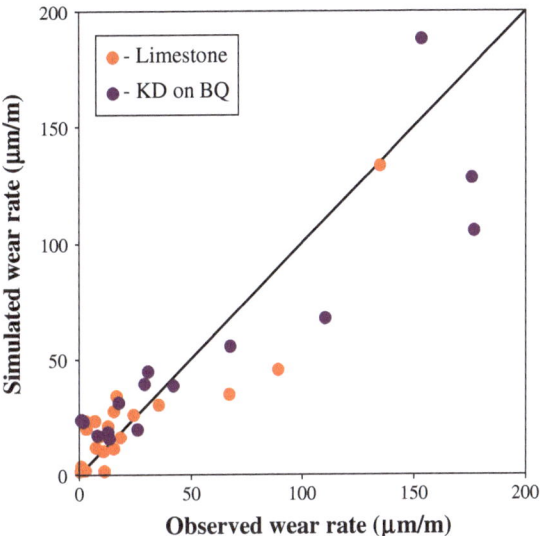

Figure 5

Model simulated versus experimentally observed wear-rates for the two sets of experiments shown in Fig. 4

model, but at a different slope that depends on the steady-state friction. The slopes marked I–III in Fig. 3c correspond to the representative friction values of $\mu_S = 0.75$ and 0.5 for the regimes I and II and is zero (no wear) for $\mu_S = 0.25$ in regime III.

3.3. Comparison of Experimental Results and Model Predictions

3.3.1 Carbonate Faults

The experimentally measured wear-rates are now compared with the predicted trends of the model.

Figure 4 shows measured wear-rates reported by BONEH *et al.* (2013) for Dover limestone samples (Fig. 4a) and for samples containing an upper Kasota dolomite and a lower Blue quartzite (Fig. 4b). Experimentally measured wear-rates were fitted with the model Eq. (3a) using least-squares by changing two parameters, material constant F and critical friction coefficient μ_{cr}. The best fitting model parameters are: $F = 98.4$ µm m^{-1} MPa^{-1}, $\mu_{cr} = 0.68$ for Dover limestone, $F = 49.8$ µm m^{-1} MPa^{-1}, $\mu_{cr} = 0.35$ for Kasota dolomite on Blue quartzite. Experimental results for each rock type are plotted in three groups according to the observed steady-state friction value. The modeled trend lines (dashed lines with the same color as markers representing specific group) are the calculated linear relations between the experimental wear-rate and normal stress for the given friction group. The systematic slope decrease with decreasing friction coefficients fits the model predictions. Comparison of the experimentally observed wear-rates with the model's calculated wear-rates (Eq. 3a) in Fig. 5 indicates that the simplified 1-D model reproduces the general trends of the wear-rate changes with both normal stress and frictional strength.

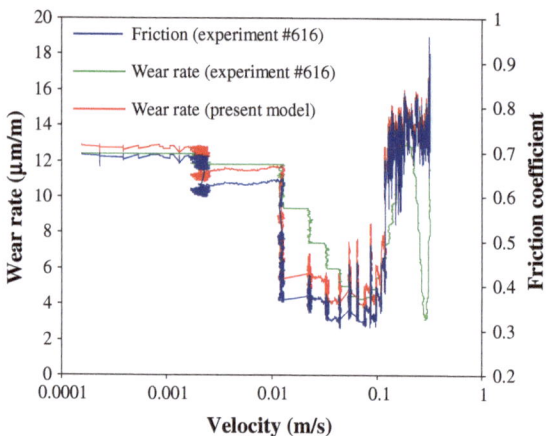

Figure 6
Wear-rate and friction coefficient during rotary shear experiment (#616 in RECHES and LOCKNER 2010) plotted with wear-rate predictions of the present model. The experiment was conducted on a tonalite (Sierra White granite) sample at normal stress of 3.05 MPa with total slip of 65 m at 30 slip velocity steps (see text)

3.3.2 Tonalite Fault

An extensive series of friction experiments on samples of tonalite, which is better known by its commercial name Sierra White granite (SWG), were reported by RECHES and LOCKNER (2010). We apply the present model to their experiment #616 that was conducted at a constant normal load of 3.08 Mpa and stepping velocity. This experiment was conducted on the same rotary apparatus and similar sample geometry as BONEH et al. (2013), but differs in two central parameters: rock composition and loading configuration. The #616 experimental fault was made of tonalite, an igneous rock dominated by quartz and feldspar (RECHES and LOCKNER 2010), whereas BONEH et al.'s (2013) experiments were conducted on limestone and dolomite samples. Also, experiment #616 was subjected to a continuous sequence of 30 slip velocity steps from $V = 0.0015$ to 0.32 m/s, and the slip distance during each velocity step was 2.1 m; on the other hand, the reported experiments in Fig. 4 are single velocity runs. The experimental results of #616 (Fig. 6) are shown by friction coefficient (blue line) and wear-rate (green line) as a function of the changing slip velocity. We used Eq. (3a) to calculate the expected wear-rates for the measured friction coefficient, and the red line in Fig. 6 is the calculated

curve for $F = 6.5\ \mu\mathrm{m}\ \mathrm{m}^{-1}\ \mathrm{MPa}^{-1}$, $\mu_{\mathrm{cr}} = 0.25$. Instead of the best-fit procedure, these values were chosen to reproduce high wear-rate values during the initial experimental stages with low slip velocities (up to 0.01 m/s), and the low wear-rate values for the velocity range between 0.01 and 0.1 m/s. The present model based on steady-state conditions is not expected to fit experimental results of transient conditions which are intrinsic for stepping velocity experiments. Therefore, the least-squares procedure was not applied to the measured data.

In spite of the transient nature of #616 experiment, the present model, provides a reasonable fit to the experimental data, and predicts the main features of the observations: (1) abrupt wear-rate decreases as the friction coefficient drops from ~ 0.7 to ~ 0.4 (slip velocity ~ 0.01 m/s). Yet, the measured wear-rate decreases gradually, whereas the model's calculated wear-rate drops abruptly. This difference is probably associated with transient change of the wear-rate that under changing velocities does not reach the steady-state of the presented model. (2) At higher velocities (above 0.1 m/s), the experimental friction coefficient abruptly increases (due to temperature rise, RECHES and LOCKNER 2010; SAMMIS et al. 2011), as well as the experimental and model-derived wear-rates. Note again a delay of the rise of the experimental wear-rate. We also note that the model fails to predict the abrupt drop of wear-rate at the high-velocity of V 0.22–0.29 m/s). Nevertheless, the successful model simulation of the wear-rates in experiment #616 further indicate the strong correlation between frictional strength and wear-rate even under transient slip conditions.

4. Discussion

4.1. Relations to Archard's Wear Model

We derived the model for wear-rate of rock faults under steady-state conditions with constant slip velocity and loading stresses motivated mainly by the limited ability of the well-known ARCHARD (1953) model to explain recent experimental observations of rock wear (HIROSE et al. 2012; BONEH et al. 2013). ARCHARD's (1953) model was based on wear at contacting asperities (Fig. 1a), implying that the wear-rate is proportional to a strength ratio of [normal stress]/[hardness of the softer surface] (Eq. 1). The mechanical conditions at the contacting asperities (e.g., gouge presence, roughness, and lubrication) were integrated into one free parameter, K. In contrast, our model is derived for slip in a three-body mode (Fig. 1b) with a gouge layer separating the blocks; experimental (RECHES and LOCKNER 2010; BONEH *et al.* 2014) and field observations (e.g., CHESTER and CHESTER 1998; KATZ *et al.* 2003) indicate that this is the realistic mode for steady-state slip. In the model, the wear process is associated with microfracturing and gradual damage increase in the solid rock.

We analyze the wear process by using a non-local continuum damage mechanics connecting wear-rate with the rate of a propagating damage front. The evolving material damage in space and time is defined by the reaction–diffusion equation with non-linear source term (Eqs. 7, 8) known as Fisher-KPP equations. Using the general properties of the mathematical solution of Fisher-KPP equations, we can quantify the process dynamics without calculating local state and motion of every single crack in the system. The derivations lead to a simple expression, Eqs. (3a, b), connecting the wear-rate with the frictional strength of a fault. Comparison with experimental data demonstrates that our wear model is capable of predicting the wear-rates under a wide range of slip velocities and normal stresses (Figs. 4, 6) (Eqs. 3a, b). The model depends on the critical strength of the host rock, τ_{srt}, or its equivalent, the critical friction coefficient, μ_{crt}. Archard's model appears as a specific case in which the friction coefficient is constant and larger than the critical friction coefficient, μ_{crt}, (as the wear-rate vanishes when $\mu \leq \mu_{\mathrm{crt}}$). Moreover, we showed in the "Appendix" that the model is valid and even simpler for wear in relatively large confining pressures and significant degradation of the shear modulus (Fig. 7), which are the conditions assumed for natural faulting, but until now, have never been taken into account in wear models. Considering these intrinsic limitations of Archard's model, its application to field cases of natural faulting is limited, as shown below.

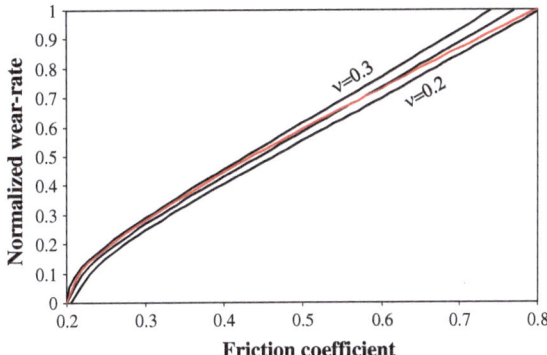

Figure 7
Normalized wear-rate versus friction coefficient calculated using (14) with Poisson ratio, $v = 0.2$, 0.25 and 0.3 (*black lines*). *Red line* shows suggested approximation (15) used in this study

4.2. Wear, Gouge Thickness and Fault Slip Distance

The present analysis provides an important insight on gouge accumulation along faults. It is assumed here that at steady-state, the front of fault wear propagates into the host rock by micro-cracking damage. Based on this assumption, the wear-rate has simple relations to the friction coefficient ("Appendix", Fig. 3, Eq. 3a) or shear-stress (Eq. 3b). Now, when the shear-stress drops, e.g., due to a friction coefficient drop at high slip velocity, the wear-rate also drops (Fig. 6), or even vanishes when $\mu_S < \mu_{cr}$ (Figs. 3, 4, Eqs. 3a, b). An obvious deduction from these results is that a fault may slip over long distances without ongoing wear and thickening of the gouge zone if the shear stress (or the equivalent friction coefficient) is low enough. This conclusion leads to a new view on gouge-zone thickness (discussed below).

As fault gouge is the main wear product of fault slip, it is commonly envisioned that gouge continuously forms and accumulates during fault slip (e.g., SCHOLZ 1987). It was thus suggested that the total gouge thickness, G_T, along a given fault is proportional to the total slip distance, D, of the fault. SCHOLZ (1987) showed that this relation is a natural result of Archard's model, and used compiled field data to demonstrate that G_T is linearly proportional to the total fault slip with $G_T/D = 0.1$–0.001 over about seven orders of magnitude of fault slip. Interestingly, based on fractal fault geometry and transient wear mechanism, POWER *et al.* (1988) also suggested that wear zone thickness of natural faults depends linearly on displacement, because the size of the asperities increases in an approximately linear relationship to displacement. On the other hand, based on fault surface roughness measurements and contact mechanics, BRODSKY et al. (2011) predicted that average wear-rate is a weak function of D (proportional to D^{-1}); therefore, even under infinite fractal roughness the wear-rate is not constant. According to BRODSKY et al. (2011), the predicted wear-rate vanishes, or $G_T \sim$ constant, for faults that already accumulated large slip distances. These works were based on an asperity failure concept (Fig. 1a), and considered the effects of roughness, lithology and normal stresses on wear-rate (SCHOLZ 1987; WANG and SCHOLZ 1994; POWER et al. 1988; BRODSKY et al. 2011), but implicitly ignored the intensity of the fault-parallel shear-stress. On the other hand, our model considered the three-body configuration (Fig. 1b) in which the wear-rate depends on the shear-stress intensity at the gouge-rock contact (Eqs. 3a, b).

Our model results can now be applied to evaluate fault wear at seismic conditions. Experimental works have shown that the dynamic friction during high slip velocity may be as low as 0.1–0.3, as shown in many high-velocity friction experiments (RECHES and LOCKNER 2010; DI TORO et al. 2011; CHANG et al. 2012) or even vanish at seismic velocities (HAN et al. 2010). Application of these experimental observations to the field indicates that earthquake slip could generate very low wear, and thus the long-term gouge accumulation along a seismically active fault is not necessarily related to its total slip (SHIPTON et al. 2006). For example, CHESTER et al. (1993) analyzed two major branches of the San Andreas system, the north branch San Gabriel and Punchbowl faults, with a total slip of 22 and 44 km, respectively. They showed that almost the entire slip along these faults was localized within a 1 m thick zone of cataclasite (=cohesive gouge). According to SCHOLZ's (1987) compilation of $G_T/D = 0.1–0.001$, the minimum expected gouge thickness of these faults is on the order of tens of meters. Thus, our model, which predicts negligible wear for faults with low to vanishing dynamic friction, provides a suitable interpretation for the observed gouge thickness along

active faults such as the north branch San Gabriel and Punchbowl, California.

Our model predictions may also hold for faults that are intrinsically weak due to their composition. For example, LOCKNER et al. (2011) analyzed the strength of core materials collected from the San Andreas fault-zone, California. The core was retrieved from a depth of 2.7 km in the creeping segment of the host fault. The strength measurements conducted at in situ conditions revealed a very weak gouge, with a friction coefficient of ~ 0.15, due to the presence of saponite, which is an extremely weak phyllosilicate mineral. According to our model, faults with such low static frictional strength will undergo negligible wear and consequently will have anomalously thin gouge zones.

The present model is a simplified one that fits well the simple configuration of the experimental faults. However, the wear of natural faults can also be affected by additional parameters such as pre-faulting damage, variable slip velocity, complex geometrical, lithological and rheological relations (e.g., KATZ et al. 2003; BRODSKY et al. 2011; NIELSEN et al. 2010).

5. Summary and Conclusions

The present wear model is based on a continuum damage-breakage rheological model that provides a general quantitative description of fault mechanics (LYAKHOVSKY and BEN-ZION 2014b). This model considers a transition from a solid phase with distributed fracturing and evolving elastic moduli to a granular phase referred to as pseudo-liquid. The model of LYAKHOVSKY and BEN-ZION (2014b) reproduces central features of a fault-zone structure, including slip localization within a narrow zone, gouge formation (three-body mode), and transition between slow motion associated with high steady-state friction to a dynamic regime with low friction. The present wear model is a 1-D simplification of the general formulation ("Appendix") that leads to a general expression for wear-rate under steady-state slip (Eqs. 3a, b).

The present analysis leads to the following conclusions:

1. The theoretical wear-rate of a fault during steady-state slip (Eqs. 3a, b) is linked to the shear stress along the slipping fault: It is relatively high when

the shear stress, τ_s, is much higher than the rock strength, τ_{str}, it drops as the shear stress drops, and the wear-rate vanishes when the shear stress is lower than the rock strength.

2. The model successfully predicted the wear-rate intensities in two sets of friction experiments (BONEH et al. 2013; RECHES and LOCKNER 2010). These experiments were conducted on different rocks (limestone, dolomite and tonalite), a wide range of slip velocities (0.001–0.97 m/s), and two types of loading histories (constant slip velocity or stepping velocity).

3. This successful application to experimental observations indicates that the central model concept of damage-driven wear in a "three-body" configuration (Fig. 1b) is a reasonable approach for wear under steady-state slip. We conclude that this concept is more realistic than the asperities contact model of bare rock surfaces (Fig. 2a).

4. The present wear model suggests that the thickness of gouge (cataclasite) layers along a natural fault strongly depends on their frictional strength, and thus faults with low frictional resistance (static and/or dynamic) may slip with low to negligible wear.

Acknowledgments

The manuscript benefitted from useful comments by Y. Ben-Zion (editor), W. Ashley Griffith (reviewer), and an anonymous reviewer. The study was supported by the NSF, Geosciences, Equipment and Facilities, Grant No. 0732715, with partial support of NSF, Geosciences, Geophysics, Grant No. 1045414, and ConocoPhillips Foundation grant. VL acknowledges support by the US–Israel Binational Science Foundation (Grant 2008248); AS acknowledges support by the Israel Science Foundation (Grant 929/10).

Appendix

Damage Rheology Model

Key derivations of the damage rheology model leading to the estimate of the speed of the propagating damage front are presented in this "Appendix". Following thermodynamic balance relations and the ONSAGER (1931) principle, LYAKHOVSKY et al. (1997) developed a kinetic equation for the damage state variable, α (weakening and healing) which is a function of the progressive deformation. Non-linear elasticity that connects the effective elastic moduli to a damage variable and loading conditions allows accounting for the transition from damage accumulation to healing. This transition is controlled by the strain invariants ratio $\xi = I_1/\sqrt{I_2}$, where $I_1 = \varepsilon_{kk}$ and $I_2 = \varepsilon_{ij}\varepsilon_{ij}$ are the invariants of the elastic strain tensor ε_{ij}. The ξ value is a conjugate quantity to the ratio between shear and normal stress expressed in terms of strains, instead of stresses. The rate of damage/healing accumulation is given by LYAKHOVSKY et al. (1997):

$$\frac{d\alpha}{dt} = \begin{cases} C_d I_2(\xi - \xi_0) & \text{for } \xi > \xi_0 \\ C_1 \exp\left(\frac{\alpha}{C_2}\right) I_2(\xi - \xi_0) & \text{for } \xi < \xi_0 \end{cases}. \quad (4)$$

where the coefficient C_d is the rate of positive damage evolution (material degradation) that is constrained by laboratory experiments (LYAKHOVSKY et al. 1997; HAMIEL et al. 2004, 2006). The value $\xi = \xi_0$ controls the onset of damage accumulation or transition from material healing to weakening associated with microcrack nucleation and growth. The value of the critical strain invariants ratio also called "modified internal friction" (see Fig. 3 in LYAKHOVSKY et al. 1997) is explicitly related to the internal friction angle of Byerlee's law (BYERLEE 1978) and Poisson ratio of the intact rock.

The rate of damage recovery (healing) is assumed to depend exponentially on α and produces logarithmic healing with time in agreement with the behavior observed in laboratory experiments with rocks and other materials (e.g., DIETERICH and KILGORE 1996; SCHOLZ 2002; MUHURI et al. 2003; JOHNSON and JIA 2005). LYAKHOVSKY et al. (2005) showed that the local damage model reproduces the main phenomenological features of the rate- and state-dependent friction, and constrained the healing parameters C_1, C_2 by comparing the model calculations with empiric parameters of the slow-rate frictional sliding (e.g., DIETERICH 1972, 1979; MARONE 1998).

The damage accumulation under constant stress in a simplified 1-D model with effective elastic moduli degrading proportionally to $(1 - \alpha)$ follows a power law solution (e.g., BEN-ZION and LYAKHOVSKY 2002; TURCOTTE et al. 2003):

$$\alpha(t) = 1 - \left(1 - 3\frac{C_d\sigma^2}{G_0^2}t\right)^{1/3}. \qquad (5)$$

where G_0 is the elastic moduli of the intact rock, σ, the applied stress, and C_d the damage rate coefficient. This solution allows us to introduce a time scale, t_f, which is the time-to-failure when $\alpha = 1$ (total damage) (e.g., PATERSON and WONG 2005) that becomes

$$t_f = \frac{G_0^2}{3C_d\sigma^2}. \qquad (6)$$

This parameter controls the time scale of all processes associated with accumulated damage, including the growth rate of narrow fracture zones.

Recently, LYAKHOVSKY et al. (2011) developed a gradient-type damage rheology formulation that incorporates non-local behavior by enriching the local constitutive relations with a gradient of the damage state variable. This addition modifies the kinetic equation for the damage evolution. In addition to the source term controlling the damage growth in Eq. (4), the damage accumulation for $\xi > \xi_0$ in non-local formulation includes damage diffusion term with a coefficient D:

$$\frac{\partial\alpha}{\partial t} = C_d I_2(\xi - \xi_0) + D\nabla^2\alpha. \qquad (7)$$

The non-local gradient-type damage kinetic Eq. (7) has the form of a Fisher-KPP reaction–diffusion type equation for which the general one-dimensional form is

$$\frac{\partial u}{\partial t} = \frac{\partial^2 u}{\partial x^2} + f(u) \qquad (8)$$

The solution of the Fisher-KPP Eq. (8) exhibits a traveling wave or fronts switching between equilibrium states $f(u) = 0$ (see text). Travelling fronts propagating with the speed c exist in homogeneous media when $c \geq c_* \equiv 2\sqrt{f'(0)}$. We calculate the source term $f(\alpha) = C_d I_2(\xi - \xi_0)$ for $\alpha = 0$ under loading condition that mimics the experimental set-up with the normal stress σ_n and shear stress τ, and then estimate the speed of the propagating damage front.

Adopting damage model with $(1 - \alpha)$ reduction of the elastic moduli, the relations between stress and strain components (axial, ε_a, and transversal, ε_τ) are (E_0, G_0, Young and shear moduli of the intact rock):

$$\varepsilon_a = \frac{\sigma_n}{E_0(1 - \alpha)}; \; \varepsilon_\tau = \frac{\tau}{2G_0(1 - \alpha)} \qquad (9)$$

Using relations (9), the second strain invariant, I_2, and strain invariant ratio, ξ, are,

$$I_2 = \varepsilon_a^2 + 2\varepsilon_\tau^2 = \frac{1}{(1 - \alpha)^2}\left[\frac{1}{E_0^2}\sigma_n^2 + \frac{1}{2G_0^2}\tau^2\right] \qquad (10)$$

$$\xi = \frac{\varepsilon_a}{\sqrt{\varepsilon_a^2 + 2\varepsilon_\tau^2}} = \frac{-1}{\sqrt{1 + 2\left(\frac{E_0\tau}{2G_0\sigma_n}\right)^2}} \qquad (11)$$

The minus sign in (11) implies that the compaction strains are negative. Note that without any shear loading ($\tau = 0$), the strain invariants ratio is $\xi = -1$, which is slightly below typical ξ_0 range (LYAKHOVSKY et al. 1997). This implies that damage is not accumulated at loading by normal stress alone without shear loading. Substituting (10, 11) into equation for the damage accumulation (7), leads to the following relation for the source term:

$$f(\alpha) = C_d\frac{1}{(1 - \alpha)^2}\left[\frac{1}{E_0^2}\sigma_n^2 + \frac{1}{2G_0^2}\tau^2\right]$$
$$\times \left[\frac{-1}{\sqrt{1 + 2\left(\frac{E_0\tau}{2G_0\sigma_n}\right)^2}} - \xi_0\right] \qquad (12)$$

The speed of the travelling damage front, which in the present work is defined as the wear-rate (text), is controlled by the value $c_* = 2\sqrt{D \cdot f'(0)}$ equal to

$$c_* = 2\sqrt{2C_d D}\frac{\sigma_n}{E_0}$$
$$\times \sqrt{-\xi_0\left[1 + 2\left(\frac{E_0\tau}{2G_0\sigma_n}\right)^2\right] - \sqrt{1 + 2\left(\frac{E_0\tau}{2G_0\sigma_n}\right)^2}} \qquad (13)$$

Taking $\mu_S = \tau/\sigma_n$ for the steady-state friction coefficient, Eq. (13) is rearranged to:

$$c_* = 2\sqrt{2C_\mathrm{d}D}\,\frac{\sigma_\mathrm{n}}{E_0}\sqrt{-\xi_0\left(1 + A\mu_\mathrm{S}^2\right) - \sqrt{1 + A\mu_\mathrm{S}^2}}$$

$$(14)$$

where $A = 2(E_0/2G_0)^2 = 2(1 + v)^2$, and v is the Poisson ratio. Note that typical values of the strain invariant ratio, ξ_0, controlling the onset of damage accumulation in (4) vary between $\xi_0 = -1.2$ and $\xi_0 = -0.6$. These values are taken from Fig. 3 of LYAKHOVSKY et al. (1997) for friction angle 30°, $v = 0.2$ and for friction angle 40°, $v = 0.3$. The steady-state friction coefficient should be above certain critical or threshold values ($\mu_\mathrm{S} \geq \mu_\mathrm{cr}$) related to the material strength to give positive expression under the radical in (14). A negative value for $\mu_\mathrm{S} < \mu_\mathrm{cr}$ implies that the applied shear stress is below the level needed for the onset of damage accumulation and the source term in damage growth Eq. (7) is negative; no any damage is accumulated, and no wear is anticipated.

The Poisson ratio for most rocks varies within the small range $v = 0.2$–0.3. Variation of the steady-state friction, μ_S, is also limited. It decreases from static friction values about 0.6–0.8 to dynamic values about 0.2–0.4. Accounting for this limited range of the material properties, the speed of the travelling damage front or wear-rate calculated using (14) only slightly depends on the Poisson ratio (black lines in Fig. 7). Hence, the wear-rate may be approximated by much simpler equation (red line in Fig. 7):

$$\text{wear\_rate} \propto \sigma_\mathrm{n}\sqrt{\mu_\mathrm{S}^2 - \mu_\mathrm{cr}^2} \qquad (15)$$

The units of the speed of the travelling damage front (13, 14) are length per time, while in the laboratory experiments wear-rate measured the thinning of the sample per unit slip. As the area of the surface is constant (BONEH et al. 2013), this thinning is proportional to the wear volume generation. Such unit conversion could be done under steady-state conditions with constant slip rate.

We note here that Eq. (14) was derived for the loading conditions of experimental set-up. More appropriate conditions for deep seismogenic zones should account for relatively large confining pressure, small damage-induced change of the bulk modulus and significant degradation of the shear modulus.

Similar derivations lead directly to the more simple form (15) instead of (14) for the natural conditions.

REFERENCES

ALLIX, O. and HILD, F. (2002) Continuum damage mechanics of materials and structures, Elsevier, 396 pp.

ARCHARD, J.F. (1953) Contact and rubbing of flat surfaces, J. Appl. Phys., 24, 981–988.

ARCHARD, J.F., HIRCH, W. (1956) Wear of metals under unlubricated conditions. Proc. R. Soc. Lond. A Math. Phys. Sci. 236, 397–410.

BAZANT, Z.P. (1991) Why continuum damage is nonlocal: Micromechanics arguments, J. Eng. Mech., 1070–1087.

BAZANT, Z.P. (2005) Scaling of structural strength. Elsevier, 327 pp.

BAZANT, Z.P. and JIRASEK, M. (2002) Nonlocal integral formulations of plasticity and damage: Survey of progress, J. Eng. Mech., 128, 1119–1149.

BEN-ZION, Y., LYAKHOVSKY, V., (2002), Accelerating seismic release and related aspects of seismicity patterns on earthquake faults. Pure Appl. Geophys. 159, 2385–2412.

BERCOVICI, D. and RICARD, Y. (2003) Energetics of a two-phase model of lithospheric damage, shear localization and plate boundary formation, Geophys. J. Int., 152, 581–596.

BERCOVICI, D., RICARD, Y., and SCHUBERT, G. (2001) A two-phase model for compaction and damage, part 1: general theory, J. Geophys. Res., 106, 8887–8906.

BONEH, Y, CHANG, J.J., LOCKNER, D.A., and RECHES, Z. (2014) Fault evolution by transient processes of wear and friction, Pure. Appl. Geoph. (this volume).

BONEH, Y. (2012) Wear and gouge along faults: Experimental and mechanical analysis, Thesis, Univ. Oklahoma.

BONEH, Y., SAGY, A., and RECHES, Z. (2013) Frictional strength and wear-rate of carbonate faults during high-velocity, steady-state sliding, Earth Planet. Sci. Lett., 381, 127–137.

BRODSKY, E.E., GILCHRIST, J.J SAGY, A., and COLLETTINI, C. (2011) Faults smooth gradually as a function of slip, Earth Planet. Sci. Lett., 302, 185–193.

BYERELEE, J.D. (1967) Frictional characteristics of granite under high confining pressure, J. Geophys. Res., 72, 3639-3648.

BYERLEE, J.D., (1978), Friction of rocks. Pure Appl. Geophys., 116, 615–626.

CHANG, J.C., LOCKNER, D.A., RECHES, Z. (2012) Rapid acceleration leads to rapid weakening in earthquake-like laboratory experiments, Science, 338, 101, doi:10.1126/science.1221195.

CHESTER, F.M., and CHESTER, J.S. (1998) Ultracataclasite structure and friction processes of the San Andreas fault, Tectonophysics, 295, 199–221.

CHESTER, F.M., EVANS, J.P., BIEGEL, R.L. (1993) Internal structure and weakening mechanisms of the San Andreas fault, J. Geophys. Res., 98, 771–786.

CHESTER, F.M., LOGAN, J.M. (1987) Composite planar fabric of gouge from the Punchbowl fault, California, J. Struct. Geol., 9, 621–634.

DI TORO, G., HAN, R., HIROSE, T., DE PAOLA, N., NIELSEN, S., MIZOGUCHI, K., FERRI, F., COCCO, M., SHIMAMOTO, T. (2011) Fault lubrication during earthquakes. Nature, 471, 494–498.

DIETERICH, J.H. and KILGORE, B.D., (1996), *Imaging surface contacts; power law contact distributions and contact stresses in quarts, calcite, glass, and acrylic plastic*, Tectonophysics, *256*, 219–239.

DIETERICH, J.H., (1972) *Time-dependent friction in rocks*. J. Geophys. Res., *77*, 3690–3697.

DIETERICH, J.H., (1979) *Modeling of rock friction 1. Experimental results and constitutive equations*, J. Geophys. Res., *84*, 2161–2168.

ENGELDER, J. T. (1974), *Cataclasis and the generation of fault gouge*. Geol. Soc. Am. Bull., *85*, 1515–1522.

ERINGEN, A.C. (1966) *A unified theory of thermomechanical materials*, Int. J. Energ. Sci., *4*, 179–202.

FILLOT, N., IORDANOFF, I., and BERTHIER, Y. (2007) *Wear modeling and the third body concept*, Wear, *262*, 949–957.

FISHER, R. (1937), *The wave of advance of advantageous genes*, Ann. Eugenics, *7*, 355–369.

GODET, M. (1984) *The third-body approach: A mechanical view of wear*. Wear, *100*, 437–452.

GRINDROD, P. (1996) The theory and applications of reaction-diffusion equations: Patterns and waves, Oxford Applied Mathematics and Computing Science Series, The Clarendon Press, Oxford Univ. Press, New York, 2nd ed., 275 p.

HAMIEL, Y., LIU, Y., LYAKHOVSKY, V., BEN-ZION. Y. and LOCKNER, D., (2004), *A visco-elastic damage model with applications to stable and unstable fracturing*. J. Geophys. Int. *159*, 1155–1165.

HAMIEL, Y., KATZ, O., LYAKHOVSKY, V. RECHES, R. and FIALKO, Y. (2006) *Stable and unstable damage evolution in rocks with implications to fracturing of granite*. Geophys. J. Int. *167*, 1005–1016, doi:10.1111/j.1365-246X.2006.03126.x.

HAN, R., HIROSE, T., SHIMAMOTO, T. (2010) *Strong velocity weakening and powder lubrication of simulated carbonate faults at seismic slip rates*. J. Geophys. Res., *115*, B03412, doi:10.1029/2008JB006136.

HANSEN, N.R. and SCHREYER, H.L. (1994) *A thermodynamically consistent framework for theories of elasticity coupled with damage*, Int. J. Solids Struct., *31*, 359–389.

HEESAKKERS, V., MURPHY, S., and RECHES, Z. (2011a) *Earthquake rupture at focal depth, Part I: Structure and rupture of the Pretorius fault, TauTona mine, South Africa*, Pure Appl. Geophys., doi:10.1007/s00024-011-0354-7.

HEESAKKERS V., MURPHY. S., LOCKER, D.A., and RECHES, Z. (2011b) *Earthquake rupture at focal depth, Part II: Mechanics of the 2004 M2.2 earthquake along the Pretorius fault, TauTona mine, South Africa*, Pure Appl. Geophys., doi:10.1007/s00024-011-0355-6.

HIROSE, T., MIZOGUCHI, K., SHIMAMOTO, T. (2012) *Wear processes in rocks at slow to high slip rates*. J. Struc. Geol., doi:10.1016/j.jsg.2011.12.007.

HOFF, N.J. (1953) *The necking and rupture of rods subjected to constant tensile loads*, J. Apl. Mech., *20*, 105–108.

JOHNSON, P.A. and JIA, X., (2005), *Non-linear dynamics, granular media and dynamic earthquake triggering*, Nature, *437*, 871–874.

KACHANOV, L.M. (1958) *On the time to rupture under creep conditions*, Izv. Acad. Nauk SSSR, *OTN 8*, 26-31 (in Russian).

KACHANOV, L.M. (1986), Introduction to Continuum Damage Mechanics, Martinus Nijhoff Publishers, 135 p.

KACHANOV, M. (1994) *On the concept of damage in creep and in the brittle-elastic range*, Int. J. Damage Mech., *3*, 329–337.

KARRECH, A. REGENAUER-LIEB, K., and POULET, T. (2011) *Continuum damage mechanics for the lithosphere*. J. Geophys. Res., *116*, B04205.

KATO, K., ADACHI, K. (2000) Wear mechanisms, In: Bhushan, B. (ed.), Modern Tribology Handbook. CRC Press, Boca Raton, Florida, 273–300.

KATZ, O., RECHES, Z. E. and BAER, G. (2003) *Faults and their associated host rock deformation: Part I. Structure of small faults in a quartz–syenite body, southern Israel*, J. Struct. Geol., *25*, 1675–1689.

KOLMOGOROV, A.N., PETROVSKII, I.G., and PISCOUNOV, N.S. (1937) *A study of the diffusion equation with increase in the amount of substance, and its application to a biological problem*, Bull. Moscow Univ., Math. Mech. *1*, 1–25. Translated by V.M. Volosov in V.M. Tikhomirov, editor, Selected Works of A. N.

KRAJCINOVIC, D. (1996) Damage Mechanics, Amsterdam, Elsevier. 774 p.

KRONER, E., (1968) *Elasticity theory of materials with long-range cohesive force*. Int. J. Solids Struct., *3*, 731–742.

LEMAITRE, J. (1996) A Course on Damage Mechanics, Springer-Verlag, Berlin.

LEVY, A. V., and JEE, N. (1988) *Unlubricated sliding wear of ceramic materials*, Wear, *121*, 363–380.

LIFSHITZ, E.M. and PITAEVSKII, L.P. (1981) Physical Kinetics, Course of Theoretical Physics, L.D. Landau and E.M. Lifshitz, Vol. 10, Elsevier, 452 p.

LOCKNER, D.A., MORROW, C., MOORE, D. and HICKMAN, S. (2011) *Low strength of deep San Andreas fault gouge from SAFOD core*, Nature, 472, p. doi:10.1038/nature09927.

LOCKNER, D.A., RECHER, Z., MOORE, D.E. (1992) Microcrack interaction leading to shear fracture. In Proc. 33rd U.S. Symposium on Rock Mechanics (ed. W. Wawersik), A.A. Balkema Rotterdam.

LU, K., BRODSKY, E.E., and KAVEHPOUR, H.P. (2007) *Shear-weakening of the transitional regime for granular flow*, J. Fluid Mech., *587*, 347–372.

LYAKHOVSKY, V., BEN-ZION, Y. and AGNON, A., (2005) *A visco-elastic damage rheology and rate- and state-dependent friction*. Geophys. J. Int. *161*, 179–190.

LYAKHOVSKY, V., and BEN-ZION Y., (2014a), *Damage-Breakage rheology model and solid-granular transition near brittle instability*, J. Mech. Phys. Solids, *64*, 184–197.

LYAKHOVSKY V. and BEN-ZION Y. (2014b) *Continuum damage-breakage model for faulting accounting for solid-granular transition*, Pure Appl. Geophys. (this volume).

LYAKHOVSKY V., HAMIEL, Y. and BEN-ZION, Y. (2011) *A non-local visco-elastic damage model and dynamic fracturing*. J. Mech. Phys. Solids, *59*, 1752–1776. doi:10.1016/j.jmps.2011.05.016.

LYAKHOVSKY, V., BEN-ZION, Y., and AGNON, A. (1997) *Distributed damage, faulting, and friction*, J. Geophys. Res., *102*, 27635–27649.

MA, S. K. (2000), Modern theory of critical phenomena, Westview press, 561 p.

MARONE, C., (1998), *Laboratory-derived friction laws and their application to seismic faulting*, Annu. Rev. Earth Planet. Sci., *26*, 643–649.

MUHURI, S.K., DEWERS, T.A., SCOTT, T.E., and RECHES, Z. (2003) *Interseismic fault strengthening and earthquake-slip stability: Friction or cohesion?* Geology, *31*(10), 881–884.

MURRAY, J. D. (2002) Mathematical Biology: I. An Introduction, 3rd ed., Springer, 551 p.

NEWMAN W.I. and PHOENIX, S.L. (2001) *Time-dependent fiber bundles with local load sharing*, Phys. Rev. E., *63*(2), 021507, doi:10.1103/PhysRevE.63.021507.

NIELSEN, S., DI TORO, G. and GRIFFITH, W.A. (2010), *Friction and roughness of a melting rock surface.* Geophys. J. Int. *182*, 299–310.

NOLEN, J., ROQUEJOFFRE, J.M., RYZHIK, L. and ZLATOŠ, A. (2012) *Existence and non-existence of Fisher-KPP transition fronts*, Arch. Rational Mech. Anal., *203*, 217–246, doi:10.1007/s00205-011-0449-4.

ONSAGER, L., (1931), *Reciprocal relations in irreversible processes.* Phys. Rev., *37*, 405–416.

PATERSON, M.S. and WONG, T.-F. (2005) Experimental Rock Deformation—The Brittle Field. Berlin, Heidelberg, New York: Springer-Verlag, 348 pp.

PETIT, J.P. (1987) *Criteria for the sense of movement on fault surfaces in brittle rocks*, J. Struct. Geol., *9*, 597–608.

POWER, W.L., TULLIS, T.E. and WEEKS, J.D. (1988) *Roughness and wear during brittle faulting*, J. Geophys. Res. *93*, 15268–15278.

QUEENER, C.A., SMITH, T.C., and MITCHELL, W.L. (1965) *Transient wear of machine parts*, Wear, *8*, 391–400.

RABINOWICZ, E., DUNN, L.A., and RUSSELL, P.G. (1961) *A study of abrasive wear under three-body conditions*, Wear, *4*, 345–355.

RABOTNOV, Y.N. (1959) A mechanism of a long time failure, in: Creep problems in structural members, 5–7, USSR Academy of Sci. Publ., Moscow.

RABOTNOV, Y.N. (1988) *Mechanics of Deformable Solids, Moscow,* Science, 712 p.

RECHES, Z. and LOCKNER, D.A., (1994) *Nucleation and growth of faults in brittle rocks.* J. Geophys. Res., *99*, 18159–18173, doi:10.1029/94JB00115.

RECHES, Z., and LOCKNER, D.A. (2010) *Fault weakening and earthquake instability by powder lubrication*, Nature, *467*, 452–456, doi:10.1038/nature09348.

RICARD, Y. and BERCOVICI, D. (2009) *A continuum theory of grain-size evolution and damage*, J. Geophys. Res., *114*. doi:10.1029/2007JB005491.

ROBINSON, E.L. (1952) *Effect of temperature variation on the long-term rupture strength of steels*, Trans. Am. Soc. Mech. Eng., *174*, 777–781.

SAGY, A. and BRODSKY, E.E. (2009) *Geometric and rheological asperities in an exposed fault zone*, J. Geophys. Res. *114*, B02301. doi:10.1029/2008JB005701.

SAMMIS, C., LOCKNER, D., RECHES, Z., 2011. *The role of adsorbed water on the friction of a layer of submicron particles.* Pure Appl. Geophys. *168*, 2325–2334.

SCHOLZ, C.H. (1987) *Wear and gouge formation in brittle faulting*, Geology, *15*, 493–495.

SCHOLZ, C.H., (2002) The Mechanics of Earthquakes and Faulting, 2nd ed. Cambridge University Press, 471 p.

SHCHERBAKOV, R. and TURCOTTE, D.L. (2003) *Damage and self-similarity in fracture*, Theor. Appl. Fracture Mech., *39*, 245–258.

SHIPTON, Z.K., EVANS, J.P., ABERCROMBIE, R.E., and BRODSKY, E.E. (2006) The missing sinks: slip localization in faults, damage zones, and the seismic energy budget, In: Abercrombie, R. (eds.) Earthquakes: Radiated Energy and the Physics of Faulting, 217–222. Washington, DC.

SIBSON, R.H. (1977) *Fault rocks and fault mechanisms.* J. Geol. Soc. *133*, 191–213.

TURCOTTE, D.L., NEWMAN, W.I., and SHCHERBAKOV, R. (2003) *Micro and macroscopic models of rock fracture*, Geophys. J. Int., *152*, 718–728.

WANG, W. and SCHOLZ, C.H. (1994) *Wear processes during frictional sliding of rock: a theoretical and experimental study.* J. Geophys. Res., *99*, 6789–6799.

WIBBERLEY, C.A., YIELDING, G., and DI TORO, G. (2008) *Recent advances in the understanding of fault zone internal structure: A review.* Geol. Soc., London, Special Publ., *299*, 5–33.

WILSON, B.T., DEWERS, T., RECHES, T. and BRUNE, J.N. (2005) *Particle size and energetics of gouge from earthquake rupture zones.* Nature, *434*, 749–752.

ZANG, A., WAGNER, F., STANCHITS, S., JANSSEN, C. and DRESEN, G., 2000. *Fracture process zone in granite*, J. Geophys. Res., *105*, 23651–23661.

(Received August 28, 2013, revised January 29, 2014, accepted January 30, 2014, Published online February 26, 2014)

Pure Appl. Geophys. 171 (2014), 3159–3174
© 2014 Springer Basel
DOI 10.1007/s00024-014-0835-6

| Pure and Applied Geophysics

The Fluid Dynamics of Solid Mechanical Shear Zones

E. Veveakis[1] and K. Regenauer-Lieb[2]

Abstract—Shear zones in outcrops and core drillings on active faults commonly reveal two scales of localization, with centimeter to tens of meters thick deformation zones embedding much narrower zones of mm-scale to cm-scale. The narrow zones are often attributed to some form of fast instability such as earthquakes or slow slip events. Surprisingly, the double localisation phenomenon seem to be independent of the mode of failure, as it is observed in brittle cataclastic fault zones as well as ductile mylonitic shear zones. In both, a very thin layer of chemically altered, ultra fine grained ultracataclasite or ultramylonite is noted. We present an extension to the classical solid mechanical theory where both length scales emerge as part of the same evolutionary process of shearing the host rock. We highlight the important role of any type of solid-fluid phase transitions that govern the second degree localisation process in the core of the shear zone. In both brittle and ductile shear zones, chemistry stops the localisation process caused by a multiphysics feedback loop leading to an unstable slip. The microstructural evolutionary processes govern the time-scale of the transition between slow background shear and fast, intermittent instabilities in the fault zone core. The fast cataclastic fragmentation processes are limiting the rates of forming the ultracataclasites in the brittle domain, while the slow dynamic recrystallisation prolongs the transition to ultramylonites into a slow slip instability in the ductile realm.

Key words: Cataclastic fault zone, mylonitic shear zone, chemical reactions, state variables.

1. Introduction

The geologist in the field is often confronted with two scales of localisation when investigating shear or fault zones (Chester and Chester 1998; Ben-Zion

---
[1] CSIRO Earth Science and Resource Engineering and School of Mathematics and Statistics, University of Western Australia, ARRC, 26 Dick Perry Avenue, Kensington, WA 6151, Australia. E-mail: manolis.veveakis@csiro.au
[2] School of Earth and Environment, University of Western Australia and CSIRO Earth Science and Resource Engineering, ARRC, 26 Dick Perry Avenue, Kensington, WA 6151, Australia. E-mail: klaus.regenauer-lieb@csiro.au

and Sammis 2003). Brittle fault zones show striking examples of extremely localized slip events, occurring within a thin shear zone, <1–5 mm thick, called the principal slipping zone (PSZ) (Sibson 2003). This localized PSZ lies within a finely granulated fault zone of typically tens to hundreds millimeter thickness. These fault zones are either cataclastic in the brittle regime or mylonitic in the ductile regime.

Exhumed field examples for the morphology of these brittle fault zones can be found in the North Branch San Gabriel fault (Chester et al. 1993), the Punchbowl fault of the San Andreas system in southern California (Chester and Chester 1998), the Median Tectonic Line fault in Japan (Wibberley and Shimamoto 2003), and in the Hanaore fault in southwest Japan (Noda and Shimamoto 2005). Active faults have been intersected in wells in, e.g., the Aigion system, central Greece, where the fault zone of clay size particles consists of finely crushed radiolarites, extended to about 1 m. This localized zone of deformation was found to be intercepted by a "fresh" distinct slip surface of sub-millimiter size (Sulem 2007; Cornet et al. 2004). Other wells that have been specifically designed to intersect active faults are the completed San Andreas Fault project (Holdsworth et al. 2011) and the ongoing Alpine Fault drilling project (Townend 2009). The San Andreas Fault project revealed a fault zone of 1–2.5 m width at 3 km depth, with several ultralocalized PSZ's.

Similarly, in the ductile field the dynamic recrystallization of the matrix minerals is telltale for crystal plastic or diffusion creep, forming shear zones of vast thickness interrupted by ultralocalized anastomosing patterns of extreme grain size reduction in the ultramylonite. Exhumed thrusts in creeping carbonates around the world like the Naukluft thrust in Namibia (Rowe et al. 2012), the McConnell thrust in Alberta Canada (Kennedy and Logan 1997) or the

Glarus Thrust in Switzerland (HERWEGH *et al.* 2008), all present a common structure of a meter-wide (1–5 m) zone which accomodates several thin (mm–cm at most) veins of ultralocalized deformation and chemical alteration.

The brittle field features micro-mechanisms, governed by grain breakage, rolling, cleavage and brittle fragmentation processes. The ductile field features dislocation, diffusion and dissolution mechanisms. Although these mechanisms are completely different, they surprisingly exhibit he same dual pattern of localisation: a broad meter wide shear zone with ultra localised PSZ's in its middle.

Each of the different phenomena can be explained independently but to date no comprehensive theory has been suggested that can investigate the reason for the double degree localisation process and the potential commonalities between brittle and ductile localisation. This is because the dynamic process of the extreme localisation and the quasi static response of the background flow of the shear zone are treated as two different fundamental theories. The former instability is described in solid mechanics and the latter flow in fluid dynamics.

While solid mechanics is capable of accurately describing the conditions for failure it does not emphasise the long term post-failure behaviour. This is due to the nature of the solid materials that maintain a high degree of structural integrity after failure and are not significantly affected by the elapsed time. The consequence of this is that a quasi static approach is preferred, thus, making it possible to cast the weak rate effects into approximate laboratory determined hardening laws.

Fluid Dynamics on the opposite does not have the concept of failure, it knows nothing about the processes that happen prior to failure and is totally concerned with the description of the rate of flow of the material under an applied stress. Since most mantle convection codes are based on the fluid dynamical approach, the geophysical community has over the recent years expended a lot of effort into incorporating the theory of Solid Mechanics.

The fact that both theories apply to geology has been universally acknowledged and the after dinner contribution of Reiner on the "Deborah Number" (REINER 1964) is well worth reading. The number is

defined as the ratio of the stress relaxation time over the time of observation. Reiner's quote "the mountains flowed before the lord" implies that given enough time one may smoothly transition from one to the other theory. This can be easily conceptionalized in a Newtonian (linear viscous) viscoelastic framework, however, the inclusion of non-linear temperature dependence of the creep processes gave rise to the Time-Temperature Superposition (TTS) in polymer sciences (TOBOLSKY and ANDREWS 1945). The TTS principle implies that the temperature dependent behaviour of materials can be substituted by appropriate experiments at different time-scales for a reference temperature. It is thereby assumed that the time evolution described in the temperature equation is the only time-scale relevant for the long term mechanical behaviour of solids. This simple concept allows us to directly extend it for the non-linear viscous behaviour of solids within a single theoretical framework incorporating elasticity, plastic failure and viscous post failure evolution.

In the present work, we summarise a theory that couples the solid and fluid-like behaviour. This theory must be able to provide the framework for modeling both the formation of faults (refer, for more in depth reading, to REGENAUER-LIEB *et al.* 2013) and their post-failure evolution. We show that the formation and post failure evolution of faults depend strongly on the coupled multiphysical effects affecting pressure, temperature and chemical conditions. We, thus, aim at providing in this paper a comprehensive guide through the energetics of faults and explain how they apply to the different types of failure.

2. Time-Independent Formation of Shear Zones from Solid Mechanics

The theory describing the onset of localized failure from uniform deformation has been a direct extension of the classic Mohr's theory of the strength of materials, and is applied to geomaterials using the so-called Thomas–Hill–Mandel shear-band model (HILL 1962; MANDEL 1966), which was introduced in the early 1960s. A seminal paper was contributed by RUDNICKI and RICE 1975. More recently, the mathematical formulation of bifurcation and post-bifurcation phenomena and related instabilities were summarized by VARDOULAKIS and SULEM (1995) to form the basis of an improved continuum theory of failure of geomaterials.

This theory defines failure as a stationary elastic wave and does not consider time evolution as a degree of freedom. While velocity dependent solutions are sought, they are independent of their evolution in true time. This is known as quasi-static deformation. The fundamental stress and strain solutions, therefore, degenerate into a geometric problem where failure lines appear with a finite width d.

These concepts of shear banding as material bifurcation of an elasto-plastic skeleton have led to the identification of a material length scale defining the width of shear bands. The critical stress, as well as the orientation and the thickness of the localized shear failure planes (shear zones) are calculated through the eigenvalues of the elastoplastic stiffness modulus C_{ijkl}^{ep} of the material (RUDNICKI and RICE 1975) obeying a rate-independent rheology, $\dot{\sigma}_{ij}' = C_{ijkl}^{ep}\dot{\varepsilon}_{kl}$ (the prime denoting effective stress). Within the framework of the solid mechanical instabilities the shear (fault) zone thickness emerges as a solution depending on the microstructure (MUHLHAUS and VARDOULAKIS 1987; PAPANICOLOPULOS and VEVEAKIS 2011).

This rate-independent regime is typically used as an adequate description of brittle processes, thus placing the formation of the shear zones (onset of localization) near the maximum deviatoric stress (RUDNICKI and RICE 1975). For example, in linear elastic fracture mechanics brittle fracture occurs without thermal activation when a critical stress level is reached to split the bonds. At a critical energy threshold, an elastodynamic fast time scale instability ensues where a variety of dissipative processes kick in such as grain/particle rotations, which release heat in extremely fast timescales. The fast time scale instability simplifies the processes as temperature and fluids do not have time to diffuse. This regime is called undrained-adiabatic and is characterising an extremely fast co-seismic slip (RICE 2006; SULEM et al. 2011) of brittle failure events.

Within this brittle, solid mechanical framework, recent studies (SULEM et al. 2011; VEVEAKIS et al. 2012, 2013) derived different levels of localisation corresponding to different energy (temperature) regimes. This result is a direct generalisation of the extension of the visco-elastic Deborah number concept to different temperatures using the time temperature superposition principle (TOBOLSKY and ANDREWS 1945) for elasto-plastic media. This approach is generalising the concept of visco-elastic relaxation experiment at different temperatures to that of nonlinear visco-elasto-plastic solutions where quasi static solutions are sought for different temperatures.

In order to so, we first freeze time by adopting classical, rate-independent, elasto-plastic theory for a fault zone material incorporating chemical, hydraulic and thermal sensitivity. In the brittle field these fault zones are characterised by a damage zone surrounding the fault zone with an embedded PSZ as shown in Fig. 1a. In the brittle crust the width of this hierarchical damage zone can be up to 3–5 km wide (BEN-ZION and SAMMIS 2003; ALLAM and BEN-ZION 2012), containing in principle multiple localization zones (FAULKNER et al. 2003). In order to keep the mathematical model tractable, we restrict the present work to the study of a single fault zone.

We, therefore, calculate the extent (thickness) of the brittle fault zone at different temperature regimes, marked by: (1) the temperature inferred from geothermal consideration for the boundary of the fault zone, and (2) the activation temperature of the dominant chemical reaction observed in the PSZ. The details of the approach can be found in SULEM et al. 2011, VEVEAKIS et al. 2012, 2013 and the results are summarised in Fig. 1b, c.

2.1. Shear Zone Thickness at Boundary Temperature Conditions

The general solution for a quasi static shear zone considers the average of the background particle size d_{50}, the fluid pressure and the dominant chemical reaction at a given temperature. Both fluid pressure and reaction rate are strongly affected by the temperature. At ambient geothermal temperatures of the shear zone, the chemical reaction is inactive and

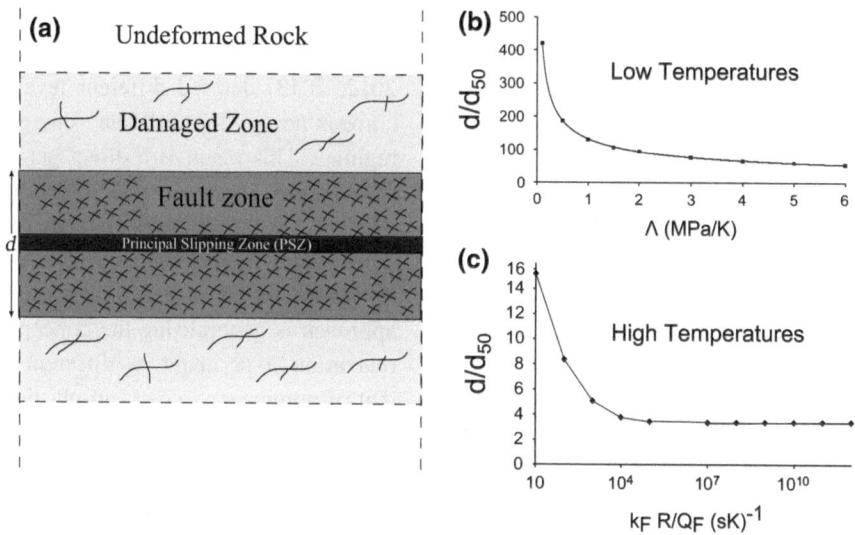

Figure 1

a Conceptual model of the internal structure of a fault described by two scales: the principal slip zone (PSZ) in the *centre* surrounded by the fault zone, **b** using the geothermal temperature as a thermal condition for the boundary of the fault zone we obtain the thickness of the fault zone as a function of its internal structure (average particle size) and the thermal pressurisation, **c** using the higher temperature required for chemical reactions as recorded in the PSZ a much smaller length scale down to three times the average grain/particle size is obtained

the thickness of the shear zone is predominantly controlled by the thermal-elastic pressurisation of the poro-elastic skeleton (Fig. 1b), characterized by the pressurization coefficient $\Lambda = \frac{\alpha_f - \alpha_s}{K_s + K_f}$ (MPa/K). In this expression, α is the thermal expansion coefficient and K the compressibility, with subscripts s, f denoting solid and fluid phases, respectively.

The solid mechanical thickness of the shear zone can be calculated directly as function of the average particle size d_{50}, and the thermal pressurisation Λ as well as a multiplier C representing the elasto-plastic material:

$$\frac{d}{d_{50}} = \sqrt{\frac{C}{\Lambda}}. \tag{1}$$

For a Drucker–Prager, Cosserat material it was shown by VEVEAKIS *et al.* (2013) that $C = 17.5 \times 10^3 \, \mathrm{K\,MPa^{-1}}$. For typical values of Λ between 10^{-4} (MPa/K) (CECINATO *et al.* 2011) and 1 (MPa/K) (RICE 2006), with d_{50} of the order of 1 mm, the shear zone width ranges between 0.1 and 13 m.

2.2. *Shear Zone Thickness at Elevated Temperature*

The effect of chemical reaction can be incorporated by considering the reaction rate to follow Arrhenius kinetics.

In this framework the reaction rate depends on the activation enthalpy Q_F and the reaction rate frequency k_F. The reaction is activated at elevated temperatures, near the activation temperature of the reaction at given pressure conditions. At these temperature conditions, and considering calcite decomposition as a typical reaction for a carbonate host rock (FAMIN *et al.* 2008; SULEM and FAMIN 2009), the thickness of the PSZ is several orders of magnitude smaller than the low temperature shear zone as shown in Fig. 1c.

This result is reconciling field observations from two different temperature regimes but has no predictive power on how these temperatures may have been achieved. The traditional concept of TTS is not extendable to obtain an extended solution for non-linear solids. We, therefore, need a non-linear superposition that allows fault zone evolution through time. The non-linear concept in this work is introduced by a Taylor expansion of the plastic strain rate, accounting for higher-order terms describing the non-linearities.

3. *Time-Dependent Evolution of Shear Zones*

To this end, we assume a smooth function of the effective stress, temperature T and additional internal variables ξ, expressing the constitutive behaviour of the material

$$\dot{\epsilon}^p_{ij} = f(\sigma'_{ij}, T, \xi) \,. \tag{2}$$

The temperature and ξ obey their own evolution laws, namely the energy balance for T and experimentally deduced evolution laws for ξ. By expanding Eq. (2) around the effective yield stress σ'_Y we obtain

$$\dot{\epsilon}^p_{ij} = f' \left(\frac{\bar{\sigma}_{ij}}{\sigma'_n} \right) + \sum_{m \geq 2} f^{(m)} \left(\frac{\bar{\sigma}_{ij}}{\sigma'_n} \right)^m , \tag{3}$$

where $\bar{\sigma}_{ij} = \sigma'_{ij} - \sigma_Y$, σ'_n a reference stress and $f^{(m)} = \frac{1}{m!} \left| \frac{d^m f(\sigma'_{ij}, T, \xi)}{d\sigma'^m_{ij}} \right|_{\sigma'_{ij} = \sigma'_Y}$.

The first order of the Taylor expansion corresponds to the time independent solid mechanical regime presented earlier and describes the formation of the fault near initial yield. Once the fault has been formed after initial yield, it continues to deform plastically (irreversible deformation) upon continuation of loading. This restricts our modelling efforts to the behaviour of the material inside the solid mechanical shear zone obtained from the first order expansion. Therefore, by considering explicitly the rate sensitivity emerging from the higher order terms of the expansion, we study the non-linear (fluid-like) evolution of the mechanisms inside the solid mechanical fault zone solution.

The higher order terms of Eq. (3) give a natural extension of the classical TTS in the nonlinear viscoplastic space. If the solid mechanical yield is considered as a large scale reference state of our model we can describe the higher order terms in the nonlinear fluid dynamic space by the so-called overstress $\bar{\sigma}_{ij} = \sigma'_{ij} - \sigma'_Y$, defined as the stress that follows the evolution post the initial yield point (PERZYNA 1966), hence, $\bar{\sigma}_{ij} > 0$ is always satisfied. The rate of plastic strain is, therefore, expressed as

$$\dot{\epsilon}^p_{ij} = \sum_m f^{(m)} \left(\frac{\bar{\sigma}'_{ij}}{\sigma'_n} \right)^m \quad \text{when } \bar{\sigma}_{ij} > 0 \,, \tag{4}$$

where $f^{(m)}$ has now dimensions of $(s)^{-1}$, being thus a reference strain rate.

In this sense the nonlinear TTS expresses a straight forward extension of the linear TTS with the difference that the time scale now plays an explicit role. In the classical TTS the time scale is simply

derived from relaxation experiments returning the system near to its equilibrium whereas in the nonlinear TTS the system is always driven far from equilibrium. In the nonlinear framework the relaxation experiments, therefore, must be replaced by far from equilibrium energy considerations, which define the time scales over which the rate of processes exchange energy inside the fault zone (REGENAUER-LIEB et al. 2009).

3.1. Energy Considerations

We introduce the Helmholtz free energy ψ, being a function of the elastic strain, temperature and the internal variables ξ. The energy balance equation, together with the second law of thermodynamics and Fourier's law of diffusion provide the local form of the entropy production equation (ROSAKIS et al. 2000; REGENAUER-LIEB et al. 2009):

$$\frac{\partial T}{\partial t} + v^m_k \frac{\partial T}{\partial x_k} = c_{\text{th}} \frac{\partial^2 T}{\partial x_k^2} + \frac{q_e}{(\rho C)} + \frac{\Phi_m}{(\rho C)} + \frac{L}{(\rho C)} \,, \tag{5}$$

where v^m_k is the barycentric velocity of the system, $c_{\text{th}} = \frac{k_{\text{th}}}{(\rho C)}$ is the thermal diffusion coefficient of the mixture (usually depending weakly on temperature and strongly on the microstructure, here considered constant), $C = -T \frac{\partial^2 \psi}{\partial T^2}$ is the specific heat capacity of the mixture under constant volume, and $q_e = \rho_m T \frac{\partial^2 \psi}{\partial T \partial \epsilon^e_{ij}} \dot{\epsilon}^e_{ij}$ is the thermoelastic rate of heating. Note that in the framework of solid mechanics the advective terms are neglected, thus, $v^m_k = 0$.

The term $L = \rho_m T \frac{\partial^2 \psi}{\partial T \partial \xi} \dot{\xi} = \Delta h \, r$ represents the latent energy produced or absorbed during a higher order energy transition (Δh is the enthalpy of the energy transition and r its rate). This includes any microstructural changes such as grain size reduction and damage in the case that the state variable is a grain size or a damage parameter (LYAKHOVSKY and BEN-ZION 2014a, b), respectively. In the case where the state variable refers to the chemical constituents, it corresponds to the well known latent heat of a phase change. Finally, Φ_m is the local dissipation, i.e., the mechanical power that is converted into heat in the system. In generalised thermodynamics the dissipation is defined by the product of a thermodynamic force times a thermodynamic flux. In mechanics the thermodynamics force is the stress σ'_{ij} and the thermodynamic flux is the plastic strain rate $\dot{\epsilon}_{ij}^p$. In the generalised framework, when additional state variables/processes are considered, the force becomes a generalised force such as a damage force E_ξ and the thermodynamic flux is the rate of change of the state variable $\dot{\xi}$:

$$\Phi_m = \sigma_{ij}' \dot{\epsilon}_{ij}^p - E_\xi \dot{\xi} = \chi \sigma_{ij}' \dot{\epsilon}_{ij}^p. \tag{6}$$

with $E_\xi = \rho_m \frac{\partial \psi}{\partial \xi}$ the energy dual of the internal variable ξ. The Taylor–Quinney ratio χ expresses the amount of the mechanical energy converted into heat (Taylor and Quinney 1934) and is in principle a history dependent quantity rather than a constant $(0 < \chi < 1)$. Its importance is discussed in the following section.

3.2. The Taylor–Quinney Coefficient

The Taylor–Quinney coefficient is a function of the elastic strain, the temperature and the internal variables:

$$\chi = \chi(\epsilon_{ij}^e, \xi, T) = 1 - \frac{E_\xi \dot{\xi}}{\sigma_{ij}' \dot{\epsilon}_{ij}^p}. \tag{7}$$

The Taylor–Quinney coefficient is unity if all deformation work is converted into heat. For values smaller than unity and larger than zero, it describes the portion of deformation work that is stored in microstructure. The coefficient can be derived through thermographic deformation experiments (Rosakis et al. 2000). The Taylor–Quinney coefficient as derived from thermography often starts at initial yield being close to zero and progressively evolves in the transient deformation regime towards a steady state value of 0.9 for most materials (Chrysochoos and Belmahjoub 1992). Its evolution between these two values is, therefore, prescribed by the development of the internal state variables. As such, it has a fundamental meaning for discriminating deformation mechanisms.

If the dissipation process is for instance a rapid fracture, the evolution time of the Taylor–Quinney is extremely fast and obtained from the bond energy between the covalent or electronic bonds of the atoms. This equates macroscopically to an elastic interaction potential. In linear elastic fracture

mechanics, brittle fracture occurs at explosive time-scales (of the order of seconds). The Taylor–Quinney coefficient prior to fracture is zero as all mechanical work is stored elastic deformation. At the critical threshold, an elastodynamic fast time scale instability ensues where a variety of dissipative processes kick in such as grain/particle rotations which release heat therefore increasing the Taylor Quinney coefficient. This is the typical mechanism for brittle fracture.

Similarly, in ductile deformation processes, the Taylor–Quinney coefficient is zero at initial yield, however, in contrast to brittle deformation the failure is thermally activated. A series of micro-mechanical processes such as dislocation and diffusion mechanisms kick in that feature a time scale that is relatively slow compared to the elastodynamic time scale. The thermal activation is often described by an Arrhenius temperature activation. This rate-dependent failure mechanism is known as ductile (or creep) fracture:

$$\Phi_m = \Phi_0 \, \chi(\xi) \left(\frac{\bar{\sigma}_{ij}}{\sigma_n'} \right)^{m+1} e^{-\frac{T_0}{T}}, \tag{8}$$

where $\Phi_0 = \sigma_n' \dot{\epsilon}_0$ is the reference dissipation, σ_n' is a reference stress value and $T_0 = Q_d/R$ the thermal sensitivity (activation temperature) of dissipation.

3.3. Chemical Reactions

From the analysis of the brittle regime (Sect. 2), we obtained that chemical reactions have a critical role in the system, defining the ultimate width of localization (the PSZ thickness). In this work we will emphasize fluid release reactions, due to their fundamental importance in earth systems. Fluid-release reactions occur when either a hydrous mineral such as a clay mineral, serpentinite, mica, gypsum, etc., loses its water at a critical activation enthalpy or when a mineral that is made of fluid phase and a solid constituent breaks down such as carbonate breaking to lime plus carbon dioxide upon critical activation enthalpy (Famin et al. 2008; Sulem and Famin 2009).

We treat these shear zone minerals as generalized solids characterised by the bonded chemical species A and B forming the solid composite AB. We assume B to represent fluid species filling the porous matrix of AB at initial conditions. At high temperatures the solid AB breaks down, producing excess B fluid, and increasing the fluid pore pressure through a general fluid-release reaction of the form $v_1AB_{(s)} \rightleftharpoons v_2A_{(s)} + v_3B_{(f)}$.

The kinetics of this reaction are assumed to obey the standard Arrhenius dependency on temperature (see Appendix). Following these considerations, the rates of the forward (r_F) and reverse (r_R) first order reactions (and for $v_1 = v_2 = v_3 = 1$) can be calculated to be

$$r_F = \frac{\rho_{AB}}{M_{AB}}(1 - \phi)(1 - s)k_F e^{-Q_F/RT}$$

$$r_R = \frac{\rho_A \rho_B}{M_A M_B}(1 - \phi)s\Delta\phi_{\text{chem}}k_R e^{-Q_R/RT} \qquad (9)$$

The total reaction rate is, subsequently,

$$r = \left[(1 - s) - s\Delta\phi_{\text{chem}}\frac{\rho_A\rho_B}{\rho_{AB}^2}\frac{M_{AB}^2}{M_A M_B}K_c^{-1}e^{\Delta h/RT}\right]$$
$$\times (1 - \phi)\rho_{AB}k_F e^{-Q_F/RT} \qquad (10)$$

where $K_c = k_F/k_R$ and $\Delta h = Q_R - Q_F$. In these expressions k_F, k_R, Q_F, Q_R are the pre-exponential factors and activation enthalpies of the forward and reverse reaction, respectively, and M_i and ρ_i is the molar mass and density of the ith constituent. The porosity ϕ consists of an initial value ϕ_0 and the new interconnected pore volume created from the reaction $\Delta\phi_{\text{chem}}$, $\phi = \phi_0 + \Delta\phi_{\text{chem}}$. The partial solid ratio s is the volume ratio of the produced solid A in the solid matrix. The expressions for the dependency of ϕ and s on the reaction kinetics are given in ALEVIZOS et al. (2014) and are summarized here in Appendix. This formulation is essentially a damage mechanics formulation with porosity being the damage parameter being controlled by the physics of the chemical reactions.

4. Post Failure Evolution of a Shear Zone

We focus on the post failure evolution of a shear zone with thickness d, formed by material bifurcation (Fig. 2). Due to the small thickness of the shear zone compared to the thickness of the overburden, its momentum balance prescribes a constant stress profile across the shear zone, thus, $\sigma_{yx} = \tau_n(t)$ and

$\sigma_{yy} = \sigma_n(t)$ (RICE 2006; VEVEAKIS et al. 2010). In addition, in the presence of a fluid, the stress can be decomposed according to Terzaghi's principle (VARDOULAKIS and SULEM 1995) to $\sigma'_{ij} = \sigma_{ij} + p_f\delta_{ij}$, with δ_{ij} the Kronecker's delta, and $p_f = p_n + \Delta p$ the pore fluid pressure, consisting of a hydrostatic part p_n at the boundary and the excess pore pressure Δp. The final system of equations is obtained once the energy equation is coupled with the mass balance equation (VEVEAKIS et al. 2010, 2014; ALEVIZOS et al. 2014):

$$\frac{\partial \Delta p^\star}{\partial t^\star} = \frac{\partial}{\partial y^\star}\left[\frac{1}{Le}\frac{\partial \Delta p^\star}{\partial y^\star}\right] + \Lambda\frac{T_c}{\sigma'_n}\frac{\partial \theta}{\partial t^\star}$$
$$+ (1 - \phi)(1 - s)\mu_r e^{\frac{Ar\theta}{1+\theta}}$$

$$\frac{\partial \theta}{\partial t^\star} = \frac{\partial^2 \theta}{\partial y^{\star 2}} + \delta\Big[Gr(1 - \Delta p^\star)^m e^{\frac{aAr}{1+\theta}}$$
$$- \left((1 - \phi)(1 - s) + (1 - \phi)s\Delta\phi_{\text{chem}}\eta K_c^{-1}e^{\frac{xAr}{1+\theta}}\right)\Big]e^{\frac{Ar\theta}{1+\theta}}, \qquad (11)$$

where the fields were normalised with the help of a reference temperature T_c:

$$t^\star = \frac{c_{\text{th}}}{(d/2)^2}t, \ y^\star = \frac{y}{d/2}, \ \theta = \frac{T - T_c}{T_c}, \ \Delta p^\star = \frac{\Delta p}{\sigma'_n}. \qquad (12)$$

The kinetics of the reactions are normalised using the Arrhenius scaling

$$Ar = \frac{Q_F}{RT_c}, \ a = 1 - \frac{Q_d}{Q_F}, \ x = 1 - \frac{Q_R}{Q_F}. \qquad (13)$$

The remaining dimensionless quantities appearing in the system (11) are defined as

$$\delta = \frac{|\Delta h|(d/2)^2}{k_{\text{th}}T_c}k_F\rho_{AB}e^{-Ar}$$

$$\eta = \frac{\rho_A\rho_B}{\rho_{AB}^2}\frac{M_{AB}^2}{M_A M_B}$$

$$\mu_r = \left(\frac{\rho_{AB}}{\rho_B}\frac{M_B}{M_{AB}}\right)\frac{(d/2)^2}{c_{\text{th}}\sigma'_n}\frac{k_F}{(\beta_f + \beta_s)}e^{-Ar} \qquad (14)$$

$$Le = \frac{c_{\text{th}}\mu_f(\beta_f + \beta_s)}{k}$$

$$Gr = \frac{\chi(\xi),\Phi_0}{k_F|\Delta h|(\rho_{AB}/M_{AB})}$$

where $k_{\text{th}} = c_{\text{th}}(\rho C)$, the thermal conductivity of the system. The system of Eq. (11) is fully described when the eight dimensionless numbers (a, x, μ_r, Le, δ,

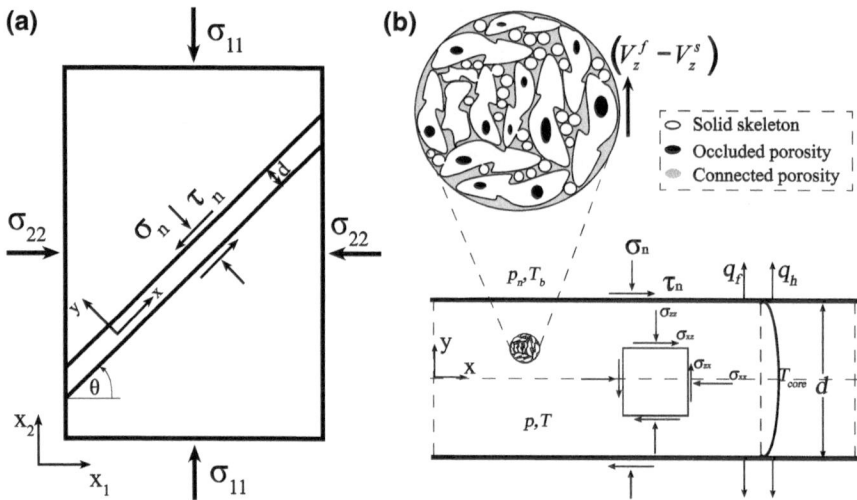

Figure 2

a Sketch of the continuum mechanical concept of shear failure as a material bifurcation (RUDNICKI and RICE 1975). At a critical stress ratio of the original stresses (in the coordinate system $x_1 - x_2$) a conjugate pair of shear zones emerge, having a thickness d and dipping at angles $\pm\theta$ thereby defining the rotated coordinate system of failure, $x - y$. For visual purposes only the shear zone dipping at θ is shown here. **b** Processes inside the shear zone of (**a**). The loading conditions and the filter velocity $v_z^f - v_z^s$, along the content of a saturated rock, are depicted. We assume that any chemical reaction is taking place at the solid-pore interface. Hence, all the produced fluid contributes to the interconnected pore volume and is concentrated on the grain boundaries. We also assume that the produced solid is added to the skeleton, establishing a common velocity field v_z^s with the reactant solid

Gr, K_c, and the boundary temperature θ_b since Ar is implicitly considered in these groups) are determined. However, the system has a much lower dimensionality, which can be revealed through a comprehensive analysis of its steady state and transient responses. These analyses has been presented in ALEVIZOS et al. (2014), VEVEAKIS et al. (2014), identifying the dominant parameters of the system. Of critical importance is the Gruntfest number Gr, as it incorporates the energetics of the microstructure through the Taylor–Quinney coefficient, as well as the mechanical loading and the characteristics of the chemical reaction. High values of Gr correspond to a regime where the mechanical input is sufficient to trigger the chemical reaction. Due to its fundamental nature, Gr will be treated as a bifurcation parameter.

The post-failure evolution of the shear zone is summarized in Fig. 3. The first panel (a) illustrates the fundamental three phase stability S-curve typical to all systems of the generalized type discussed above (Eq. 11) (YUEN and SCHUBERT 1977). The first phase is illustrated in details in panel (b) showing a steady state creep solution acting as a global stable material response without instabilities. The panel (c) is the

solution for the phase of elevated Gruntfest numbers but below the critical point B of Fig. 3a. For any initial conditions below line BC of (a) the system relaxes back to the stable branch AB. For any initial condition above line BC of (a) the system features a solitary oscillation. Panel (d) of Fig. 3 illustrates the behaviour in domain III where a stable oscillator emerges. This stable oscillator is self-sustained and has a fundamental role in the post-failure evolution of shear zones. This area, where $Gr > Gr_B$, is the domain where localisation of dissipation occurs owing to critical energy transitions. This means that the latent energy term L of Eq. 5 is triggered and energy is released. This domain does not depend on the exact micro mechanism, as shown in VEVEAKIS et al. (2010), but describes a generalised material instability where the changes of the state variable form the localisation phenomenon. The state variable can be for instance a grain size, a damage or porosity change as discussed in this study.

ALEVIZOS et al. (2014) provided asymptotic criteria for the area of oscillations to be admitted. This occurs when the following inequalities simultaneously hold:

$$Gr > \left(1 + \frac{e^{-Ar\theta_b}}{Ar\delta}\right)e^{-aAr}, \ \mu_r Le < Ar^{-3/2}, \ \log(K_c) \gg 1$$

$$(15)$$

These inequalities correlate most of the dimensionless groups of the system, implying that in environments where strong (irreversible) endothermic reactions are favoured (hence, $\log(K_c) \gg 1$ is always satisfied) any two of the remaining groups can always be selected as control over oscillatory instabilities.

4.1. Analysis of the System's Response

The solution discussed in Fig. 3 is obtained under the assumption of a constant loading stress and chemistry as well as internal microstructure, thus, giving a constant Gruntfest number. In the more general case either of these parameters can evolve in time, thus, allowing the system to cross all three areas of the phase diagram. This leads potentially to a multiplicity of chaotic responses marked by phases of self-organised chaos, transition from aseismic creep to seismic slip and back. For example, if the system follows the hysteretic loop ABB'CA of Fig. 3 we find all three different system responses in one and the same shear zone.

Of particular interest for plate tectonic loading is the particular attractor depicted in Fig. d where a constant plate velocity would lead to a constant shear stress over time, allowing for a steady state evolution of the microstructure and chemistry. The periodic instability that emerges in this regime corresponds to a stick-slip type of instability, as illustrated in Fig. 4 where the limit cycle is plotted in the logarithmic strain rate versus normalised time space. The limit cycle is characterised by two distinctly different time scale phenomena, the first being a long time scale of slow creep interrupted by a short time scale fast response owing to fluid of the release reaction.

In addition to the two timescales characterizing the system's time evolution, its spatial manifestation also comprises two length-scales. Figure 5 shows profiles of strain rate (red line), porosity (blue line) and solid product (green line) inside the solid mechanical shear zone, at their maximum points in the limit cycle. Starting from a flat initial profile when the solid mechanical shear zone is established, the strain rate localizes in an ultrathin core zone during the fast timescale. The extreme localisation of strain rate in the centre of the solid mechanical shear zone is a robust outcome for any chemical reaction. The thickness of this ultralocalized shear band inside the global mechanical shear zone depends on the activation enthalpy of the forward reaction. For large enthalpy reactions the localised zone is ultra-sharp and broadens with reducing enthalpy.

We expect from this result two important outcomes. The first is that the shear zone will show at least two different scales of localisation, linked to the corresponding timescales. The broad solid mechanical zone acts as a vessel for the long creeping timescale, whereas the sharp fluid-dynamic length-scale accompanies the fast pressurization timescale. The second outcome is that the width of the fluid dynamic localisation can be used to identify the activation energy of the chemical reaction involved. The spatial extent of the fluid release reaction is shown in the blue porosity curve in Fig. 5, which broadens with reducing enthalpy of the reaction. The same trend applies to the solid constituent of the reaction. This additional information may be used in the field as an added constraint for identification of activation enthalpies.

4.2. Timescales of the System

We expand the temperature field with respect to $\epsilon = 1/Ar$ (FOWLER 1997). At leading order, the chemical reaction is inactive, thus, reducing the system to a single equation, known as the Frank–Kamenetskii limit:

$$\frac{\partial\theta}{\partial t^\star} = \frac{\partial^2\theta}{\partial y^{\star 2}} + \lambda e^\theta,$$

$$(16)$$

where $\lambda = \delta(Gre^{aAr} - 1)$. Equation (16) is correspondingly the Frank–Kamenetskii equation (FUJITA 1969), known to have the semi-analytic solution (VEVEAKIS et al. 2007):

$$\theta = \theta_{\text{core}} - ln\left[\lambda(t_I - t^\star) + \frac{y^{\star 2}}{4(c_1 - \ln(t_I - t^\star))}\right],$$

$$(17)$$

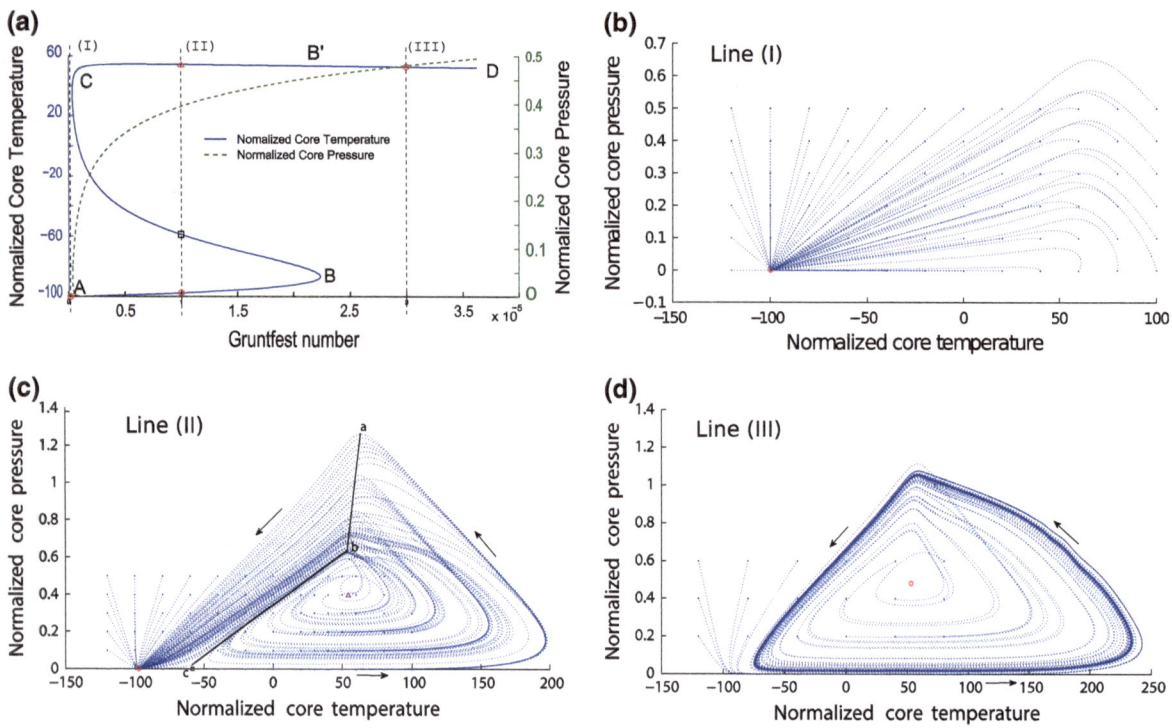

Figure 3

Summary of the system's response when Eq. (15) are satisfied. **a** Steady state response of the dimensionless core temperature (*solid line*) and the dimensionless core excess pore pressure (*dashed line*). The *three lines* annotated correspond to the three areas of interest, providing the phase diagrams of: **b** a stable, aseismic creep behavior (*line I*), where all the initial conditions in the $\Delta p - T$ phase plane end up at the stable node depicted as a *circle*, **c** non-periodic events (*line II*), where the *line a–b–c* separates the linear paths towards the lower stable node (*circle*) from the homoclinic orbits tending to the *circle* via the unstable spiral orbit of the upper solution (*triangle*), and **d** periodic instabilities (*line III*), appearing in the $\Delta p - T$ phase plane as a stable limit cycles around the unstable upper steady state solution (*circle*). The magnitude of the cycle increases with decreasing Gr, obtaining its maximum amplitude at point B, where a homoclinic bifurcation takes place, with the periodic orbit colliding with the saddle point B. Note that values of Δp larger than 1 indicate possible hydraulic fracturing in the vicinity of the fault's core, limiting the present model

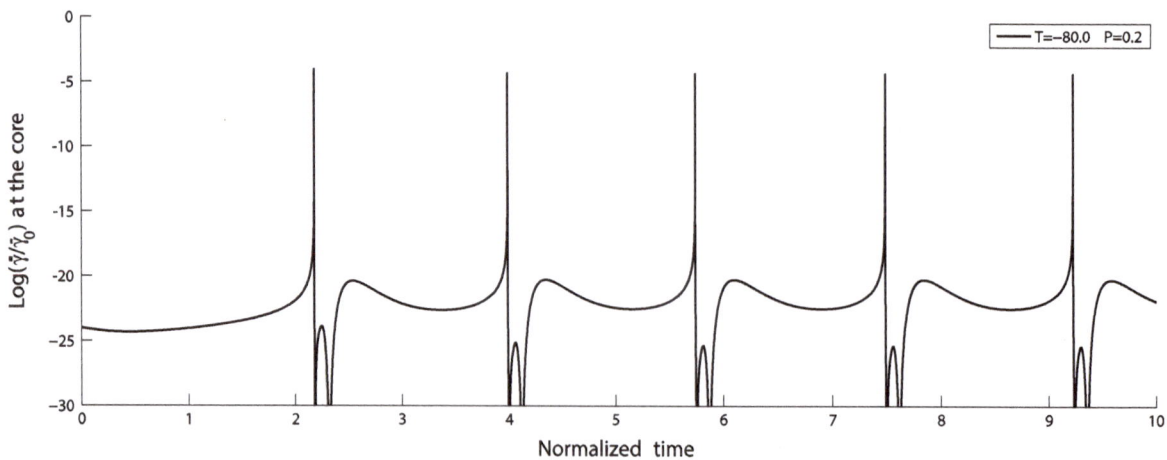

Figure 4

Evolution of the core strain rate on line III of Fig. 3a. Irrespective of the initial conditions, the system in this area undergoes cycles of abrupt acceleration followed by prolonged relaxation

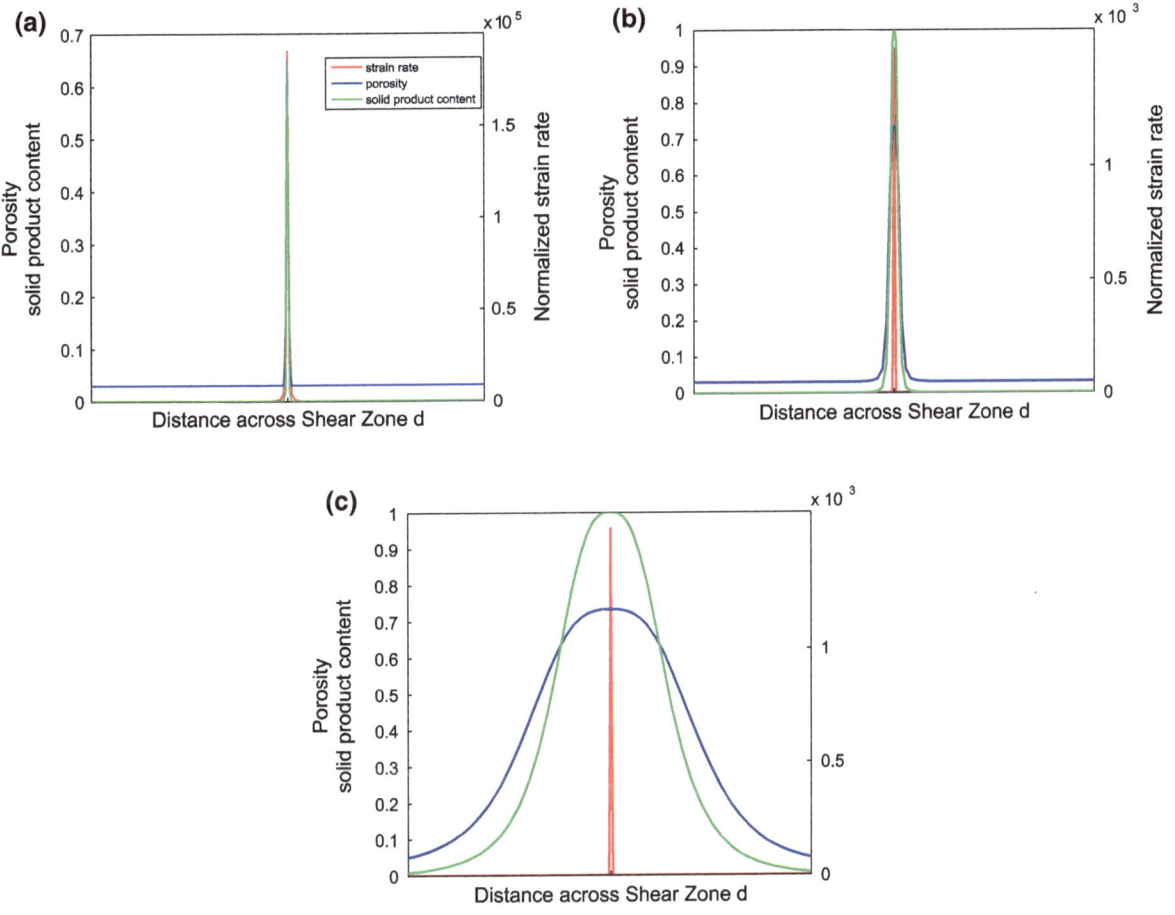

Figure 5
Influence of the characteristics of the reaction. Snapshots of the profiles of porosity ϕ, solid product content and strain rate near the time of their maximum values (the point of maximum temperature in the limit cycle of Fig. 3d). **a** For $\Delta h = Q_F$, **b** for $\Delta h = 3/4 Q_F$, **c** for $\Delta h = Q_F/2$. We notice that as Δh decreases, the reaction broadens its extent inside the shear zone d. For $\Delta h \approx Q_F$ porosity and s are weakly produced at the core of the shear zone, and the strain rate is smaller in magnitude (**a**). Once s and ϕ reach their maximum values (in this case $s = 1$ and $\phi = 0.75$) the reaction takes place in an increasingly broader zone as Q_R increases (**b**, **c**).

where θ_{core} is the initial temperature at the core (maximum), $t_I = 1/\lambda$ is the time that temperature presents a singularity (blow-up time), and the constant c_1 is determined by the boundary conditions (VEVEAKIS *et al.* 2007). We note that the initial condition should be the temperature at which the fault is initially formed. For flat, isothermal profile inside the shear zone, $\theta_{\text{core}} = \theta_b$.

As discussed in VEVEAKIS *et al.* (2007), past a critical strain rate achieved at $t \approx 0.88 t_I$, the analytical solution of Eq. (17) indicates that dissipation (strain rate) localizes towards the center of the shear band while abruptly increasing. When reaching the critical value of temperature to trigger chemical

pressurization, then the system enters the pressurization regime and excess pore pressure is being generated from the reaction in an undrained adiabatic setting. The time at which the temperature at the center of the fault zone ($y = 0$) reaches the activation temperature of the reaction, θ_{cr}, is approximately the time-scale of the frictional (stable creep under zero overpressures) process,

$$t_{\text{cr}}^{\star} = \frac{1}{\lambda} \left[1 - e^{(\theta_{\text{core}} - \theta_{\text{cr}})} \right]. \qquad (18)$$

Thus, the time at which chemical pressurization will set in is a function of the initial configuration of the system θ_{core}, of the pressurization temperature θ_{cr} and of all the material, chemical and loading parameters of the problem, incorporated into λ. Recalling that Gr (hence, λ) incorporates the Taylor–Quinney coefficient, $t_{cr}^{\star}$ is directly influenced by the evolution of the internal variables and ,thus, of the microstructure.

Note that when $\theta_{cr} = \theta_{core}$, then $t_{cr}^{\star} = 0$, meaning that the fault will enter the pressurization regime directly, without admitting any period of creep. On the other hand, when $\theta_{cr} \gg \theta_{core}$, then $t_{cr}^{\star} = 1/\lambda$ and the fault will admit all its creeping capability. Once the chemical reaction is triggered, it evolves in a fast timescale, estimated by the higher order of the expansion, where undrained-adiabatic conditions are established. In this regime, the timescale is inversely proportional to μ_r (VEVEAKIS *et al.* 2010)

$$t_p^{\star} \sim \frac{1}{\mu_r}. \tag{19}$$

5. Comparison to Field Observations

We have identified in the above theoretical considerations two fundamentally different processes (rate/temperature independent and rate/temperature dependent) with similar outcomes. Both processes lead to two scales of localisation. The large scale is associated with the solid mechanical solution and the small scale with the fluidised fault zone material inside the master shear zone.

In the rate/temperature independent solution the ultralocalised PSZ appears without a creeping phase. From Eq. (18) we conclude that the system is already in a critical state and does not require shear heating to be brought to criticality. In physical terms this could be seen as a transition that is equivalent to allowing phase changes, or grain size reduction, or damage to start at a negligible input of energy. This leads to the identification of the fast time-scale elastodynamic instability, where the energy released in the PSZ causes extreme localisation with an internal time scale governed by the energy change process. In our example, we postulated a fluid release reaction and obtained the pressurisation time scale of Eq. (19). Other mechanisms, such as damage or grain-size, impose different time scales for instability based on their energetics. This dual localisation is, therefore, the hallmark of brittle shear zones as illustrated in Fig. 6a.

In the case where the system is not close to criticality and is creeping, the equivalent ductile localisation mechanism can emerge. Upon a finite time after release of deformational work into heat inside the creeping shear zone, the system can reach the critical point for a fast energy transition. At this point the creeping zone forms a PSZ upon which the micro mechanical or chemical changes of the fast energy process occur. We have discussed the example of a chemical breakdown reaction which results into an accelerated slip instability forming the PSZ. A number of other microstructural processes are often activated in the course of this instability leading to a rich microstructure inside the PSZ. This could be dynamic recrystallisation, fluid release, dissolution–precipitation, etc. The complex nature of these instabilities leads to complex geometries and time series such as slow slip/earthquake signatures. The dual feature of creeping zone and ultralocalized PSZ, therefore, is also a hallmark of ductile shear zones as illustrated in Fig. 6b.

To juxtapose the outcomes of the theoretical approach with field evidence, we compare our solutions with observation from brittle and ductile shear zones. Figure 6 shows in (a) the famous Punchbowl cataclastic fault described by CHESTER and CHESTER (1998). The fault shows a cataclastic fault zone of the order of meter thickness with an ultralocalized, pulverised PSZ. The mineralogy of the host rock and PSZ indicate a transition from quartz and feldspar dominated to clay mineralogy in the cataclastic fault zone and ultimately higher ordered smectite and quartz in the PSZ (CHESTER and CHESTER 1998). This indicates a series of mineral dissolution–precipitation reactions that are typical for the ingression of water under lower temperature environments. The addition of water can decrease the activation temperature and thereby render the system critical at ambient conditions.

The same style of deformation is also recorded in the UNESCO world heritage Glarus fault that shows a meter-thick, chemically altered tectonite (known as Lochsite tectonite), blending the hanging wall and footwall minerals. The tectonite is deformed and folded by ductile deformation and has an ultralocalised PSZ (Fig. 6b) in its middle. Field evidence include a multiplicity of PSZ's inside the tectonite,

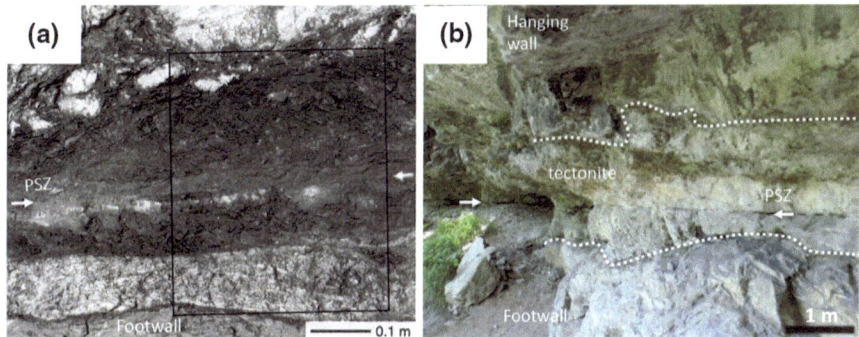

Figure 6
a Punchball fault from CHESTER and CHESTER (1998) featuring two scales of localisation, the cataclastic fault zone and the ultracatalastic PSZ annotated by *arrows*. **b** The Glarus Thrust at the Lochsite, featuring two scales of localisation, the creeping zone forming the carbonate rich tectonite and the ultralocalized PSZ in its *centre*

implying repetitive fast events interrupting ductile creep. The key chemical reaction controlling both the tectonite formation and the thickness of the PSZ was identified to be carbonate decomposition and precipitation (HERWEGH *et al.* 2008). For the ductile localisation instability, additional heat is required as the carbonate decomposition reaction happens at a much larger temperature than the boundary of the thrust zone. The addition of temperature is readily available from the long-term dissipation of the creep process, which becomes a prerequisite for the ductile instability. Because the shear zone has a characteristic background temperature, the chemical reaction has a critical temperature to be triggered and the microstructure has equilibrated over time, to a quasi-steady state, the ductile instability is inferred to have periodic signature in time. This implies a regular stick-slip type of behaviour for the ductile PSZ whereas the brittle PSZ is expected to be chaotic in time. The reason is that brittle instabilities are always at criticality while ductile instabilities require time to be brought to criticality.

6. Discussion

In this contribution we link the observed double localization patterns in fault zones with a discussion on their possible driving energetics and timescales of formation. We have shown that the thick fault zone is a result of a solid mechanical instability, acting as a vessel for the ultralocalized instability. In the case of

brittle failure negligible additional energy is required as the system behaves in a near temperature-independent, close to criticality, regime. In the case of ductile failure the shear zone is not necessarily at criticality through time but the fluid-like creep motions raise the local dissipation to eventually trigger an energy transformation in an ultralocalized PSZ. Under the assumption of a microstructural steady state of the shear zone, the latter instability is periodic.

We have presented in Fig. 6 a direct comparison of brittle and ductile/creep fractures in shear zones. In the brittle case it is known from field and laboratory observations that the PSZ represents an ultralocalised process zone, consisting of ultrafine particles that have usually undergone mechano-chemical degradation (gelification, decarbonation and dehydration reactions, melting, as thoroughly discussed in TORO *et al.* 2011). Therefore, although the mechanism of the fault zone formation is the material bifurcation from homogenous deformation (MUHLHAUS and VARDOULAKIS 1987), the formation of PSZ's should be considered the effect of the chemical–mechanical deformation following the onset of the initial structure.

The ductile shear zone in Fig. 6 features a chemical transition from host rock to hanging rock and a strong grain size reduction towards the PSZ (HERWEGH *et al.* 2008). The mechanism of localisation, although different from the brittle case, has the same style of energetic transition where microstructural changes are captured by their latent heat effects

309

inside the PSZ. The main difference in terms of energetics is the time-scale of deformation, which we have shown to be governed by the energy equation and lastly the thermal activation.

In the brittle zone thermal activation is not considered important. However, a similar energetic transition may be taking place. The activation process can equally be triggered by a lowering of the critical threshold to near ambient condition through the access of fluids interacting with the host rock. It can be argued that in the brittle case the thermal time scale control is replaced by the mechanism of fluid transfer during fault slip. These mechanisms can be low temperature pressurisation effects (SIBSON 1973; LACHENBRUCH 1980; WIBBERLEY and SHIMAMOTO 2005; SULEM and VARDOULAKIS 2005; RICE 2006) or high temperature chemical effects such as dehydration of minerals or decomposition of carbonates, theoretically studied recently (SULEM and FAMIN 2009; VEVEAKIS et al. 2010; BRANTUT et al. 2010, 2011) and reported to take place in real faults (HIRONO 2007; FAMIN et al. 2008) and experiments at laboratory conditions (HAN et al. 2007; FERRI et al. 2010; BRANTUT et al. 2011; PAOLA et al. 2011; COLLETINI et al. 2013). Additional mechanisms that include dry phase transitions such as breakage can be formulated in a similar thermodynamic framework (EINAV 2007a, b; LYAKHOVSKY and BEN-ZION 2014a, b). Therefore, the above described framework illustrates an energy based formulation for instabilities of solid matter with emergent length scales of localisation.

This fundamental outcome is illustrated in the present approach by two scales of localisation naturally emerging and corresponding to two different energy regimes. These are the traditional Solid Mechanical localization during failure leading to a finite width shear band that can be described by internal variables and the Fluid Dynamical post-failure localization related to a phase change that counterintuitively leads to ultra thin shear zones within the master Solid Mechanical containment. We identify these fundamental outcomes to be a plausible explanation for the two scale localization phenomena described in this work.

Appendix: Poro-Chemical Model

At high temperatures the solid AB breaks down, producing excess B fluid, and increasing the fluid pore pressure through a general fluid-release reaction of the form:

$$v_1 AB_s \rightleftharpoons v_2 A_s + v_3 B_f. \tag{20}$$

We assume the following relations for the partial molar reaction rates for the species,

$$r_{AB} = -\left[\frac{\rho_{AB}}{M_{AB}}(1-\phi)(1-s)\right]^{v_1} k_F \exp(-Q_F/RT)$$

$$r_A = \left[\frac{\rho_A}{M_A}(1-\phi)s\right]^{v_2} k_R \exp(-Q_R/RT)$$

$$r_B = \left[\Delta\phi_{chem}\frac{\rho_B}{M_B}\right]^{v_3} k_R \exp(-Q_R/RT). \tag{21}$$

From the stoichiometry of the considered reaction, Eq. (20), it should hold that:

$$-\frac{r_{AB}}{v_1} = \frac{r_A}{v_2} = \frac{r_B}{v_3}. \tag{22}$$

From Eqs. (21–22), and for $v_1 = v_2 = v_3 = 1$ we derive the poro-chemical model

$$\Delta\phi_{chem} = A_\phi \frac{1-\phi_0}{1+\frac{\rho_B}{\rho_A}\frac{M_A}{M_B}\frac{1}{s}},$$

$$s = \frac{\omega_{rel}}{1+\omega_{rel}}, \quad \text{and} \tag{23}$$

$$r_{rel} = \frac{\rho_{AB}}{\rho_A}\frac{M_A}{M_{AB}}K_c \exp\left(\frac{\Delta h}{RT}\right).$$

In Eq. (23), $K_c = k_F/k_R$ is the ratio of the pre-exponential factors of the Arrhenius reaction rates and $\Delta h = Q_R - Q_F$ the difference of the forward and reverse activation energies. The parameter A_ϕ is a coefficient that determines the amount of the interconnected pore-volume (porosity) created due to the reaction. We assume that all the fluid generated contributes to the interconnected pore volume, and, thus, set $A_\phi = 1$.

Following these considerations, the rates of the forward (ω_F) and reverse (ω_R) first order reactions can be calculated to be

$$r_{\text{F}} = r_{AB} \frac{\rho_{AB}}{M_{AB}} (1 - \phi)(1 - s)k_{\text{F}}\text{e}^{-Q_{\text{F}}/RT} . \quad (24)$$

$$r_{\text{R}} = r_A r_B = \frac{\rho_A \rho_B}{M_A M_B} (1 - \phi)s\Delta\phi_{\text{chem}}k_{\text{R}}\text{e}^{-Q_{\text{R}}/RT} . \quad (25)$$

Note that, for simplicity, we have assumed in Eq. (21) that the two products are produced with the same pre-exponential factor and activation energies. If this is not the case, the above model should be modified accordingly. The net reaction rate would then be $r = r_{\text{F}} - r_{\text{R}} \frac{M_{AB}}{\rho_{AB}}$ (the reverse reaction rate was normalized with the reference concentration $\frac{\rho_{AB}}{M_{AB}}$ for dimensional purposes), which however would be essentially irreversible $(r_{\text{F}} \gg r_{\text{R}})$ in the case $K_c = k_{\text{F}}/k_{\text{R}} \gg 1$.

REFERENCES

ALEVIZOS, S., POULET, T., VEVEAKIS, E.: Thermo-poro-mechanics of chemically active creeping faults. 1: Theory and steady state considerations. J. Geophys. Res. pp. In Press, 2014. doi:10.1002/2013JB010070.

ALLAM, A.A., BEN-ZION, Y.: Seismic velocity structures in the southern california plate-boundary environment from double-difference tomography. Geophys. J. Int. 190, 1181–1196 (2012). doi:10.1111/j.1365-246X.2012.05544.x.

BEN-ZION, Y., SAMMIS, C.G.: Characterization of fault zones. Pure Appl. Geophys. 160, 677–715 (2003).

BRANTUT, N., HAN, R., SHIMAMOTO, T., FINDLING, N., SCHUBNEL, A.: Fast slip with inhibited temperature rise due to mineral dehydration: evidence from experiments on gypsum. Geology 39(1), 59–62 (2011).

BRANTUT, N., SCHUBNEL, A., CORVISIER, J., SAROUT, J.: Thermo-chemical pressurization of faults during coseismic slip. J. Geophys Res. 115, B05,314 (2010).

BRANTUT, N., SULEM, J., SCHUBNEL, J.: Effect of dehydration reactions on earthquake nucleation: Stable sliding, slow transients and unstable slip. J. Geophys Res. 116, B05,304 (2011). doi:10.1029/2010JB007876.

CECINATO, F., ZERVOS, A., VEVEAKIS, E.: A thermo-mechanical model for the catastrophic collapse of large landslides. International Journal for Numerical and Analytical Methods in Geomechanics 35(14), 1507–1535 (2011). doi:10.1002/nag.963.

CHESTER, F., CHESTER, J.: Ultracataclasite structure and friction processes of the punchbowl fault, san andreas system, california. Tectonophysics pp. 199–221 (1998).

CHESTER, F.M., EVANS, J.P., BIEGEL, R.L.: Internal structure and weakening mechanisms of the san andreas fault. Journal of Geophysical Research: Solid Earth 98(B1), 771–786 (1993). doi:10.1029/92JB01866..

CHRYSOCHOOS, A., BELMAHJOUB, F.: Thermographic analysis of thermomechanical couplings. Archives Mechanics 44(1), 55–68 (1992).

COLLETTINI, C., VITI, C., TESEI, T., MOLLO., S.: Thermal decomposition along natural carbonate faults duing earthquakes. Geology p. 2013, doi:10.1130/G34421.1.

CORNET, F., DOAN, M., MORETTI, I., BORM, G.: Drilling through the active aigion fault: the aig10 well observatory. Comptes Rendus Geosciences 336(4–5), 395–406 (2004).

EINAV, I.: Breakage mechanics–part I: theory. Journal of the Mechanics and Physics of Solids 55(6), 1274–1297 (2007a). doi:10.1016/j.jmps.2006.11.003

EINAV, I.: Breakage mechanics–part ii: Modelling granular materials. Journal of the Mechanics and Physics of Solids 55(6), 1298–1320 (2007b). doi:10.1016/j.jmps.2006.11.004

FAMIN, V., NAKASHIMA, S., BOULLIER, A.M., FUJIMOTO, K., HIRONO, T.: Earthquakes produce carbon dioxide in crustal faults. Earth. Plan. Sci. Let. 265, 487–497 (2008).

FAULKNER, D.R., LEWIS, A.C., RUTTER, E.H.: On the internal structure and mechanics of large strike-slip fault zones: field observations of the carboneras fault in southeastern spain. Tectonophysics 367, 235–251 (2003). 1951, doi:10.1016/S0040-(03)00134-3.

FERRI, F., DITORO, G., HIROSE, T., SHIMAMOTO, T.: Evidence of thermal pressurization in high-velocity friction experiments on smectite-rich gouges. Terra Nova 22(5), 347–353 (2010). doi:10.1111/j.1365-3121.2010.00955.x.

FOWLER, A. (ed.): Mathematical Models in the Applied Sciences, 2 edn. Cambridge University Press (1997).

FUJITA, C.: On the non-linear equations $du + e^u = 0$ and $v_t = dv + e^v$. Bull. Am. Math. Soc. 75, 132–135, (1969).

HAN, R., SHIMAMOTO, T., HIROSE, T., REE, J., ANDO, J.: Ultralow friction of carbonate faults caused by thermal decomposition. Science 316, 878–881 (2007).

HERWEGH, M., HURZELER, J., PFIFFNER, O., SCHMID, S., ABART, R., EBERT, A.: The glarus thrust: excursion guide and report of a field trip of the swiss tectonic studies groups. Swiss Journal of Geosciences 101(2), 323–340 (2008).

HILL, R.: Acceleration waves in solids. J. Mech. Phys. Solids 10, 1–16 (1962).

HIRONO, T., et al.: A chemical kinetic approach to estimate dynamic shear stress during the 1999 taiwan chi-chi earthquake. Geophys. Res. Lett. 34, L19,308 (2007). doi:10.1029/2007GL030743.

HOLDSWORTH, R., VAN DIGGELEN, E., SPIERS, C., DE BRESSER, J., WALKER, R., BOWEN, L.: Fault rocks from the SAFOD core samples: Implications for weakening at shallow depths along the san andreas fault, California. Journal of Structural Geology 33(2), 132–144 (2011). doi:10.1016/j.jsg.2010.11.010

KENNEDY, L., LOGAN, J.: The role of veining and dissolution in the evolution of fine -grained ylonites: the mcconnell thrust, Alberta. J. Struct. Geology 19(6), 785–797 (1997).

LACHENBRUCH, A.: Frictional heating, fluid pressure and the resistance to fault motion. J. Geophys. Res. 85, 6097–6112 (1980).

LYAKHOVSKY, V., BEN-ZION, Y.: A continuum damage-breakage faulting model accounting for solid-granular transitions. Pure Appl. Geophys. p. Submitted (2014a).

LYAKHOVSKY, V., BEN-ZION, Y.: Damage-breakage rheology model and solid-granular transition near brittle instability. J. Mech. Phys. Solids 64, 184–197 (2014b). doi:10.1016/j.jmps.2013.11.007..

MANDEL, J.: Conditions de stabilite et postulate de drucker. Rheology and Soil Mechanics pp. 58–67 (1966).

MUHLHAUS, H., VARDOULAKIS, I.: *Thickness of shear bands in granular materials*. Geotechnique 37(3), 271–283 (1987).

NODA, H., SHIMAMOTO, T.: *Thermal pressurization and slip-weakening distance of a fault: an example of the hanaore fault, southwest Japan*. Bull. Seism. Soc. Am. 95(4) (2005).

PAOLA, N.D., HIROSE, T., MITCHELL, T., TORO, G.D., TOGO, T., SHIMAMOTO, T.: *Fault lubrication and earthquake propagation in thermally unstable rocks*. Geology 39(1), 35–38 (2011).

PAPANICOLOPULOS, S., VEVEAKIS, E.: *Sliding and rolling dissipation in cosserat plasticity*. Granular Matter 13(3), 197–204 (2011).

PERZYNA, P.: *Fundamental problems in viscoplasticity*. Adv. Appl. Mech. 9, 243–377 (1966).

REGENAUER-LIEB, K., VEVEAKIS, M., POULET, T., WELLMANN, F., KARRECH, A., LIU, J., HAUSER, J., SCHRANK, C., GAEDE, O., TREFRY, M.: *Multiscale coupling and multiphysics approaches in earth sciences: applications*. Journal of Coupled Systems and Multiscale, Dynamics, 1(3), 2013, doi:10.1166/jcsmd.2013.1021.

REGENAUER-LIEB, K., VEVEAKIS, M., POULET, T., WELLMANN, F., KARRECH, A., LIU, J., HAUSER, J., SCHRANK, C., GAEDE, O., TREFRY, M.: *Multiscale coupling and multiphysics approaches in earth sciences: theory*. Journal of Coupled Systems and Multiscale Dynamics, 1(1), 49–73 (2013).

REGENAUER-LIEB, K., YUEN, D., FUSSEIS, F.: *Landslides, ice quakes, earthquakes: a thermodynamic approach to surface instabilities*. Pure. Appl. Geophys, 166(10–11), 1885–1908 (2009).

REINER, M.: *The deborah number*. Physics Today pp. 152–153 (1964).

RICE, J.R.: *Heating and weakening of faults during earthquake slip*. J. Geophys. Res. 111, B05311 (2006). doi:10.1029/2005JB004006.

ROSAKIS, P., ROSAKIS, A., RAVICHANDRAN, G., HODOWANY, J.: *A thermodynamic internal variable model for the partition of plastic work into heat and stored energy in metals*. J. Mech. Phys. Solids 48, 581–607 (2000).

ROWE, C., FAGERENG, A., MILLER, J., MAPANI, B.: *Signature of coseismic decarbonation in dolomitic fault rocks of the naukluft thrust, Namibia*. Earth Plan. Sci. Let. 333–334, 200–210 (2012).

RUDNICKI, J.W., RICE, J.R.: *Conditions for the localization of deformation in pressure sensitive dilatant materials*. J. Mech. Phys. Solids 23, 371–394 (1975).

SIBSON, R.: *Interaction between temperature and pore-fluid pressure during earthquake faulting—a mechanism for partial or total stress relief*. Nature 243, 66–68 (1973).

SIBSON, R.: *Thickness of the seismic slip zone*. Bull. Seism. Soc. Am. 93, 1169–1178 (2003).

SULEM, J., FAMIN, V.: *Thermal decomposition of carbonates in fault zones: slip-weakening and temperature-limiting effects*. J. Geophys. Res. 114, B03309 (2009). doi:10.1029/2008JB006004.

SULEM, J., LAZAR, P., VARDOULAKIS, I.: *Thermo-poro-mechanical properties of clayey gouge and application to rapid fault shearing*. Int. J. Num. Anal. Meth. Geomechanics 31(3), 523–540 (2007).

SULEM, J., STEFANOU, I., VEVEAKIS, E.: *Stability analysis of undrained adiabatic shearing of a rock layer with cosserat microstructure*. Granular Matter 13(3), 261–268 (2011). doi:10.1007/s10035-010-0244-1.

SULEM J. VARDOULAKIS I., O.H., PERDIKATSIS, V.: *Thermo-poro-mechanical properties of the aigion fault clayey gouge - application to the analysis of shear heating and fluid pressurization*. Soils and Foundations 45, 97–108 (2005).

TAYLOR, G., QUINNEY, H.: *The latent energy remaining in a metal after cold working*. Proc. R. Soc., Ser. A. 143, 307–326. (1934).

TOBOLSKY, A., ANDREWS, R.: *Systems manifesting superposed elastic and viscous behaviour*. J. Chem. Phys. 13(1), 3–27 (1945).

TORO, G.D., HAN, R., HIROSE, T., DEPAOLA, N., NIELSEN, S., MIZOGUCHI, K., FERRI, F., COCCO, M., SHIMAMOTO, T.: *Fault lubrication during earthquakes*. Nature 471, 494–498 (2011).

TOWNEND, J.: *Drilling, sampling, and monitoring the alpine fault: Deep fault drilling project–alpine fault, New Zealand; Franz Josef, New Zealand, 22–28 March 2009*. Eos, Transactions American Geophysical Union 90(36), 312–312 (2009). doi:10.1029/2009EO360004

VARDOULAKIS, I., SULEM, J. (eds.): Bifurcation Analysis in Geomechanics. Blankie Acc. and Professional (1995).

VEVEAKIS, E., ALEVIZOS, S., VARDOULAKIS., I.: *Chemical reaction capping of thermal instabilities during shear of frictional faults*. J. Mech. Phys. Solids 58, 1175–1194 (2010). doi:10.1016/j.jmps.2010.06.010.

VEVEAKIS, E., POULET, T., ALEVIZOS, S.: *Thermo-poro-mechanics of chemically active creeping faults. 2: transient considerations*. J. Geophys. Res. pp. In Press, 2014. doi:10.1002/2013JB010071.

VEVEAKIS, E., STEFANOU, I., SULEM, J.: *Failure in shear bands for granular materials: thermo-hydro-chemo-mechanical effects*. Geotechnique Let. 3(2), 31–36 (2013).

VEVEAKIS, E., SULEM, J., STEFANOU, I.: *Modeling of fault gouges with cosserat continuum mechanics: Influence of thermal pressurization and chemical decomposition as coseismic weakening mechanisms*. J. Struct. Geology 38, 254–264 (2012). doi:10.1016/j.jsg.2011.09.012.

VEVEAKIS, E., VARDOULAKIS, I., DITORO., G.: *Thermoporomechanics of creeping landslides: The 1963 vaiont slide, northern Italy*. J. Geophys. Res. 112, F03026 (2007). doi:10.1029/2006JF000702.

WIBBERLEY, C., SHIMAMOTO, T.: *Earthquake slip weakening and asperities explained by thermal pressurization*. Nature 426(4), 689–692 (2005).

WIBBERLEY, C.A.J., SHIMAMOTO, T.: *Internal structure and permeability of major strike-slip fault zones: the median tectonic line in mid prefecture, southwest Japan*. J. Struct. Geol. 25, 59–78 (2003).

YUEN, D., SCHUBERT, G.: *Asthenospheric shear flow: thermally stable or unstable?* Geophys. Res. Lett 4(11), 503–506 (1977).

(Received August 17, 2013, revised March 13, 2014, accepted March 13, 2014, Published online May 3, 2014)